Quaternary Palaeoecology

Quaternary Palaeoecology

H. J. B. Birks and Hilary H. Birks

Botany School, University of Cambridge

Reprint of the 1980 Edition by Edward Arnold (Publishers) Limited

Quaternary Palaeoecology

ISBN-10: 1-930665-56-3
ISBN-13: 978-1-930665-56-9

Library of Congress Control Number: 2004111028

THE BLACKBURN PRESS
P. O. Box 287
Caldwell, New Jersey 07006 U.S.A.
973-228-7077
www.BlackburnPress.com

Preface

The idea for this book originated in a lecture course entitled 'Quaternary Palaeoecology' to final-year undergraduate students at the University of Cambridge. It was evident that no book covered this rapidly developing and interdisciplinary subject. Therefore, we have tried to write a fairly comprehensive introduction, which we hope will be of use to undergraduate and research students with varying backgrounds in botany, zoology, geology, geography, and archaeology. We also hope that it will be of use to research workers in the Quaternary, by drawing their attention to related and perhaps relevant aspects of their subject, and as a general reference book which covers a wide variety of interests.

This book covers a broad spectrum of subjects which can be usefully drawn together in various combinations for the reconstruction of past environments. This broad coverage inevitably leads to superficial treatment of some aspects, particularly those in which we ourselves have not been directly involved during our research. However, we hope that the outline and the references given will stimulate the reader, and will give a useful lead into more detailed accounts.

Because of the interdisciplinary and international nature of Quaternary palaeoecology, the literature is vast and diverse. Therefore, we have listed references at the end of each chapter, and arranged them according to subject. References on a particular topic are thus grouped together for easy use by the student. Inevitably, some references appear more than once. We have also included references to several important and influential publications to which a student should refer, even though they are not mentioned in the chapters. New papers are constantly being published, and therefore this book was out of date as soon as it reached the publisher's hands. This is an unavoidable problem, and we have had to take our cut-off point as 1 March 1979. No papers which have appeared since then are included.

This book does not cover marine Quaternary palaeoecology in any detail. We have restricted ourselves almost entirely to terrestrial and freshwater environments, as they are the ecological settings of our personal research experience. The Quaternary palaeoecology of marine environments is a rapidly developing field, which has great potential, particularly when it can be linked to terrestrial events. It would warrant a separate book to itself.

Palaeoecology, consisting of the reconstruction of past biota, communities, and environments, is the over-riding theme of this book, and we have therefore made little attempt to describe in detail the methods used in palaeoecological investigations or the identification of the various types of fossils. These can be found readily in other reference works mentioned in the text.

We could not have conceived or written this book without the benefit of contact over several years with numerous other Quaternary palaeoecologists. We are grateful to them all for stimulating and inspiring us along various research paths by their friendly discussions and free interchanges of ideas. Our greatest debts are due to Professor Sir Harry Godwin, F.R.S., and Professor R. G. West, F.R.S., who introduced us to the subject, and who have constantly encouraged us. Our research interests and experience have also been particularly influenced by Prof. W. Tutin, F.R.S., Professor H. E. Wright Jr., and Dr. E. J. Cushing, each of whom have added a new dimension to our palaeoecological outlook. Other colleagues and friends have generously discussed their ideas and approaches with us, particularly Dr. S. T. Andersen, Professor B. E. Berglund, Professor M. B. Davis, Dr. B. Huntley, the late Dr. J. Iversen, Dr. C. R. Janssen, Professor R. A. Reyment, Dr. K. Rybniček, Professor W. A. Watts, and Dr. T. Webb III. We owe a great debt to Dr. A. D. Gordon, for his work with H.J.B.B. on the development of numerical methods applicable to Quaternary palaeoecology. Dr. A. J. Stuart introduced us to the fascinating world of Quaternary vertebrates, and we thank him for reading this section of the manuscript for us. Numerous undergraduates and research students have enthusiastically acted as guinea pigs for ideas and techniques which we would never had the time to try out for ourselves.

We are grateful for assistance in the preparation of this book to Mrs. R. Hockaday, Mrs. A. Bennett, and Mrs. A. Ansell, who typed the manuscript so carefully, and to Mrs. S. Peglar and Mrs. S. Dalton who drafted some of the figures. The conversion of the manuscript into a book was cheerfully and efficiently conducted by the editorial staff of Edward Arnold.

Acknowledgements

This book includes material reproduced from original sources in several journals and other publications. We are grateful to the many authors, learned societies, and publishers who granted permission for reproduction. The sources of these figures are as follows:
American Scientist: 1.1, 11.4; Danmarks Geologiske Undersøgelse: 9.4, 10.1, 10.2, 10.3; Ecological Monographs: 11.14; Journal of Animal Ecology: 7.5; Journal of Ecology: 11.15; Limnology and Oceanography: 9.9, 9.10; Los Angeles County Museum Science Series: 7.14, 7.15; Nature: 4.4, 4.5, 4.6, 12.6; New Phytologist: 11.6; New Scientist: 4.7; Quaternary Plant Ecology: 5.4, 5.7, 10.4, 10.19; Quaternary Research: 9.8, 10.17, 10.18, 12.14; Philosophical Transactions of the Royal Society of London: 7.17, 7.18, 7.19; Proceedings of the Linnean Society of London: 4.15; Proceedings of the Prehistoric Society: 4.8, 4.9; Proceedings of the Royal Society of London: 11.20; Review of Palaeobotany and Palynology: 10.8, 11.1; Science: 9.5, 12.17, 12.18, 12.19; Textbook of Pollen Analysis: 8.5.

Quotations are reprinted with permission from 'Textbook of Pollen Analysis' by K. Faegri and J. Iversen © 1975 Munksgaard, Copenhagen, and from 'Theoretical Ecology' edited by R. M. May © 1976 Blackwell Scientific Publications, Oxford.

All sources of material are acknowledged in the text.

Cambridge H.J.B.B.
1980 H.H.B.

Contents

1
Ecology and palaeoecology

Introduction

Palaeoecology is the ecology of the past. It is strongly linked to both biology and geology. It can be studied in any period of earth's history in which there was life. Palaeoecology's main link and relevance to modern or neo-ecology is during the most recent geological interval, the Quaternary. The Quaternary covers the last 1–2 million years and it is unique in earth's history for its oscillating climates, alternating in the latitude of Europe and North America between temperate so-called interglacial phases and cold phases within which glaciation commonly occurred, and also for the fact that man evolved during the Quaternary.

We must define the terms ecology and palaeoecology, in order to see the differences and similarities between them, to delimit the sphere of the subject of this book, and to establish the close interaction between ecology and palaeoecology. ***Ecology*** can be defined as the study and understanding of the complex relationships between living organisms and their present environment. Ideally, ***Palaeoecology*** could be defined as the study and understanding of the relationships between past organisms and the environment in which they lived. In practice, however, palaeoecology is largely concerned with the reconstruction of past ecosystems. To do this, all the available evidence, both biological and geological, is used to reconstruct the past environment. Therefore, it is difficult to deduce the relationships between organisms and their environment in the past if the evidence of the organisms had already been used to reconstruct the environment. Independent lines of evidence for environmental reconstruction are required before organism–environment relationships can be assessed.

Although, in theory, ecology and palaeoecology have similar aims and invoke many of the same biological principles, in practice they have different concepts and working methods. These differences arise for two main reasons. Firstly, past ecosystems cannot be observed directly. The biotic and abiotic components of the ecosystem must be inferred from the fossils and the sediments in which the fossils are found. Palaeoecology is thus limited to the study of past organisms whose fossils are preserved. Secondly, the fossil record on which all palaeoecology depends can be seriously distorted due to the processes of transportation, diagenesis, and redeposition. Due to transportation, evidence from one area may be mixed with and indistinguishable from that from other systems. Therefore the organisms which are preserved as fossils cannot necessarily be assumed to have lived in the system within which the fossils are found today. Due to diagenesis, fossils and other environmental evidence may be modified or even destroyed by geological processes operating in the ecosystem surrounding the fossils at contemporary or subsequent time intervals. Due to redeposition, evidence that originated at one point in time and space may be deposited and preserved with evidence derived from different points in time (either earlier or later) and space. There is thus little control in palaeoecology over what can be observed and over the range in time and space that the palaeoecological evidence occupies. Fortunately, in many instances, the limitations imposed on palaeoecology by this lack of spatial and temporal control are not insurmountable, as the processes of transportation, diagenesis, and redeposition can often be identified and their effects evaluated and allowed for.

There are six major differences in approach to the study of ecology and palaeoecology. These important differences influence the working methods used in palaeoecology.

1. An ecologist can select the organisms and the physical and chemical variables to be studied. A palaeoecologist is restricted to studying those organisms preserved as fossils, and he must use evidence from them and from the associated sediments to reconstruct the past environment.

2. An ecologist must establish and operate within defined boundaries of space and time. These boundaries, defined implicitly or explicitly, delimit the ecosystem of study. A palaeoecologist has little control over the limits in space and time represented by his fossils, due to the processes of transportation, diagenesis, and redeposition. A palaeoecologist must accept the evidence where it can be found, and attempt to decide what it represents and from where it originated.

3. An ecologist can usually plan a series of observations and/or experiments, and with care, he can make repeatable observations. A palaeoecologist can usually never repeat an observation, except perhaps for taxonomic revisions or re-examination of his samples. A palaeoecologist makes a once-only investigation of his material. Any anomalies or unusual features that are found can usually only be checked by comparison with other samples from a different part of the system of interest or from other systems nearby. Such a set of new observations is not an analogy of a repeatable experiment or recording in ecology.

4. An ecologist generally makes his observations at one or a few points in time; more rarely observations may be made over a short period (20–50 years) of time. A palaeoecologist makes observations that cover long periods of time (100–10 000 years or more). Each sample studied may represent many years (often 10–200 years), and a set of samples collected in stratigraphical order may cover many thousands of years. The time dimension, although often measured with less precision than in ecology, is much more important in palaeoecology than in ecology.

5. An ecologist is usually not directly concerned with evolutionary, migrational, and other biogeographical processes. A palaeoecologist is usually very concerned about such processes, and in some cases evolution and/or migration may be the main purpose of the study.

6. An ecologist has a wide range of strategies available for sampling ecosystems. A palaeoecologist is restricted in his sampling strategy, because the palaeoecosystem of interest is dead, partially or wholly decayed, and mixed and changed by processes of diagenesis and transportation operative up to the time of sampling.

What is the fascination, relevance, and value of studying palaeoecology at the present day?

Despite many of the limitations imposed by the nature of the fossil record, palaeoecology can provide reconstructions of past ecosystems which appear, in many instances, to be valid and useful. Inevitably such reconstructions are rather gross and frequently unsophisticated. The reconstructions do, however, enable comparison to be made with ecosystems from other periods of time, including the present, so that possible causes and mechanisms of biological change with time can be sought (Deevey, 1965, 1969). Many processes which are important in understanding modern ecosystem composition, structure, and dynamics operate over long periods of time, and thus cannot be studied within a single human lifetime. Modern ecology and biogeography can benefit directly from the results of palaeoecology and from the historical and evolutionary perspectives that palaeoecology can uniquely provide. For example, in considering the succession of organisms through time, Gould (1976) emphasized the importance of palaeoecology to modern ecological theory by saying, 'Palaeoecology can provide the only record of complete *in situ* successions. The framework of classical succession theory (probably the most well known and widely discussed notion of ecology) rests largely upon the inferences from separated areas in different stages of a single, hypothetical process (much like inferring phylogeny from the comparative anatomy of modern forms). Palaeoecology can provide direct evidence to supplement ecological theory.'

A historical, palaeoecological perspective on the development and structure of modern ecosystems and on ecological processes acting through the geological past, can provide a basis for the formulation of ecosystem models in which predictions about the future effects of environmental change can be made.

Classification of ecology and palaeoecology

There are several broad approaches to the study of modern ecology, depending on the aims and interest of the investigator. Similarly, palaeoecology can also be approached in several ways, depending upon the questions asked by the palaeoecologist.

Ecology

1. Descriptive ecology, in which the ecologist aims to describe the features of an ecosystem. He seeks simplifications of the real world by classifying and generalizing observations on specific ecosystems. Hypotheses can be tested against new, independent observations.

2. Deductive ecology. In this approach the ecologist constructs generalized dynamic models of an ecosystem that simulate the relationships between organisms and their environment. This approach of 'systems analysis' tests the value and relevance of generalized models with real observations from descriptive ecology (see Clymo, 1978, for a model of peat growth).

3. Experimental ecology. This approach attempts to simplify nature by controlling as many environmental and biotic factors as possible and by varying one or a few factors at a time in order to study their influence on an ecosystem under controlled conditions.

These three approaches are complementary and essential. Ideally an ecologist should use all three in his study of an ecosystem. In practice, ecologists tend, however, to concentrate on one approach only.

Palaeoecology

Palaeoecology can similarly be subdivided into descriptive, deductive, and experimental approaches.

1. *Descriptive palaeoecology.* This is the dominant approach in much of palaeoecology, because the reconstruction and description of past ecosystems is usually difficult and time consuming, as well as being of such intrinsic interest. In such an approach, the palaeoecologist faces sampling problems that the modern ecologist can largely avoid. Besides selecting where and how to sample and finding the relevant evidence, the palaeoecologist must decide what the evidence represents and where the fossils came from. These difficulties have inevitably led to simplifications, and the palaeoecologist uses the present to model the past by extending observations about processes within present ecosystems backwards in time. There is thus a close interaction between descriptive ecology and descriptive palaeoecology.

2. *Deductive palaeoecology.* A few attempts have been made recently to develop mathematical models to simulate palaeoecological systems (see Reyment, 1968; Harborough and Bonham-Carter, 1970). For example, Craig and Oertel (1966) presented deterministic models (i.e. models in which the random variation factor is ignored) of living and fossil populations of animals involving growth rates and death rates.

One of the few examples of the deductive approach in Quaternary palaeoecology is that of Martin (1973) and Mosimann and Martin (1975). They considered the problem of the possible cause for the sudden and dramatic extinction of many species of large mammals in North America at the end of the Pleistocene about 10–12 000 years ago. One hypothesis to account for this extinction involved the arrival and expansion of man into the previously uninhabited North American continent; man was directly responsible for overkill and extinction. It has been suggested that the early North American men were highly skilled predators with thousands of years of Palaeolithic experience in Asia. The mammalian prey was, it is proposed, unable to develop suitable defensive mechanisms within the relatively short time available.

Mosimann and Martin (1975) constructed a mathematical model to simulate how an initially small human population could increase sufficiently rapidly to cause the extinction of the large fauna. They represented the arrival of man across the Bering Land Bridge at about 12 000 years ago by starting the model with 100 men and women at Edmonton, Alberta. If this population doubled every 30 years, the model predicts that a wave of 300 000 humans would have reached the Gulf of Mexico in about 300 years, having populated an area of 780×10^6 ha (3 million square miles). Such an advancing front of humans would have been large enough to kill a biomass of large mammals comparable to 42×10^9 kg (93×10^9 pounds), which would have reflected an animal density similar to a modern African gamepark. Such a human population explosion could have resulted in massive predation, or overkill of the native mammal fauna, leading to its rapid extinction in about 300 years. As the food supply dwindled behind the advancing front, the model predicts a decline in the human population to a level in equilibrium with the environment. The model proposes a mechanism to explain the observations of a rapid extinction of large mammals in North America, and the paucity of archaeological sites of this age where both human artifacts and animal remains have been found together.

Mosimann and Martin inserted different values for population growth, animal density and biomass, etc. into their model. Even with figures well below the theoretical maximum, the result was always an extinction of the fauna in a relatively short time (see Fig. 1.1).

The deductive approach shows considerable promise, and although the numerical and computer techniques can be rather complex, there is already a wealth of experience in this type of approach in ecology.

3. *Experimental palaeoecology.* This approach involves controlled experiments using either living organisms or scale models of fossils to investigate the effect of processes and factors that are recognizable at the present day and that were almost certainly operative in the past.

Reyment (1971, 1973) has devised ingenious experiments with scale models of shells of ammonites and nautiloids in tanks of water. The experiments were designed to estimate their necroplanktonic dispersal properties, for example what shell sizes and shapes floated or sank after death. The results obtained help to explain the composition of some fossil assemblages which, due to processes of transportation, do not reflect the life assemblages of the different cephalopod shells. Other types of palaeoecological experiments involve observations in the laboratory on palaeoecologically important living organisms. For example, Reyment and Bränn-

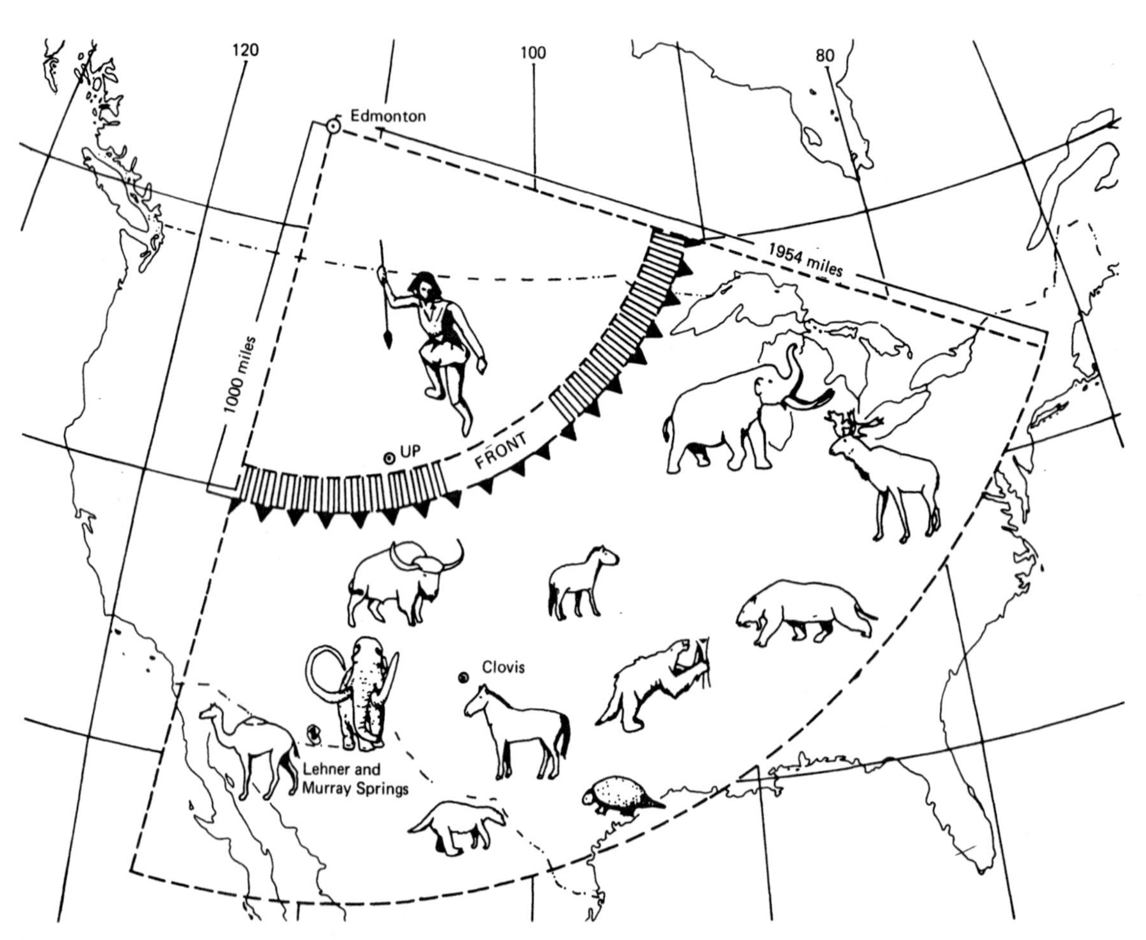

Fig. 1.1 A hypothetical model for the spread of man and the overkill of large mammals in North America. Upon arrival the population of hunters reached a critical density, and then moved southwards in a quarter-circle front. One thousand miles south of Edmonton, the front is beginning to sweep past radiocarbon-dated Palaeoindian mammoth kill sites, which will be overrun in less than 200 years. By the time the front has moved nearly 2000 miles to the Gulf of Mexico, the herds of North America will have been hunted to extinction. For further explanation, see text. (After Mosimann and Martin, 1975.)

ström (1962) studied the growth reaction of the carapace of the ostracod *Cypridopsis vidua* to various environmental factors such as calcium carbonate and water aeration.

Examples of the experimental approach in Quaternary palaeoecology will be described in various parts of this book, for example the study of the dispersal, deposition, and preservation of pollen, spores, and seeds in different sedimentary environments. For example, laboratory studies under controlled experimental conditions have been made on the differential preservation of pollen with high sporopollenin content, and on the mixing of pollen and sediment in a lake by burrowing worms.

Experimental palaeoecology is an extremely important approach and one which merits a great deal more attention than it has received up to now. It differs from experimental ecology in that the experiments are generally designed to explain particular features of the fossil assemblage relevant to the reconstruction of the past ecosystem. Ecologists can experiment directly on living organisms, to study and quantify the organism's response to environmental factors.

Palaeoautecology and palaeosynecology

The bulk of palaeoecology falls within the general framework of descriptive palaeoecology. This in turn can usefully be subdivided into palaeoautecology and palaeosynecology, just as modern ecology is conveniently divided into autecology and synecology.

Palaeoautecology

Autecology considers the ecology of the individual organism or species, and it is primarily concerned with life histories, behaviour, adaptive morphology, and ecological tolerance. Palaeoautecology is the palaeoecological study of individual fossils or species of fossils, and, as in autecology, the emphasis is on behaviour, adaptive morphology, and life histories. A good example of a palaeoautecological study is that by Gould (1974) on the Giant Irish Elk (*Megaloceros giganteus*). This huge deer was common in Ireland at the end of the last glaciation, and its enormous antlers, weighing about 40 kg (90 pounds), have frequently been found in lake and bog sediments of this age in Ireland. Gould studied the morphology and structure of the antlers, and suggested that their function was in display and courtship rather than in fighting. The Giant Irish Elk became extinct at the beginning of the Holocene, possibly because the antlers were very cumbersome, and because they could not be seen in a wooded environment.

Palaeosynecology

Synecology considers the ecology of groups of organisms which are associated with each other as a functional unit, either as a population, a community, or an ecosystem. Palaeosynecology is the study of groups of fossils, so-called fossil assemblages, and the reconstruction of past environments. Such studies are also concerned with populations, communities, or ecosystems. It is more informative to talk of community palaeoecology, population palaeoecology, etc. rather than the broad term palaeosynecology.

Descriptive palaeoecology

Just as descriptive ecology is often subdivided on the basis of habitat type (marine, freshwater, terrestrial) or a taxonomic basis (vascular plants, birds, mammals, insects), descriptive palaeoecology can usefully be subdivided in various ways, according to habitat, taxonomy or geological age.

a) *Habitat.* Fossils can be studied in sediments formed in marine, freshwater, or terrestrial habitats.
b) *Taxonomy.* Different groups of organisms can be studied as fossils, such as diatoms, vascular plants, vertebrates, foraminifera, mollusca, etc.
c) *Geological age.* Fossils can be studied from sediments of different geological age; Palaeozoic, Mesozoic, Tertiary, Quaternary. In this book we shall restrict ourselves to the study of Quaternary fossils, preserved in continental rather than marine environments.

These subdivisions are often sharp within descriptive palaeoecology, because of the nature of the evidence and because of the uneven distribution of the fossil record. For example, the bulk of the fossiliferous rocks of the Palaeozoic and Mesozoic now preserved were sedimented in the sea. Hence their palaeoecology is concerned mainly with the marine environment and with those marine organisms with readily preservable hard parts, such as shells of brachiopods, molluscs, and echinoderms. In contrast, the vast majority of fossiliferous sediments of Quaternary age that are accessible and easily studied were formed in terrestrial or freshwater habitats. Their palaeoecology is thus largely concerned with terrestrial or freshwater organisms with readily preservable hard parts, such as vertebrates (including man), insects, vascular plants (pollen, spores, seeds), and diatoms. Quaternary marine sediments are beginning to be studied in detail (see, for example, Imbrie and Kipp, 1971), but compared with terrestrial and freshwater Quaternary palaeoecology, marine Quaternary palaeoecology is in its infancy.

Approaches to palaeoecology

Despite the numerous subdivisions of palaeoecology, the approach to the subject is the same in all branches of palaeoecology. The topic of primary concern to the palaeoecologist is the palaeoecosystem (Fig. 1.2). In contrast to the modern ecosystem studied by an ecologist, a palaeoecologist has

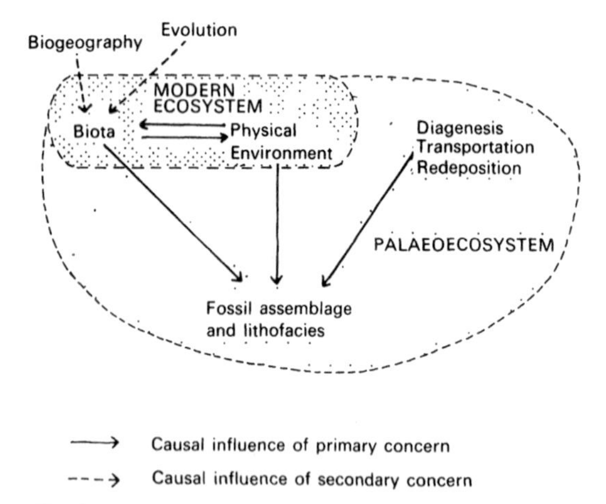

Fig. 1.2 Representation of a palaeoecologist's approach to the study of the fossil record.

to consider not only the ecosystem, but also the processes of diagenesis, transportation, and redeposition, along with the influences of biogeography and evolution. This palaeoecological viewpoint can be contrasted with the approach of a historical biogeographer and of a palaeontologist (Fig. 1.3). Both gather their primary evidence from the fossil record. Their main interests are the biota and the resulting fossil record, whereas the influence of the physical environment on the fossil record and the effects of diagenesis and transportation are usually of less interest (see Imbrie and Newell, 1964).

Various methodological approaches to palaeoecology are possible (Fig. 1.2), and in this book particular attention is given to (1) biological approaches, ranging from interpretations of individual species to broader studies of entire fossil assemblages, (2) sedimentary approaches involving the study of sediment lithology and chemistry, and (3) statistical approaches.

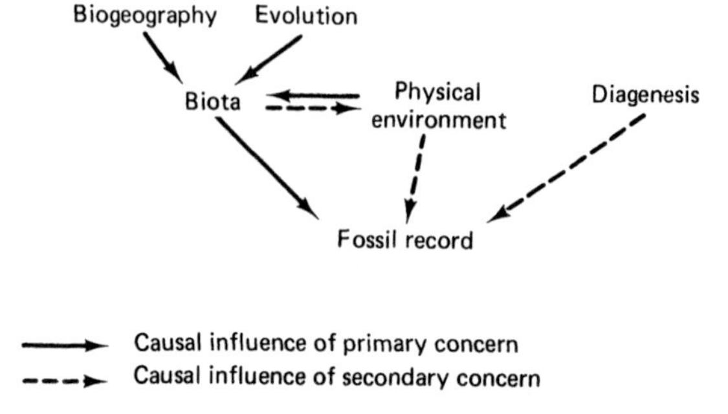

Fig. 1.3 Representation of a historical biogeographer's and a palaeontologist's approach to the study of the fossil record.

Philosophical principles of palaeoecology

Since palaeoecology has a very close relationship with geology, many of the principles and philosophies of geology are applicable to palaeoecology. Watson (1969), Simpson (1970), and Albritton *et al.* (1967) provide useful insights into the philosophy of geology. Because a whole book could be written solely on the philosophy of palaeoecology, we must restrict ourselves to a brief mention of the philosophical principles of the subject. Rudwick (1972) and Albritton (1963, 1975) provide comprehensive accounts of the philosophy of geology and palaeontology, to which the interested reader is referred.

There are seven major features of palaeoecological philosophy.

1. Palaeoecology is a descriptive, historical science, and as such it depends largely on inductive inferences and reasoning. Inductive reasoning is the basic mode of reasoning in empirical science (Hempel, 1966), where one observation leads to another, and where extrapolations are made in an attempt to present generalizations about nature.
2. The method of multiple working hypotheses (see Chamberlin, 1965). This method forces one to consider as many explanations of a phenomenon as possible. Such explanations are likely to be nearer

the correct explanation than if only one explanation was considered, which would then become a tentative theory or working hypothesis, and finally a ruling theory into which all subsequent observations are fitted, often without due regard for the evidence. In presenting the method, Chamberlin said 'the effort is to bring up into view every rational explanation of new phenomena and to develop every tenable hypothesis regarding their cause and history'. Multiple hypotheses encourage the seeking of new evidence that will lead to the rejection of, hopefully, all but one working hypothesis. The method thus contributes directly to the planning and design of new investigations.

3. Simplicity. The general scientific principle called Occam's razor, derives from the saying, 'It is vain to do with more what can be done with fewer' (William of Occam, *ca* 1300–49). In other words, let the simplest explanation suffice until more evidence is available which necessitates more complicated explanations (see Anderson in Albritton, 1963, for an analysis of the principle of simplicity in historical geology).

4. Taxonomy and evolution. Palaeoecology deals with fossil organisms. Consequently a sound taxonomy and an appreciation of evolutionary processes is essential in any palaeoecological study. There is thus inevitably some preoccupation with taxonomy, especially at the species and genus levels. It is ecologically more meaningful to consider fossil assemblages as communities of species rather than from the viewpoint of abstract stratigraphic units or constituents of particular lithological units.

5. Language. The terms and vocabulary of palaeoecology are primarily those of biology and geology.

6. Data. Palaeoecological data are frequently quantitative and invariably complex, consisting of many observations and many variables. For example, many different types of fossils may be counted in a sample, which may be one of a whole series of samples related to each other in time or space, or both. Such data are called 'multivariate data' and may be too complex for the palaeoecologist to sort efficiently and to synthesize fully. The use of multivariate mathematical methods for data analysis, such as those described by Reyment (1969, 1972) can be of considerable assistance to the palaeoecologist. Such methods can deal with large amounts of complex data and can process them in precise and repeatable ways. With the primary data simplified and synthesized in this way, the palaeoecologist can then devote himself to the interpretation of the data in as critical and as meaningful a way as possible.

7. Uniformitarianism. This as a basic assumption and philosophical principle of palaeoecology. It was first formulated by James Hutton in 1788, but more fully defined and discussed by Charles Lyell in 1830 (see Simpson (1970) for a full and penetrating analysis of uniformitarianism). It is the principle of the uniformity of nature, generally regarded as the philosophical foundation upon which historical geology is based. Briefly it can be stated as 'the present is the key to the past'. However, although an attractive cliché, this is an unsatisfactory statement of uniformitarianism in philosophical terms (see Gould, 1965; Scott, 1963).

Uniformitarianism was developed explicitly by Charles Lyell in 1830–33 in his *Principles of Geology*, in response to 'catastrophism' and religious views of divine intervention in the history of the earth. Catastrophism was developed during the eighteenth century, and invoked a series of catastrophies, such as the Great Flood of Biblical times, in order to explain geological features such as the presence of shells on mountain tops, and also to fit in with Bishop Ussher's estimated origin of the earth at 4004 B.C. Lyell, who is commonly regarded as the founder of modern geology, subtitled the first edition of his *Principles of Geology* 'An attempt to explain the former changes of the earth's surface by reference to causes now in operation'. Such uniformitarian ideas were later extended into biology by Charles Darwin. Although there have been arguments about uniformitarianism ever since, Lyell was originally intending to exclude divine intervention in the processes of geology, since all the features of the earth could be explained by processes which still operated at the present day (see Albritton, 1975).

Gould (1965), in a valuable essay, distinguished between 'substantive uniformitarianism', where the rates of geological processes are said to have been constant through the past, and 'methodological uniformitarianism' or 'actualism', which states that the nature of the processes are the same, but that they may occur at different rates at different times. In other words, catastrophies do occur, such as volcanic eruptions, glaciations, floods etc., but they do involve and obey the laws of nature because the properties of matter and energy are invariant with time. Hence these laws can be extended back in time, and are thus applicable to the explanation of past events. Methodological uniformitarianism is basically an extension of the laws of physics, but primarily of those laws relevant to geological processes. It is the basic logic and methodology by which the past can be reconstructed. There is no way to prove methodological uniformitarianism, but alternatively, there is no way to reject it.

Frequently, methodological uniformitarianism merges with, and indeed represents, the simplest approach to reconstructing the past, and thus it is also the geological, and hence palaeoecological, formulation of the logical principles of simplicity (see Albritton *et al.*, 1967; Gould, 1965; Cushing and Wright, 1967) and of induction. Uniformitarianism and its basic postulate that the laws of nature are invariant with time is not unique to geology, but is now a common denominator of all science (Hubbert in Albritton *et al.*, 1967). Hubbert proposes the following definition of history: 'History, human or geological, represents our hypothesis, couched in terms of past events devised to explain our present-day observations.'

The nature of palaeoecological evidence

The observations made by palaeoecologists are of two main types. Fossils are remains of organisms, and can be called ***biotic*** evidence, in contrast to ***abiotic*** evidence, which includes the physical and chemical characteristics of the sediments.

Biotic evidence

Fossils can be defined as the remains or indications of past biota. By indication, we mean such fossils as animal tracks, so-called trace fossils, and leaf impressions. There are five main types of fossils (see Krasilov, 1975).

1. *Original material preserved.* This type would include hard parts of organisms such as shells, plant cuticles, bones, and the exines of pollen grains and spores.

2. *Impressions and films.* Carbonized films found on bedding planes of rocks are the commonest example of this type of fossil. The volatile organic components of plant leaves and of animals with a chitinous exoskeleton such as arthropods have gradually disappeared until only a film of carbon is left. As more and more carbon is lost, the film becomes an impression.

3. *Petrifications and replacements.* Fossils, whether calcareous or carbonaceous, may be altered by the effects of water percolating through the rocks. In the simplest case, pore spaces originally filled by organic matter may be infilled with precipitated mineral matter. In petrification complete replacement takes place, the original hard parts being replaced by silica, by iron compounds, or by phosphate compounds.

4. *Moulds and casts.* Percolating water, instead of replacing the organic material, may dissolve it away. If the walls of the cavity so produced are strong enough, a mould of the original fossil is left.

5. *Trace fossils.* These are markings and structures found in sedimentary rocks resulting from the activities of animals moving on or through the sediment during its deposition. Footprints, tracks, and coprolites are the commonest types of trace fossils.

In Quaternary deposits, the commonest type of fossil is the first, namely preserved organic material. The reasons why some organic compounds are preserved, but others are not, are complex. In general, those compounds with a low free-energy in relation to the depositional environment are stable, and are thus more resistant to decay. There are relatively few such compounds that are sufficiently stable to be preserved, with the result that the fossil record is biased towards those organisms with preservable parts. In the animal kingdom, such compounds include calcium carbonate deposited either as calcite or aragonite in shells, silica, and chitin. In the plant kingdom such compounds include calcium carbonate, silica, cutin, lignin, and sporopollenin.

The animal groups producing fossils useful in the study of Quaternary palaeoecology are chiefly vertebrates, molluscs, arthropods, testaceous rhizopods, and foraminifers. Plant fossils most commonly studied are pollen and spores, cuticles, wood, leaves, seeds and fruits, phytoliths (siliceous cell thickenings), mosses, fungal hyphae and spores, diatoms, other algae such as *Pediastrum*, and algal cysts from dinoflagellates, chrysophytes, and charophytes.

Fossils may be deposited in the place where the organism died, in which case they are termed ***autochthonous***. If, however, the dead remains are transported to another locality by any agency, they are termed ***allochthonous***. A collection of dead remains ready to be preserved is called a death assemblage, or ***thanatocoenose***, and it is usually different from the life assemblage, or ***biocoenose***, because of the addition of allochthonous material. These processes are summarized in Fig. 1.4.

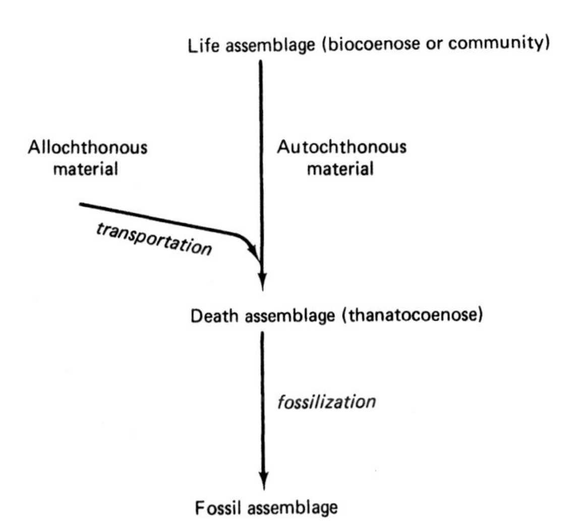

Fig. 1.4 Summary of the processes leading to the formation of the fossil assemblage.

The proportion of autochthonous and allochthonous fossils in a death assemblage varies with the environment of deposition. For example, in a raised bog, the material forming the peat is closely related to the living community of the bog surface, and there is little allochthonous addition, apart from windblown particles. In a lake, however, the material deposited at a certain point may be largely allochthonous, consisting of matter originating in another part of the lake, and transported by water currents, of matter washed in by streams from the catchment area, and matter blown on to the lake surface. Therefore, in reconstructing a past ecosystem, it is important to bear in mind the mode of formation of the death assemblage, or taphonomy (see West, 1973; Lawrence, 1968, 1971).

Abiotic evidence

Abiotic evidence is derived from the physical and chemical characteristics of the sediments. For example, the size of the particles of sediment provides information about the energy of the environment of deposition. In high energy environments, there is much kinetic energy available, and thus only large particles will be deposited, the smaller ones being carried away. For example, shingle beaches are formed at points of strong wave action at the edge of the sea or a lake, whereas sands and muds are deposited in quieter, deeper waters with less movement. In general, deposition occurs in low-energy environments, and erosion in high-energy environments. Thus the situations for fossil preservation tend to be low-energy environments, such as in still water.

Other properties of sediments are also important. For example, chemical characteristics can lead to diagenesis after deposition. Redeposition and mixing of sediments of different origin may also be detected by examination of the physical characteristics of the sediments. For example, the presence of silt or sand particles in sediments of a deep lake suggests that there has been inwashing of this material from the surrounding landscape, and thus the environment at the time of deposition led to erosion of soils. Much of the sediment of a lake with active inflows and outflows may be allochthonous, whereas the organic sediments of peat bogs and swamp forests are mainly autochthonous.

Quaternary sediments may be divided into those which are inorganic in origin, and those which are primarily organic. The main types of inorganic sediments are:

glacial – till, outwash, stratified drift
eolian – wind-blown silt, or loess
alluvial – stream deposits
colluvial – solifluction deposits, hillwash

The main types of organic sediments are:

limnic – lake sediments (mainly allochthonous)
telmatic – deposited at or very near water level (mainly autochthonous)
terrestrial – deposited above water level (mainly autochthonous)

Sediments originating from other environments may be a mixture of organic and inorganic components, for example marine deposits from the sea bottom and estuaries; soils; cave earths; and spring deposits such as tufa. West (1977) gives a detailed account of the full range of Quaternary sediments.

Space and time in palaeoecology

Now that we have discussed the nature of palaeoecological evidence we can construct a conceptual model of a palaeoecological investigation. An ecologist can readily delimit boundaries of space and time in his study of the present day, but a palaeoecologist has great difficulties in defining either, and indeed, the evidence he studies in his fossil assemblage may have originated in several different points in space and time. He can usually define his basin of deposition as a lake, bog, or the sea, but its spatial boundaries may have changed through time, for example, the lake gradually became filled with sediment. Similarly his time scale may not be linear with depth of sediment sampled, if, for example, there have been changes in the rate of deposition, which in an extreme case may lead to a period of no deposition, or even erosion of sediment. The palaeoecologist must also limit his point of observation. Clearly he cannot study all the sediments in a lake or peat bog, but must concentrate his efforts on what he hopes is a representative sample from one small area of space and time. Therefore his reconstruction of past events in the whole area, such as a lake's catchment area, over a period of time may be unknowingly biased because an atypical sample was selected. Finally, the palaeoecologist publishes the results of his observations and his inferences from them, in the hope that fellow palaeoecologists, and perhaps ecologists too, may find them of interest and perhaps also of relevance to their own studies.

The interrelations of these processes with space and time, and how they affect a palaeoecological investigation are shown in Fig. 1.5, based on a model devised by E. J. Cushing (personal communication).

During the time of the life assemblage, three processes may occur to alter its composition (the numbers refer to processes marked on Fig. 1.5).

1. Some organisms may migrate out of the defined area of interest, either to different parts of the defined region of space, or beyond it.
2. Some organisms may migrate into the defined area of interest from other parts of the defined region of space.
3. Some organisms may migrate into the defined area of interest from an undefined region of space beyond it.

Several processes may then affect the death assemblage as it is converted to the fossil assemblage.

4. After death, the soft parts of the organisms are usually decomposed, and their bodies may be

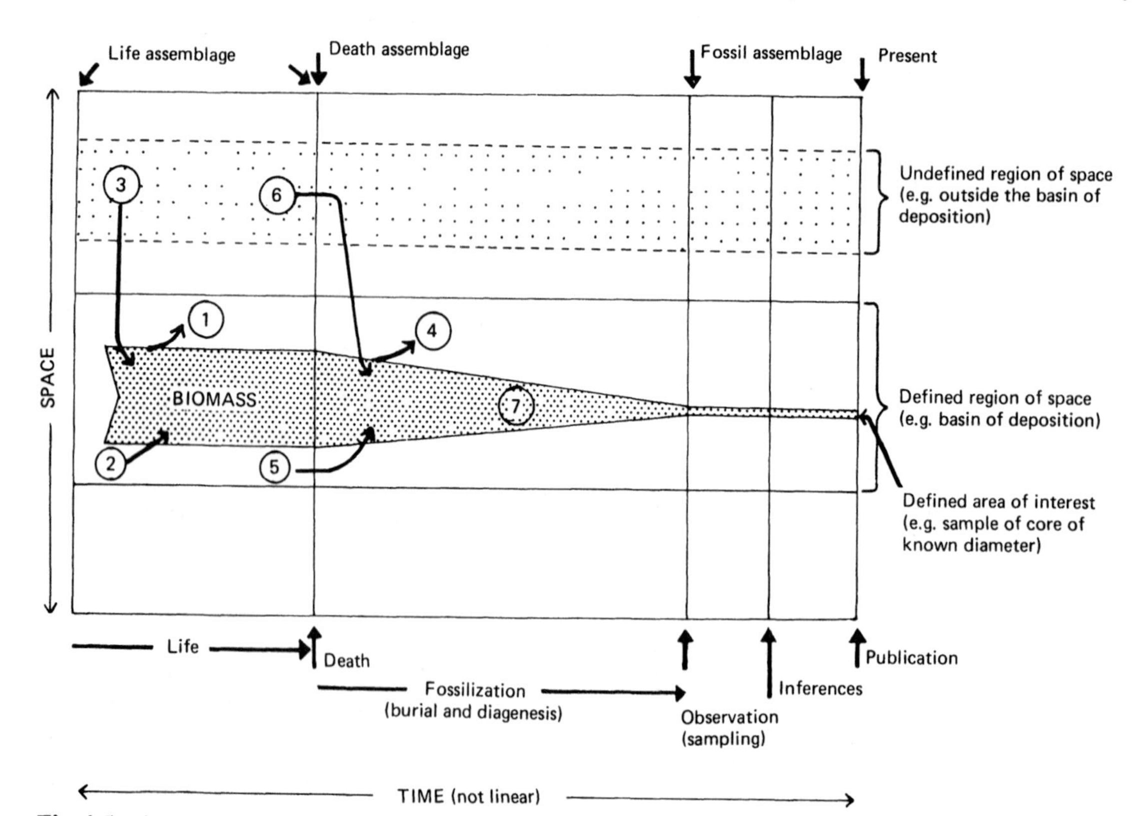

Fig. 1.5 A model of the processes in space and time leading to the formation of a fossil assemblage, and its study by a palaeoecologist. For explanation of numbers, see text. (Based on an unpublished diagram by E. J. Cushing.)

removed by scavengers or erosive physical forces. There is therefore a rather rapid loss of information about the death assemblage soon after its formation due to diagenesis and removal.

5. Redeposition of fossils of the same or different ages from elsewhere in the basin of deposition may alter the composition of the death assemblage.

6. Redeposition of fossils of the same or different age from beyond the basin of deposition may also alter the composition of the death assemblage.

7. The over-all result is a reduction in total biomass due to a reduction in the number of taxa, the quantity of each taxon, and the loss of soft parts.

The relative importance of these various processes is different in different environments of deposition and fossilization. They also have different relative effects on different types of fossils. For example, in pollen analysis of a lake sediment, processes 3 and 6 are probably the most important, namely the addition of living and/or dead material from outside the basin of deposition (see West, 1973). Processes 1, 2, and 5 involving movement of material within the depositional basin also occur but they are generally of secondary importance. In a study of plant macrofossils or diatoms from the same lake, processes 1, 2, 4, and 5 are most important, because few seeds and fruits or diatoms originate from beyond the basin of deposition. In any palaeoecological study it is valuable to be able to recognize the relative importance of these various processes before attempting any interpretations of the observed fossil assemblages. One of the basic aims of experimental palaeoecology is to study and to quantify these processes at the present day, in an attempt to assess their significance in the past.

References

General aspects of palaeoecology

CLOUD, P. E. (1959). Palaeoecology – retrospect and prospect. *J. Paleontol.*, **33**, 926–62.

CRAIG, G. Y. (1966). Concepts in palaeoecology. *Earth Sci. Rev.*, **2**, 127–55.

CUSHING, E. J. AND WRIGHT, H. E. (1967). Introduction. In *Quaternary Paleoecology* (eds E. J. Cushing and H. E. Wright). Yale University Press.

DEEVEY, E. S. (1965). Environments of the geologic past. *Science*, **147**, 592–4.

DEEVEY, E. S. (1969). Coaxing history to conduct experiments. *Bioscience*, **19**, 40–3.

GOULD, S. J. (1976). Palaeontology plus ecology as palaeobiology. In *Theoretical Ecology, Principles and Applications* (ed. R. M. May). Blackwell, Oxford.

IMBRIE, J. AND KIPP, N. G. (1971). A new micropaleontological method for quantitative paleoclimatology: application to a Late Pleistocene Caribbean core. In *The Late Cenozoic Glacial Ages* (ed. K. K. Turekian). Yale University Press.

IMBRIE, J. AND NEWELL, N. (1964). Introduction: the viewpoint of paleoecology. In *Approaches to Paleoecology* (eds J. Imbrie and N. Newell). J. Wiley and Sons.

KRASILOV, V. A. (1975). *Paleoecology of Terrestrial Plants.* J. Wiley and Sons (especially Parts I, II, and IV).

REYMENT, R. A. (1969). Biometrical techniques in systematics. In *Systematic Biology*. National Academy of Sciences, Washington D.C.

REYMENT, R. A. (1971). *Introduction to Quantitative Paleoecology*. Elsevier.

REYMENT, R. A. (1972). Application of multivariate morphometrics in paleontology. *Proc. 24th Int. Geol. Congr.*, **7**. 238–45.

WEST, R. G. (1973). Introduction. In *Quaternary Plant Ecology* (eds H. J. B. Birks and R. G. West). Blackwell, Oxford.

WEST, R. G. (1977). *Pleistocene Geology and Biology* (2nd edition). Longman.

Models in palaeoecology

CLYMO, R. S. (1978). A model of peat bog growth. In *Production Ecology of British Moors and Montane Grasslands* (eds O. W. Heal and D. F. Perkins). Springer-Verlag.

CRAIG, G. Y. AND OERTEL, G. (1966). Deterministic models of living and fossil populations of animals. *Q. Jl. geol. Soc. Lond.*, **122**, 315–55.

HARBAUGH, J. W. AND BONHAM-CARTER, G. (1970). *Computer Simulation in Geology*. J. Wiley and Sons (especially Chapter 10).

MARTIN, P. S. (1973). The discovery of America. *Science*, **179**, 969–74.

MOSIMANN, J. E. AND MARTIN, P. S. (1975). Simulating overkill by paleoindians. *Amer. Sci.*, **63**, 304–13.

RAUP, D. M. AND SEILACHER, A. (1969). Fossil foraging behaviour: computer simulation. *Science*, **166**, 994–5.

REYMENT, R. A. (1968). Systems analysis in paleoecology. *Geol. Fören. Förhandl. Stock.*, **89**, 440–7.

Experimental palaeoecology

JOHNSON, R. G. (1957). Experiments on the burial of shells. *J. Geol.*, **65**, 527–35.

KUMMEL, B. AND LLOYD, R. M. (1955). Experiments on relative streamlining of coiled cephalopod shells. *J. Paleontol.*, **29**, 159–70.

MENARD, H. W. AND BOUCOT, A. J. (1951). Experiments on the movement of shells by water. *Am. J. Sci.*, **249**, 131–51.

REYMENT, R. A. (1958). Factors in the distribution of fossil cephalopods. *Stockholm Contrib. Geol.*, **1**, 97–184.

REYMENT, R. A. (1973). Factors in the distribution of fossil cephalopods. 3. Experiments with exact models of certain shell types. *Bull. Geol. Inst. Univ. Upsala*, N.S. **4**, 7–41.

REYMENT, R. A. AND BRÄNNSTRÖM, B. (1962). Certain aspects of the physiology of *Cypridopsis* (Ostracoda, Crustacea). *Stockholm Contrib. Geol.*, **9**, 207–42.

RUDWICK, M. J. S. (1961). The feeding mechanism of the Permian brachiopod *Prorichthofenia*. *Palaeontology*, **3**, 450–71.

Palaeoautecology

ANDREWS, H. E. *et al.* (1974). Growth and variation in *Eurypterus remipes* DeKay. *Bull. Geol. Inst. Univ. Upsala*, N.S. **4**, 81–114.

BROWER, J. C. (1973). Ontogeny of a Miocene pelecypod. *J. Math. Geol.*, **5**, 73–90.

GOULD, S. J. (1969). An evolutionary microcosm; Pleistocene and recent history of the land snail *P* (*Poecilozonites*) in Bermuda. *Bull. Museum. Comp. Zool. Harvard.*, **138**, 407–532.

GOULD, S. J. (1973). Positive allometry of antlers in the 'Irish Elk', *Megaloceros giganteus*. *Nature*, **244**, 375–6.

GOULD, S. J. (1974). The origin and function of 'bizarre' structures: antler size and skull size in the 'Irish Elk', *Megaloceros giganteus*. *Evolution*, **28**, 191–220.

MALMGREN, B. A. (1974). Morphometric studies of planktonic foraminifers from the type Danian of southern Scandinavia. *Stockholm. Contrib. Geol.*, **29**, 1–126.

Uniformitarianism and the philosophy of palaeoecology

ALBRITTON, C. C. (1963). *The Fabric of Geology*. Addison-Wesley.

ALBRITTON, C. C. (1975). *Philosophy of Geohistory: 1785–1970*. Dowden, Hutchinson and Ross.

ALBRITTON, C. C. *et al.* (1967). Uniformity and simplicity. *Geol. Soc. Amer. Spec. Paper*, **89**, 99pp.

CHAMBERLIN, T. C. (1965). The method of multiple working hypotheses. *Science*, **148**, 754–9.

GOULD, S. J. (1965). Is uniformitarianism necessary? *Am. J. Sci.*, **163**, 223–8.

HEMPEL, C. G. (1966). *Philosophy of Natural Science*. Prentice Hall.

LAWRENCE, D. R. (1971). The nature and structure of paleoecology. *J. Paleontol.*, **45**, 593–607.

MANN, C. J. (1970). Randomness in nature. *Bull. Geol. Soc. Amer.*, **81**, 95–104.

RUDWICK, M. J. S. (1972). *The Meaning of Fossils. Episodes in the History of Palaeontology*. Macdonald.

SCOTT, G. H. (1963). Uniformitarianism, the uniformity of nature, and paleoecology. *New Zealand J. Geol. Geophys.*, **6**, 510–27.

SIMPSON, G. G. (1970). Uniformitarianism. An inquiry into principle, theory, and method in geohistory and biohistory. In *Essays in Evolution and Genetics in honour of Theodosius Dobzhansky* (eds M. K. Hecht and W. C. Steere). North Holland.

WATSON, R. A. (1969). Explanation and prediction in geology. *J. Geol.*, **77**, 488–94.

Palaeoecological evidence

BOUCOT, A. J. (1953). Life and death assemblages among fossils. *Am. J. Sci.*, **251**, 25–40.

CRAIG, G. (1953). Fossil communities and assemblages. *Am. J. Sci.*, **251**, 547–8.

IMBRIE, J. (1955). Biofacies analysis. *Geol. Soc. Amer. Spec. Paper*, **62**, 449–64.

IMBRIE, J. (1955). Quantitative lithofacies and biofacies study of Florena Shale (Permian) of Kansas. *Bull. Amer. Ass. Petrol. Geol.*, **39**, 649–70.

LAWRENCE, D. R. (1968). Taphonomy and information losses in fossil communities. *Bull. Geol. Soc. Amer.*, **79**, 1315–40.

WEST, R. G. (1973). Introduction. In *Quaternary Plant Ecology* (eds H. J. B. Birks and R. G. West). Blackwell, Oxford.

2

Principles of palaeoecology

Introduction

As discussed in Chapter 1, palaeoecology involves both biological and geological principles and working methods. In this chapter we will discuss these principles as they relate to Quaternary palaeoecological investigations. Initially the principles of geological stratigraphy and correlation, the units of stratigraphy, and the approaches to sampling geological material are considered. The various biological stages in a palaeoecological investigation are then discussed in terms of the reconstruction of past organisms, populations, communities, and ecosystems.

Detailed accounts of the principles and methods in palaeoecology, particularly of pre-Quaternary deposits, are given by Ager (1963), Imbrie and Newell (1964), and Reyment (1971).

Principles of Geological Stratigraphy

Palaeoecology is a historical science, in which the time dimension is an essential feature. In this sense, palaeoecology is part of geology and the principles of geological stratigraphy are applicable and important to palaeoecologists, whether their interests are Quaternary or pre-Quaternary palaeoecology. The temporal relationships of past events and of geological evidence such as fossils are of paramount importance in any palaeoecological study. Correlation in geology (including palaeoecology) deals with the relationships of events in time and with the temporal equivalence of these events. In principle two events are correlated if they occurred 'simultaneously'. In a classic essay on geological time Kitts (1966) critically examines the bases for our conception of geological time. He argues on both theoretical and practical grounds that in the correlation of geological events only approximate simultaneity is ever possible and he concludes that geological correlation is 'a method which permits the ordering of spatially separated historical–geological events in the relation "earlier than" – "later than"'. Geological and stratigraphical evidence usually allows the palaeoecologist to infer whether events coincide or overlap in time and can thus be correlated, or if the events do not coincide or overlap so that no correlation can be made. Figure 2.1 illustrates the types of correlation that

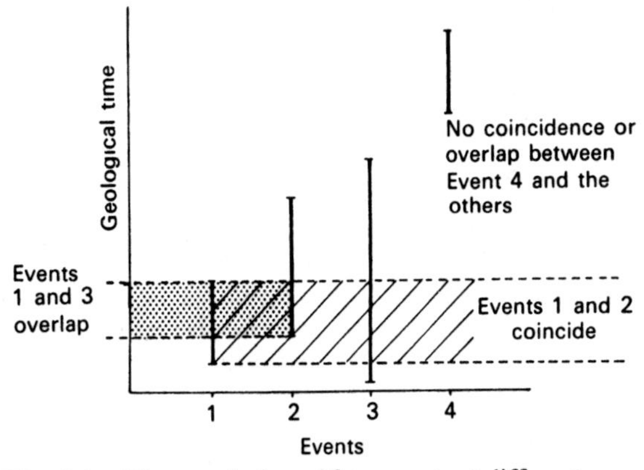

Fig. 2.1 The correlation of four events at different places which cover different periods of geological time.

can be made. Four events at different points in space are considered, each covering a certain period of geological time. Events at points 1 and 2 coincide in time and can thus be correlated. Events at points 1 and 3 overlap in time and can also be correlated. Events at points 1 and 4 do not coincide or overlap in time and thus no correlation can be made.

There are three broad approaches to geological correlation.

1. *Cause-and-effect relationships.* If event 1 caused event 2, then event 2 must be younger than event 1. In this way, a chain of events can be established, and the events can thus be ordered in time and correlations suggested.
2. *Stratigraphy of sections.* This approach involves correlating an event in one stratigraphical sequence

at one place with another event in another sequence at another place. This is the commonest method of establishing correlations when no independent time-scale is available. The lithological (rock or sediment type) or biological (fossil content) features of the section are used to compare sequences. The law of superposition is essential in any stratigraphical study. It states that overlying sediments are younger than the underlying sediments, in the absence of any evidence for sediment disturbance or reworking.

Hence if two sections from different areas contain the same rocks or sediments, the two sequences are correlated even though they may not match throughout. Figure 2.2 illustrates the type of

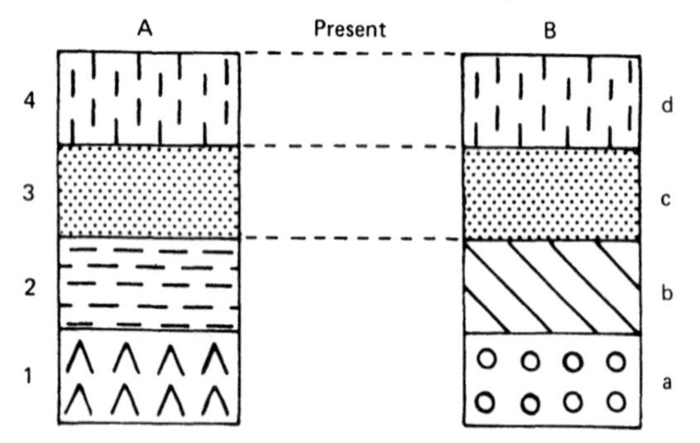

Fig. 2.2 The correlation of two stratigraphic sequences, A and B. The symbols may be taken to represent lithological units or fossil assemblages.

correlation which can be made between two sections A and B. The top of both sections is assumed to be the present day. Stratum 4 is the same lithologically and palaeontologically as stratum d, and they are thus correlated. Their lower boundaries are also correlated. Similarly with strata 3 and c. Below the base of strata 3 and c, the sequences differ and thus cannot be correlated further. However, we can say that stratum 3 is younger than strata 1 and 2 and also, by indirect correlation, strata a and b. This approach to correlation assumes that the event leading to the deposition of sediments 3 and c was the same, and acted at similar times at places A and B. This is not necessarily true in detail, especially if A and B are widely separated in space, although a *broad* correlation may be made on the grounds that the event was likely to have occurred at the two places during the same broad time period.

In Quaternary stratigraphy, for example, boulder clay deposited by glaciers is taken as evidence of glaciation after the retreat of the glaciers. Glaciation on a large scale is assumed to be a world-wide phenomenon. Thus extensive boulder clays in Europe and North America, although not directly related in cause, are the products of the same event, in this case glaciation, and hence are correlated in time. Similarly any underlying or overlying deposits, such as solifluction deposits or organic lake deposits accumulating in hollows in the boulder clay can be ordered in time and hence correlated in relation to the underlying boulder clay. Fossiliferous deposits such as silts, lake muds, and peats can also be correlated by their fossil content, both in terms of the species present and the relative abundances of the different species. One of the most successful attempts at combining biotic and abiotic evidence for correlation purposes is that of the East Anglian Quaternary sequence elucidated by West (1963, 1967, 1977). Professor West has worked out the number and sequence of glaciations and the associated interglacials during the Pleistocene of East Anglia, England, as well as the sequence of alternating cold and temperate stages prior to the first Pleistocene glaciation. A similar approach to stratigraphy and to correlation has been made by Zagwijn (1963) and Van der Heide and Zagwijn (1967) in The Netherlands. In general the two stratigraphies can be correlated by a combination of lithological, palaeontological, and inferred synchroneity of glacial events.

In pre-Quaternary palaeoecology, sequences can be correlated by the presence of particular types of fossil organisms found in the rocks. Fossils that are similar morphologically are considered to have reached the same evolutionary stage, and are thus used for correlation purposes. There are several problems associated with this type of approach. The organism may have died out in one area due to unfavourable environmental conditions before it could colonize a second area. The stratigraphical range of the organism in the two areas would resemble events at points 1 and 4 in Fig. 2.1, and the two sequences would not be correlated. In less extreme circumstances, the relative abundance of the organism in the first area may be changed by local habitat factors, whereas such changes did not occur at the second area. Although the two sequences may be of the same age, their observed fossil composition is different, and they might not be correlated. A further problem arises because different habitats occur simultaneously in an area, and different fossil assemblages may be characteristic of these different habitats. Although the assemblages are in fact contemporary, they may not be recognized as such unless the whole palaeoenvironment is reconstructed and ecological and causal relationships are assumed. This problem commonly arises when the lateral variation in rocks is examined. The variation from limestones through limestone rubble to reef limestones may represent an ecological gradient from a lagoon, through a beach, to an offshore reef. All three rock types would contain different fossil assemblages, and correlations are only possible when the inferred environment is considered and the ecological relationships are assumed. As in Quaternary palaeoecology, correlation of pre-Quaternary rocks must be done with considerable care and with an open but critical mind. Any palaeoecological study requires a reliable stratigraphy and a sound taxonomy.

3. *'Absolute dating'*. Correlations can also be made by 'absolute dating' of the sediments or the rock strata in which particular geological events are referred to an independent event, namely the present day or to a standard sequence of events. This is usually done by one of several methods of radiometric dating. These methods measure the amount of an isotope produced by radioactive decay, or the amount of the radioactive isotope itself. By assuming that the proportion of the radioactive to the stable isotope (e.g. $^{14}C:^{12}C$ for carbon) was the same when the sediment was laid down as it is today and that no subsequent addition or dilution has occurred, the proportion of the radioactive isotope or its product remaining at the present-day is a function of the time that it has had to decay at its constant known rate (so-called 'half-life'). Carbon-14 is the most commonly used radiometric dating technique in Quaternary palaeoecology, as it has a suitable decay rate (conventionally its half-life is 5568 years), which allows datings to be made back to about 40 000 years ago. Details of radiocarbon dating and some of its problems and limitations are given by West (1977), Shotton (1967), Olsson (1968), and Burleigh (1974). Successful attempts have been made to extend the range of radiocarbon dating back to 75 000 by Stuiver *et al.* (1978) and in the future datings involving the use of cyclotrons as a high-energy mass spectrometer (Muller, 1977) may extend the technique back even further to 100 000 years. At present, however, 40–75 000 years takes us back to the middle of the last glaciation only, so the absolute age of previous glaciations and interglacials cannot be measured by radiocarbon methods at present. Other methods, such as potassium/argon dating can be used to date older rocks. The decay rate of potassium-40 to argon-40 is so slow that it shows a minute change only during the 1–2 million years of the Quaternary, thus again making the age of the interglacials indeterminable. This problem has not yet been satisfactorily overcome, and estimates of the duration of some interglacials have been made by other 'absolute dating' methods such as counting annual laminations in interglacial lake sediments (Müller, 1974a,b; Shackleton and Turner, 1967), or by assuming constant rates of sedimentation in the oceans.

Despite the widespread notion of 'absolute dating' and 'absolute time', Kitts (1966) urges the abandonment of the term 'absolute time' in geology as pretentious and misleading. It is important to realize in any palaeoecological study that in the correlation of geological events *approximate* simultaneity is only possible, and that 'absolute time' is an unattainable goal.

Units of stratigraphy

As correct stratigraphy is an essential prerequisite in palaeoecology, we will now consider the different types of stratigraphical units recognized by geologists, with particular reference to the Quaternary. Hedberg (1972a,b, 1976) and Van Eysinga (1970) provide valuable guides to stratigraphical classification, terminology and nomenclature, and usage.

Stratigraphy, literally the 'descriptive science of strata', deals, in geology, with the form, distribution, chronological succession, classification, and relationships of rock strata. The main object of stratigraphy *per se* is to construct a stratigraphic column representing a systematic and geochronological picture of the sequence of stratigraphical events which occurred during earth's history. To a palaeoecologist the main objective of stratigraphy is to subdivide stratigraphical sequences for the description, discussion, comparison, correlation, and ecological interpretation of the sequences investigated. Stratigraphical sequences can be subdivided in several different ways, and it is essential to use the correct units for the different bases of subdivision (see Hedberg, 1972a,b, 1976; Van Eysinga, 1970, for details and definitions).

There are four basic types of stratigraphical unit (see Table 2.1): observed, measured, inferred, and assumed.

Table 2.1 Classification of stratigraphic units (with special reference to the Quaternary).

Measured	Observed*		Inferred	Assumed	
Absolute time units (radiometric years before present defined as 1950 A.D.)	Lithostratigraphic units (group, formation, member, bed)	Biostratigraphic units (zone‡, subzone, zonule)	Geologic-climatic units (glaciation, interglacial, stadial, interstadial)	Geochronologic units† (era, period, epoch, age, chron)	Chronostratigraphic units† (erathem, system, series, stage, chronozone)

See: American Commission on Stratigraphic Nomenclature (1961), Van Eysinga (1970), Hedberg (1972b, 1976).

* May also have soil-stratigraphic units, biochemical units, etc.
† Geochronologic units also called ***geologic-time*** or chronomeric units. Chronostratigraphic units also called ***time-stratigraphic***, chronostratic, or stratomeric units.
‡ There can be a variety of biostratigraphic zones, e.g. assemblage zone, range zone, peak or acme zone, interval zone; see Van Eysinga (1970) and Hedberg (1972b). Biostratigraphic zones are strictly termed biozones, the time equivalent of which is a biochron.

1. *Observed units.* These are defined and delimited on the basis of observations made on the rocks or sediments and on their contained fossil content.

a) Lithostratigraphic units. These are defined by Hedberg (1972b) as 'a body of rock strata which is unified by consisting dominantly of a certain lithologic type, or combination of lithologic types, or by possessing some impressive and unifying lithologic features. It may consist of sedimentary, or igneous, or metamorphic rocks, or, in some cases, of intricate interbedding of two or more of these. It is a three-dimensional body and its concept must be based on its character as a unit through its full extent, both vertically and laterally'. The smallest formal unit of lithostratigraphy is the bed. These can be united into members, formations, and finally groups. Lithostratigraphy is commonly used in pre-Quaternary and in Pleistocene geology.

b) Biostratigraphic units. These are defined by Hedberg (1972b) as 'a body of rock strata which is unified by its fossil content or paleontologic character and thus differentiated from adjacent strata'. The sequence is divided into biozones of defined fossil content and if necessary the biozones can be subdivided into subzones and zonules. Of the various types of biostratigraphic unit recognized by geologists, the most commonly used type in Quaternary palaeoecology is the ***assemblage zone*** defined by Hedberg (1972b) as 'a body of strata whose content of fossils, or of fossils of a certain kind, taken in its entirety, constitutes a natural assemblage or association which distinguishes it in biostratigraphic character from adjacent strata'. As Hedberg (1972b) comments, assemblage zones are particularly significant as indicators of past environments. An example of a Quaternary pollen assemblage biozone (or zone) would be the *Betula-Corylus* regional pollen assemblage zone commonly found in western Scotland in sediments deposited at the beginning of the Flandrian stage or post-glacial period (see Birks, 1973). The characteristic or diagnostic taxa need not occur throughout the assemblage zone or be restricted to it.

Other types of biostratigraphic zones recognized by geologists include the ***taxon-range-zone***, which is a body of strata representing the total horizontal and vertical range of occurrence of specimens of a particular taxon, ***concurrent-range-zone*** (overlap or range-overlap zone), which is a body of strata defined by those parts of the ranges of two or more selected taxa which are concurrent or coincident, ***acme-zone*** (peak zone), which is a body of strata representing the acme or maximum development of a taxon but not its total range, and ***interval-zone***, which comprises strata between two distinctive biostratigraphical horizons, but does not itself represent any distinctive biostratigraphic assemblage, acme, or range.

2. *Measured units.* These are 'absolute' time units measured by some independent, often radiometric method. In Quaternary palaeoecology, the commonest measure of 'absolute time' is radiocarbon dating. The results are expressed conventionally as radiocarbon years before present (B.P.), where present is defined as 1950 A.D.

3. *Inferred units.* The American Commission on Stratigraphic Nomenclature (1961) introduced geologic-climatic units for use in the Quaternary. These units of glaciation, interglaciation, stade, and interstade reflect glacial and non-glacial events which are assumed to result from widespread, often global, and synchronous climatic change.

4. *Assumed units.* These are abstract, conceptual chronostratigraphic units (*sensu* Hedberg, 1972b) which attempt to organize systematically the sequence of rock strata into named units (chronostratigraphic or time-stratigraphic units) which correspond to intervals of geological time (geochronologic or geological-time units). Such units may facilitate time-correlation and age-determination of strata and serve as a reference system for recording events in geological history. A chronostratigraphic unit is a specific body of rock strata formed during a particular interval of geological time at a specific locality where the upper and lower boundaries are observable. For each chronostratigraphic unit or interval of rock-strata, there is a corresponding geochronologic unit or interval of geological time. The terms for chronostratigraphic units (erathem, system, series, stage, chronozone) are different from the terms for their corresponding geochronologic unit (era, period, epoch, age, chron), even though the two are given the same proper name, e.g. Permian System (chronostratigraphic) and Permian Period (geochronologic). Thus the Flandrian Age is the interval of geological time during which Flandrian Stage sediments were deposited.

In establishing any stratigraphic unit, the unit must be defined carefully and described thoroughly. Such a description should include the following (see Hedberg, 1972b):

a) Name
b) Type locality and section
c) Kind and rank of unit – a statement of the kind of unit is very important
d) Description – distinctive features, fossil content
e) Boundaries and relationships to adjacent units and to units based on other criteria
f) Thickness and lateral extent
g) Age and correlation
h) General concept – reasons for establishing the unit
i) Other occurrences and facies
j) Prior usage, previous nomenclature, and literature references.

Stages in a palaeoecological study

In an attempt to reconstruct the life assemblage from a fossil assemblage and the past environment in which the life assemblage lived the palaeoecologist must go through several stages of operations (Fig. 2.3). First, he must decide what the problem is that he wishes to investigate. He must then design a suitable procedure for sampling the fossil assemblage so that the results obtained are relevant to the original problem. Before suitably preparing the samples, he must decide upon the organisms or parts of organisms he is going to observe, record, and count. To do this he needs a thorough knowledge of the morphological features and taxonomy of these fossils, and he must be able to reconcile the morphological entities he can identify with whole organisms, and, particularly in the Quaternary, with living taxa. If the palaeoecologist quantifies his observations, he can potentially reconstruct the fossil population, and then, if sufficient evidence is available, he can go a certain way in reconstructing the community of organisms which gave rise to the fossil assemblage. From there he can reconstruct features of the past environment from the known ecological preferences of his fossils, combined with a study of the sediment that was surrounding them (see Imbrie, 1955).

Sampling procedures

Because of the nature of the fossil record, a palaeoecologist has much less choice about sampling than does a modern ecologist. An ecologist can select almost any point in a chosen ecosystem to sample in a variety of different ways. A palaeoecologist's material is limited to places where fossil assemblages are preserved, and the fossil assemblage has been modified considerably from the life assemblage, or

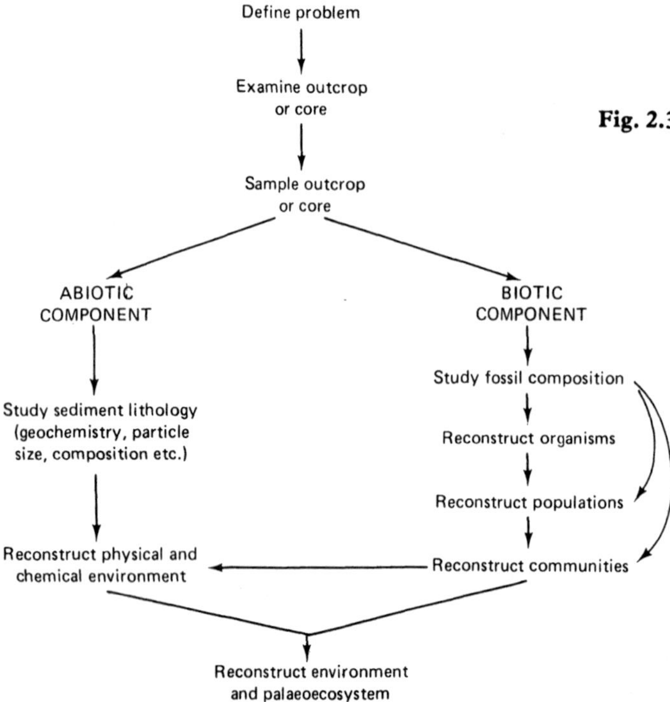

Fig. 2.3 Flow chart of a palaeoecological investigation.

community which formed it (see Fig. 1.5). Therefore a palaeoecologist must take his evidence where he can find it, and do as best he can with its incomplete nature. Indeed, this is part of the challenge of palaeoecology – how much can you get out of the unsatisfactory material available?

The most usual aim of a palaeoecological study is the reconstruction of the life assemblage from the preserved fossil assemblage. A life assemblage cannot be directly studied and is not preserved intact anywhere within its geographical range. Moreover, the palaeoecologist can only reconstruct such a life assemblage where the fossil assemblage is preserved at a particular outcrop, lake, or other fossiliferous locality. In Quaternary palaeoecology, basins of deposition are studied which have collected and accumulated fossil material from the surrounding and largely undefined catchment to a greater or lesser extent (see Fig. 1.5). When sampling is so restricted, the palaeoecologist must be careful with any generalizations about other areas not available for sampling.

In palaeoecology the life assemblage corresponds to Krumbein and Graybill's (1965) ***target population***, which is a conceptual population containing all that the palaeoecologist is interested in, and wishes to sample. In many geological situations most of the individuals are not available for sampling, hence a second kind of population, the ***sampled population*** is defined as that population which includes each element available for sampling. If the target population is fully accessible and is sampled, the target and the sampled populations are the same (Fig. 2.4). Most commonly the target population is not all accessible, so inferences about the target population must be based on samples of the sampled population (central column of Fig. 2.4). The nature and reliability of such inferences will depend on how the sampled population is sampled.

In Quaternary palaeoecology a familiar example of sampling problems is provided by studying a lake basin and its associated sediments. The aim of the study may be to reconstruct the vegetation of the lake and its surrounds using pollen analysis. The target population is thus the vegetation of the catchment and of the lake. This is not fully preserved anywhere, except for what is represented in the lake sediments. These therefore contain the sampled population, i.e. the fossils available to be sampled. Because the palaeoecologist cannot sample

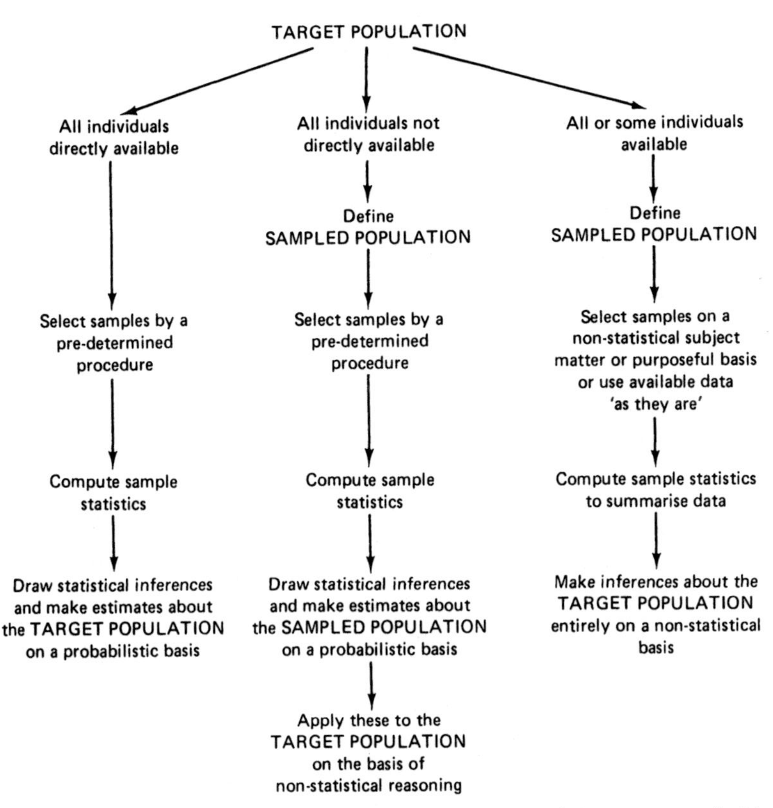

Fig. 2.4 Routes towards the statistical reconstruction of the target population in a palaeoecological study. (After Krumbein and Graybill, 1965.)

all the sediments in the lake, he takes a sample (usually a small-diameter vertical core) of the sampled population (see Fig. 2.4 centre column). Clearly the palaeoecologist should be aware of these stages of sampling in his generalizations and reconstructions of the past vegetation, the target population. Alternatively the aim of the study could have been the reconstruction of the diatom flora of the lake. In this case the lake sediments contain the target population, and the sampled population is therefore the same as the target population (see Fig. 2.4 left-hand column). A third alternative, represented by the right-hand column of Fig. 2.4, arises when the palaeoecologist selects his observational data solely on his judgement and on a subject-matter basis without defining a target or a sampled population, for example, by analysing whatever fossils can be found in a sediment core from anywhere within the lake basin.

As in ecology, the sampling strategy adopted by a palaeoecologist should be related to the problem in hand, and should be designed before the project is begun. Before discussing sampling strategies, we should first consider the objectives of any sampling in a palaeoecological study.

1. *Quantitative description of an assemblage.* A common objective is to describe the fossil content of a sediment quantitatively and to compare it with similar descriptions of assemblages from elsewhere, including outside the defined area of study. The scale of the comparison is important. For example, if large areas are involved, any assemblage is heterogeneous with respect to geography and coarse samples will suffice for comparison, for example, in comparing pollen assemblages from northern boreal forest with spectra from subtropical forests. If the area of interest is small, the palaeoecologist is interested in fine-scale differences, and accurate, quantitative samples are required, for example, for comparing pollen spectra from different types of forest in western Scotland. When a small area has been studied, the investigator can summarize the area and compare it with other areas, and hence build up a picture of large areas.

2. *Quantitative variation within an area.* In this a palaeoecologist is interested in similarities and differences between samples from the same area of interest; in other words the objective is to quantify the variance within an area. A common example involves studying the pollen composition at several points within a bog or lake to provide a basis for estimating the within-site variability, before considering variability between sites.

3. *Quantitative biofacies and lithofacies analyses.* This involves the quantitative comparison of biotic variables such as frequencies of different fossils with abiotic, lithological variables in an attempt to reconstruct the past environment or palaeoecosystem at the sampling site. Such analyses require quantitative analyses of both the contained fossils and the sediment lithologies.

Having determined the objectives of the study, the palaeoecologist must now decide how to sample the available material. There are three main classes of sampling strategy (Krumbein and Graybill, 1965).

1. *Search sampling.* In this the aim is to find out what material is available and to see what fossil assemblages are present and how they change with time. Search sampling is often used to locate particular stratigraphical horizons or time planes in a sequence for subsequent, more detailed study.

2. *Purposeful sampling.* Once the object of the search sampling is found, samples can be collected for specific purposes from particular areas and/or times.

3. *Statistical or probability sampling.* This strategy allows for the selection of one or more samples from a population in such a manner that each individual in the population has a known probability of being found in the sample. In palaeoecology, this sampling strategy is often difficult because of lack of access to some individuals in the population. Statistical sampling can be implemented in several ways (see Krumbein and Graybill, 1965).

a) Random sampling. In this samples are taken from randomly selected points in space and/or time along the outcrop or core of interest (see Fig. 2.5) using a table of random numbers to position the sampling points. This strategy provides an unbiased estimate of the mean and variance of the population, but in practice the strategy may be difficult and expensive to implement.

b) Systematic sampling. In this approach, samples are collected systematically, for example, at equidistant intervals down a core (Fig. 2.5). Such an approach provides an estimate of the population mean but not of the variance.

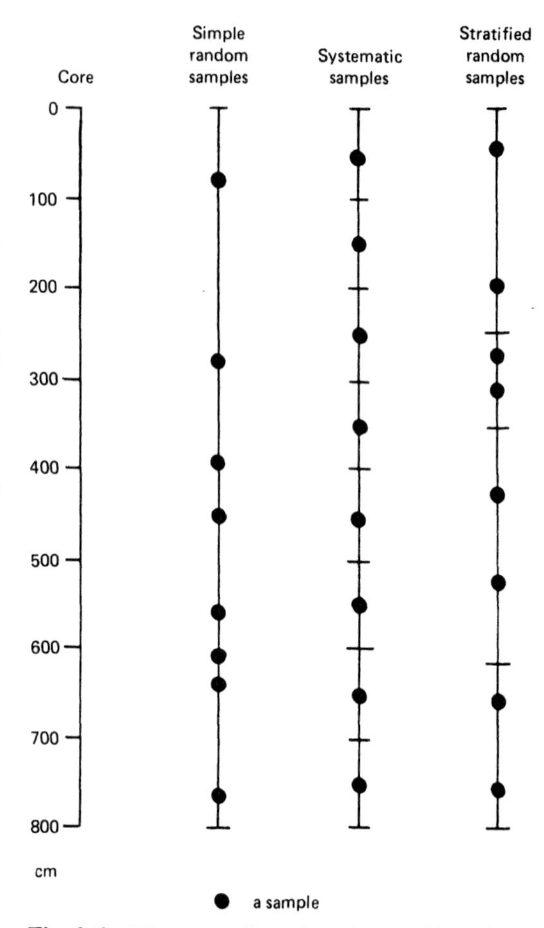

Fig. 2.5 Three sampling plans for an 800 cm-long sediment core. For explanation, see text.

c) Stratified sampling. In some studies, it is desirable to divide the area or sequence of interest into subpopulations, using prior information, and then to take random samples from each subpopulation. Estimates of the subpopulation means and variances can be derived. In Fig. 2.5 the core is divided into four subpopulations on the basis of sediment lithology, and two random samples are collected from each subpopulation.

In addition to the positioning of the samples, there is the question of how many individuals must be counted in the sample. The sample size depends on the aims of the study which in turn dictate how accurately the sample means and variances need be estimated. The question of sample size in relation to pollen analysis is discussed in Chapter 8. De Caprariis *et al.* (1976), Heck *et al.* (1975), and Raup (1975) discuss suitable methods for deriving the optimal sample size when the aim of the study is estimation of taxonomic diversity. Stanton and Evans (1972) consider the importance of sample size in studies concerned with the reconstruction of the structure and composition of the life assemblages.

In Quaternary pollen analysis the most commonly used sampling strategy is search sampling. Samples are taken from different levels in a vertical sedimentary sequence, and the fossil pollen in each of these samples are counted to see what assemblages are present and how they vary with depth. These assemblages can be interpreted in terms of past vegetation and vegetational changes with time. Stratified sampling has been used in some recent Quaternary pollen analytical studies (e.g. Brubaker, 1975; Birks, 1973). In these studies the area of interest was divided into smaller areas or subpopulations on the basis of geology and climate, and randomly positioned sampling points selected within each area. In practice, the sampled populations, such as lake sediments, are never randomly distributed within an area, so the sampling is 'pseudo-random'. Occasionally, purposeful sampling is used in Quaternary palynology. A particular pollen analytical event such as the marked decline of *Ulmus* pollen values may be chosen and investigated at many sites throughout the area of interest (e.g. Birks *et al.*, 1975). Systematic sampling is also used occasionally, in which sampling sites are positioned at regular intervals along transects (e.g. Turner, 1970; Caseldine and Gordon, 1978).

Clearly, as in any descriptive science the sampling strategy used in palaeoecology should be related to the types of questions being studied, and the sampling strategy should be designed before the project is initiated. Any generalizations that can be made about the target population in question will be a function of the sampling strategy adopted, and the type of sampling method adopted should always be borne in mind in assessing and evaluating any interpretations of the data derived from the samples studied.

Reconstruction of the organism

Reconstruction of the individual

In general, Quaternary fossils can be referred to modern living counterparts, and thus the individual can usually be reconstructed readily. This is because relatively little morphological evolution has occurred during the Quaternary Period in most groups of organisms. The major exception is the mammals, which have shown extensive adaptive radiation, evolution, and extinction. However, the extinct forms can all be related to living forms fairly easily. For example, the find of a single hominid jaw gives a large amount of information about the whole animal, how it may have differed from modern man, and how it may perhaps fit into his ancestral evolutionary line. Subsequent finds of other skeletal parts allow more detailed reconstructions to be made, for example, how the hominid animal walked. Similarly, finds of bones of other mammals indicate that many of them differed in certain respects from their modern descendents. However, it is still fairly easy for a skilled palaeontologist to say that a certain bone belongs to an extinct species of horse or hippopotamus.

Classification of the reconstructed individual

The basic unit of biological classification is the 'species'. We therefore have to decide 'what is a fossil species?' in palaeoecological terms. Imbrie (1957), Weller (1961), McAlester (1962), and others have discussed the nature of fossil species and how they may be usefully defined. A systematist bases his opinions on living species, emphasizing either their morphological characteristics (typological species) or their breeding relationships (biological or genetic species). In practice, evidence of both sorts is used to define and delimit modern species, and, in general, species defined by either method coincide. Occasionally, however, situations are found where genetic characters prevent interbreeding between two morphologically indistinguishable groups of organisms, making the definition of a 'species' difficult. A stratigraphical palaeontologist bases his opinions on the morphological characteristics of his fossils. However, he may also take into account features of their biogeography, biostratigraphy, and palaeoecology, in which case the fossil species may well reflect interbreeding populations at that point in geological time. It is relatively easy for a Quaternary palaeoecologist to reconstruct his organisms by reference to living material, and so his concept of a species approaches that of the modern systematist. However, he has to contend with the processes of fossilization, such as decay of non-preserved parts, erosion, and poor preservation of the hard parts. The exact specific identity of the fossil in modern terms cannot always be established. In general, palaeoecologists like ecologists are not very interested in taxonomy for its own sake, although they should always try to identify their fossils to the lowest taxonomic level. They tend to accept the palaeontologist's or taxonomist's judgement, perhaps rather uncritically at times.

Recognition of fossil species

The classification of fossils into taxonomic categories provides special problems not applicable to a modern taxonomist. Usually only part of the organism is preserved, and the state of preservation may be critical in its identification. It is often difficult to separate the fossilizable parts of two modern species which are quite distinct when the whole organisms are present. For example, the pollen grains of nearly all grasses are extremely similar, and only a few can be separated on morphological characteristics, often using refined techniques such as scanning electron microscopy. In palaeoecology we deal primarily with morphological characteristics, and two general approaches to species recognition can be distinguished.

Intrinsic classification

A morphological feature can be determined for each fossil within an assemblage of similar fossils and the results plotted graphically as in Fig. 2.6. In (a) the graph is unimodal, and therefore, in terms of the character measured, there is only one taxon present. In (b) the graph is bimodal, suggesting that there are two taxa present. However, there is an area of overlap shaded in the diagram, so although two taxa are present, they are not perfectly distinct, and cannot always be separated in terms of the morphological character measured. Some other character needs to be found which will differentiate between the two taxa in the area of overlap.

When several morphological characters are measured, the situation becomes 'multivariate', with individuals being defined by many characters. Multivariate methods of data analysis can be used to simplify the situation into a few dimensions. The main types of multivariate methods relevant to this problem are cluster analysis (Everitt, 1974), similar to that used in numerical taxonomy (see Sneath and Sokal, 1973; Kaesler, 1969), and ordination or scaling procedures (such as principal components analysis).

Cluster analysis produces results in the form of a dendrogram illustrated in Fig. 2.7. Each individual is linked to the individual most similar to it, or to the group of individuals most similar to it. If a cut-off point is taken at a degree of similarity marked by a dashed line in Fig. 2.7, the result is three groups of individuals marked *a*, *b*, and *c*. Whether these three groups really represent three taxa is up to the investigator's choice, but cluster analysis can provide a guide to any discontinuities within the data.

Ordination, for example, by the method of principal components analysis, represents the multidimensional data in a few dimensions by positioning each individual in relation to the principal component axes in such a way that the new distances between individuals reflect, as accurately as possible, the original distances or dissimilarities between the individual specimens. It is a property of

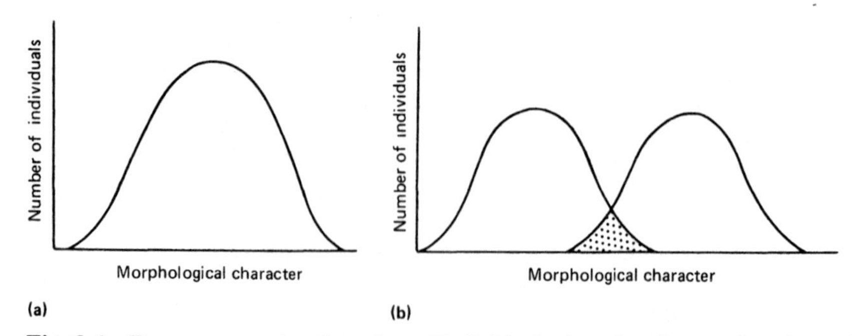

Fig. 2.6 Frequency graphs of numbers of individuals plotted against a selected morphological character, such as size. In (**a**) the graph is unimodal, suggesting that only one taxon is present. In (**b**) the graph is bimodal, suggesting the presence of two taxa, which overlap in the shaded area.

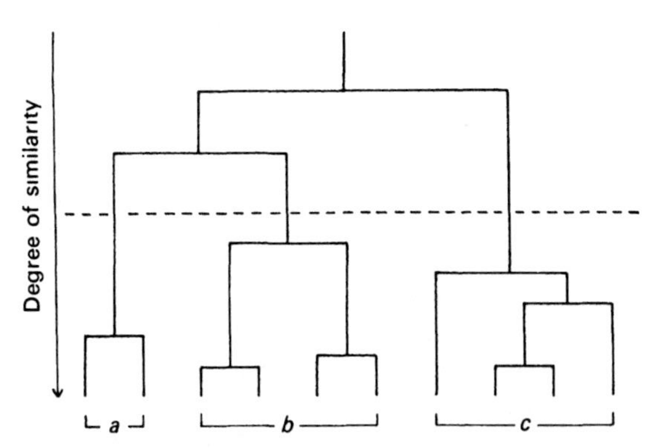

Fig. 2.7 Hypothetical dendrogram. If the degree of similarity at the level of the dashed line is taken as the cut-off point, the individuals can be grouped into three clusters, *a*, *b*, and *c*.

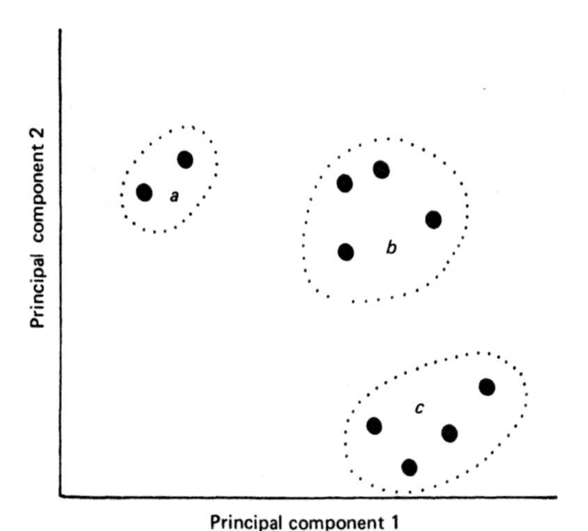

Fig. 2.8 Results of a principal components analysis of the hypothetical data used in Fig. 2.7. When plotted on the first two principal component axes, the three clusters, *a*, *b*, and *c*, can be distinguished.

principal component axes that they are selected along the major directions of variation within the data. An ordination plot is illustrated in Fig. 2.8 of the data represented in Fig. 2.7.

The mathematics of principal components analysis is complex. Clear and relatively simple accounts of the method are given by Blackith and Reyment (1971), Davis (1973), and Gittins (1969). More advanced accounts are presented by Jöreskog *et al.* (1974) and Seal (1964). There are numerous applications of these and related techniques in morphological studies of fossils, particularly of pre-Quaternary fossils (see, for example, Christensen (1974), Hills *et al.* (1974), Malmgren (1974), Reyment (1963, 1966b)).

The general approach of deriving an intrinsic classification based solely on the observable morphological characters in a set of fossils is most valuable in pre-Quaternary studies, where the fossils are frequently not referable to living taxa. In Quaternary palaeoecology, however, an extrinsic classificatory approach is generally adopted.

Extrinsic classification

In Quaternary palaeontology, the investigator accepts a pre-existing modern taxonomy as a standard, and attempts to identify his fossils by comparison with the standard. If the fossil is similar to a taxonomic category of modern reference material, but dissimilar to all the other categories, the name of the living taxon is applied to the fossil. This approach is different from the grouping or intrinsic classification approach, in that individuals are added to a pre-existing or extrinsic classification. Mathematically this approach corresponds with discrimination or identification methods.

The technique of discriminant analysis seeks to distinguish between two groups of individuals (e.g. taxon A and B) which have been defined *a priori* on any number of characters in such a way that individuals of unknown identity can be assigned (or identified) to one or other of the two groups. Klovan and Billings (1967), Reyment (1961), and Davis (1973) give excellent, non-mathematical accounts of the method, and Reyment (1960, 1963, 1966b), and Blackith and Reyment (1971) discuss numerous applications of it to palaeoecology. If many *a priori* groups (e.g. taxa A, B, C, D, E, F, etc.) are involved, the method can be extended to discriminate between many groups and to assign unknown individuals to one of the groups. This method is known as multiple discriminant analysis or canonical variates analysis, and Blackith and Reyment (1971) and Oxnard (1973) provide useful introductions to the method. Palaeoecological applications of it include those by Buzas (1966), Malmgren (1974), Reyment (1963, 1966b), Oxnard (1972), and Ashton *et al.* (1957).

Although, in practice, few palaeontologists explicitly use such numerically defined procedures for distinguishing between taxa and for assigning new fossils to the existing taxonomy, discriminant analysis and its relative canonical variates analysis are mathematical analogues for the mental processes of 'distinguishing', 'matching', and 'identifying' unknown biological samples (see Gower, 1975) in relation to a pre-existing taxonomy.

Other aids have been made to the identification of fossils, such as the use of dichotomous keys, polyclaves, and pattern recognition either visually or automatically by machine.

Nomenclature of fossils

In palaeoecological studies, the taxonomic identification of an individual fossil or of a group of fossils may be in doubt in the mind of the investigator. The confidence an investigator feels in his identification will depend on a variety of factors. For example, (1) the investigator may lack sufficient competence or experience with the taxonomic group concerned, (2) he may lack the time or the equipment required to study the fossils in sufficient detail, (3) the fossils observed may lack some essential diagnostic characters because of poor preservation, (4) the fossil may belong to a taxonomic group that is inadequately understood or described within the region of his study, and (5) the extent and the observed variability of the reference material with which the fossils are being compared may limit the taxonomic precision to which the fossils can be identified.

The degree to which the investigator attempts to overcome these problems will depend on his judgement of the degree of taxonomic precision required for his study. Because he cannot know to what uses his data may be put by others, however, it is essential that the degree of confidence in the identifications be clearly stated and defined in his notes and in subsequent publications. If this is done, equal weight in interpretation will not be given to both certain (i.e. beyond reasonable doubt) and doubtful identifications. It is better to admit doubt than to mislead others with false confidence.

It is also essential to cite the taxonomic treatment followed for the taxonomic nomenclature used in a palaeoecological study. If the treatment is cited, it is not then necessary to give the authorities for each taxonomic entity.

Several systems of conventions have been devised to indicate the certainty of identification. Perhaps the most widely used and most easily understood is the one used throughout this book, which is illustrated by the following examples using the plant family Plantaginaceae.

1. Plantaginaceae Family determination certain; types, genera, or species undetermined or indeterminable.
2. *Plantago* Genus determination certain; types or species undetermined or indeterminable.
3. *Plantago lanceolata* Species determination certain.
4. *Plantago* cf. *P. lanceolata* Genus determination certain; species determination less certain because of poor preservation, inadequate reference material, or close morphological similarity of the fossil with those of other taxa.
5. *Plantago lanceolata*/*P. maritima* One fossil type present; only two taxa are considered probable alternatives, but further distinctions are not possible on the basis of fossil morphology alone.
6. *Plantago*-type One fossil type present; three or more taxa are possible alternatives, but further distinctions are not possible on the basis of fossil morphology alone. The taxonomic composition of the type should be stated.
7. Plantaginaceae undifferentiated (undiff.) Family determination certain, some morphological types distinguished and presented separately. The fossil type represents fossils that were not or could not be separated beyond family level.
8. *Plantago* undiff. Genus determination certain, some morphological types distinguished and presented separately. The fossil type represents fossils that were not or could not be separated beyond genus level.

Reconstruction of populations

The population of an organism is the number living at any one time. Usually population studies are restricted to the population of one particular area. It is important to a palaeoecologist to know whether the organism he is interested in is abundant or rare, as this is an important aspect of the community he is trying to reconstruct, and whether the size of the population changes with time. Quaternary palaeoecologists have not been particularly interested in the population dynamics of any organism in detail. Such studies are more frequently undertaken by pre-Quaternary palaeontologists. However, population studies in the Quaternary may well be a fruitful line of research (Watts, 1973), as evolution by natural selection acts upon populations rather than individuals. Natural selection acts most intensively at weak points in the life cycle, for example at birth, and thus in a population study it is necessary to know something

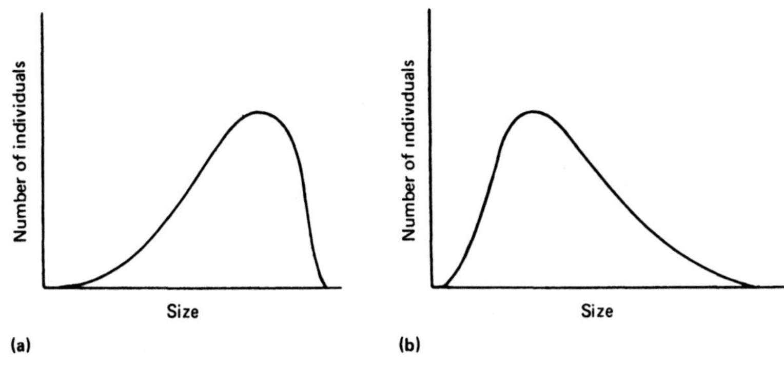

Fig. 2.9 Graphs of numbers of individuals against size. If size is an indication of age, the maxima of the curves show the point in the life cycle where mortality was highest.

of the age structure of the population, the number of individuals, the mortality, the birth rate, etc. (see Reyment, 1971). These features can be combined in a life table (see Chapter 7). Population studies have usually been made on organisms which produce one fossil per individual. These include the majority of animals and some algae. It is more difficult to reconstruct populations of other plants because each individual may produce many seeds and pollen grains or spores. The production and representation of the fossil parts must be known in order to calculate the number of individuals which produced them. The only way to find these values is by studying modern individuals, deriving modern representation factors, and transferring the data to the fossil situation. We shall discuss how this has been done in Chapter 10.

In animal population studies, size is often taken as an index of age of the individual. The use of size measurements in reconstructing fossil populations has been extensively discussed by Olson (1957). Examples of typical curves of numbers of individuals in a fossil assemblage plotted against size are shown in Fig. 2.9. Curve (a) shows that larger fossils are more numerous, and therefore that the highest mortality was amongst older organisms. Curve (b), on the contrary, shows that the highest mortality rate was amongst the smaller, and hence the younger individuals and that relatively few attained large size and hence maturity.

As we know, the fossil assemblage may not reflect faithfully the death or life assemblages, so care has to be taken in interpreting such curves. There may well have been differential movement of different sized material to the final site of fossilization. If possible, other indices besides size should be used in the estimate of age of a population, such as growth rings in corals and fish scales, layers of dentine and amount of wear in teeth, and degree of ossification of bones. Amongst plants, tree rings are useful for estimating the age of temperate trees.

Besides differential transport problems, there are other limitations on the use of size to estimate age. Many species have a sexual dimorphism, and unless the fossils can be sexed, a bimodal curve may result. An amusing example of sexual dimorphism misleading palaeontologists was described by Kurtén (1958). A museum collection of cave bear bones contained 90% male individuals and 10% females. The museum collectors had had first choice of a set of bones from a bear cave, and had chosen the largest and most impressive. As the male bears are much larger than the females, the result was a highly biased and spurious sample.

Another problem in the use of size measurements occurs in those organisms with irregular growth rates, such as the larval growth stages in ostracods, and the larval instars of insects. Figure 2.10 illustrates this situation for an ostracod. The larval stages 1–5 are discrete size groups and the adult stage 6 shows sexual dimorphism. Kurtén (1964), Chapter 29 in Blackith and Reyment (1971), and Reyment (1966b) expand this discussion of reconstruction of fossil animal populations (see also Chapter 7).

As mentioned above, it is much more difficult to reconstruct fossil plant populations, because of the numerous and unknown number of fossils produced by one individual. An accurate population study is only really possible when the life and death

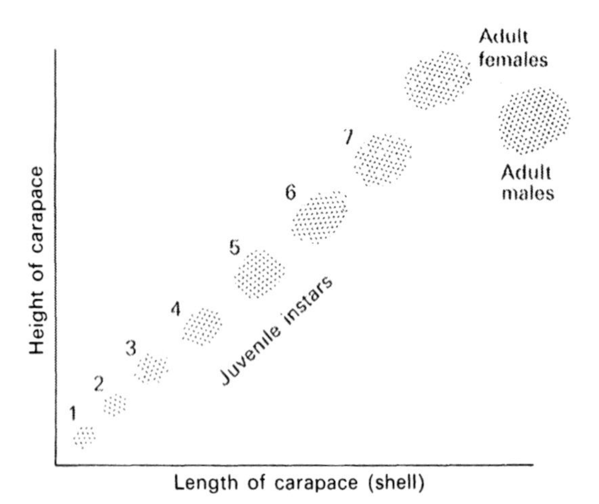

Fig. 2.10 Growth series for a Palaeocene cytherid ostracod, showing the discontinuous increase in size for each larval stage, and sexual dimorphism in the adult. (After Blackith and Reyment, 1971.)

assemblages are preserved *in situ*, and the whole population can be studied. Such a situation was found at Terneuzen in The Netherlands, where a layer of pine stumps and trunks about 4300 years old was preserved at one level in a peat bog. The investigations by Munaut were redescribed by Godwin (1968). Munaut demonstrated by tree-ring analysis (dendrochronology) and matching ring sequences that the trees started growth within 100 years and died some 250 years later over a period of about 40 years as a result of a gradual rise in the water table and the initiation of raised bog growth. He was thus able to derive a picture of the population growth and death for this short interval of geological time.

In favourable situations, the fossil assemblage may be sufficiently well preserved to allow the palaeoecologist to assemble data suitable for the construction of life tables (see Reyment (1971) and Chapter 7). A life table is a means of expressing facts about mortality in terms of probabilities. From such a table, survivorship curves showing the number of individuals living at particular ages plotted against age can be constructed, along with the curves showing mortality at different ages. Life tables permit comparisons to be made between two or more populations, as the tables are structured with respect to age. The tables also provide a means of making population estimates for different ages, and hence of suggesting possible causes of death.

Reconstructions of past populations of two or more species that may have interacted, such as in predator–prey relationships, or in competition for resources, allow the palaeoecologist to study population dynamics of two or more species varying through time. Such studies are rare, no doubt due to the considerable technical difficulties. Reyment (1966a,b) has considered the frequency and magnitude of predation by drilling gastropods, called 'drills', on pelecypods, ostracods, other gastropods, etc. (see also Reyment, 1971). He was able to estimate the numbers of predators indirectly from the number of drilled individuals of ostracods and the numbers of prey directly from the number of ostracods and to study the variation in time of the predators and the prey in long stratigraphical sequences in Nigeria. In general the frequency of predators was closely related to the frequency of prey. These estimates can now be tested against theoretical models for predator–prey relationships such as the Lotka–Volterra equations (see Hutchinson (1978) and Wilson and Bossert (1971) for useful introductions to population dynamics relevant to palaeoecology). Although palaeopopulation ecology is in its infancy, it is a field of very considerable potential, particularly as palaeoecologists can study population changes over long time periods which cannot be readily studied by modern ecologists (Van Valen, 1969; Deevey, 1969; Watts, 1973).

Reconstruction of communities

A community consists of an assemblage of organisms living together in space and time within defined boundaries of both. Communities most often include animals and plants, although ecologists and palaeoecologists frequently study only one or the other, and talk of either the ***animal community*** or the ***plant community***. In general, the community approach has been studied most by terrestrial plant ecologists and marine benthonic ecologists, whereas other types of ecologists are generally more interested in populations rather than communities. It is useful to distinguish at the outset between 'abstract' and 'concrete' communities. A ***concrete*** community is represented, for example, by particular stands or plots of vegetation in the field that can be visited, described, and sampled. An ***abstract*** community or ***community type*** is the result of classifying several 'concrete' communities together into an abstract, descriptive unit, such as is commonly done in phytosociology.

The reconstruction of past communities from the fossil record represents a major step towards the reconstruction of the past ecosystem. The biotic component, represented by communities, is the most complex part of the ecosystem, and when they have been reconstructed, inferences can be made about the environment of the past ecosystem. Such reconstructions require knowledge about the ecological requirements and tolerances of the species and communities involved.

The ecological requirements of an organism fall into two levels. Firstly, it occupies a certain habitat, with certain attributes. Secondly, within the habitat, it forms part of a community, and the factors limiting its position within the community can be said to define its ***niche***. Hutchinson (1978) suggests that the niche can be defined as the range of environmental tolerances of a species to which the species is adapted, and that the niche can be represented in a *m*-dimensional hyperspace where each of the *m* axes represents an environmental variable (see Green, 1971, for an attempt at representing the niches of freshwater molluscs using multivariate techniques). The potential niche of the species in question is represented in the hyperspace by the range of ecological tolerances of the species along each environmental axis. Each species is tolerant of certain levels of environmental variables, but its tolerances may overlap with other species. In this case, there is interspecific competition. If their tolerances coincide, one species will become extinct (competitive exclusion principle). If their tolerances do not completely overlap, the potential niches of each species will become restricted (actual niche) within the community. The community can also occupy a niche, which is the range of environmental factors common to all the organisms comprising the community.

For palaeoecological interpretations, the presence of a species of known tolerances will provide a certain spectrum of information about the past environment. However, if a past community, or even part of a community can be reconstructed, more precise information will be gained, because the niche of the community is more restricted than that of its individual components, as illustrated in Fig. 2.11.

It may be quite difficult to determine a past community from a fossil assemblage, because of all the processes affecting the death assemblage as it becomes fossilized (see Chapter 1 and Johnson, 1960; Fagerstrom, 1964). Certain situations may

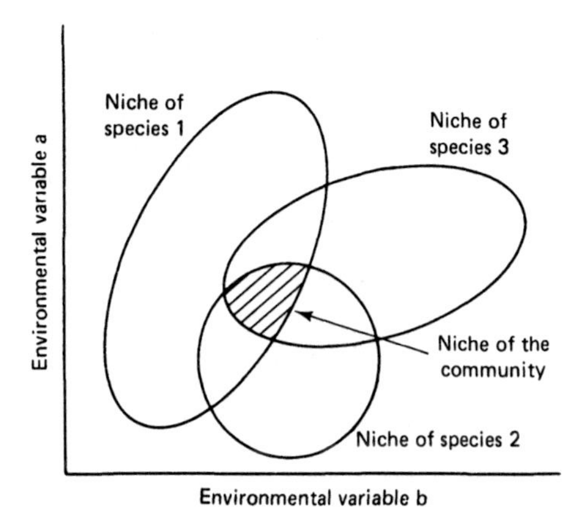

Fig. 2.11 The niches of three species as defined by two environmental variables. The niche of the community containing all three species is restricted to the area where the niches overlap.

contain communities fossilized more or less as they were in life. Examples of these are autochthonous sediments, such as peat deposits, with very little input of material into the death assemblage, and little loss of material during fossilization. Sudden catastrophes may also preserve a community intact, such as a sudden flood resulting in burial under silts, or burial by volcanic ash, of which Pompeii and Herculaneum are excellent examples.

If the community is not preserved intact or nearly so, it has to be reconstructed from available evidence. After the identification of the fossils, taxonomic analogy with modern communities may be possible. It is unlikely that all the species of a modern community will be preserved, recovered, or recognized in the fossil assemblage. However, certain species will be identified which, because of their known modern niches and the ways in which they are associated, can, by inference, be used to indicate the past occurrence of a particular community. Such species are called ***indicator species***.

If an assemblage is recovered which cannot be recognized as resembling any known modern community or which is derived from several communities, a statistical approach may be adopted. Fossils which are consistently found together can be delimited as recurrent groups (Johnson, 1962; Valentine and Mallory, 1965; Valentine and Peddicord, 1967), and interpreted as having formed a community in the past. The indicator species approach to community reconstruction is most commonly used in Quaternary palaeoecology, but the use of statistical and numerical analysis is becoming more important for analysing large and complex data sets, and providing a preliminary framework upon which more sophisticated interpretations can be based. Chapter 7 in Reyment (1971) provides a good discussion of the use of statistical and numerical methods in community reconstruction, so only a brief mention will be made here in order to provide a background for more detailed discussions in later chapters.

In a situation where the sediment is largely autochthonous and most of the fossils lived within the basin of deposition, any association in time between the fossils is likely to reflect associations between the organisms in the life assemblage. The presences and absences of fossils from particular samples can be compared numerically in order to determine whether any of the fossils occur together more frequently than would be expected by chance, and hence are associated in a statistical sense. Care must be taken to ensure that the fossils to be compared are comparable. This is particularly important in relating the observed numbers of fossils to the likely number of individuals in the community. Brachiopods, for example, present few problems, as one entire fossil represents one individual. Seeds and fruits, however, are more difficult, as plants produce variable numbers of seeds and fruits, with the result that the numbers of fossils are not comparable between different types of plants or even between species of plants. In general, the presence or absence of particular fossils can provide a basis for reconstructing past communities, because many modern communities are defined in terms of their taxonomic composition rather than their quantitative composition.

In delimiting recurrent groups of fossils using presence/absence data, measures of interspecific association are required to quantify the strength of association between the different taxa. Chi-squared statistics provide a convenient way of testing the null hypothesis that the fossils in question are independent, and that there is no association between them (see Chapter 7 in Reyment (1971) and Johnson (1962)). If many species are considered, an association matrix can be constructed and an association net or plexus can be drawn to show the patterns of association between the fossil taxa. Taxa which are most associated together (recurrent groups) can be viewed as possibly having lived together in the past as a community.

If the majority of the fossils preserved in a stratigraphical sequence is derived from beyond the basin of deposition (e.g. pollen grains and spores), the recurrent-group approach using the presence or absence of particular fossils is not appropriate. For example, in pollen analysis, some pollen types such as pine, are abundantly produced, are widely dispersed by the wind, and are thus deposited great distances from their parent plant. Pine pollen is therefore usually present in every sample, with the result that presence/absence data are of little value. Quantitative data such as proportions of the various pollen types are, however, appropriate, as in general the observed proportions of a fossil type increase with increasing abundance of the plants in the area or with the nearness of the plants to the site of deposition. Plants whose pollen show similar quantitative changes may thus have grown together in life.

Interspecific correlations between all pairs of taxa can be calculated using frequency data (Clapham, 1970), and the results can be presented as a correlation matrix or as a correlation net or plexus. Such nets or plexuses are multidimensional in character and are thus difficult to represent accurately in two dimensions. A useful way of simplifying the correlation matrix is to subject it to principal components analysis, and to plot the loadings of the various taxa on the first few principal components. In general, taxa that are correlated will be positioned near each other (Reyment, 1963). Due to the complex processes of pollen dispersal (see Chapter 9) the fossil pollen assemblage is derived from an indefinable source area. There is thus no simple spatial relationship between the fossil pollen assemblage and the life assemblage or plant communities that produced it. Correlated groups of fossil pollen can thus only refer to 'correlations' in time, whereas community reconstructions involve the patterns between taxa in both time and space.

Reconstruction of past environments and ecosystems

The reconstruction of ecosystems follows very closely on the reconstruction of the communities represented by the fossil assemblage, because it involves reconstructing the environment in which the fossil organisms lived. There are three main approaches to reconstructing the fossil ecosystem: (1) using modern analogues from known environments to compare with the fossil assemblages, (2) analysing the quantitative patterns between the fossil organisms and determining and interpreting any gradients in the data, and (3) using information from the abiotic component (sediment lithology, chemistry, etc.).

Analogues with living organisms

This approach assumes that there has been little or no change in the ecological requirements and niches of organisms and communities since the time the fossil assemblage was laid down. It is essentially an 'indicator species' approach, carrying on further from the reconstruction of the community, and is the main approach used in interpreting Quaternary palaeoecological data.

Some organisms have well-defined adaptations to environmental conditions, and a good example is the marine planktonic foraminifer *Globigerina pachyderma*. The shell of this species is coiled, and the direction of coiling varies with the sea surface temperature where it lives. In the arctic, south to about 10° or 20° latitude, the shell coils to the right. In warmer water it coils to the left. The shells are preserved in cores of marine sediment, and the direction of coiling can be used to interpret the sea surface temperatures of the past (see Fig. 2.12 and Bandy, 1960).

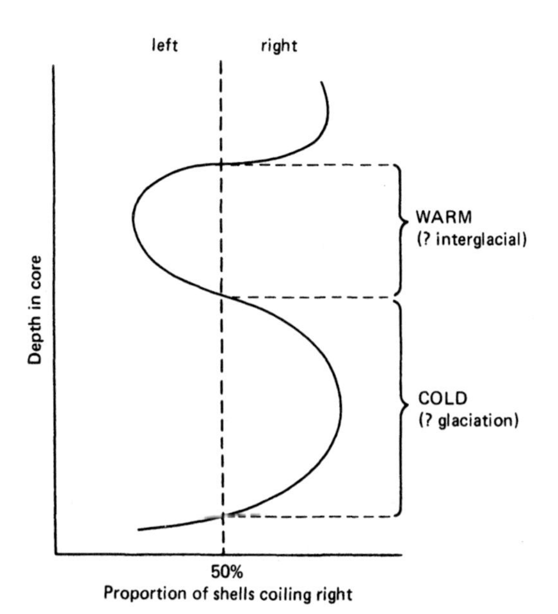

Fig. 2.12 The direction of shell-coiling in the foraminifer *Globigerina pachyderma* down a core of marine sediment, and its interpretation in terms of sea-surface temperatures.

Another example of morphological adaptation to environment is found in angiosperms. The morphology of the leaves of tropical rain forest species, irrespective of taxonomic position is characteristically entire with a long thin tip, called a drip tip. In temperate climates, the drip tips are absent, and many leaves have a lobed or serrated margin (see Fig. 2.13 and Krassilov, 1975).

In fossil leaf floras, particularly from Tertiary deposits, a change from a predominantly entire leaf shape with drip tips to a predominantly lobed or serrated leaf shape is taken to indicate a change from tropical to temperate climatic conditions. A good account of such studies from the Western United States is given by Wolfe (1971).

However, these neat situations are uncommon, and the palaeoecologist has to rely on comparison with modern floras and faunas. He may rely entirely on his own intuition and judgement, which, if he has a good working knowledge of the modern ecology of the group he is studying, will probably provide an acceptable interpretation. However, there are always the unknown factors acting during deposition of the fossils and during their fossilization. Hence some method of testing the representation of the modern taxa in a situation similar to the fossil locality is necessary. To do this, modern surface samples have been taken and

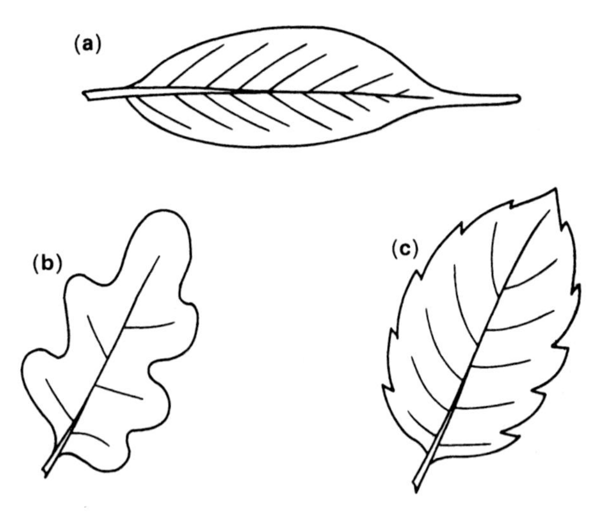

Fig. 2.13 Different leaf shapes. (**a**) has a drip tip, characteristic of tropical environments. (**b**) and (**c**) are lobed and serrated, shapes characteristic of temperate trees.

analysed. They can be directly compared to the life assemblage which produced them, and factors of representation can be assessed. Surface samples have been analysed for many fossilizable organisms, but the most extensive studies have been on the representation of vascular plants by pollen and spores (see Chapter 11).

Recently attempts have been made to derive quantitative calibration or 'transfer' functions that relate entire modern assemblage of organisms to present-day environments. By assuming that these transfer functions are invariant in time and space, they can be applied to fossil assemblages of the same organisms to derive quantitative estimates of the past environment (see Chapter 12).

By means of a variety of multiple regression and correlation techniques (Webb and Clark, 1977) it is possible to derive modern transfer functions T_m that express the quantitative correlations between a modern assemblage X_m and the modern environment E_m,

i.e. $X_m T_m = E_m$

Given a fossil assemblage X_f, and using T_m, we can derive estimates of E_f, the past environment,

i.e. $X_f T_m = \hat{E}_f$

Imbrie and Kipp (1971) developed this approach using modern assemblages of foraminifers to reconstruct past sea-surface temperatures in the winter and the summer and salinities from deep-sea cores. Similar approaches have been applied to coccoliths, radiolarians, pollen, ostracods, tree-rings, and diatoms. The approach is discussed more fully in Chapter 12.

Numerical analysis of fossil data

Changes in the proportions of different fossils may occur through a stratigraphical sequence. Multivariate analysis of the palaeontological data can often help to detect gradients of biological change that may reflect changing environments (see Reyment, 1966c). Principal components analysis is a suitable technique.

Suppose we have four hypothetical species (A, B, C, D) whose percentage frequencies have been determined in 24 samples in a borehole.

Sample no.	Species A	B	C	D
1	45	52	1	2
2	52	45	2	1
3	50	47	1	2
4	44	52	2	2
5	50	45	3	2
6	49	49	1	1
7	48	49	1	2
8	47	50	2	1
9	1	25	50	24
10	2	24	48	26
11	1	23	49	27
12	1	26	50	23
13	1	25	48	26
14	1	24	50	25
15	2	23	51	24
16	2	24	52	22
17	1	2	44	53
18	2	2	42	54
19	1	1	48	50
20	1	1	49	49
21	1	1	50	48
22	2	1	47	50
23	2	2	45	51
24	1	1	46	52

(depth ↓ — samples 9–19)

If we calculate the correlations between the four species we get the following matrix of correlations:

A	1.0			
B	0.865	1.0		
C	−0.993	−0.839	1.0	
D	−0.843	−0.992	0.802	1.0
	A	B	C	D

indicating that species A and B are strongly and positively correlated, that C and D are also positively correlated together, and that A and C, A and D, B and C, and B and D are all strongly negatively correlated.

If we now perform a principal components analysis of these data and extract the eigenvalues and eigenvectors of this correlation matrix, we get the following results

	Principal Component (eigenvectors)			
	1	2	3	4
Species A	0.505	−0.444	0.519	0.528
Species B	0.504	0.449	−0.575	0.462
Species C	−0.496	0.551	0.418	0.525
Species D	−0.496	−0.546	−0.473	0.482
Eigenvalues	3.667	0.321	0.012	0.001
Percentage of total variance	91.67	8.03	0.29	0.001
Cumulative percentage of total variance	91.67	99.70	99.99	100.00
Total variance (trace) = 4.0				

The first principal component reflects the positive correlation of species A and B and the negative correlation with C and D, whereas component two indicates a positive correlation between B and C and negative correlation between A and D. The other two components are so small that we can neglect them.

We can now position the 24 samples on the first and second components, to obtain the configuration shown in Fig. 2.14. We can now see that there are three groups of samples with similar fossil compositions – samples 1–8, 9–16, and 17–24. From other studies it is known that taxa A and B are only found in fine-grained silts and clays, whereas taxa C and D are characteristic of coarse organic muds. These observations suggest that A and B are indicative of deep, still water whereas C and D reflect shallow, organic-rich water. By applying these ecological inferences to the data displayed in Fig. 2.14 we can now suggest that the first principal component reflects a gradient from shallow water to deep water and possibly from inorganic to organic sedimentation. The second component may also reflect a gradient from deep to intermediate-depth water. Independent support for these interpretations comes from examining the sediment lithology of the sediment samples. Samples 1–8 are from fine-grained, laminated silts and clays of low organic content, 17–24 are from coarse organic muds, and 9–16 are from fine-detritus organic silty muds.

Burnaby (1961) analysed the frequencies of benthonic foraminifers in 27 samples from the Chalk Marl (Cretaceous) in Cambridgeshire. By analogues with present-day foraminifers, Burnaby was able to suggest that some of the samples represented deep-water and some represented shallow-water conditions. By regarding these two groups of samples as *a priori* groups, Burnaby calculated a discriminant function between the two groups (see p. 23). By computing the position of the other known samples on this function he was able to suggest a relative measure of the water depth of the Chalk Marl sea.

Analysis of the abiotic, sediment component

The sedimentary component in a palaeoecological study can provide valuable information about the past environment at the time of its deposition. For example, the particle size of an aquatic sediment reflects the amount of water movement, as fine particles are only deposited in still water. The amount of inorganic matter in a lake sediment may reflect the degree of soil erosion from the surrounding landscape and inwashing by the inflow streams. The reason for soil disturbance could either be a cold climate limiting plant cover, or disturbance by man of an otherwise stable soil by tree felling and cultivation. The type of peat laid down in a peat bog may impart information on the wetness of the surface, which may or may not be related to climatic conditions.

If the characteristics of the sediment are analysed, the data can be subjected to multivariate analysis in a way similar to the biotic component, and the two types of data can be linked in an analysis to determine any association of fossils with particular sediment types. This type of study was done, for example, by Fox (1968) on Upper Ordovician sediments in Indiana. He determined various recurrent faunal groups by interspecific association, and

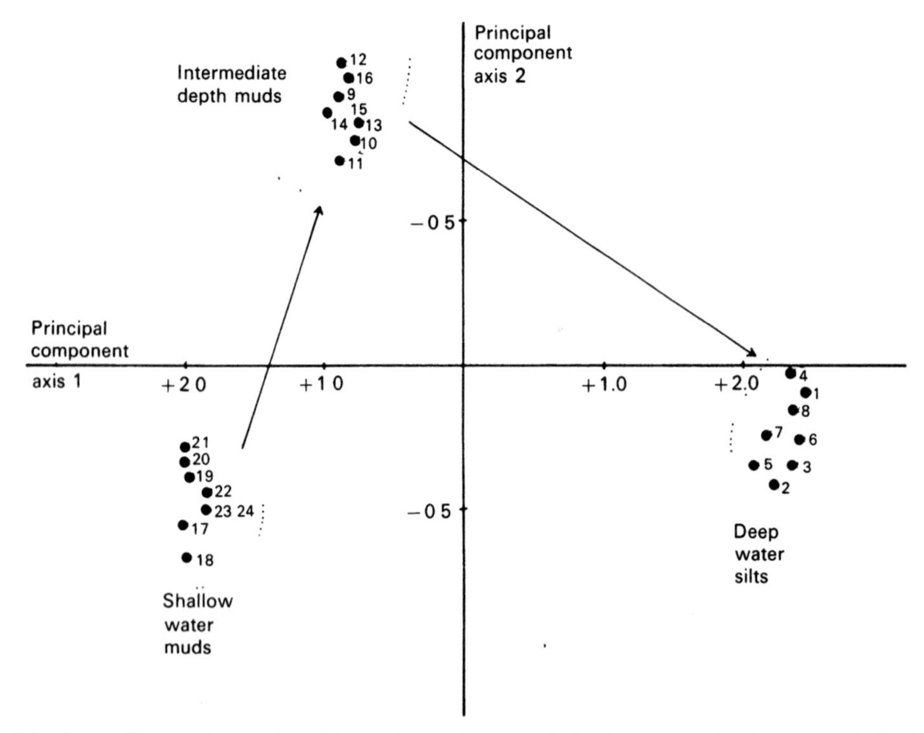

Fig. 2.14 The position of the 24 samples in the example in the text on the first two principal component axes. For further explanation, see text.

related them to particular sedimentary environments using factor analysis, a method closely related to principal components analysis. This enabled him to suggest that the major changes in his faunal assemblages over time were caused by transgressions and regressions of the sea.

Geochemical characteristics of the sediments can also be used in environmental reconstructions. For example, Reyment (1976) related the frequencies of ostracods and foraminifers in late Cretaceous boreholds from western Nigeria to 14 geochemical variables (Si, Fe, Mg, Ca, Na, K, Ti, P, Mn, V, Mo, Sr, Pb, Zn) using canonical correlation analysis. This mathematically complex method is the multivariate extension of linear correlation analysis between two variables in which correlations are calculated between two *sets* of variables rather than between two individual variables. The first canonical correlation indicates a positive correlation between Si, Fe, K, Mn, V, Mo, Pb, and Zn with *Gaboriella elongata* and *Valvulineria* (both foraminifers) and a negative correlation between Mg, Ca, and Sr with ostracods. This pattern can be interpreted as contrasting clastic sediments containing foraminifers with carbonate-rich sediments containing ostracods, perhaps related to a changing position of the shoreline during the late Cretaceous.

In any palaeoecological reconstruction of a past ecosystem, it is essential to utilize all the available evidence, both biotic and abiotic. Information from the sediment matrix will often provide different insights and information relevant to the environmental reconstruction from those obtained by studying the fossil assemblages alone. Multivariate methods of data analysis are useful aids in the handling and simplification of complex sets of palaeoecological data (see Imbrie, 1964). These methods do not interpret the results, but they can often sort the data in such a way as to suggest hypotheses which can be tested by further investigation.

Conclusions

In this chapter we have outlined how palaeoecological investigations proceed. Obviously, different problems will require different techniques, but we have tried to present a survey of the various approaches possible at different stages in an investigation. We can summarize the whole process in the flow-chart of Fig. 2.3.

At all stages of an investigation, modern analogues are inevitably involved, particularly in Quaternary palaeoecology where the fossils are the same as or closely related to living organisms. The rest of this book will be devoted to enlarging upon the various stages involved in palaeoecological studies of Quaternary animals, plants, and environments.

References

General principles of palaeoecology

AGER, D. V. (1963). *Principles of Palaeoecology*. McGraw-Hill.

IMBRIE, J. AND NEWELL, N. (1964). *Approaches to Paleoecology*. J. Wiley and Sons.

REYMENT, R. A. (1971). *Introduction to Quantitative Paleoecology*. Elsevier.

Stratigraphical principles

AMERICAN COMMISSION ON STRATIGRAPHIC NOMENCLATURE (1961). Code of stratigraphic nomenclature. *Amer. Assoc. Petrol. Geol. Bull.*, **45**, 645–65.

HARLAND, W. B. *et al.* (1972). A concise guide to stratigraphical procedure. *Q. Jl. geol. Soc. Lond.*, **128**, 295–305.

HEDBERG, H. D. (1972a). Introduction to an international guide to stratigraphic classification, terminology, and usage. *Boreas*, **1**, 199–211.

HEDBERG, H. D. (1972b). Summary of an international guide to stratigraphic classification, terminology, and usage. *Boreas*, **1**, 213–39.

HEDBERG, H. D. (1976). *International Stratigraphic Guide*. J. Wiley and Sons.

VAN EYSINGA, F. W. B. (1970). Stratigraphic terminology and nomenclature; a guide for editors and authors. *Earth-Sci. Rev.*, **6**, 267–88.

Stratigraphical applications in the Quaternary

MANGERUD, J., ANDERSEN, S. T., BERGLUND, B. E. AND DONNER, J. J. (1974). Quaternary stratigraphy of Norden, a proposal for terminology and classification. *Boreas*, **3**, 109–28.

MITCHELL, G. F., PENNY, L. F., SHOTTON, F. W. AND WEST, R. G. (1973). A correlation of Quaternary deposits in the British Isles. *Geol. Soc. Lond., Special Report*, No. **4**, 99 pp.

VAN DER HEIDE, S. AND ZAGWIJN, W. H. (1967). Stratigraphical nomenclature of the Quaternary deposits in the Netherlands. *Medel. Geologische Sticht.*, N.S. **18**, 23–9.

WEST, R. G. (1963). Problems of the British Quaternary. *Proc. Geol. Assoc.*, **74**, 147–86.

WEST, R. G. (1967). The Quaternary of the British Isles. In *The Quaternary* Vol. 2 (ed. K. Rankama). Interscience Publishers.

WEST, R. G. (1977). *Pleistocene Geology and Biology* (2nd edition). Longman (Chapters 11, 12).

ZAGWIJN, W. H. (1963). Pleistocene stratigraphy in the Netherlands, based on changes in vegetation and climate. *Verh. Kon. Ned. Geol. Mijnb. Gen. Geol.*, Serie **21–2**, Jub. Conv. pt. 2, 173–96.

Geological time and radiometric dating

BURLEIGH, R. (1974). Radiocarbon dating: some practical considerations for the archaeologist. *J. Arch. Sci.*, **1**, 69–87.

KITTS, D. B. (1966). Geologic time. *J. Geol.*, **74**, 127–46.

MÜLLER, H. (1974a). Pollenanalytische Untersuchungen und Jahresschichtenzählungen an der holstein-zeitlichen Kieselgur von Munster-Breloh. *Geol. Jb.*, **A21**, 107–40.

MÜLLER, H. (1974b). Pollenanalytische Untersuchungen und Jahresschichtenzählungen an der eem-zeitlichen Kieselgur von Bispingen/Luhe. *Geol. Jb.*, **A21**, 149–69.

MULLER, R. A. (1977). Radioisotope dating with a cyclotron. *Science*, **196**, 489–94.

OLSSON, I. U. (1968). Modern aspects of radiocarbon datings. *Earth-Sci. Rev.*, **4**, 203–18.

OLSSON, I. U. (1974). Some problems in connection with the evaluation of C^{14} dates. *Geol. Fören. Förhandl. Stock.*, **96**, 311–20.

SHACKLETON, N. J. AND TURNER, C. (1967). Correlation between marine and terrestrial Pleistocene successions. *Nature*, **216**, 1079–82.

SHOTTON, F. W. (1967). The problems and contributions of methods of absolute dating within the Pleistocene period. *Q. Jl. geol. Soc. Lond.*, **122**, 356–83.

STUIVER, M., HEUSSER, C. J., AND YANG, I. C. (1978). North American Glacial History extended to 75,000 Years Ago. *Science*, **200**, 16–21.

WATKINS, T. (1975). *Radiocarbon: Calibration and Prehistory*. Edinburgh University Press.

WEST, R. G. (1977). *Pleistocene Geology and Biology* (2nd edition). Longman (Chapter 9).

Sampling procedures

DE CAPRARIIS, P., LINDEMANN, R. H. AND COLLINS, C. M. (1976). A method for determining optimum sample size in species diversity studies. *J. Math. Geol.*, **8**, 575–81.

HECK, K. L., VAN BELLE, G. AND SIMBERLOFF, D. (1975). Explicit calculation of the rarefaction diversity measurement and the determination of sufficient sample size. *Ecology*, **56**, 1459–61.

IMBRIE, J. (1955). Biofacies analysis. *Geol. Soc. Amer. Spec. Paper*, **62**, 449–64.

KOCH, G. S. AND LINK, R. F. (1970). *Statistical Analysis of Geological Data*. J. Wiley and Sons (Chapters 3, 7, and 8).

KRUMBEIN, W. C. AND GRAYBILL, F. A. (1965). *An Introduction to Statistical Models in Geology*. McGraw-Hill (Chapter 7).

RAUP, D. M. (1975). Taxonomic diversity estimation using rarefaction. *Paleobiology*, **1**, 333–42.

STANTON, R. J. AND EVANS, I. (1972). Community structure and sampling requirements in palaeoecology. *J. Paleontol.*, **46**, 845–58.

Sampling strategies in Quaternary pollen analysis

BIRKS, H. J. B. (1973). *Past and Present Vegetation of the Isle of Skye – a Palaeoecological Study*. Cambridge University Press.

BIRKS, H. J. B., DEACON, J. AND PEGLAR, S. M. (1975). Pollen maps for the British Isles 5000 years ago. *Proc. R. Soc.* B, **189**, 87–105.

BRUBAKER, L. B. (1975). Postglacial forest patterns associated with till and outwash in northcentral Upper Michigan. *Quat. Res.*, **5**, 499–527.

CASELDINE, C. J. AND GORDON, A. D. (1978). Numerical analysis of surface pollen spectra from Bankhead Moss, Fife. *New Phytol.*, **80**, 435–53.

TURNER, J. (1970). Post-Neolithic disturbance of British vegetation. In *Studies in the Vegetational History of the British Isles* (eds D. Walker and R. G. West). Cambridge University Press.

The fossil species

IMBRIE, J. (1957). The species problem with fossil animals. In *The Species Problem* (ed. E. Mayr). *Amer. Assocn. Advan. Sci. Pub.*, **50**, 125–53.

MCALESTER, A. L. (1962). Some comments on the species problem. *J. Paleontol.*, **36**, 1377–81.

SYLVESTER-BRADLEY, P. C. (1956). The species concept in palaeontology. *Systematics Association Publ.*, **2**, 145 pp.

WELLER, J. M. (1961). The species problem. *J. Paleontol.*, **35**, 1181–92.

Methods of classifying and identifying fossils

BLACKITH, R. E. AND REYMENT, R. A. (1971). *Multivariate Morphometrics*. Academic Press (Chapters 3, 7, 8, 12, 29).

DAVIS, J. C. (1973). *Statistics and Data Analysis in Geology*. J. Wiley and Sons (Chapter 7).

EVERITT, B. (1974). *Cluster Analysis*. Heinemann.

GITTINS, R. (1969). The application of ordination techniques. In *Ecological Aspects of the Mineral Nutrition of Plants* (ed. I. H. Rorison). Blackwell.

GOWER, J. C. (1975). Relating classification to identification. In *Biological Identification with Computers* (ed. R. J. Pankhurst). Academic Press.

JÖRESKOG, K. G., KLOVAN, J. E. AND REYMENT, R. A. (1974). *Geological Factor Analysis*. Elsevier.

KAESLER, R. L. (1969). Numerical taxonomy in paleontology: classification, ordination and reconstruction of phylogenies. *Proc. N. Amer. Paleontol. Convention* B, 84–100.

KLOVAN, J. E. AND BILLINGS, G. K. (1967). Classification of geological samples by discriminant-function analysis. *Bull. Canad. Petrol. Geology*, **15**, 313–30.

OXNARD, C. E. (1973). *Form and Pattern in Human Evolution*. University of Chicago Press.

REYMENT, R. A. (1961). A note on geographical variation in European *Rana*. *Growth*, **25**, 219–27.

REYMENT, R. A. (1973). The discriminant function in systematic biology. In *Discriminant Analysis and Applications* (ed. T. Cacoulos). Academic Press.

SEAL, H. (1964). *Multivariate Statistical Analysis for Biologists*. Methuen.

SNEATH, P. H. A. AND SOKAL, R. R. (1973). *Numerical Taxonomy*. W. H. Freeman.

Applications of multivariate techniques to classifying and identifying fossils

ASHTON, E. H., HEALY, M. J. R. AND LIPTON, S. (1957). The descriptive use of discriminant functions in physical anthropology. *Proc. R. Soc.* B, **146**, 552–72.

BUZAS, M. A. (1966). The discrimination of morphological groups of *Elphidium* (foraminifer) in Long Island Sound through canonical analysis and invariant characters. *J. Paleontol.*, **40**, 585–94.

CHRISTENSEN, W. K. (1974). Morphometric analysis of *Actinocamax plenus* from England. *Bull. geol. soc. Denmark*, **23**, 1–26.

HILLS, L. V., KLOVAN, J. E. AND SWEET, A. R. (1974). *Juglans eocinerea* n.sp., Beaufort Formation (Tertiary), southwestern Banks Island, Arctic Canada. *Can. J. Bot.*, **52**, 65–90.

KLOVAN, J. E. (1970). Numerical classification of *Stictostroma* Parks from the Devonian of southern Ontario, Canada. *Bull. Geol. Inst. Univ. Upsala*, N.S. **2**, 29–40.

MALMGREN, B. A. (1974). Morphometric studies of planktonic foraminifers from the type Danian of southern Scandinavia. *Stockholm Contrib. Geol.*, **29**, 1–126.

MALMGREN, B. A. AND KENNETT, J. P. (1972). Biometric analysis of phenotypic variation: *Globigerina pachyderma* (Ehrenberg) in the South Pacific Ocean. *Micropaleontol.*, **18**, 241–8.

OXNARD, C. E. (1972). Some African fossil foot bones: a note on the interpolation of fossils into a matrix of extant species. *Am. J. Phys. Anthrop.*, **37**, 3–12.

REYMENT, R. A. (1960). Studies on Nigerian Upper Cretaceous and Lower Tertiary Ostracoda. I. Senonian and Maestrichtian Ostracoda. *Stockholm Contrib. Geol.*, **7**, 1–238.

REYMENT, R. A. (1963). Studies on Nigerian Upper Cretaceous and Lower Tertiary Ostracoda. II. Danian, Paleocene, and Eocene Ostracoda. *Stockholm Contrib. Geol.*, **10**, 1–286.

REYMENT, R. A. (1966b). Studies on Nigerian Upper Cretaceous and Lower Tertiary Ostracoda. III. Stratigraphical, paleoecological and biometrical conclusions. *Stockholm Contrib. Geol.*, **14**, 1–151.

Reconstruction of populations

DEEVEY, E. S. (1969). Specific diversity in fossil assemblages. In *Diversity and Stability in Ecological Systems* (eds G. M. Woodwell and H. H. Smith). Brookhaven.

GODWIN, H. (1968). Terneuzen and buried forests of the East Anglian Fenland. *New Phytol.*, **67**, 733–8.

KURTÉN, B. (1958). Life and death of the Pleistocene cave bear, a study in paleoecology. *Acta Zool. Fennica*, **95**, 1–59.

KURTÉN, B. (1964). Population structure in paleoecology. In *Approaches to Paleoecology* (eds J. Imbrie and N. Newell). J. Wiley and Sons.

OLSON, E. C. (1957). Size-frequency distributions in samples of extinct organisms. *J. Geol.*, **65**, 309–33.

REYMENT, R. A. (1966a). Preliminary observations on gastropod predation in the Western Niger delta. *Palaeogeography, Palaeoclimatology, Palaeoecology*, **2**, 81–102.

REYMENT, R. A. (1966b). Studies on Nigerian Upper Cretaceous and Lower Tertiary Ostracoda. III. Stratigraphical, paleoecological and biometrical conclusions. *Stockholm Contrib. Geol.*, **14**, 1–151.

REYMENT, R. A. (1971). *Introduction to Quantitative Paleoecology*. Elsevier (Chapter 5).

SYLVESTER-BRADLEY, P. C. (1958). The description of fossil communities. *J. Paleontol.*, **32**, 214–35.

VAN VALEN, L. (1969). Variation genetics of extinct animals. *Amer. Natur.*, **103**, 193–224.

WATTS, W. A. (1973). Rates of change and stability in vegetation in the perspective of long periods of time. In *Quaternary Plant Ecology* (eds H. J. B. Birks and R. G. West). Blackwell.

Principles of population ecology

GREEN, R. H. (1971). A multivariate statistical approach to the Hutchinsonian niche: Bivalve molluscs of central Canada. *Ecology*, **52**, 543–56.

HUTCHINSON, G. E. (1978). *An Introduction to Population Ecology*. Yale University Press.

WILSON, E. O. AND BOSSERT, W. H. (1971). *A Primer of Population Biology*. Sinauer.

Reconstruction of communities

BRIDEAUX, W. W. (1971). Recurrent species groupings in fossil microplankton assemblages. *Palaeogeography, Palaeoclimatology, Palaeoecology*, **9**, 101–22.

CLAPHAM, W. B. (1970). Nature and paleogeography of Middle Permian floras of Oklahoma as inferred from their pollen record. *J. Geol.*, **78**, 153–71.

FAGERSTROM, J. A. (1964). Fossil communities in paleoecology: their recognition and significance. *Bull. Geol. Soc. Amer.*, **75**, 1197–216.

JOHNSON, R. G. (1960). Models and methods for analysis of the mode of formation of fossil assemblages. *Bull. Geol. Soc. Amer.*, **71**, 1075–86.

JOHNSON, R. G. (1962). Interspecific associations in Pennsylvanian fossil assemblages. *J. Geol.*, **72**, 32–55.

JOHNSON, R. G. (1964). The community approach to paleoecology. In *Approaches to Paleoecology* (eds J. Imbrie and N. Newell). J. Wiley and Sons.

REYMENT, R. A. (1963). Multivariate analytical treatment of quantitative species associations: an example from palaeoecology. *J. Anim. Ecol.*, **32**, 535–47.

REYMENT, R. A. (1971). *Introduction to Quantitative Paleoecology*. Elsevier (Chapter 7).

VALENTINE, J. W. AND MALLORY, B. (1965). Recurrent groups of bonded species in mixed death assemblages. *J. Geol.*, **73**, 683–701.

VALENTINE, J. W. AND PEDDICORD, R. G. (1967). Evaluation of fossil assemblages by cluster analysis. *J. Paleontol.*, **41**, 502–7.

Reconstructions of environments and ecosystems

BANDY, O. L. (1960). The geologic significance of coiling ratios in the foraminifer *Globigerina pachyderma* (Ehrenberg). *J. Paleontol.*, **34**, 671–81.

BURNABY, T. P. (1961). The palaeoecology of the foraminifera of the Chalk Marl. *Palaeontology*, **4**, 599–608.

FOX, W. T. (1968). Quantitative paleoecologic analysis of fossil communities in the Richmond Group. *J. Geol.*, **76**, 613–40.

IMBRIE, J. (1964). Factor analytic model in paleoecology. In *Approaches to Paleoecology* (eds J. Imbrie and N. Newell). J. Wiley and Sons.

IMBRIE, J. AND KIPP, N. G. (1971). A new micropaleontological method for quantitative paleoclimatology: application to a late-Pleistocene Caribbean core. In *The Late Cenozoic Glacial Ages* (ed. K. K. Turekian). Yale University Press.

KRASILOV, V. A. (1975). *Paleoecology of Terrestrial Plants*. J. Wiley and Sons (Part IV).

REYMENT, R. A. (1966c). Illustrative constructed example for quantitative ecologic reconstructions. *Stockholm Contrib. Geol.*, **14**, 138–42.

REYMENT, R. A. (1976). Chemical components of the environment and Late Campanian microfossil frequencies. *Geol. Fören. Förhandl. Stock.*, **98**, 322–8.

WEBB, T. AND CLARK, D. R. (1977). Calibrating micropaleontological data in climatic terms: a critical review. *Annals New York Academy of Sciences*, **288**, 93–118.

WOLFE, J. A. (1971). Tertiary climatic fluctuations and methods of analysis of Tertiary floras. *Palaeogeography, Palaeoclimatology, Palaeoecology*, **9**, 27–57.

3

Sampling and description of organic sediments

Coring methods

After defining the problem to be studied, the first step is to select and sample the sediments containing the fossils. In the Quaternary, fossils are most commonly studied from unconsolidated organic sediments, such as lake muds or peats. However, particularly during glacial phases, fossils may be preserved in minerogenic silts, clays, sand, and gravels. In addition, fossils can be found in ancient soils and cave earths, and also, of course, in marine sediments.

Each sediment deposit contains a sample population, which occasionally may also represent the target population, but not very often. Such a wide variety calls for a wide variety of sampling strategies and techniques. The three most common types of sampling situations are:

1. *Open sections.* Examples include drainage ditches and peat haggs in bogs, river terraces, gravel pits, sea cliffs, cave earths, soil profiles, etc. These sections may be sampled directly in the field, or monoliths can be dug, and a complete section removed to the laboratory for further analysis.
2. *Bogs and fens.* These are usually sampled by walking on to the bog or fen, drilling with some sort of coring equipment and extracting a vertical core from through the deposit.
3. *Lakes.* These can be sampled by using coring equipment operated from a boat or raft, or by a Mackereth deep-water corer (Mackereth, 1958), which is operated by compressed air. If the lakes freeze sufficiently, they can be cored from the ice in a way similar to bogs and fens (see Cushing and Wright, 1965).

If very deep or stiff sediments are to be sampled, motor-driven corers are necessary. Such a method was used, for example, by West (1977) to obtain a long core through the Crag sediments of the lower Pleistocene at Ludham in East Anglia. However, Quaternary sediments are usually reasonably accessible to hand-operated coring equipment, and deep boreholes are rarely required. In order to sample ocean sediments, special ships are equipped to take long continuous cores from beneath great depths of water (see Kullenberg, 1955).

There is a great variety of suitable coring equipment available (see West, 1977; Faegri and Iversen, 1975; Wright *et al.*, 1965; and Mott, 1966, for details). Each is suited to a particular type of sediment and site. They fall into three main categories:

1. *Peat corers.* These corers are operated by pushing them to the level required and then twisting them to cut out and remove a segment of sediment. The most commonly used corer of this type is the Swedish Hiller peat borer (see Fries and Hafsten, 1965). Various modifications have been made to this by, for example, Thomas (1964) to allow 50 cm long cores of peat to be removed intact from the corer. Another type of corer commonly used is the so-called Russian corer (Jowsey, 1966). Both this and the Hiller corer are ideal for sampling coarse, fibrous sediments such as peats. Smith *et al.* (1968) have designed a useful corer for taking long (1.5 m) large-diameter (12 cm) cores of peat that are required to provide sufficient material for close-interval radiocarbon dates.
2. *Piston corers.* The commonest type of piston corer used is the Livingstone sampler designed by D. A. Livingstone (1955), and subsequently modified and improved by Vallentyne (1955), Rowley and Dahl (1956), Livingstone (1967), Walker (1964), Cushing and Wright (1965), and Wright (1967). The most useful variant of the original Livingstone sampler is the one described by Wright (1967) in which the corer is a metal tube fitted with a moveable piston whose diameter can be varied to alter the tightness of fit. A square-rod fitted with a pair of splines rests on top of the piston whilst the corer is inserted to the depth required. This square-rod prevents the piston at the bottom of the core tube being pushed up while the corer is

inserted. When the required depth is reached, the piston is then secured, the square-rod raised, and the core tube is pushed down past the piston, thereby taking a core of sediment the length of the empty core tube. The whole corer is then raised, and the core extruded by pushing it out with the piston. Further details about the use of the Livingstone sampler in the field are given by Wright, Livingstone, and Cushing (1965) and by Deevey (1965). The Livingstone sampler is ideal for sampling non-fibrous sediments such as lake muds, marls, clays, and silts. Its one disadvantage is that the cores are only one metre long and gaps may exist between cores. With care, sets of overlapping cores can be obtained by sampling alternately from adjacent boreholes.

Large-diameter (10 cm) cores of lake sediments that are required to provide samples of sufficient size for radiocarbon dating, quantitative macrofossil analysis, and sediment chemistry can be obtained using an enlarged Livingstone corer (Cushing and Wright, 1965), or by specially designed samplers (Digerfeldt, 1965; Digerfeldt and Lettevall, 1969). With large-diameter corers, there are problems in inserting and removing the corer and in lifting the corer when it is full of sediment. Drive-frames and high-geared chain hoists (Cushing and Wright, 1965) are often essential aids for obtaining large-diameter cores from lakes and fens.

3. *Surface mud corers.* The uppermost 5 or 10 cm of lake mud are invariably very soft, flocculent, and water saturated, which leads to difficulties in raising an undisturbed core. A variety of methods exist, the most useful of which are described by Wright (1969) and Hongve (1972). These include a simple modification of the Livingstone corer using a perspex tube carefully lowered to a point about 20 cm above the sediment surface. The piston is then secured, and the plastic tube is pushed into the soft sediment until the piston has risen to near the top of the tube. A core of undisturbed surface mud overlain by about 20 cm of water is thus obtained.

More sophisticated methods for sampling uppermost muds include the freeze-corer or 'frozen finger' (Shapiro, 1958; Swain, 1973), an electrically triggered piston corer (Davis and Doyle, 1969), and a small mini-version of the Mackereth corer (Mackereth, 1969).

Sediment sampling

Obviously a sequence containing the most complete section through the deposit is required. This can often be easily selected from an open section, although it should be borne in mind that only one plane is visible in the face. In a bog or lake, the most complete sections are usually found in the deepest part of the basin, which is usually near the middle. It may be necessary to level in a series of test borings to determine the three-dimensional stratigraphy of the deposit and thus select the most appropriate site for sampling.

Depending upon the problem being investigated, the next step after obtaining the cores or monoliths is to sample them for fossil and sediment analysis. The surface of the sediment should be carefully scraped clean, to avoid contamination. Small samples, about 1–2 cm^3 are required for analysis of small abundant fossils such as pollen grains or diatoms. Larger samples of about 100–400 cm^3 are required for analysis of bigger fossils such as seeds and fruits, molluscs, and beetles. Radiocarbon dates also usually require samples of this magnitude. The cores may either be sampled in the field, or they may be transported back to the laboratory. The latter is preferable, as contamination is much more likely to occur in the field, especially if it is raining. As Faegri and Iversen (1975, p. 100) say, 'In rain, field work is impossible, or nearly so. The note book becomes wet, the pencil tears holes in the paper, rain drops into the sample and splashes the substance all over, and everything becomes hopelessly dirty in a short time. In such circumstances it is almost impossible to work with sufficient precision and to obtain pure samples .... *No method can compensate for contamination of samples!*'

Field work is an extremely important part of any palaeoecological investigation. If the site selected is unsatisfactory, or the samples are badly taken, no amount of sophisticated laboratory techniques will compensate for the loss or confusion of information. When sampling the sediment in the field, it is often helpful to make notes on the surrounding environment. Depending on the study to be undertaken, the geology and geomorphology of the surrounding area may affect the interpretation, and a knowledge of the vegetation around the area may also be helpful. When taking samples for pollen analysis in the field, it is important to note any flowers nearby, in case any modern pollen from them lands on the sediment.

Sediment description

It is useful to describe the composition and structure of the sediments in as much detail as possible, so that the maximum amount of environmental information can be extracted from the abiotic component. Unless a special problem is being studied, the sediments are usually described in the field, and the field descriptions can be supplemented by laboratory observations if necessary. Inorganic sediments (e.g. sands, gravels) are relatively straightforward to describe. Organic sediments are more complex, and a comprehensive field description system was presented by Troels-Smith (1955). Although it is rather complicated at first sight, no other more logical or versatile descriptive method has yet been devised. Many other systems for describing organic sediments are genetic in their character, (e.g. Osvald's system in West, 1977), and tend to require the investigator to be able to reconstruct the origin of the sediment prior to its description. The great advantages of the Troels-Smith system are (1) that it is basically a descriptive approach, and (2) that it recognizes that sediments are frequently mixtures of elements.

The Troels-Smith sediment description system

There are three major properties of the sediment which should be described.

1. Physical properties (colour, dryness, stratification etc.).
2. Humification (the degree of decomposition of organic material).
3. Composition (lake mud, silt, *Sphagnum* peat etc.).

Sediments are usually a mixture of components, and Troels-Smith's system estimates the abundance of each component on a 5-point scale.

0 = absent
1 = up to $\frac{1}{4}$
2 = $\frac{1}{4}$ to $\frac{1}{2}$
3 = $\frac{1}{2}$ to $\frac{3}{4}$
4 = whole
+ = trace, less than $\frac{1}{8}$

1. *Physical properties*

a) **nig** – nigror, is the degree of darkness, ranging from the lightest quartz sands or lake marls at nig 0 to the darkest most humified peats at nig 4.

A note of the actual colour can be made by comparison with the standard Munsell soil colour code. Sometimes the sediment changes colour on exposure to air, as it becomes oxidized. Any variations in colour due to banding or mottling should also be recorded.

b) **strf** – stratificatio, is the degree of stratification. Stratification may be structural, in that the deposit readily splits horizontally, or visual, in that different coloured layers have been deposited with otherwise similar composition. Strf 0 indicates a homogeneous sediment, and strf 4 a sediment consisting of many minor layers.

c) **elas** – elasticitas, is the degree of elasticity. The elasticity of a sediment is its ability to regain its shape after deformation. Some freshwater swamp peats and algal muds have great powers of recovery, whereas silts and plastic clays have none. Elas 0 indicates absence of elasticity, and elas 4 is the highest degree of elasticity.

d) **sicc** – siccitas, is the degree of dryness. Sediments differ in their water content. Sicc 0 is pure water, and sicc 4 indicates an air-dry sediment. We usually deal with sediments between these two extremes; sicc 1 indicates a very soft sediment with such a high water content that it runs or spreads (for example surface muds); sicc 2 indicates a sediment saturated with water, which slumps and will not retain its shape; sicc 3 indicates a sediment not saturated with water, which retains its shape but can be deformed by applying pressure. Sediments of sicc 1 are impossible to sample accurately by normal sampling methods. As discussed above, special samplers are needed to take cores through them without disturbance, and the core is sampled by careful vertical extrusion and scooping off the sediment a certain amount at a time.

e) In addition to the above physical features of a sediment, notes should also be made on the structure of the sediment (granular, fibrous etc.) and on the nature of the contacts with adjoining sediment types (**lim** – limes). The boundary can be recorded as very sharp (over <0.5 mm), sharp (over <1 mm and >0.5 mm), gradual (over <2 mm and >1 mm), and very gradual (over <1 cm and >2 mm). A very diffuse boundary occurs over more than 1 cm. The calcareousness (**calc**) and sulphide (**sulp**) content can also be estimated by the addition of dilute acid, and the amount of resulting effervescence of carbon dioxide or hydrogen sulphide recorded.

2. *Humification*

Humo – humositas, is the degree of humification. This is the degree of decomposition of the organic material into humic acids. These are brown, and they can be estimated by squeezing the sediment and assessing the colour of the solution squeezed out. Humic acids are soluble in dilute alkali, such as NaOH or KOH, and a more accurate way of estimating humification is to extract the acids by treatment with hot dilute alkali and then assess the colour intensity of the solution colorimetrically, for example with an EEL colorimeter. Detailed studies of this type have been done by Bahnson (1968) and Aaby and Tauber (1975). The method can detect small changes in humification in a profile from a peat bog. The more humified layers correspond to areas of slower peat growth, so-called 'retardation layers' or 'recurrence surfaces', which may be a result of drier conditions, perhaps climatically controlled.

3. *Sediment composition*

There are six basic components of sediments (see Table 3.1).

a) Substantia humosa (**Sh**; humous substance). This is completely disintegrated organic material, and it is blackish and homogeneous with no apparent structure. It dissolves in dilute alkali to give a very dark liquid. The material is so completely decomposed that it is not possible to ascertain whether the sediment, or part of it, is limus, turfa, or detritus.

b) Turfa (peat) consists of macroscopic plant remains, such as roots, stems, trunks, mosses etc. It is formed mainly from underground parts of plants, and plant parts connected to the root system. The underground parts are not exposed to oxidation, as they are generally rooted in a waterlogged environment.

Turfa is divided into three types, depending on whether it is made up of mosses, wood, or herbaceous material.

Turfa bryophytica (**Tb**; moss peat). This is peat composed of moss remains, which are often identifiable. The main types of moss peat are Turfa Sphagni, *Sphagnum* peat, Turfa hypnacea, hypnaceous moss peat, and Turfa Polytrichi, *Polytrichum* peat.

Turfa lignosa (**Tl**; wood peat). This is peat composed of wood remains. Large pieces of wood may be classed as an accessory element (see p. 41). Sometimes large stumps may be preserved. Often only small twig or bark fragments remain. It is often possible to identify the type of wood by microscopic examination of thin sections.

Turfa herbacea (**Th**; herbaceous peat). This is peat composed largely of the roots of herbaceous plants, but also rhizomes and above-ground stems attached directly to the roots. It may be further classified on the species of plant which produced it, such as Turfa Phragmitis, *Phragmites* peat, Turfa Dryopteridis, fern peat etc.

The type of peat can be difficult to distinguish if the degree of humification is high, and some peats, for example, from some British blanket bogs, are almost completely homogeneous. It is often useful in such cases to wash out some of the peat in water or dilute alkali and examine any macroscopic remnants with a microscope in the laboratory to try to determine the original components. Fragments of *Sphagnum* leaves, *Calluna* stems, or *Juncus* seeds are often recognizable, and there is considerable scope for the careful identification of epidermal fragments, many of which have a unique cell pattern. Gross-Brauckmann (1972, 1974) has produced comprehensive descriptions and keys of commonly found epidermis types.

c) Detritus. In contrast to Turfa, detritus consists of above ground parts of plants not directly attached to the roots. There are three major types of detritus.

Detritus lignosus (**Dl**) – fragments of wood and bark >2 mm in size. They can range from small fragments to trunks and branches.

Detritus herbosus (**Dh**) – fragments of herbaceous plants >2 mm in size, such as stems, leaves, and seeds.

Detritus granosus (**Dg**) – fragments of woody and herbaceous plants <2 mm but >0.1 mm in size. Small animal fragments may also be included.

Detritus is easy to determine when it is found in a lake sediment, but it is often impossible to separate from Turfa in peats unless microscopic examination of the sediment is made in the laboratory.

d) Limus. This is aquatic mud, composed of particles or colloids <0.1 mm in size. It is often called by its Danish name 'gyttja'. It is composed of very small parts of plants and animals, micro-organisms and the breakdown products of these, together with marl and other precipitates. There are four main types of Limus, although other types may be infrequently distinguished.

Limus detrituosus (**Ld**) – is the commonest form of lake mud. It is usually a highly elastic, non-greasy, non-sticky substance, composed mainly of more or less decayed micro-organisms or <0.1 mm plant fragments. When it is very humified (Ld^4 or 'dy') it is much less elastic, and more greasy and sticky, as it consists mostly of humous substance. It can be distinguished from highly humified Turfa ($\mathbf{Tb}^4$) by the absence of root fragments. It is not easy to distinguish from Substantia humosa (**Sh**) except by its origin below water rather than above it. This is usually evident either from the situation of the deposit and the adjacent sediment types, or from examination of the microfossils. Remains of obligate aquatic organisms will be found in Limus, which will be absent from Substantia humosa.

Limus siliceus organogenes (**Lso**) – consists of the siliceous skeletons of plants or animals. When pure, it is pale and non-elastic, and dries to a very light, easily powdered substance. It is often called diatomite. It is often impossible to distinguish in the field when mixed with Limus detrituosus, and again microscopic examination is helpful.

Limus calcareus (**Lc**) – is marl. When pure, this is composed almost entirely of calcium carbonate, and is very pale or white, and dense in comparison with diatomite. Small quantities in a deposit are easily detected by effervescence with dilute hydrochloric acid. However, it can be difficult to distinguish from calcareous silt mixed with limus detrituosus.

Limus ferrugineus (**Lf**) – consists of precipitated iron oxides or sulphides. In the reduced state the deposit is blackish, but when oxidized, for example, by exposure to air, it is reddish or yellowish.

e) Argilla. This consists of mineral particles <0.06 mm in size. In comparison with Limus it is heavier when dry, and when moist it is plastic and not elastic. There are two types of Argilla.

Argilla steatodes (**As**) – is composed of clay particles <0.002 mm in size, and it is very plastic when wet.

Argilla granosa (**Ag**) – is silt composed of mineral particles between 0.06 and 0.002 mm in size. It is scarcely plastic in comparison with clay and it can be rubbed to a dust when it is dry, rather than drying into a hard lump like clay.

f) Grana. Grana consists of mineral particles >0.06 mm in size. When it is chewed, the grit can be felt between the teeth. The types of grana are divided according to size.

Grana arenosa (**Ga**) – is fine sand, with particles between 0.06 and 0.6 mm in size.

Grana suburralia (**Gs**) – is coarse sand, with particles between 0.6 and 2 mm in size.

Grana glareosa (**Gg**) – is small to medium gravel, 2 mm in size. If it is <6 mm it is termed G. glareosa minora (Gg(min.)), if >6 mm, G. glareosa majora (Gg(maj.)).

In addition to the main components, the Troels-Smith system allows for additional 'accessory elements'. These do not occur commonly, and often do not make up a substantial proportion of the sediment. Their quantity is estimated outside the sum of the main elements. For example, in a marly clay containing mollusc shells the proportions of the components by volume may be marl 1, clay 1, shell 2. Because the sum of the main components must be 4, marl is scored as 2 and clay as 2. Shells make up half of the total sediment and are therefore given the value 2. The notation for the sediment is therefore Lc2, As2, [test.(moll.)2].

The main accessory elements are: Testae molluscorum – test. (moll.) which are entire mollusc shells; Particulae testarum molluscorum – part. test. (moll.), which are broken mollusc shells; Stirpae, trunci et rami, which are stumps, trunks and branches of trees whose identity may be specified thus; stirpes alni (alder stump), stirpes betulae (birch stump), and stirpes pini (pine stump); Rudimenta culturae, or archaelogical remains. Troels-Smith delimits many different sorts of cultural remains which may be found in bogs. However, they are not usually found unless a special archaeological investigation is being made. Perhaps the most useful in other investigations is anthrax, charcoal, which may or may not be associated with human activities.

Although the Troels-Smith system appears to be complicated at first sight, it is in fact simple and logical to use, and it is designed for use in the field. Quick descriptions of the sediments can be made which accurately reflect their composition. The system is also useful in that the investigator is forced to examine the sediment carefully. A superficial description would result in the recording of 'peat' or 'mud'. However, if the components have to be identified and estimates of their abundance made, much more information of considerable palaeoecological value is retained, and with practice, the use of the system becomes quite fast. There are a few difficulties, as mentioned in the description of the system, where it is difficult if not impossible to distinguish elements in the field. A major stumbling block is a very humified and decayed deposit, which may be peat or mud. The category Substantia humosa is very useful for describing this sort of sediment. Many of the more difficult components can be distinguished in the laboratory if the sediment is suspended in water and examined microscopically. Much more information on the items composing the sediment is gained this way. If necessary, a special microscopic study of the sediment can be made, and rough estimates made of the abun-

Table 3.1 Scheme for the field determination of the composition of lake and bog sediments, modified from J. Troels-Smith ('Karakterisering af løse jordarter', Danmarks Geologiske Undersøgelse, IV Raekke, Bd, **3**, Nrd. 10, 1955).

Class	Symbol	Element	Description
Turfa	Tb^{0-4}	T. bryophytica	Mosses, +/– humous substance
	Tl^{0-4}	T. lignosa	Stumps, roots, intertwined rootlets, of ligneous plants +/– trunks, stems, branches, etc., connected with these. +/– humous substance
	Th^{0-4}	T. herbacea	Roots, intertwined rootlets, rhizomes, of herbaceous plants +/– stems, leaves, etc., connected with these. +/– humous substance
Detritus	Dl	D. lignosus	Fragments of ligneous plants >2 mm
	Dh	D. herbosus	Fragments of herbaceous plants >2 mm
	Dg	D. granosus	Fragments of ligneous and herbaceous plants, and sometimes of animal fossils <2 mm > *ca.* 0.1 mm
Limus	Ld^{0-4}	L. detrituosus	Plants and animals or fragments of these; particles <*ca.* 0.1 mm. +/– humous substance
	Lso	L. siliceus organogenes	Diatoms, needles of sponge, siliceous skeletons, etc., of organic origin, or parts of these. Particles <*ca.* 0.1 mm
	Lc	L. calcareus	Marl, not hardened like calcareous tufa. Particles <*ca.* 0.1 mm
	Lf	L. ferrugineus	Iron oxide. Particles <*ca.* 0.1 mm
Argilla	As	Clay	Mineral particles <0.002 mm
	Ag	Silt	Mineral particles 0.002 to 0.06 mm
Grana	Ga	Fine sand	Mineral particles 0.06 to 0.6 mm
	Gs	Coarse sand	Mineral particles 0.6 to 2 mm
	Gg	Gravel	Mineral particles >2 mm
Substantia humosa	Sh	Humous substance	Completely disintegrated organic substances and precipitated humic acids

dance of such items as mineral particles, pyrite, charcoal fragments, bryophyte remains, fungal hyphae, algae, insect remains, cladocera, diatoms etc. A five-point scale of abundance is useful; abundant, frequent, occasional, rare, absent.

The Troels-Smith system uses a shorthand for describing the sediments. Each component has an abbreviation, as indicated in the descriptions of the components. In addition, the degree of humification of peat (turfa) and lake mud (limus) is shown by a superscript. Th^3 is very humified herbaceous peat, whereas Ld^1 is slightly humified detritus lake mud.

A description of unhumified marly lake mud containing some fine detritus and small quantities of coarse detritus and silt would be

$$\underbrace{Ld^0 2,\ Lc1,\ Dg1,}_{\text{Adds up to 4}}\ \underbrace{Dh+,\ Ag+}_{\text{traces } (<\frac{1}{8})}$$

Similarly, a very humified *Sphagnum* peat with herbaceous fragments and a few *Calluna* twigs would be

Tb Sphagni^{3}2, Th^{3}1, Sh1, Tl Callunae +

A more complete description of a sediment would include some estimate of its physical properties in addition to its composition. For example:

Marly fine detritus mud with shells	Medium olive, no colour change on exposure to air; Nig 2, strf 0, elas 3, sicc 2, calc 4; humo 1; Ld^{1}2, Lc1, Dg1, Dh +, [test.(moll.)2]. Lower boundary sharp over 0.5 mm.
Highly humified fibrous peat with charcoal fragments	Dark brown, darkens on exposure to air; Nig 3, strf 0, elas 2, sicc 2–3, calc 0; humo 3; Tb Sphagni^3 2, Th^{3}1, Sh1, Tl Callunae +, [anth.1]. Lower boundary gradual over 1.5 mm.

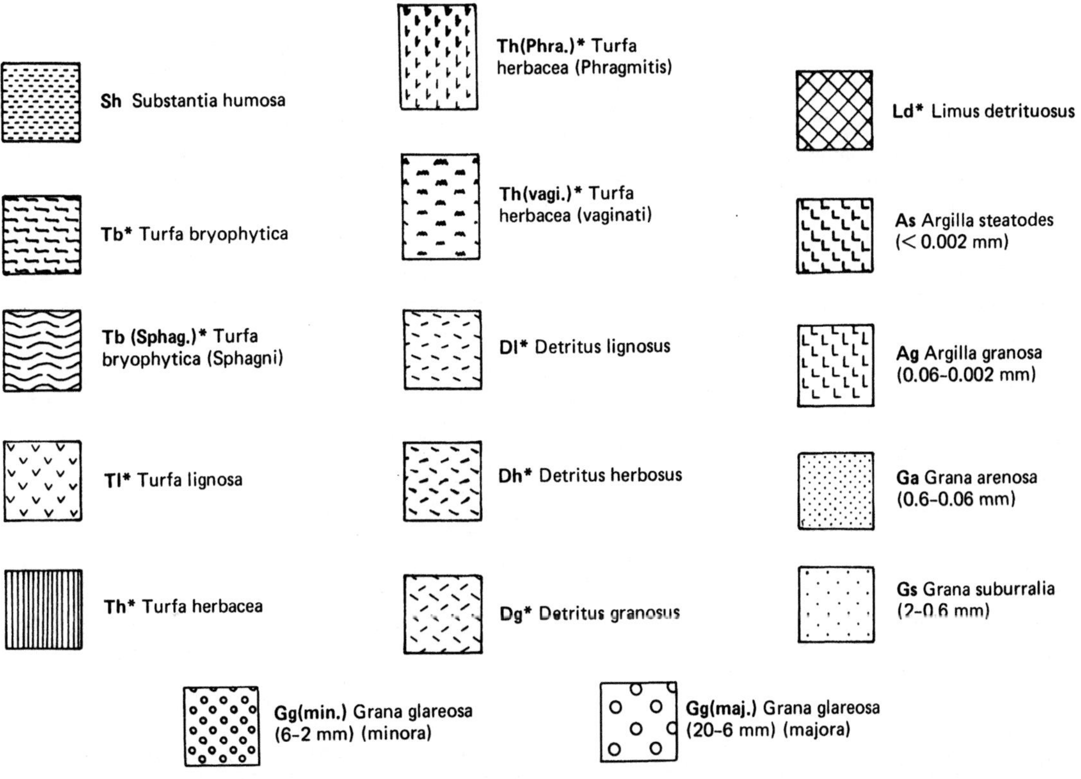

Fig. 3.1 Examples of the Troels-Smith representation of sediment types. Each sediment type is represented here at density 2 on the Troels-Smith scale. Those elements which can be humified are indicated by a star. For explanation of sediment types, see text. Diagrammatic representations of the whole range of sediment types is found in Troels-Smith (1955).

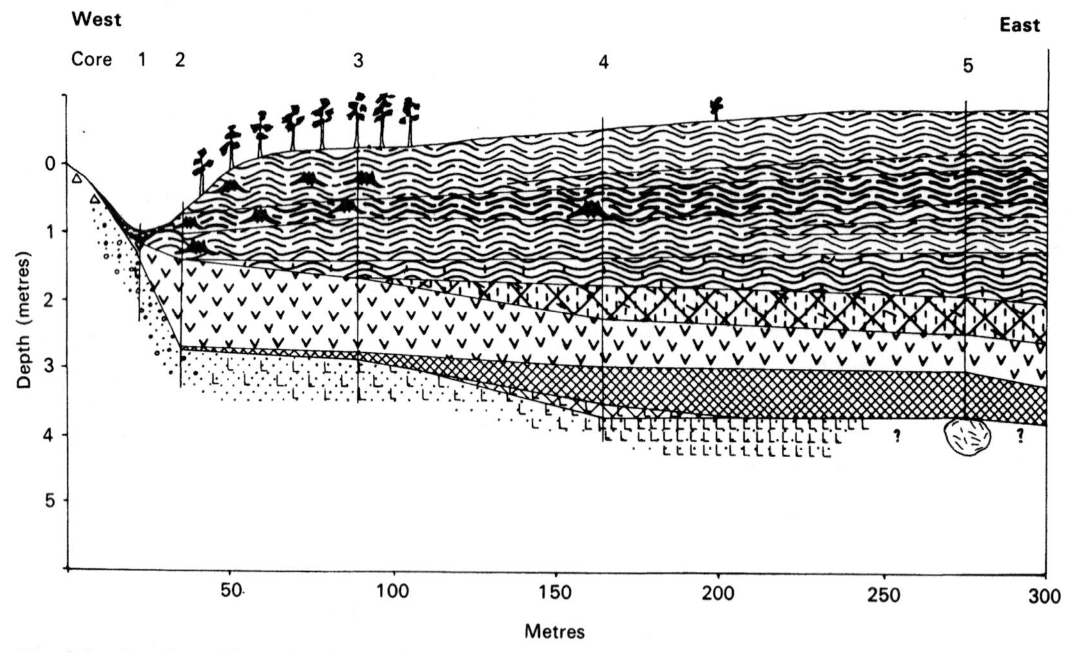

Fig. 3.2 Stratigraphic section through part of a Swedish raised bog, with the sediments depicted in the Troels-Smith notation. The bog started development as a lake, which became overgrown by fen wood. The wood peat was replaced by reedswamp peat, followed by *Sphagnum* peat of varying humification, which contains pine stumps towards the rand, where pine trees still grow today.

Table 3.1 is a summary of the Troels-Smith sediment description system.

Troels-Smith also designed a means of representing a sediment diagrammatically, so that a sediment column could be put beside a pollen diagram, or a section could be drawn across a deposit. He gave each component a symbol, and varied the density of the symbol with the proportion of the component. The major symbols are illustrated in Fig. 3.1. The thickness of the lines is varied for those components which can be humified. The different symbols for the appropriate quantity of each component are drawn over each other. Some examples are illustrated in Fig. 3.2. It may appear to result in a complex diagram, but with practice it gives a quick idea of the composition of the sediment. The types of symbols ingeniously use different basic line directions, so an overall impression is immediately obvious of a mud, peat, or minerogenic deposit. Further details of the diagrammatic symbols are given in Troels-Smith (1955).

In the next chapter we will discuss the ecological situations where organic sediments are being deposited today, and some studies on sediments which illustrate the amount of palaeoecological information which can be extracted from them.

References

Coring methods

DEEVEY, E. S. (1965). Sampling lake sediments by use of the Livingstone sampler. In *Handbook of Paleontological Techniques* (eds B. Kummel and D. M. Raup). W. H. Freeman.

FAEGRI, K. AND IVERSEN, J. (1975). *Textbook of Pollen Analysis* (3rd edition). Blackwell (pp. 85–100).

MOTT, R. J. (1966). Quaternary palynological sampling techniques of the Geological Survey of Canada. *Geol. Survey Canada Paper*, **66–41**, 24 pp.

WEST, R. G. (1977). *Pleistocene Geology and Biology* (2nd edition). Longman (Chapter 6).

WRIGHT, H. E., LIVINGSTONE, D. A. AND CUSHING, E. J. (1965). Coring devices for lake sediments. In *Handbook of Paleontological Techniques* (eds B. Kummel and D. M. Raup). W. H. Freeman.

Description of corers

CUSHING, E. J. AND WRIGHT, H. E. (1965). Hand-operated piston corers for lake sediments. *Ecology*, **46**, 380–4.

DAVIS, R. B. AND DOYLE, R. W. (1969). A piston corer for upper sediment in lakes. *Limnol. Ocean.*, **14**, 643–8.

DIGERFELDT, G. (1965). A new type of large-capacity sampler. *Geol. Fören. Förhandl. Stock.*, **87**, 425–30.

DIGERFELDT, G. AND LETTEVALL, U. (1969). A new type of sediment sampler. *Geol. Fören. Förhandl. Stock.*, **91**, 399–406.

FRIES, M. AND HAFSTEN, U. (1965). Asbjørnsen's peat sampler – the prototype of the Hiller sampler. *Geol. Fören. Förhandl. Stock.*, **87**, 307–13.

HONGVE, D. (1972). En bunnhenter som er lett å lage. *Fauna*, **25**, 281–3.

JOWSEY, P. C. (1966). An improved peat sampler. *New Phytol.*, **65**, 245–8.

KULLENBERG, B. (1947). The piston core sampler. *Svensk. Hydrograf – biol. Komm. Skr. Tredje*, Ser: Hydrografi **1** (2).

KULLENBERG, B. (1955). Deep-sea coring. *Reports Swedish Deep-Sea Expeditions*, **4**, 35–96.

LIVINGSTONE, D. A. (1955). A lightweight piston sampler for lake deposits. *Ecology*, **36**, 137–9.

LIVINGSTONE, D. A. (1967). The use of filament tape in raising long cores from soft sediment. *Limnol. Ocean.*, **12**, 346–8.

MACKERETH, F. J. H. (1958). A portable core sampler for lake deposits. *Limnol. Ocean.*, **3**, 181–91.

MACKERETH, F. J. H. (1969). A short core sampler for subaqueous deposits. *Limnol. Ocean.*, **14**, 145–51.

ROWLEY, J. R. AND DAHL, A. O. (1956). Modifications in design and use of the Livingstone piston sampler. *Ecology*, **37**, 849–51.

SHAPIRO, J. (1958). The core-freezer – a new sampler for lake sediments. *Ecology*, **39**, 758.

SMITH, A. G., PILCHER, J. R. AND SINGH, G. (1968). A large capacity hand-operated peat sampler. *New Phytol.*, **67**, 119–24.

SWAIN, A. M. (1973). A history of fire and vegetation in northeastern Minnesota as recorded in lake sediments. *Quat. Res.*, **3**, 383–96.

THOMAS, K. W. (1964). A new design for a peat sampler. *New Phytol.*, **63**, 422–5.

VALLENTYNE, J. R. (1955). A modification of the Livingstone piston sampler for lake deposits. *Ecology*, **36**, 139–41.

WALKER, D. (1964). A modified Vallentyne mud sampler. *Ecology*, **45**, 642–4.

WRIGHT, H. E. (1967). A square-rod piston sampler for lake sediments. *J. Sed. Petrol.*, **37**, 975–6.

WRIGHT, H. E. (1969). Cores of soft lake sediments. Unpublished manuscript.

Description and characterization of organic sediments

AABY, B. AND TAUBER, H. (1975). Rates of peat formation in relation to degree of humification and local environment, as shown by studies of a raised bog in Denmark. *Boreas*, **4**, 1–17.

BAHNSON, H. (1968). Kolorimetriske bestemmelser af humificeringstal i højmosetørv fra Fuglsø mose på Djursland. *Medd. Dansk. Geol. Foren.*, **18**, 55–63.

FAEGRI, K. AND IVERSEN, J. (1975). *Textbook of Pollen Analysis* (3rd edition). Blackwell (pp. 72–84).

GROSSE-BRAUCKMANN, G. (1972). Über pflanzliche Makrofossilen mitteleuropäischer Torfe. 1. Gewebereste Krautiger Pflanzen und ihre Merkmale. *Telma*, **2**, 19–55.

GROSSE-BRAUCKMANN, G. (1974). Über pflanzliche Makrofossilen mitteleuropäischer Torfe. 2. Weiter Reste (Frücht und Samen, Moose u.a.) und ihre Bestimmungsmöglichkeiten. *Telma*, **4**, 51–117.

TROELS-SMITH, J. (1955). Karakterisering af løse jordater. Characterisation of unconsolidated sediments. *Danm. geol. Unders.* Ser. IV, **3** (10), 73 pp.

WEST, R. G. (1977). *Pleistocene Geology and Biology* (2nd edition). Longman (Chapter 4).

4

Organic sediments in palaeoecology

Introduction

What can organic sediments tell us about the past environment and the past ecosystem?

To attempt to answer this question we have to know something about the environments in which organic sediments are accumulating today. Then we can extend our observations backwards in time to interpret the information we have extracted from our fossil sediments. The main environments in which organic sediments are presently accumulating are bogs, fens, and lakes. A considerable amount is known about bog and fen ecology, and also about lake environments and ecosystems (limnology). However, modern ecological studies are not always directly helpful to a palaeoecologist. He is mainly concerned with the taxonomy of the organisms involved, and their autecology and community ecology. The latter two aspects are rarely known in detail for modern organisms, and the palaeoecologist often has to collect scraps of information from where he may.

Hydroseres and mire classifications

As time passes, sediment accumulates in the bottom of a lake, and gradually fills the lake basin. Plants colonize the shallow margins and speed up the filling process, so that the whole basin may become occupied by a fen community. As fen peat accumulates it may become drier and be colonized by shrubs and trees, developing eventually into a forest community. However, in cool, wet climates such as that of western Britain, the heavy rainfall tends to leach out the nutrients from the fen surface, allowing bog plants such as *Sphagnum* to colonize, and form bog peat. Unless the surface dries out for any length of time, trees will not colonize the bog. To avoid nomenclatural difficulties in exactly defining the dividing line between fens and bogs, they can be usefully amalgamated in the general term ***mire***, which has been defined by Birks (1973, p. 40) as 'an area of ground where the water table is permanently at or near the ground surface, where the vegetation (particularly the root system) is adapted to the water-logged substrata and the associated anaerobic conditions, and where organic materials (generally peat) accumulate'.

The succession from open water to mire is called a ***hydrosere***. A hydrosere can be divided into three broad stages, according to the position of the water table (see Table 4.1).

Although in principle, the Troels-Smith system of sediment description (see Chapter 3) does not take into account the origin of the sediment, the sediments fall naturally into types characteristic of limnic, telmatic, or terrestrial situations. It is therefore usually easy to tell whether we are dealing with

Table 4.1 Features of a hydrosere.

Hydrosere	Lake ——————→	Reedswamp ——————→	Other mires
Environment	Limnic	Telmatic	Terrestrial
Sediment origin	Formed below water level	Formed between low and high water	Formed at or above water table
Sediment types	Allochthonous Lake mud (limus)	Autochthonous Fen peat (Turfa herbacea)	Autochthonous Fen or bog peat (Turfa herbacea, T. bryophytica)

Table 4.2 Calcium ion and pH levels in mire soils and water. (From Ratcliffe, 1964.)

	Soil		Water	
	Ca^{++} mg 100 g^{-1}	pH	Ca^{++} (mg l^{-1})	pH
Oligotrophic	<30	<5	<4	<5.7
Mesotrophic	30–300	5–6	4–10	5.7–6.5
Eutrophic	>300	>6	>10	>6.5

a lake sediment, formed underwater, or a fen or bog sediment formed at or above water level.

Table 4.1 shows the general direction of a hydrosere. The main controlling factor is the height of the water table. However, a second factor that is very important in determining the type of plant communities at a given water level is the level of available nutrients, most particularly calcium ions, which influence the pH of the soil or water. Ratcliffe (1964a) in an excellent article on the different types of mire in Scotland, has divided the mires into three groups on the basis of their vegetation, and found that these groups corresponded closely to calcium and pH levels in the soil and water. The poorest mires are called oligotrophic, the richest are described as eutrophic, and those in between are termed mesotrophic. The calcium and pH values for these types are shown in Table 4.2.

The supply of nutrients is determined by the geology of the area and the topography, which controls how much ground-water, if any, reaches the mire. The way in which mires *receive* their nutrients has also been used to classify them.

Ombrotrophic mires

These are commonly called ***bogs***. The water table is maintained solely by atmospheric precipitation (the bog is said to be 'ombrogenous'). As a result, it is oligotrophic (poor in nutrients) with a low pH. Nutrients are deficient and the peat has a low cation saturation. Only a relatively few plants are adapted to such conditions. *Sphagnum* mosses are important on bogs. They act as a sponge and maintain a high water table, often above the surrounding water table in the mineral soil. It is interesting that insectivorous plants such as *Drosera*, *Pinguicula*, and *Sarracenia* are usually found in bog habitats, where they supplement their poor mineral diet by catching and digesting insects. In Britain, bogs are usually treeless. However, in more continental parts of the boreal region, trees, particularly conifers such as pine, larch, and spruce species typically grow on the drier parts of bogs. These continental bog types are very rare in the British Isles, with their oceanic climate.

Bogs can be subdivided on topographical criteria into two major types, ***raised bogs*** and ***blanket bogs***.

1. *Raised bogs*. In Britain these occur on flood-plains of mature river systems, and, most commonly, on alluvial deposits in estuaries. They owe their origin to the absence of slope, which would have otherwise favoured the formation of mesotrophic or eutrophic fens. As peat accumulates, the vegetation becomes isolated from and independent of the ground-water table, and an ombrogenous hydrology becomes established. Raised bogs can be very large, up to 20 km², with peat up to 12 m thick. The typical stratigraphy of a British raised bog is shown in Fig. 4.1.

Around the bog, ground-water draining from surrounding slopes and streams supplies nutrients, and the vegetation of this part, called the lagg, may be of

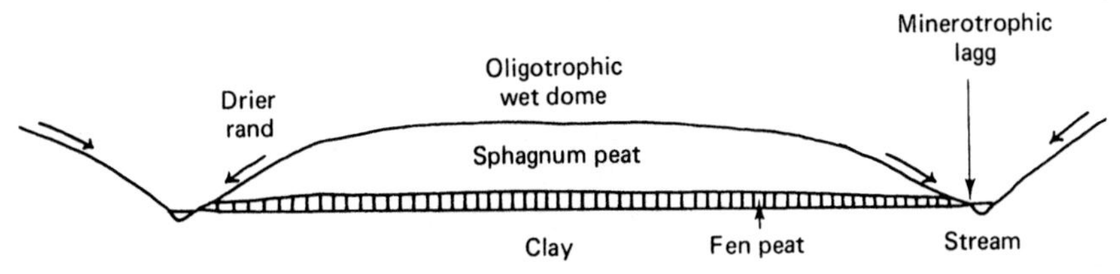

Fig. 4.1 Generalized stratigraphic section through a typical British raised bog.

the poor-fen type. Peat does not accumulate so rapidly in this situation because of greater oxidation of organic material, and there is thus a slope up to the dome, called the rand. The rand is often drier than the dome, and it may support tree growth of, for example, pine. Rates of run-off from the dome are relatively low, and hence raised bogs do not depend on such a high precipitation/evaporation ratio as do blanket bogs. Therefore, they occur in drier areas than do blanket bogs, for example at Thorne Waste, Yorkshire where the annual amount of rain is only 58 cm per year at present. Similarly, there are remnants of raised bogs at Holme Fen and Woodwalton Fen in the low rain-fall area of East Anglia. Raised bogs occur most commonly in the central Irish plain, the Solway, Morecambe, and Forth estuaries, the central England plain of Cheshire, Lancashire, and Shropshire, the Somerset Levels, and in Wales, where the most famous examples are Tregaron Bog and Borth Bog. They are rare in the wettest parts of Britain perhaps because of a lack of suitably flat topography, the most notable bog being Claish Moss, Argyll (see Ratcliffe, 1964a; Moore, 1977). Raised bogs generally occur below 200 m altitude, although the raised bog at Malham is unusual in being at an altitude of 380 m (Proctor, 1960, 1974). In Europe, raised bogs are most frequent in north France, Belgium, Holland, Denmark, and north Germany.

2. *Blanket bogs.* This is an apt descriptive name for the type of bog which develops over a whole landscape on slopes less than about 15°. It can only develop extensively in areas with a high rain-fall of more than 130 cm a year and with at least 160 wet days a year (a wet day is defined as one in which at least 1 mm of rain is recorded). This includes northern England, north and west Scotland, and western Ireland. In very wet areas of acid rocks any soil nutrients are readily leached out of the soil, which commonly becomes a podsol. Below the leached layer, the minerals are redeposited, and iron in particular is redeposited to form a characteristic layer called an ***iron pan***. If this reaches a certain thickness it impedes drainage. This allows bog species to colonize the vegetation, and the bog begins to grow. In many instances bog has grown over forest or scrub vegetation, of which traces may be found as wood remains at the base of the peat. In other instances, particularly in Ireland, the activities of prehistoric men have caused the soil fertility to deteriorate, which then allowed bog to develop over their deserted fields, and even their villages and grave mounds.

Today, many of the blanket bogs of the British Isles are being eroded, largely due again to man's activities. Moor burning and drainage to encourage pasture for sheep, atmospheric pollution, and peat cutting for fuel have often destroyed the natural vegetation. In many instances the bogs have stopped growing entirely, the surface has dried out and cracked, and the peat has started to erode. The cracks form natural drainage channels, and these are gradually enlarged and can eventually cut right through the thickness of peat to the mineral soil beneath. Such channels and remaining peat islands are called ***peat haggs***. They are most developed on the blanket peats of the Southern Pennines, where the rainfall is relatively low and human pressure is relatively high (Tallis, 1964, 1965).

The distinction between raised bogs and blanket bogs is not always sharp. For example, the bogs of the Silver Flowe in the Galloway Hills of southwest Scotland, are intermediate in type. Discrete areas of bog growth in shallow basins in the valley have coalesced by lateral spread of the blanket-bog type, until the lower, gentler slopes have become peat covered. The original bogs are still detectable because they have a wetter vegetation, often with large bog pools (p. 62). The ecology of this very interesting series of bogs is described by Ratcliffe and Walker (1958).

Blanket bog is virtually restricted to Britain and western Norway, where the climatic conditions are suitable. Vast raised or sloping peatlands have developed in more continental areas of the boreal region. They frequently have very large elongated pools, and trees growing on the intervening peat strings (Heinselman, 1963, 1970; Sjörs, 1961).

Minerotrophic mires

These are commonly called fens. The water table is maintained by atmospheric precipitation combined with ground-water; the fen is said to be geogenous. As a result it is mesotrophic or eutrophic (rich in nutrients, and with a high pH), due to the supply of nutrients dissolved in the ground water, which has percolated through mineral soil before reaching the fen. Depending upon the nutrient level, the vegetation can be very rich in species. The different fen communities are controlled by a combination of the water level, the nutrient status, the rate of water flow, the aeration of the water, and the extent of human interference.

Table 4.3 Descriptive terms used in mire ecology.

Ecological feature	Bog	Fen
Nutrient status	Oligotrophic (poor feeding)	Mesotrophic and Eutrophic (middle and rich feeding)
Method of nutrient supply	Ombrotrophic (cloud feeding)	Minerotrophic (mineral feeding)
Origin of nutrients	Ombrogenous (cloud generated)	Geogenous (ground generated)

If the fen occupies an enclosed basin in which the water table is maintained by inflow of water from run-off and from streams, and impeded drainage, the fen can be called ***topogenous*** (controlled by topography). However, fens may also occur on gently sloping ground, in which case water flows downhill through them, some feature of impeded drainage leading to a high water table. This type of fen can be called ***soligenous***. Soligenous fens do not usually occupy large areas of ground, but topogenous fens may be extensive, depending upon the size of the basin they occupy.

The words used to describe bogs and fens are somewhat confusing, so they have been tabulated in Table 4.3 for comparison of their slightly different shades of meaning.

The above division of fens is based on habitat criteria. However, the type of vegetation is also of interest to the palaeoecologist. Many fen communities have been described and classified. They can be generally divided on floristic criteria into poor-fens and rich-fens. An account of the range of fen communities and the habitat factors controlling them on the Isle of Skye, off the mainland of western Scotland, is given by Birks (1973). All the common northern and western mire types are included, and they are compared with types found in more southern areas (see also Ratcliffe, 1977).

Lakes

The plant and animal communities of a lake are chiefly controlled by water depth, lake size and shape, and nutrient status of the water. The substrate near the shore, ranging from coarse gravel to fine mud and silt, may also determine the types of plants which can become established there. Depending upon the nutrient concentration, lakes may also be termed oligotrophic, mesotrophic, or eutrophic. Calcium is important, but levels of other nutrients are equally, if not more important, particularly nitrates and phosphates. A eutrophic lake with abundant calcium may be deficient in nitrates and phosphates, and support a relatively restricted flora and fauna. The sediment deposited will be marl. However, if all nutrients, including nitrates and phosphates are plentiful, the lake is extremely productive. Submerged vegetation is rampant, and the rest of the water may be occupied by blooms of algae, particularly blue-green algae. Such lakes are rare in the natural state. They are usually the result of excessive input of nutrients through man's activities in the catchment, such as the felling of the surrounding forest, fertilization of the land for agriculture, and the disposal of sewage effluent into the lake. These lakes are generally regarded as 'polluted' (see Chapter 6).

At the other extreme, oligotrophic lakes are poor in nutrients and support a restricted flora and fauna. The water is often brownish due to dissolved humic acids washed in from surrounding peat. Productivity is low, and sediment accumulation rates are low. Many oligotrophic mountain lakes have very clear water and contain very little organic sediment, so that the bottom may be stony or silty.

Sediment type and depositional environments

We have already mentioned that lakes, fens, and bogs produce characteristic sediments; lake muds, fen peats, or bog peats. In addition we have also mentioned that lakes and mires have different vegetation, depending upon the water level and the nutrient status. We can now consider what sort of sediment is being deposited in these different environments. If we consider eutrophic and oligotrophic situations separately, we can draw up a chart (Fig. 4.2) indicating the different sediments formed with increasing nutrient status and increasing water depth.

In a lake, fine particles of silt or mud will be suspended for a relatively long time and also

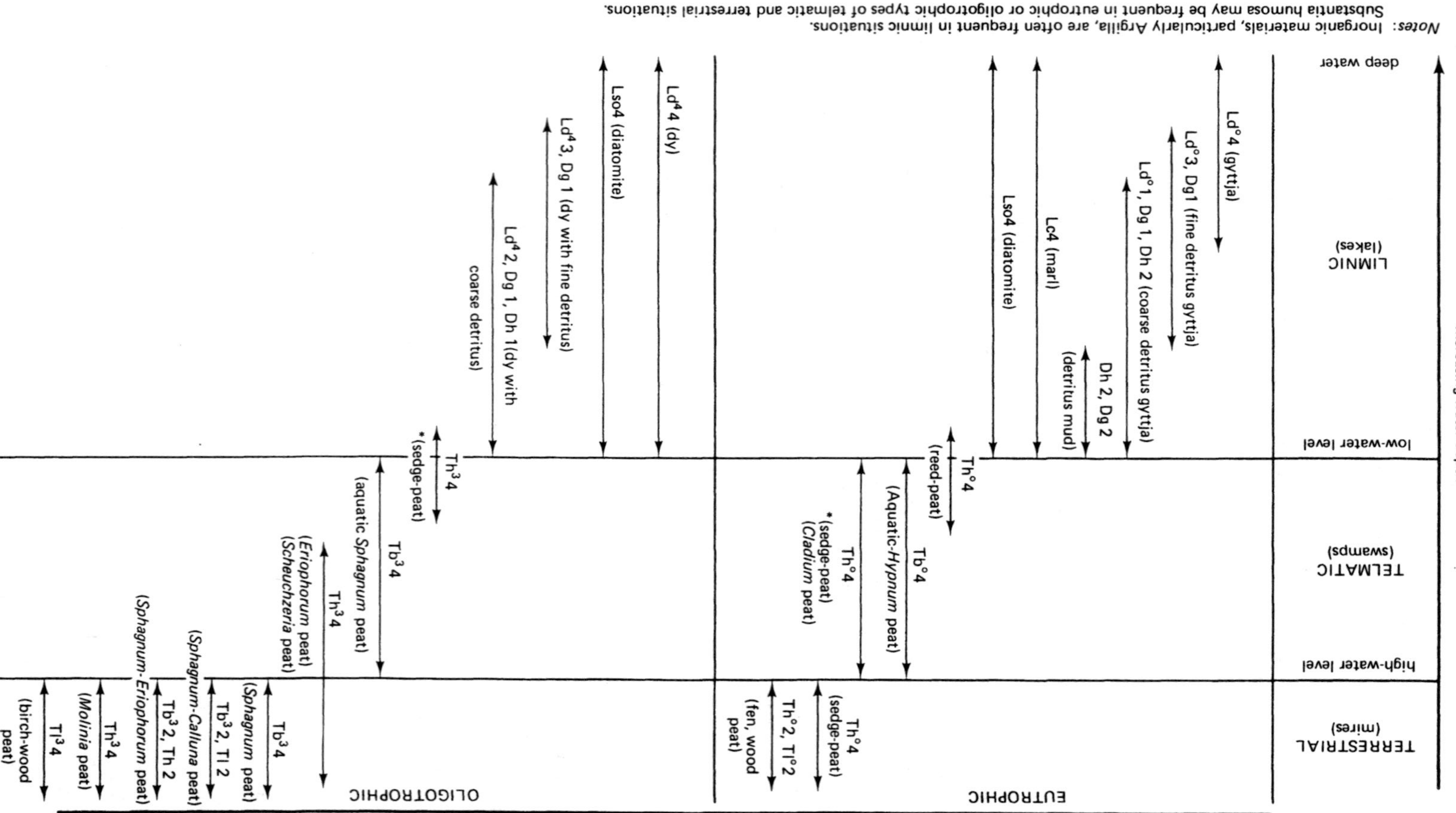

Fig. 4.2 A chart indicating the types of sediment deposited with increasing water depth under oligotrophic and eutrophic conditions. For discussion, see text.

resuspended more easily than coarser particles. Therefore, fine particles tend to be deposited eventually in the deepest water near the centre of the lake (see Chapter 9). In an oligotrophic lake the sediment may consist of humic substances and inwashed peat. The Swedish name for this dark brown jelly-like deposit is dy, and its composition is $Ld^{4}4$. Other deep-water sediments are diatomite (Lso4) which can be produced in mildly oligotrophic and eutrophic lakes, depending upon the species of diatoms, marl (Lc4) produced in lakes rich in calcium carbonate, but poor in nitrate and phosphate, and lake mud or gyttja ($Ld^{0-3}4$). Lake mud is the most common limnic deep-water deposit, and is produced in most lakes which are not exceedingly oligotrophic or rich in calcium carbonate.

As the water becomes shallower, coarser particles are deposited along with the limus particles, the coarse particles becoming more abundant with decreasing water depth, until a coarse-detritus mud may be deposited (Dh2, Dg2) under submerged or floating-leaved aquatic vegetation. Reedswamp or sedgeswamp colonizes the mud surface as it approaches the water level, and the telmatic stage is reached. The sediments now become largely autochthonous. In oligotrophic lakes the fringing fen is usually dominated by sedges (*Carex* spp.) especially *Carex rostrata* and *C. lasiocarpa*. Sedges such as *Carex paniculata* and *C. diandra* may colonize more eutrophic lakes, but these are frequently fringed by reedswamps, often dominated by *Phragmites communis*, *Phalaris arundinacea*, *Schoenoplectus lacustris*, or *Glyceria maxima*.

An oligotrophic sedgeswamp is often colonized by aquatic species of *Sphagnum*, such as *S. cuspidatum* and *S. recurvum*, and by other sedge-like plants, such as *Eriophorum angustifolium* and *Scheuchzeria palustris*. A mesotrophic or eutrophic sedgeswamp or reedswamp may be colonized by 'brown mosses', which are large creeping mosses of the family Hypnaceae, for example *Scorpidium scorpioides*, *Drepanocladus revolvens*, and *Campylium stellatum*, and by other reeds and sedges such as *Cladium mariscus*.

As the level of peat accumulates above the maximum lake water-level, the terrestrial phase begins. Mesotrophic and eutrophic sedgeswamps and reedswamps become colonized by a great diversity of species. Typical European examples include *Iris pseudacorus*, *Rumex hydrolapathum*, *Lythrum salicaria*, *Filipendula ulmaria*, *Lysimachia vulgaris*, and *Angelica sylvestris*. Such fens deposit coarse herbaceous peat, $Th^{0-4}4$.

As these fens become drier, they may be colonized by fen trees and shrubs, such as *Alnus glutinosa* and *Salix* spp., and later even by *Betula* and *Quercus*. The fen scrub is called fen carr, and it can be a dense jungle. Fen carr deposits wood peat, $Th^{0-4}2$, $Tl^{0-4}2$.

In contrast, the telmatic sedgeswamps around oligotrophic lakes are frequently colonized by *Sphagnum* spp., such as *S. palustre* and *S. plumulosum*, together with bog plants such as *Calluna vulgaris*, *Eriophorum vaginatum*, and *Molinia caerulea*. Bog peats are formed; $Th^{0-4}2$, $Tb^{0-4}2$. If the peat becomes sufficiently dry, *Betula pubescens* and even *Pinus sylvestris* may colonize it. *Sphagnum* forms *Sphagnum* peat Turfa bryophytica sphagni, Tb Sphag 4; *Calluna* remains are often woody twigs, classed as wood peat, Tl Callunae; *Eriophorum* and *Molinia* produce herbaceous peat, Th Eriophorae or Th Moliniae. The resulting raised bog peat may have the overall composition Tb $Sphag^{2}2$, Tl Callunae 1, $Th^{2}1$.

By carefully studying and describing fossil sediments, and by comparing the fossil composition with the sediments shown in Fig. 4.2, deductions about past water table and nutrient status of the site being investigated can be made.

Examples of palaeoecological studies based primarily on the study of sediments

The Blytt–Sernander scheme of climatic periods

Before pollen analysis was invented by Lennart von Post in 1916 peat stratigraphy was the main source of evidence for past changes in the vegetation and climate of northern Europe. Layers of scarcely humified *Sphagnum* peat were taken as indicators of fast peat growth and therefore of wet climatic conditions. Layers of darker, more decomposed humified peat, often with a layer of tree stumps of birch or pine, were taken to represent a drier bog surface, and hence a drier, warmer climate. A climatic sequence emerged, often called the Blytt–Sernander scheme of climatic periods. It is illustrated in Fig. 4.3.

In spite of fierce arguments in Scandinavia about the validity of the scheme at the end of the last century, it gradually became accepted as the outline

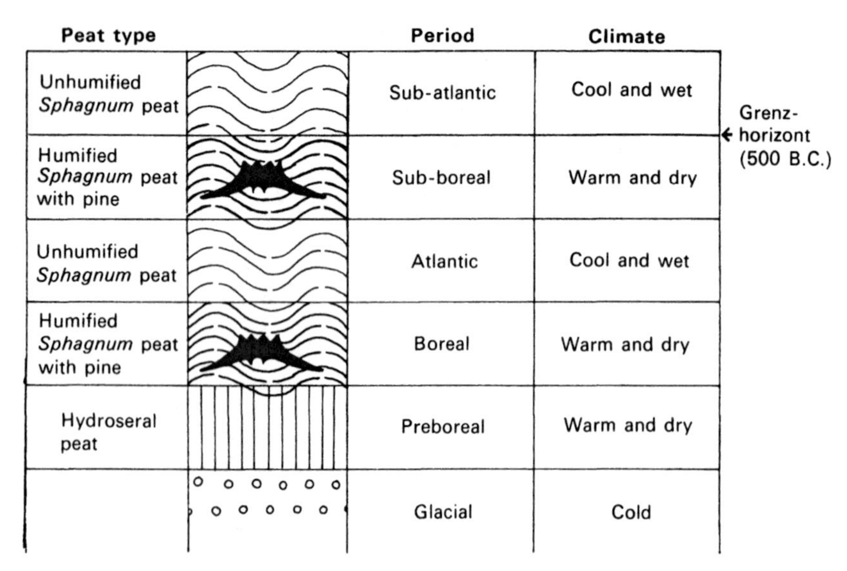

Fig. 4.3 Diagrammatic generalization of the Blytt–Sernander scheme of climatic periods during the post-glacial (Holocene, Flandrian).

of post-glacial changes in climate. When pollen analysis was developed, the main pollen zones were equated with these climatic periods. The scheme gained acceptance in Britain, largely through the work of Geikie (1877) and Samuelsson (1910), using the detailed and extensive work of F. J. Lewis (1905–11) on the blanket bogs of Scotland.

The ages of the various boundaries were not generally known, except for the boundary between the Sub-boreal and the Sub-atlantic. This is a very clear boundary in many north-west European raised bogs, fresh *Sphagnum* peat overlying dark humified peat. The relationship of archaeological finds with this layer, called the Grenzhorizont, showed that it occurred at the boundary between the Bronze Age and the Iron Age, at about 500 B.C.

Today, in the face of all the evidence from pollen analysis, macrofossil analysis of both plants and animals, and chemical studies, the Blytt–Sernander scheme cannot be totally upheld (Birks, 1975). The Grenzhorizont boundary is still accepted by some as a time of relatively sudden climatic change, but the other climatic periods are not wholly substantiated. However, the names of the periods are often used, almost colloquially, to denote periods of time in the post-glacial, for example, 'Alder immigrated at the beginning of the Atlantic period'.

A reinvestigation of some of Lewis's sites with pine-stump layers in peat bogs in Scotland was undertaken by Birks (1975). Radiocarbon dating revealed that the stumps did not fall into the accepted time limits of the Boreal and Sub-boreal periods, and the palaeoecology of the peats indicated several causes for their growth and death which were mostly related to local hydrological features and fires, which were only indirectly controlled by climatic changes. The pine stumps in north-west Scotland were exceptional in that they all grew and died between about 4500 and 4000 B.P. This horizon of climatic deterioration can also be detected in other parts of western Britain and Scandinavia. For example, Karlén (1976) has demonstrated that pine stumps extended to a maximum altitude by 5000 B.P. in northern Sweden and then began to decline at about 4000 B.P.

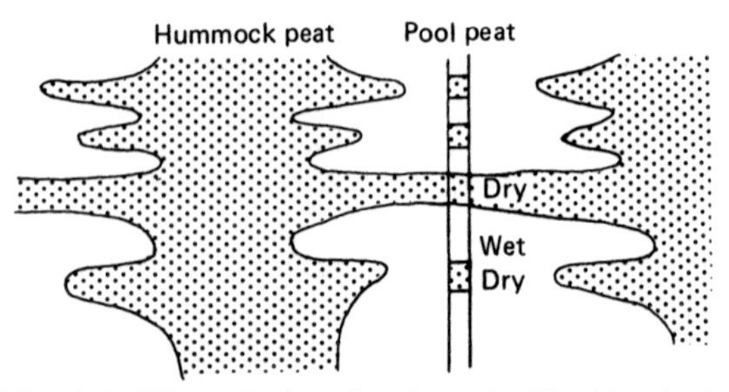

Fig. 4.4 Theoretical section through a Danish raised bog, showing that the hummock area increases in dry periods, and becomes smaller in wetter periods. (After Aaby, 1976.)

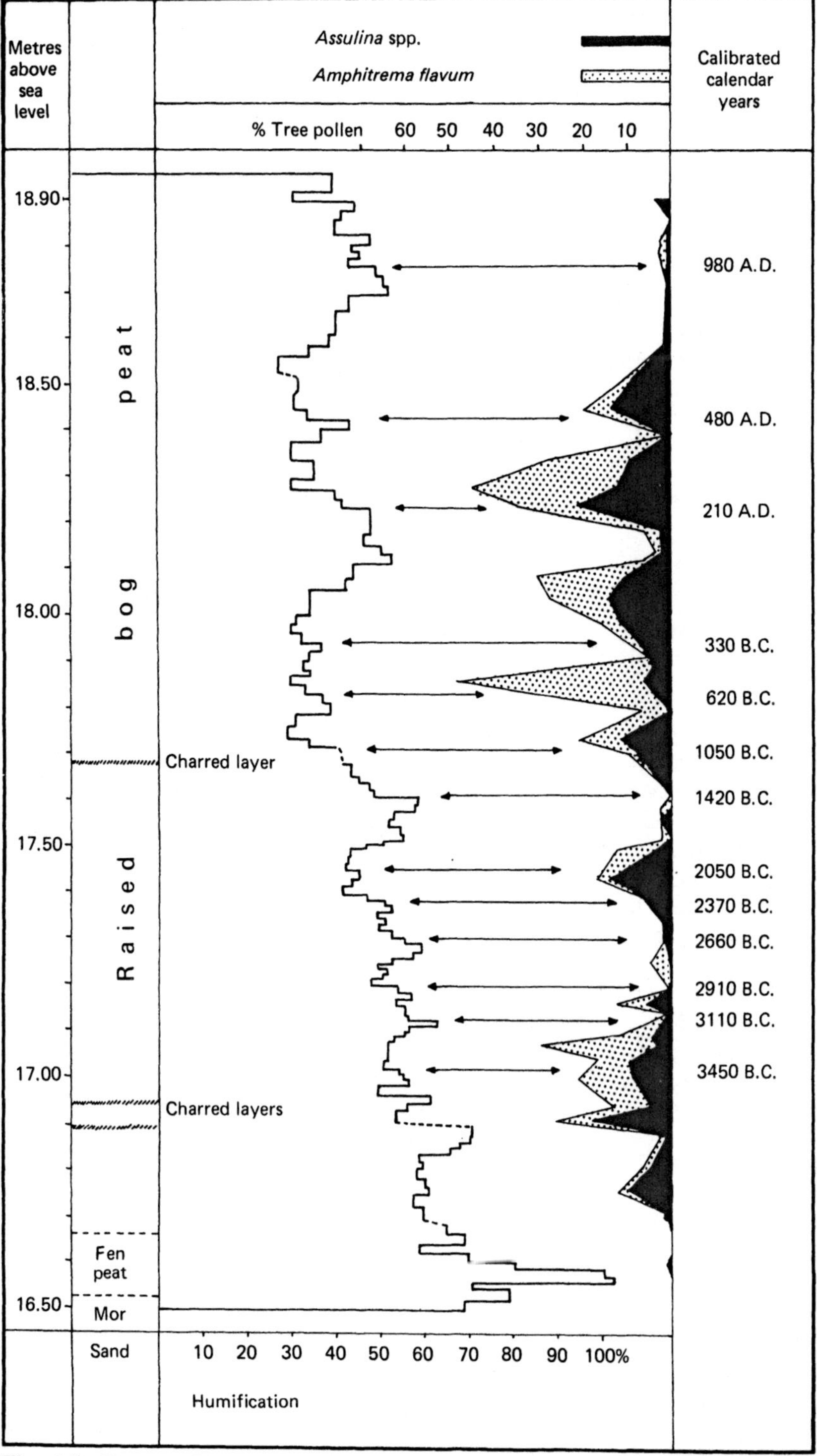

Fig. 4.5 Degree of humification and content of two rhizopod genera, *Assulina* and *Amphitrema* in a vertical peat column from Draved Mose, Denmark. The arrows indicate levels assumed to reflect climatic shifts, and their age is indicated on the right. (After Aaby, 1976.)

Peat humification studies and climatic change in raised bogs

The Grenzhorizont is only one, albeit the major one, of several recurrence surfaces in raised bogs in northern Europe. A recurrence surface is a boundary where humified peat is overlaid sharply by fresh *Sphagnum* peat, indicating regrowth of the bog surface after a period of relatively slow growth under dry conditions.

A detailed study using modern techniques was undertaken by Aaby (1976) of a section through Draved Mose, a large raised bog in south-west Denmark. Draved Mose shows many recurrence surfaces and by examining extensive peat sections, they can be related to extensions and retractions of the main drier hummocks and wetter pools (see Fig. 4.4).

Aaby analysed sections in detail with 2 cm contiguous samples for humification, the rhizopods *Assulina* and *Amphitrema flavum*, and pollen (Fig. 4.5). Humification was measured colorimetrically by the method of Bahnson (1968) (see also Aaby and Tauber, 1975). The two rhizopod taxa are characteristic of bog pools with a lawn of *Sphagnum cuspidatum*, and thus they indicate wet conditions on the bog surface.

The results from the Draved Mose profile are shown in Figure 4.5. The degree of humification is shown on the left, and the values for the rhizopods on the right. The scale of calibrated calendar years is derived from 55 radiocarbon dates. The horizontal arrows indicate points where the humification decreases and the values of rhizopods increase. These are interpreted as shifts to wetter conditions on the bog surface. The general tendency for a bog under constant climatic conditions is to dry out, as peat growth raises the surface above the water table. Hence any shift to a wetter surface probably indicates an increased climatic wetness and coolness.

Aaby found several climatic shifts over a period of about 5500 years, and more studies on other bogs showed that the events more or less coincided in time. Statistical analysis revealed that the shifts occurred regularly at intervals of 260 years, as shown in Fig. 4.6.

It should be briefly mentioned that radiocarbon years do not coincide precisely with calendar years, as seen in Fig. 4.6. Radiocarbon years have been reasonably well calibrated to calendar years by means of tree-ring counts. Tree rings are laid down annually, and by using the oldest living trees known, the bristle-cone pines of California, U.S.A.

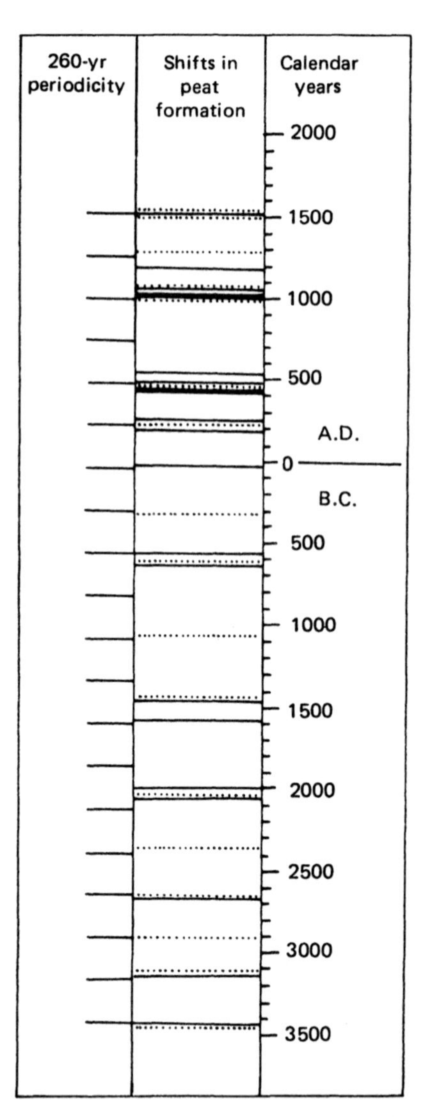

Fig. 4.6 Record of climatically conditioned shifts from dark to light peat in sections through five Danish raised bogs. (· · · · · ·) dates of changes from Draved Mose. (——) dates of changes from other bogs. The changes fit statistically a 260-year periodicity, indicated on the left. (After Aaby, 1976.)

counts reach back to about 5000 years. Radiocarbon years are then calibrated by dating wood of known calendar age (see Aaby and Tauber, 1975).

The origin of the Norfolk Broads, England

The Norfolk Broads are shallow lakes linked to the river systems of south-east Norfolk, and are much in demand by holidaymakers in the summer.

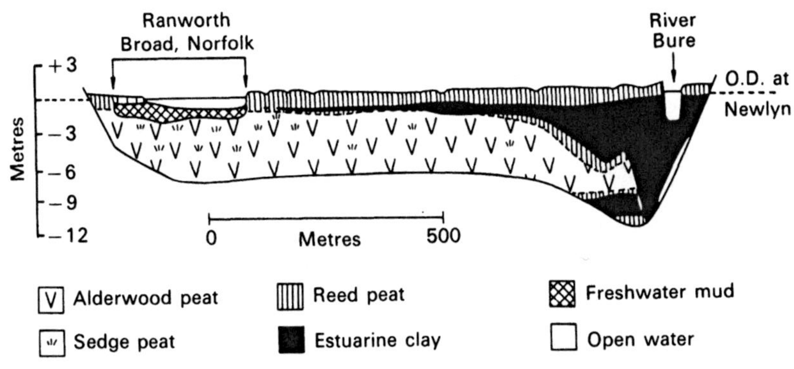

Fig. 4.7 Vertical section through Ranworth Broad to the River Bure, Norfolk, England, showing the distribution of the peat types. Note the steep sides of the broad, cut into the alder wood peat below. The estuarine clay of the marine incursion in Roman times did not reach the area of the broad at the edge of the valley. (After Lambert and Smith, 1960.)

They were originally thought to be relics of a huge lake or estuary which covered the area as a result of a sea incursion during Roman times. Lambert and Smith (1960) give a short account of stratigraphical and historical studies and these are described in considerable detail by Lambert *et al.* (1960). The Broads all lie in peat-filled river valleys, and are all relatively shallow and flat-bottomed, about 12 feet (4 m) deep. After about 2000 hand corings through the peat, stratigraphical sections were constructed, as in Fig. 4.7 for Ranworth Broad. Although the marine clay of Roman times indicates the extent of the transgression of the sea, the Broads are generally situated beyond its limit, and therefore the original theory of their formation cannot be correct. Detailed corings demonstrated that the sides of the Broads are vertical or stepped, except where they adjoin the surrounding gravels. The lake mud deposited in the Broads is sharply distinct from the surrounding fen peats.

This evidence suggests that the Broads are man-made lakes. Further research into historical records, such as tithe maps and manorial accounts revealed that the Broads were originally dug as peat cuttings in the twelfth century, and peat production continued through the thirteenth and fourteenth centuries. Turves were used locally, and exported as far as Norwich and Yarmouth. Norwich Cathedral Priory used 400 000 turves a year in its kitchens in the early fourteenth century.

Tithe maps of the time showed that the turbaries (peat cutting areas) were well defined, usually in long narrow strips. These strips are recognizable today in the alignment of the shores of the broads, and aerial photographs show how reedswamp areas tend to follow straight lines along previous boundary banks, which are now underwater.

Towards the end of the fourteenth century, sea level rose slightly, and the turf pits gradually became flooded, until peat digging became impossible. The resulting lakes were then used as fisheries, and wood replaced peat as local fuel.

The prehistoric wooden trackways of the Somerset Levels

The Somerset Levels occupy the infilled valley between the Polden and Mendip hills, which reaches the sea at Burnham-on-Sea in south-west England. The base of the valley is about 97 feet (27 m) below sea level, but it is filled up to about present sea-level by estuarine clay. The sea retreated between 6000 and 5500 B.P., leaving a freshwater marshy landscape. A reedswamp composed primarily of *Phragmites communis* and *Cladium mariscus* covered the whole valley, and laid down 1–2 m of telmatic peat. Gradually, the reedswamps were invaded by alder and birch. These fen woods deposited wood peat until about 4000 B.P. Then they were invaded by *Sphagnum* as the influence of the calcareous run-off water from the hills became progressively less important. The *Sphagnum* was associated with *Calluna*, *Eriophorum*, and other typical bog plants, and 1–2 m of humified bog peat were deposited. There is a sharp transition to fresh unhumified *Sphagnum* peat over most of the area. In the region of Shapwick Heath, this recurrence surface is accentuated by a layer of *Cladium* – brown moss peat, indicating that calcareous water must have flooded the bog surface. However, ombrogenous peat growth soon resumed. There is similar evidence

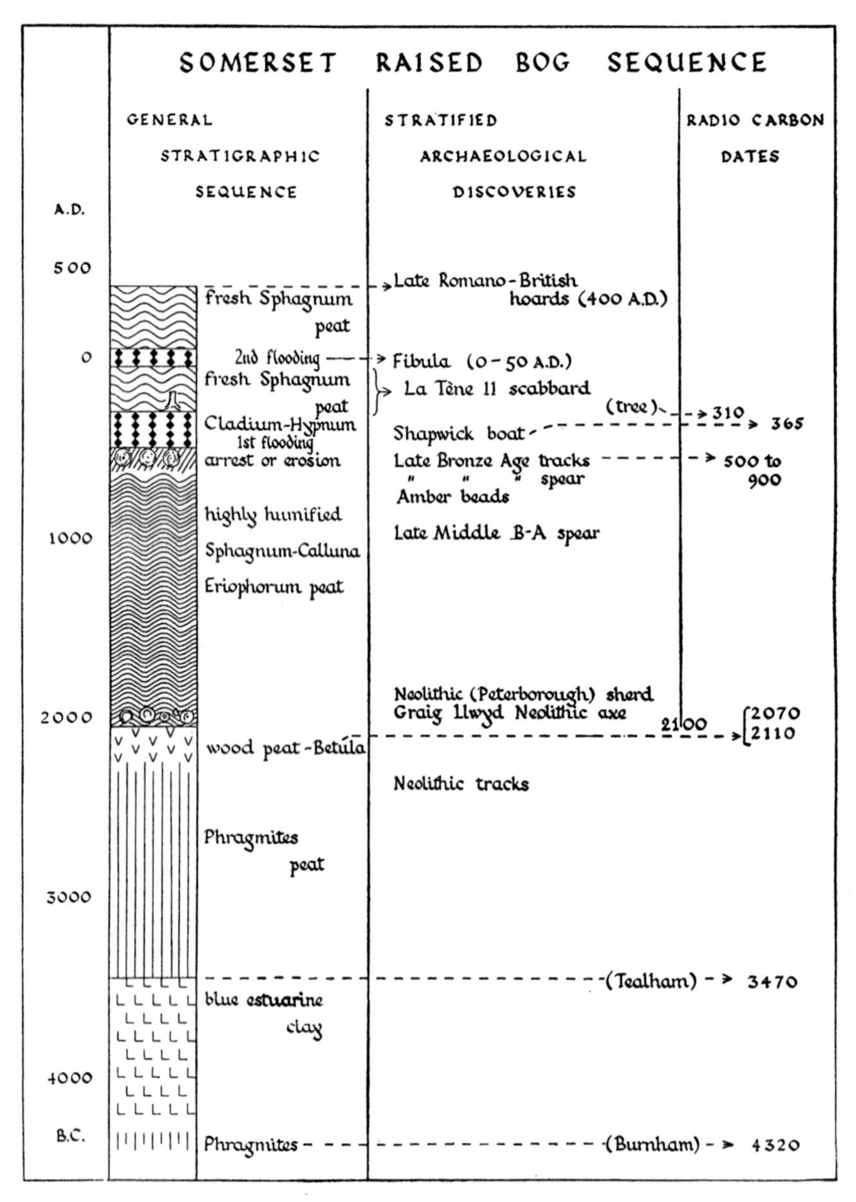

Fig. 4.8 Summary of the stratigraphic sequence in the Shapwick Heath area of the Somerset Levels, and its correlation with archaeological finds and radiocarbon dates. (From Godwin, 1960.)

for a second flooding at Shapwick Heath, where another layer of *Cladium*–brown moss peat was deposited. Elsewhere, this recurrence surface is marked by a change to unhumified *Sphagnum* peat. Once more, ombrogenous bog communities developed, until active peat growth ceased at about 1500 B.P. The stratigraphy is summarized in Fig. 4.8.

This complicated stratigraphic sequence was worked out by Professor Sir Harry Godwin, the principal instigator of Quaternary palaeoecological studies in Britain. The Somerset Levels have long been utilized for peat cutting, and this is now carried out on a commercial scale there. During the peat extraction, many finds of archaeological interest have been made, most particularly the

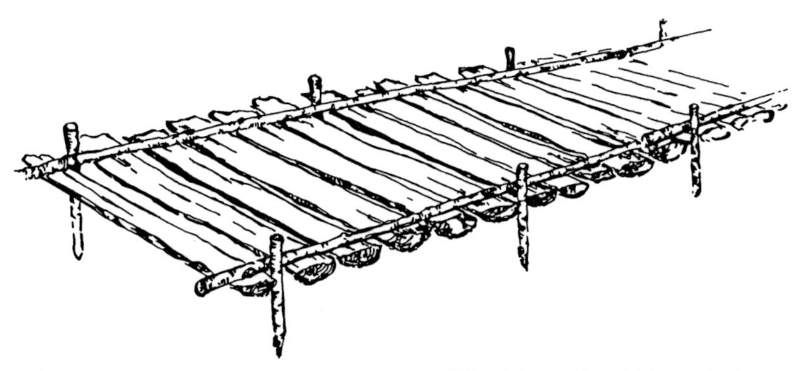

Fig. 4.9 Abbots Way: artist's reconstruction of the track with transverse timbers held by posts and lateral stringers. This type of track construction was used by both Neolithic and Bronze Age people. (After Godwin, 1960.)

remains of several wooden trackways crossing the valley. The aim of Godwin's investigation was to elucidate the age and nature of the trackways (with the help of archaeologists) and to place them into a palaeoecological context (see Dewar and Godwin, 1963; Godwin, 1960).

The trackways were of great interest in themselves, varying in construction with the different types of terrain they crossed. Some were constructed with logs placed longitudinally, and covered by transversely laid heather, brushwood, or turves. Others were constructed of transversely laid logs, held together by lateral posts driven into the peat and lateral stringers, as illustrated in Fig. 4.9.

The trackways occurrred at two stratigraphic levels. The lower ones lay above the wood peat but below the humified bog peat. The upper ones lay above the humified bog peat, but below the fresh *Sphagnum* peat or *Cladium* peat. They were roughly dated by means of associated archaeological objects and by pollen analysis, as indicated in Fig. 4.8, but much more precise dates were later obtained by radiocarbon dating both of the tracks and the associated peat. These confirmed a Neolithic age of between 4800 and 4000 B.P. for the lower tracks, and a Bronze Age date of 2500–2900 B.P. for the upper tracks (see Godwin, 1960).

Both the phases of track-building were associated with increases in wetness of the valley floor. Neolithic people could easily cross the valley in the relatively dry fen-carr. However, the onset of *Sphagnum* peat growth indicated wetter conditions. The abundance of *Sphagnum cuspidatum* (the main aquatic species) in the peat was evidence for wet pools. The people built tracks over the wettest areas which gradually extended. Eventually the tracks themselves were swamped, and presumably communications became very difficult. Later on, Bronze Age people could walk fairly easily across the rather dry bog surface, which was forming humified peat. Track building started in response to an increased wetness, probably due to climatic causes, as the parts of the bog near the high ground were eventually flooded by calcareous run-off water, allowing the development of a *Cladium*–brown moss fen community. The tracks had a limited usefulness, as they were relatively soon submerged, either by fresh *Sphagnum* peat, or by *Cladium* fen. The extent of the flooding is emphasized by the find of a primitive boat at Shapwick, radiocarbon dated to 1315 B.P. The people inhabiting the region must have had to adapt quite rapidly to their changing environment.

The second flooding episode at about 2000 B.P. is probably not due to increased climatic wetness, but to a marine transgression registered in the peat near the sea as a layer of marine clay. Drainage water would have been impounded, and hence the bogs would have become wetter than before. The over-all stratigraphical relationship of the valley to changes in sea-level is shown in Fig. 4.10.

Godwin's (1960) study is an excellent example of a coordinated approach using several techniques. The backbone of the study was the elucidation of the stratigraphy and the environmental inferences made from it. Radiocarbon dating was also very important in confirming the ages of the various layers and artifacts which had been suggested by archaeological and pollen analytical correlations. A secondary result of the study was the dating of the changes in sea-level in that area.

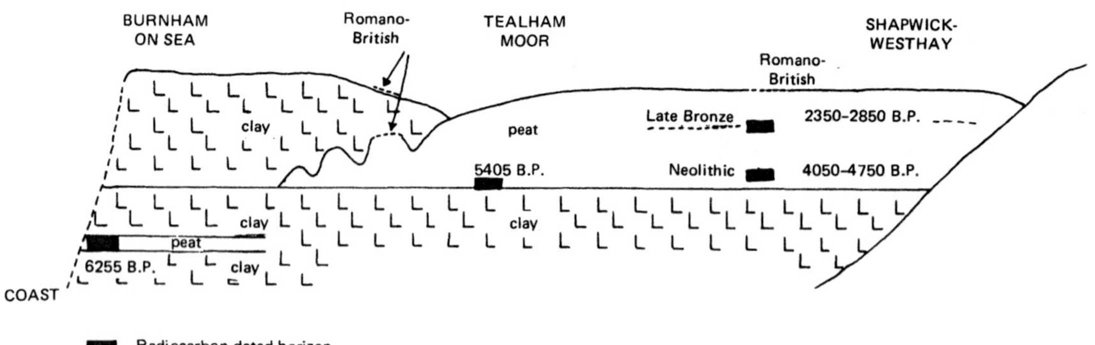

Fig. 4.10 Schematic representation of the main freshwater and marine layers in the post-glacial sediments of the Somerset Levels, showing the main archaeological horizons and radiocarbon dates. The marine transgression in Romano–British time did not cover the raised bogs, but had the effect of making them wetter due to impeded drainage. (After Godwin, 1960.)

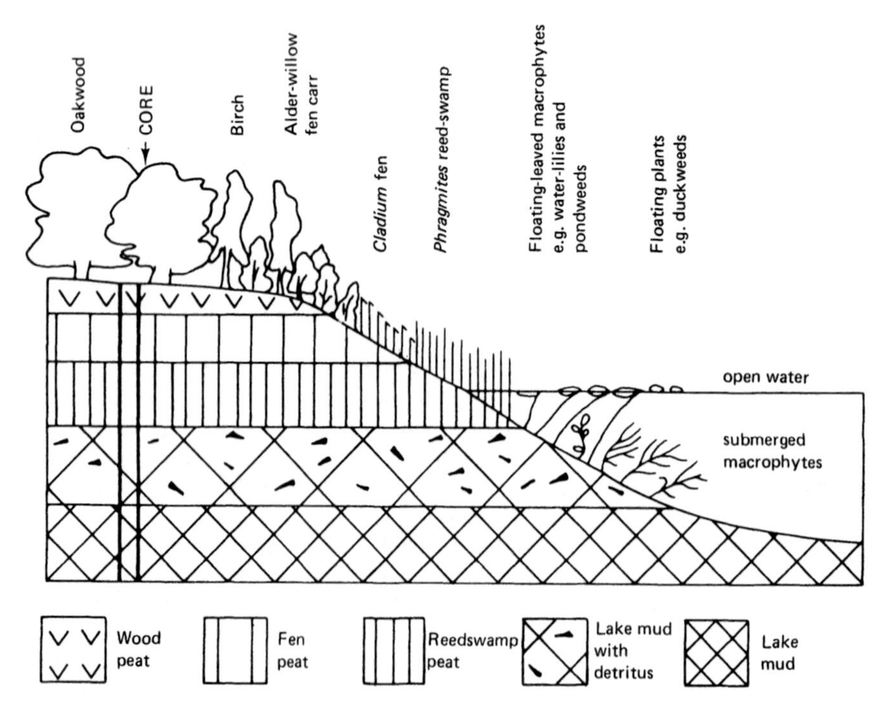

Fig. 4.11 Theoretical hydrosere, and the sediments produced by each vegetational stage. If the processes have been continuous, a core under the oakwood should produce a sequence of sediments laid down in time which is the same as the sequence being laid down in space at the present day.

Hydroseral patterns in time and space

The stratigraphy of the Somerset Levels represents the hydrosere from brackish water to fresh water, to reedswamp, fen carr, and then bog. The standard ecological concept of a hydrosere is that described originally by Tansley (1939) and illustrated in Fig. 4.11.

In the open water, fine mud is deposited, and as the water becomes shallower, coarser particles are incorporated, until the water becomes shallow enough to allow the growth of submerged and then floating-leaved aquatic plants (often called macrophytes) and then by *Phragmites* or other reedswamp species such as *Schoenoplectus lacustris.* As the reedswamp peat accumulates, fen plants colonize it, such as *Cladium mariscus*, followed by drier fen species, such as *Filipendula ulmaria.* When the fen peat becomes sufficiently dry, shrubs colonize it, forming a fen carr. These are usually willows, alder, and even birch. Eventually, the surface is dry enough for oakwood to develop. If the climate is sufficiently wet to keep the surface waterlogged, leaching and acidification allow *Sphagnum* to grow, and the fen carr may be converted to a bog, as occurred in the Somerset Levels.

Each plant community lays down its own characteristic sediment, as described earlier in the chapter. Therefore, if the zonation of the hydrosere in space reflects its development in time, a core through the sediments under the oakwood (or the bog) should produce a series of sediment types corresponding to the present-day zonation of communities.

Walker (1970) tested this theory of hydroseral development by assembling the available post-glacial stratigraphic information from Britain. In a total of 159 sites, 12 plant communities or hydroseral stages could be readily recognized from the type of sediment and information from pollen analysis. He arranged them in order 1 to 12 in the predicted order if the hydroseral concept was correct, from open water to oakwood or bog.

1. Biologically unproductive water
2. Micro-organisms in open water
3. Totally submerged macrophytes (e.g. *Potamogeton, Myriophyllum*)
4. Floating leaved macrophytes (e.g. *Nymphaea alba, Potamogeton natans*)
5. Reedswamp, rooted in the substratum below water, but with aerial shoots and leaves (e.g. *Phragmites communis, Schoenoplectus lacustris, Carex rostrata*)
6. Sedge tussock, rooted in the substratum below water, and standing in more or less perennial water (e.g. *Carex paniculata, C. acutiformis*)
7. Fen, dominated by grasses (e.g. *Phragmites communis, Molinia caerulea*) or sedges (e.g. *Carex panicea, C. nigra*) with a wide variety of acid intolerant herbs (e.g. *Potentilla palustris, Filipendula ulmaria, Valeriana officinalis*), all rooting in organic deposits which are waterlogged for the greater part of the year
8. Swamp carr formed by trees (e.g. *Salix atrocinerea, Alnus glutinosa*) growing on unstable sedge tussocks with some fen herbs, the intervening pools often harbouring thin reedswamp or floating-leaved macrophytes
9. Fen carr, dominated by trees (e.g. *Alnus glutinosa, Frangula alnus, Betula pubescens, Fraxinus excelsior*) with an undergrowth rich in fen herbs and ferns (e.g. *Thelypteris palustris*), all rooting in a physically stable peat mass
10. Aquatic sphagna, floating very near or on the water surface (e.g. *Sphagnum subsecundum, S. cuspidatum*)
11. Bog, usually distinguished by a variety of *Sphagnum* species (notably *S. palustre* and *S. imbricatum*) and acid-tolerant phanerogams (e.g. *Vaccinium oxycoccos, Erica tetralix, Eriophorum* spp., *Calluna vulgaris*) growing in an organic substratum
12. Marsh, composed of 'fen' species (e.g. *Filipendula ulmaria, Eleocharis palustris*) growing on waterlogged mineral soil

Having defined his working sedimentary units, Walker then scored the number of times stage 1 was followed by stage 2, stage 1 by stage 3, and so on in all the published stratigraphical studies in the British Isles. This resulted in the patterns shown in Fig. 4.12. The main feature of this chart is its diversity. There is no main hydroseral succession, apart from the over-all reduction of open water, build up of peat, and eventual raising of the surface above water level, either as fen or usually as bog. Reedswamp seems to be a key stage in most hydroseres, marking the transition from aquatic to terrestrial communities. The vegetation zonation around a lake is not necessarily a reflection of what has happened at a specific point over the length of time of sediment deposition. Almost all combinations of stages can follow each other; even reversals may occur, presumably as a result of flooding. However, once conditions become acid enough for the successful growth of *Sphagnum*, the route to bog develop-

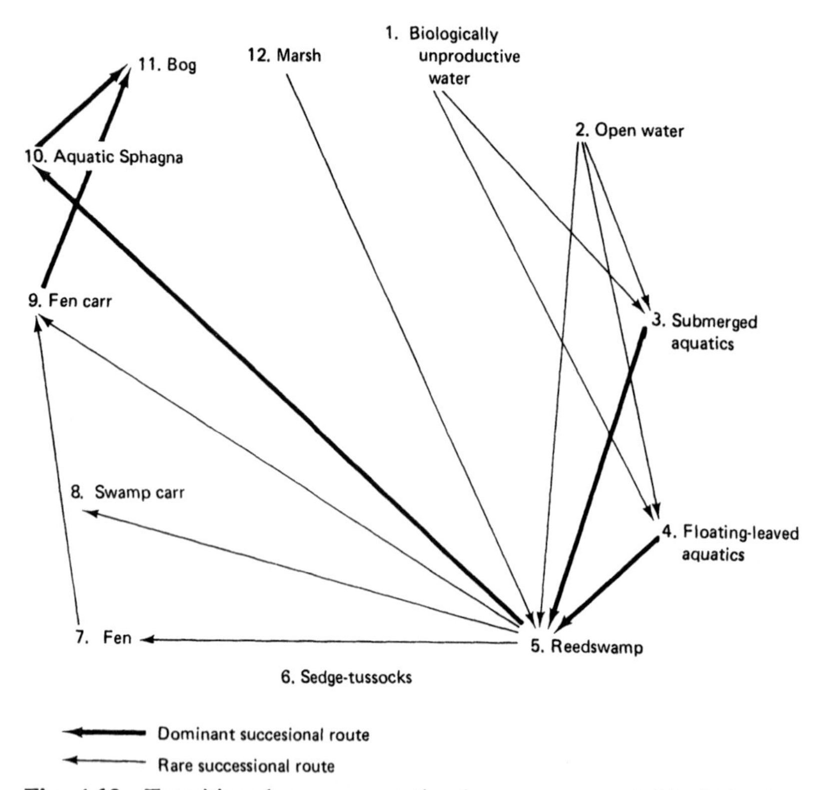

Fig. 4.12 Transitions between vegetational stages as recorded in 159 mire sites in the British Isles. (After Walker, 1970.)

ment is almost irreversible, unless there is some catastrophic flooding with calcareous water, as in the Somerset Levels. Even here, however, the reedswamp stage was relatively short-lived, and *Sphagnum* rapidly regained dominance. *Sphagnum* may invade a hydrosere at any stage from floating-leaved aquatics to fen carr. A good example of *Sphagnum* invading fen carr today can be seen at Wicken Fen, Cambridgeshire, England. Over the last few years increasing numbers of *Sphagnum* colonies have been found in the oldest, most acidic fen carr, under *Frangula alnus* and *Betula* trees. Its progress is being monitored to see how rapidly it spreads. It is likely that the end point of nearly all hydroseres throughout the British Isles, not just in the wettest areas, is bog development, and, in fact, oakwood is a rather unusual climax (see Godwin and Turner, 1933).

Because of the long time taken for hydroseral changes to occur (Spence, 1964), direct observations of hydrosere dynamics are difficult. Pigott and Wilson (1978) compared the detailed map of Esthwaite North Fen (English Lake District) prepared by W. H. Pearsall 60 years before with a new map of the fen communities. They showed that all the fen communities had moved towards the lake. However, each seemed to move at a different rate, with the result that the fen carr had broadened at the expense of the outer *Phragmites* reedswamp. The trees seemed to be able to colonize the reedswamp more quickly than the reedswamp could invade the aquatic communities, suggesting that the orderly progression lakewards of the vegetational communities did not necessarily occur. Walker's palaeoecological evidence of hydroseral development encompasses the whole 10 000 years of the Flandrian. It shows how important the time aspect is to modern ecological theory. The historical evidence of successions should be sought before theories are expounded based on the appearance of the vegetational complex at one point in time. Such palaeoecological evidence showed the inadequacy of another theoretical ecological situation, the 'regeneration complex' theory of the occurrence of pools and hummocks on a raised bog, which we shall consider next.

The 'regeneration complex' theory of bog growth

Living raised bogs have on their surfaces a pattern of pools and hummocks, often linked by *Sphagnum* lawns (see Fig. 4.13). The regeneration complex theory of bog growth was proposed by H. Osvald in 1923 to explain the role of pools and hummocks in the growth of a bog. A theoretical section through a pool and hummock complex is shown in Fig. 4.14. The open water of a pool is encroached on by growth of aquatic and lawn species of *Sphagnum*. Hummock-forming species such as *S. papillosum* and *S. imbricatum* then colonize the lawn to form a new hummock. The drier top of the hummock allows the growth of vascular plants such as *Calluna vulgaris*. The growth of the hummock slows down as it is colonized by species such as *Sphagnum rubellum* and *S. fuscum*. Eventually the top dries out, and is colonized by lichens and *Rhacomitrium lanuginosum*. It will eventually be eroded by wind and rain. By this time, new hummocks will have originated on the adjacent lawns, and they will eventually overtop the old hummock, which will subsequently become a pool, and the cycle will start again, as shown in Fig. 4.14.

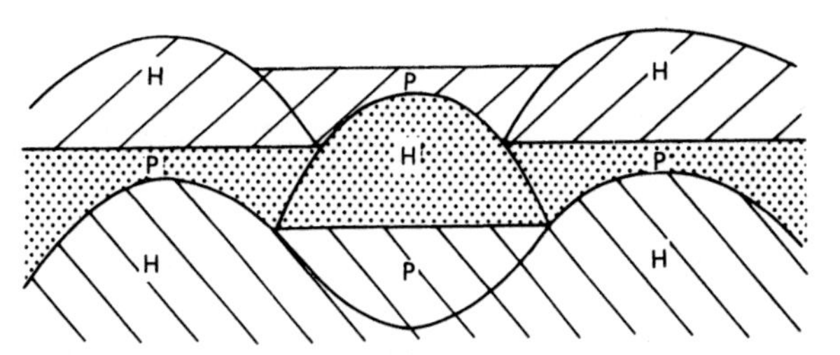

Fig. 4.14 Theoretical stratigraphic section through a pool and hummock complex on a raised bog, showing how each generation originates from pools of the previous generation.

Hummock peat should be humified and composed of *Calluna*, *Sphagnum papillosum*, and *S. imbricatum*. Pool and lawn peat should be unhumified, and be composed almost entirely of *Sphagnum cuspidatum*, and other lawn species such as *S. recurvum*, *S. subsecundum*, and *S. pulchrum*.

Walker and Walker (1961) tested the regeneration complex theory by investigating the vertical peat stratigraphy of eight Irish bogs. Their results are summarized in Fig. 4.15. Previously, the theory had been questioned, but no conclusive evidence had been presented. Walker and Walker found that the full regeneration cycle was very rare, although they could frequently find evidence that neighbouring pools and hummocks had been converted to hummocks and pools. The commonest situation was that

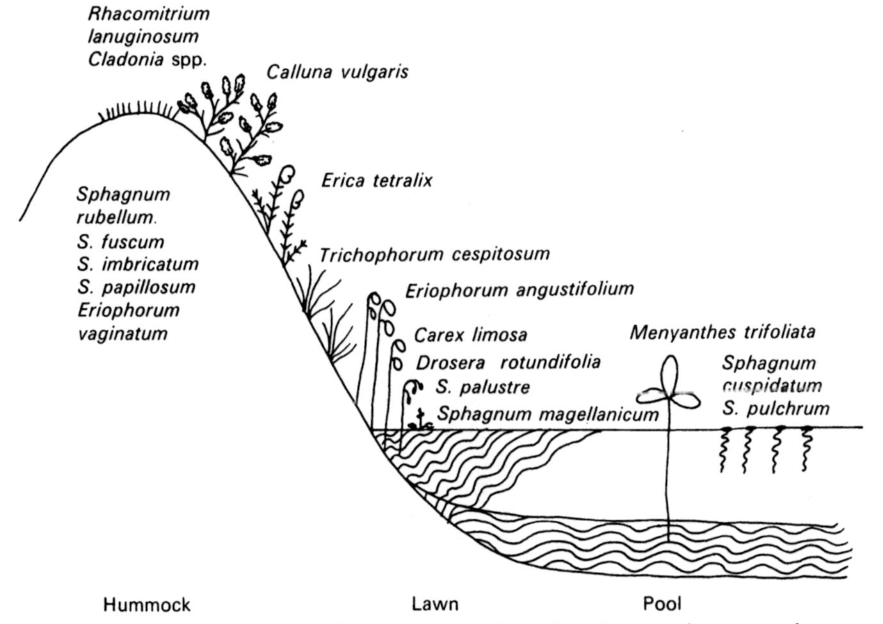

Fig. 4.13 Generalized vegetational zonation through a hummock to a pool on a British raised bog.

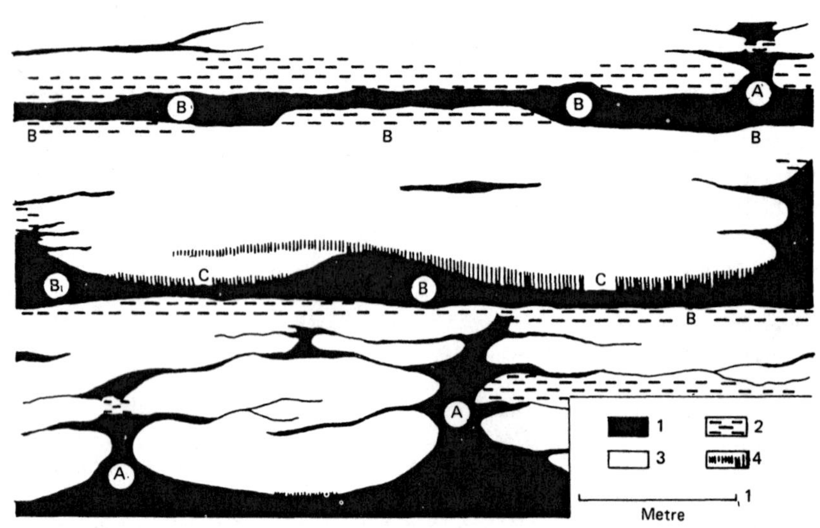

Fig. 4.15 Diagrammatical representation of some common stratigraphic features in peat faces of Irish bogs. See text for explanation of events at A, B, and C. 1, Highly humified *Calluna* peat with *Sphagnum*. 2, Medium humified *Sphagnum* peat with *Calluna*. 3, Slightly humified *Sphagnum* peat. 4, Pool mud and pool peat. (After Walker, 1961.)

a hummock had arisen from a lawn, but that it showed no sign of autonomous degeneration. In fact, as the water table was raised by growth of the lawn, the hummock showed renewed growth as well, as at 'A' in Fig. 4.15. However, all hummocks eventually disappear, either by becoming nuclei for the establishment of a *Calluna*-dominated sward, or by being overgrown by *Sphagnum*-rich communities which also form a more or less even surface. These processes are indicated by 'B' on Fig. 4.15. They also found evidence for persistent large pools usually at fairly distinct levels in the peat (C in Fig. 4.15), which suggests that they are the result of changes in the water régime of the bog leading to a phase of flooding. In other words, there was a minor recurrence surface. By careful stratigraphic study, Walker and Walker added a great deal to our knowledge of bog growth, showing that the regeneration complex was not an important mechanism. Since then, other studies have confirmed Walker and Walker's results from Irish bogs. For example, Aaby's (1976) sections showed the persistence of hummocks in Denmark (Fig. 4.4) and careful work by Casparie (1969, 1972) in Holland has shown similar results.

Aerial photographs of raised bogs often show that the pools are aligned parallel to the contours. This is well shown in Scandinavia, and is also shown by some undisturbed British bogs, such as the Silver Flowe series of bogs in the Galloway Hills, in south-west Scotland, and Claish Moss on the Argyll coast. An aerial photograph of Snibe Bog, one of the Silver Flowe bogs, is shown in Ratcliffe and Walker (1958). There have been several hypotheses proposed for the alignment of the pools (e.g. Pearsall, 1956; Ratcliffe and Walker, 1958; Boatman and Armstrong, 1968; Boatman and Tomlinson, 1977) but no truly satisfactory explanation has yet been put forward.

Ratcliffe and Walker (1958) studied the vegetation of the eight Silver Flowe bogs, and investigated the stratigraphy of one of them, Snibe Bog. They discovered that the pools had developed quite recently, over a flat layer of highly-humified uniform *Molinia*-rich *Sphagnum* peat. They postulated that the bog had been in equilibrium with the prevailing environment, and had reached an over-all fairly uniform vegetation. The development of the pools may have been caused by a flooding episode which was probably the result of increased run-off of rain water from the surrounding hills as a consequence of widespread forest clearance. This hypothesis was supported by subsequent pollen analytical studies by Birks (1972). In a short discussion on bog regeneration Walker (1961) proposed that neither any small scale regeneration complex nor the large scale system of pools and hummocks are self-perpetuating systems, independent of the hydrology of the whole bog. After a time, a uniform sward of vegetation develops in equilibrium with the prevailing climate, until some change initiates a new series of hummocks and pools. This is the response to climatic and other topographical or hydrological changes shown by raised bogs, which makes them such sensitive indicators of increased wetness rather than of increased dryness.

In contrast to the findings of Ratcliffe and Walker (1958) at the Silver Flowe bogs, Moore (1977) found that the massive aligned pools at Claish Moss, Argyll may have been present for the last 5000 years.

Although the mechanism for the formation of these impressive surface patterns on raised bogs is not fully elucidated, the work of Boatman and his colleagues (Boatman and Armstrong, 1968; Boatman and Tomlinson, 1973, 1977) suggests the following working hypothesis. The formation of hummocks is initiated by vascular plants such as *Calluna vulgaris* which provide support for mosses such as *Sphagnum rubellum* and *Rhacomitrium lanuginosum*. There is inevitably some patchiness in the distribution of these hummock-forming species, and small pools will occur wherever drainage water collects. The hydraulic conductivity of peat is so low that once a small pool is formed, the peat effectively retains and impounds the water. On many bogs the growth of *Sphagnum cuspidatum* is suboptimal, possibly due to nutrient deficiencies or competition with pool algae. Due to the poor growth of *S. cuspidatum* the pool becomes progressively lower in relation to the bog surface as the hummocks and lawns around grow upward. The pools may enlarge by flooding ridges between them. This will occur if the water level in adjacent pools is the same, and a pool will extend across the slope of the bog, parallel to the surface contours forming the characteristic concentric pattern. The pools are gradually enlarged by erosion of their banks by wind and wave action, and very large, bare pools may form, perhaps with islands.

References

Ecology of mires

BIRKS, H. J. B. (1973). *Past and Present Vegetation of the Isle of Skye – a Palaeoecological Study*. Cambridge University Press (Chapter 4).

CONWAY, V. M. (1949). Ringinglow Bog, near Sheffield. *J. Ecol.*, **37**, 148–70.

GORHAM, E. (1957). The development of peat lands. *Quart. Rev. Biol.*, **32**, 145–66.

GREEN, B. H. AND PEARSON, M. C. (1968). The ecology of Wybunbury Moss, Cheshire. I. The present vegetation and some physical, chemical and historical factors controlling its nature and distribution. *J. Ecol.*, **56**, 245–67.

GREEN, B. H. AND PEARSON, M. C. (1977). The ecology of Wybunbury Moss, Cheshire. II. Post-glacial history and the formation of the Cheshire mere and mire landscape. *J. Ecol.*, **65**, 793–814.

HEINSELMAN, M. L. (1963). Forest sites, bog processes, and peatland types in the Glacial Lake Agassiz Region, Minnesota. *Ecol. Monogr.*, **33**, 327–74.

HEINSELMAN, M. L. (1970). Landscape evolution, peatland types, and the environment in the Lake Agassiz Peatlands Natural Area, Minnesota. *Ecol. Monogr.*, **40**, 235–61.

INGRAM, H. A. P. (1967). Problems of hydrology and plant distribution in mires. *J. Ecol.*, **55**, 711–24.

MCVEAN, D. N. AND RATCLIFFE, D. A. (1962). *Plant Communities of the Scottish Highlands*. H.M.S.O. (pp. 101–36).

MOORE, P. D. AND BELLAMY, D. J. (1974). *Peatlands*. Elek Science.

PROCTOR, M. C. F. (1960). Mosses and liverworts of the Malham district. *Field Studies*, **1**, 61–85.

PROCTOR, M. C. F. (1974). The vegetation of the Malham Tarn Fens. *Field Studies*, **4**, 1–38.

RATCLIFFE, D. A. (1964a). Mires and bogs. In *The Vegetation of Scotland* (ed. J. H. Burnett). Oliver and Boyd.

RATCLIFFE, D. A. (1964b). Montane mires and bogs. In *The Vegetation of Scotland* (ed. J. H. Burnett). Oliver and Boyd.

RATCLIFFE, D. A. (1977). *A Nature Conservation Review I*. Cambridge (pp. 249–84).

SINKER, C. A. (1962). The North Shropshire Meres and Mosses: a background for ecologists. *Field Studies*, **1**, 101–38.

SJÖRS, H. (1950). On the relation between vegetation and electrolytes in North Swedish mire waters. *Oikos*, **2**, 241–58.

TALLIS, J. H. (1964). Studies on Southern Pennine peats. II. The pattern of erosion. III. The behaviour of *Sphagnum*. *J. Ecol.*, **52**, 333–44, 345–53.

TALLIS, J. H. (1965). Studies on Southern Pennine peats. IV. Evidence of recent erosion. *J. Ecol.*, **53**, 509–20.

TALLIS, J. H. (1969). The blanket bog vegetation of the Berwyn Mountains, North Wales, *J. Ecol.*, **57**, 765–87.

TALLIS, J. H. (1973). The terrestrialization of lake basins in North Cheshire with special reference to the development of a 'Schwingmoor' structure. *J. Ecol.*, **61**, 537–67.

Bog stratigraphy, buried tree remains, and post-glacial climatic change

BIRKS, H. H. (1975). Studies in the vegetational history of Scotland. IV. Pine stumps in Scottish blanket peats. *Phil. Trans. R. Soc.* B, **270**, 181–226.

GEIKIE, J. (1877). *The Great Ice Age*. Daldy, Isbister and Co.

KARLÉN, W. (1976). Lacustrine sediments and tree-limit variations as indicators of Holocene climatic fluctuations in Lappland, Northern Sweden. *Geogr. Annaler*, **58A**, 1–34.

LEWIS, F. J. (1905–11). The plant remains in the Scottish peat mosses. I–IV. *Trans. Royal Society Edinb.*, **41**, 699–723; **45**, 335–60; **46**, 33–70; **47**, 793–833.

SAMUELSSON, G. (1910). Scottish Peat Mosses. A contribution to the knowledge of the late-Quaternary vegetation and climate of North Western Europe. *Bull. Geol. Inst. Univ. Upsala*, **10**, 197–260.

Raised bog stratigraphy and peat humification studies

AABY, B. (1976). Cyclic climatic variations in climate over the past 5,500 yr reflected in raised bogs. *Nature*, **263**, 281–4.

AABY, B. AND TAUBER, H. (1975). Rates of peat formation in relation to degree of humification and local environment, as shown by studies of a raised bog in Denmark. *Boreas*, **4**, 1–17.

BAHNSON, H. (1968). Kolorimetriske bestemmelser af humificeringstal i højmosetørv fra Fuglsø mose på Djursland. *Medd. Dansk. Geol. Foren.*, **18**, 55–63.

GEEL, VAN B. (1976). A paleoecological study of Holocene peat bog sections in Germany and the Netherlands, based on the analysis of pollen, spores and macro- and microscopic remains of fungi, algae, cormophytes and animals. *Rev. Palaeobot. Palynol.*, **25**, 1–120.

GODWIN, H. AND CONWAY, V. M. (1939). The ecology of a raised bog near Tregaron, Cardiganshire. *J. Ecol.*, **27**, 313–63.

GREEN, B. H. (1968). Factors affecting the spatial and temporal distribution of *Sphagnum imbricatum* Hornsch. ex Russ. in the British Isles. *J. Ecol.*, **56**, 47–58.

MORRISON, M. E. S. (1959). The ecology of a raised bog in C. Tyrone, Northern Ireland. *Proc. R. Ir. Acad.* B, **60**, 291–308.

Peat stratigraphical studies in the Norfolk Broads

LAMBERT, J. M. (1951). Alluvial stratigraphy and vegetational succession in the region of the Bure Valley Broads. III. Classification, status and distribution of communities. *J. Ecol.*, **39**, 149–70.

LAMBERT, J. M., JENNINGS, J. N., SMITH, C. T., GREEN, C. AND HUTCHINSON, J. N. (1960). *The Making of the Broads*. Royal Geographical Society Research Series, **3**, 153 pp.

LAMBERT, J. M. AND SMITH, C. T. (1960). The Norfolk Broads as man-made features. *New Scientist*, 31 March 1960.

Peat stratigraphical studies in The Somerset Levels

CLAPHAM, A. R. AND GODWIN, H. (1948). Studies of the post-glacial history of British vegetation. VIII. Swamping surfaces in peats of the Somerset Levels; IX. Prehistoric trackways in the Somerset Levels. *Phil. Trans. R. Soc.* B, **233**, 233–73.

COLES, J. M. AND HIBBERT, F. A. (1968). Prehistoric roads and tracks in Somerset, England: I. Neolithic. *Proc. Prehist. Soc.*, **34**, 238–58.

COLES, J. M., HIBBERT, F. A. AND CLEMENTS, C. F. (1970). Prehistoric roads and tracks in Somerset, England: 2. Neolithic. *Proc. Prehist. Soc.*, **36**, 125–51.

COLES, J. M., HIBBERT, F. A. AND ORME, B. J. (1973). Prehistoric roads and tracks in Somerset: 3. The Sweet Track. *Proc. Prehist. Soc.*, **39**, 256–93.

COLES, J. M. AND ORME, B. J. (1976). The Meare Heath trackway: excavation of a Bronze Age structure in the Somerset Levels. *Proc. Prehist. Soc.*, **42**, 293–318.

DEWAR, H. S. L. AND GODWIN, H. (1963). Archaeological discoveries in the raised bogs of the Somerset Levels, England. *Proc. Prehist. Soc.*, **29**, 17–49.

GODWIN, H. (1948). Studies of the post-glacial history of British vegetation. X. Correlations between climate, forest composition, prehistoric agriculture and peat stratigraphy in Sub-boreal and Sub-atlantic peats of the Somerset Levels. *Phil. Trans. R. Soc.* B, **233**, 275–86.

GODWIN, H. (1960). Prehistoric wooden trackways of the Somerset Levels: their construction, age and relation to climatic change. *Proc. Prehist. Soc.*, **26**, 1–36.

Hydroseral patterns and temporal successions

GODWIN, H. AND TURNER, J. S. (1933). Soil acidity in relation to vegetational succession in Calthorpe Broad, Norfolk. *J. Ecol.*, **21**, 235–62.

PIGOTT, C. D. AND WILSON, J. F. (1978). The vegetation of North Fen at Esthwaite in 1967–9. *Proc. R. Soc.* B, **200**, 331–51.

SPENCE, D. H. N. (1964). The macrophytic vegetation of freshwater lochs, swamps and associated fens. In *The Vegetation of Scotland* (ed. J. H. Burnett). Oliver and Boyd.

TANSLEY, A. G. (1939). *The British Isles and Their Vegetation*. Cambridge University Press (Chapters 29–35).

WALKER, D. (1970). Direction and rate in some British post-glacial hydroseres. In *Studies of the Vegetational History of the British Isles* (eds D. Walker and R. G. West). Cambridge University Press.

Surface patterns of bogs and the origin of bog pools

BIRKS, H. H. (1972). Studies in the vegetational history of Scotland. II. Two pollen diagrams from the Galloway Hills, Kirkcudbrightshire. *J. Ecol.*, **60**, 183–217.

BOATMAN, D. J. AND ARMSTRONG, W. (1968). A bog type in North-West Sutherland. *J. Ecol.*, **56**, 129–41.

BOATMAN, D. J. AND TOMLINSON, R. W. (1973). The Silver Flowe. I. Some structural and hydrological features of Brishie Bog and their bearing on pool formation. *J. Ecol.*, **61**, 653–66.

BOATMAN, D. J. AND TOMLINSON, R. W. (1977). The Silver Flowe. II. Features of the vegetation and stratigraphy of Brishie Bog and their bearing on pool formation. *J. Ecol.*, **65**, 531–46.

CASPARIE, W. A. (1969). Bult – und Schlenkenbildung in Hochmoortorf. *Vegetatio*, **19**, 146–80.

CASPARIE, W. A. (1972). *Bog Development in Southeastern Drenthe* (The Netherlands). Dr. Junk.

MOORE, P. D. (1977). Stratigraphy and pollen analysis of Claish Moss, North-West Scotland: significance for the origin of surface pools and forest history. *J. Ecol.*, **65**, 375–97.

OSVALD, H. (1923). Die Vegetation des Hochmoores Komosse. *Svenska Växtsoe. Sallsk. Handl.*, **I**, 1–266.

PEARSALL, W. H. (1956). Two blanket-bogs in Sutherland. *J. Ecol.*, **44**, 493–516.

RATCLIFFE, D. A. AND WALKER, D. (1958). The Silver Flowe, Galloway, Scotland. *J. Ecol.*, **46**, 407–45.

SJÖRS, H. (1961). Surface patterns in boreal peatlands. *Endeavour*, 1961, 217–24.

WALKER, D. (1961). Peat stratigraphy and bog regeneration. *Proc. Linn. Soc. Lond.*, **172**, 29–33.

WALKER, D. AND WALKER, P. M. (1961). Stratigraphic evidence of regeneration in some Irish bogs. *J. Ecol.*, **49**, 169–85.

5
Plant macrofossils

Introduction

Macrofossil is a term used to describe any potentially identifiable fossil preserved in sediments which can be seen by the naked eye. Macrofossils can usually only be satisfactorily identified using a good quality binocular microscope. Only occasionally is the use of a high-powered light microscope necessary, and sometimes essential information has been obtained by the use of the transmission electron microscope or the scanning electron microscope.

Macrofossils include animal as well as plant remains, but in this chapter we shall restrict ourselves to plant macrofossils, and show how they have been successfully used in palaeoecological studies.

Plant macrofossils can consist of almost any part of the plant. Most frequently they are fruits, seeds, or megaspores, which are often well preserved in sediments, and are also more or less readily identifiable. However, other parts of plants are useful, including wood, larger fragments of which can be identified by a study of their anatomy, and also leaves, rhizomes, flower parts, etc. (e.g. Dilcher, 1974). In this vegetative category we can also place tissue fragments, such as cuticles, which usually need a high-powered light microscope for their identification (e.g. Palmer, 1976). Lower plants also produce identifiable macrofossils. Moss remains are usually characteristic and often well preserved although liverworts for some reason are very rarely preserved in sediments, and have seldom been recorded (Dickson, 1973). Among the larger algae, charophytes produce well preserved oospores. Some useful works for identification of plant macrofossils are mentioned at the end of the chapter.

Characteristics of plant macrofossils

Plant macrofossils have several characteristics which make them useful in the palaeoecological reconstruction of past environments.

1. Plant macrofossils are frequently determinable to species level. Quite precise ecological inferences can therefore be drawn from them. In contrast pollen can often only be identified to generic or family level, and hence the palaeoecological inferences which can be made are broader and less precise.
2. Because of their relatively large size, macrofossils are not usually transported very far from their point of origin. In a telmatic or terrestrial deposit, the parent plants probably grew at the point at which the macrofossils were recovered. Therefore, if we study a peat deposit we will be able to reconstruct the local community that formed the peat. In a lake sediment, the macrofossils may have originated from the point of sampling if their parents were aquatic. However, macrofossils found in lake muds have often been transported short distances from nearby aquatic and lake-margin vegetation. Hence a lake sediment usually contains an assemblage of macrofossils derived from several nearby communities. In general we can say that an assemblage of macrofossils is derived largely from local vegetation, and that macrofossils are therefore of most value in reconstructing plant communities and environments of wetland and aquatic habitats. They are generally of little value for reconstructing upland vegetation, in contrast to pollen grains. As we shall see, the information provided by macrofossils and pollen is largely complementary, and thus a study combining the analysis of both sorts of fossils can provide a more complete palaeoecological picture of the past vegetation and environment than can an analysis of either alone (Birks and Mathewes, 1978).
3. Macrofossils have an advantage over pollen in that identifiable remains are often preserved of plants which either produce very low amounts of inefficiently dispersed pollen (for example, *Dryas octopetala*), or which produce fragile pollen which is not fossilized (for example, *Najas flexilis*, *Juncus* spp., *Luzula* spp.), or which do not produce distinctive microfossils at all (for example, some

mosses and charophyte algae). Some mosses produce characteristic preservable spores (see Boros and Jarai-Komlodi, 1975), but they are often produced in such small quantity so close to the ground that they are very rarely found and recognized. The exception is *Sphagnum* spp. which produce abundant spores in favourable conditions.

Macrofossils have several characteristics which limit their value for reconstructing past environments.

1. Macrofossils are not very effectively mixed over a large area because of their limited transport. Therefore a macrofossil sample from one place in a lake will probably be quite different from a contemporaneous sample from another place in the same lake, or from another nearby lake. Therefore macrofossils are unsuitable for reconstructing regional vegetation, and cannot usually be used for stratigraphical correlations in the way that pollen analyses can be used.
2. Macrofossils are produced in relatively small quantities, and usually, therefore, at least 100 cm^3 of sediment have to be examined. This contrasts with pollen analysis, which requires 1 cm^3 or less of sediment. There are problems of sampling for macrofossils therefore, because a large diameter core is necessary or a large monolith dug from an exposed peat face. It is not always easy to obtain large diameter cores from suitable sites, especially from lakes.
3. Because of their local dispersal, the numbers of macrofossils per unit volume of sediment may fluctuate sharply in vertically adjacent samples because the local vegetation has changed slightly, or a current of water happened to transport seeds to a certain point. Therefore histograms of numbers of macrofossils deposited over time have to be interpreted with caution. Little is known at present of the relationship of the numbers of macrofossils deposited at a place with the production of macrofossils by the plants concerned.

There are many variables to consider, all of which are important in interpreting a fossil assemblage. (a) It is helpful to know the rate of macrofossil production by the plant, and how this is affected by environmental factors and the plant's performance. (b) The size, weight, and shape of the macrofossils will be important for determining their transportability. Some seeds, for example, are specially adapted for dispersal by air, water, or animal agents. (c) The distance from the source of macrofossils will be very important in determining the numbers that are deposited, because macrofossils tend to be so locally dispersed. (d) In the case of seeds, the germination rate is important. Clearly no seeds will be preserved if they all germinate. (e) Similarly, the palatability of seeds to animals is important. Seeds can be a major source of food and few may remain to be preserved. (f) The ability to be preserved is obviously important. As in most groups of fossils, some sorts of macrofossils are not readily preserved, particularly soft parts such as leaves and shoots. Seeds and fruits are, however, usually fairly resistant to decay because they are often adapted to withstand a period of dormancy. However, several kinds are rarely preserved, particularly grasses, and some very small fragile seeds such as those of orchids and some Ericaceae. (g) A final consideration must be the abundance of the parent plants in relation to the macrofossils deposited.

This is a formidable list of complications inherent in interpreting macrofossil data, and there is a wide field of research open to try to establish and quantify these relationships. However, similar considerations are inherent in interpreting any sort of fossil data, and the palaeoecologist is not usually deterred. If no information is available, it is usually assumed that the number of fossils is approximately proportional to the abundance of the organism in the environment, and on the whole, this generalization seems to hold up quite well! This should not allow palaeoecologists to become uncritical or complacent as a result. Information on the modern representation of almost all kinds of fossils is needed. The major exception is pollen, where a great deal of work has already been done, and the interpretation of pollen analyses has become fairly refined. This will be discussed in detail later in Chapters 8, 9, 10, 11, and 12.

A method of macrofossil analysis

1. Take a sample of sediment of known volume. The volume is most easily measured by displacement in a large measuring cylinder. Add a known volume of water. Then add the sediment, read off the new volume, and calculate the volume of sediment.
2. If the sediment will not break down readily in water, leave it to soak for several hours in dilute

(10%) nitric acid, or 10% sodium hydroxide. If the sediment is calcareous, beware of effervescence with the acid.

3. After the lumps have been softened sufficiently, place small portions of the sediment in a sieve and wash the finer particles through with a gentle stream of water. For coarse sediments, it may be helpful to wash the sediment through a coarse sieve (e.g. 15 mesh to the cm) and then through a fine one (e.g. 40 mesh to the cm).

4. Store the material retained by the sieve in water. Examine small portions at a time under a binocular microscope. It is usually best to use a white dish, unless light-coloured objects such as shells are being separated, in which case a dark background may be preferable. The dish should be shallow, and of such a size that the whole of it can be examined systematically. The macrofossils can be picked out either by a fine brush or by a pair of fine forceps.

5. The macrofossils are sorted, and each category is identified by comparison with accurately determined reference material. It has to be assumed that the reference material covers the range of variation of each species, and that morphological evolution has been inconsequential over the time period involved. Seed identification manuals are also valuable. Some are listed at the end of the chapter. In some cases high magnification microscopy is necessary for distinguishing closely similar taxa, and the scanning electron microscope has become a valuable tool (e.g. Conolly, 1976; Huckerby *et al.*, 1972).

6. When all the macrofossils have been identified, the numbers of each taxon are counted and recorded. It is not always possible to count the macrofossils, particularly in the case of vegetative plant parts and moss stems. They may best be recorded from a relative scale of abundance, such as abundant, frequent, occasional, rare, or present. Similarly, some seeds may be broken. A count of all the fragments would probably over-estimate the number of seeds initially present. The investigator should use common sense to decide how many fragments compose one entire seed, and the final count will obviously be a minimal estimate of the number of seeds originally present.

7. The data are tabulated.

8. The data are plotted stratigraphically as a ***macrofossil diagram*** if this will aid their interpretation. The data can be plotted in various ways.

a) The macrofossils may be plotted as present or absent (+ or –).

b) They may be plotted as estimates of abundance; e.g. abundant, frequent, occasional, rare.

c) They may be plotted quantitatively. The number of macrofossils per unit volume of sediment may be plotted. This would be an adequate representation if the sediment deposition rate had stayed constant throughout. If, however, it had changed, the concentration of macrofossils in part of the sequence may be greater because of a slower sedimentation rather than a greater abundance of plants. It is therefore preferable to calculate the number of macrofossils which have been deposited on a unit area during a unit time, for example, as numbers per cm^2 per year. This rate is called the ***macrofossil influx***. The derivation of influx figures from the original concentration figures requires measurement of the rate of sediment deposition. The most usual way of doing this is by obtaining a series of radiocarbon dates through the profile, and calculating the sedimentation rate (see Chapter 10).

d) Pollen analyses are often presented as percentages of a certain sum, which is usually the sum of all tree pollen in the sample, or the sum of all 'dry land' pollen (originated from plants growing on dry land, rather than in the lake or bog (p. 167)). Attempts have been made to present plant macrofossils as percentages (e.g. Watts and Winter, 1966). However, it is often difficult to decide what should be in the 'seed' sum, because the majority of 'seeds' have originated from marsh or aquatic plants. The numbers of 'seeds' vary tremendously, and the abundant taxa (for example, annuals such as *Najas flexilis*) comprise a consistent majority of the 'seed' sum, suppressing any potentially interesting trends in the other taxa. The advantage of percentages is that all components are less than 100%, whereas enormous numbers of certain taxa may have to be plotted on a concentration or influx diagram. However, this is a question of logistics, and the huge numbers may be written in, or even calculated and plotted logarithmically, in order to reduce the size of the final diagram.

Palaeoecological studies using plant macrofossils

As we have seen, plant macrofossils are not generally dispersed far from their source. Several researchers have taken advantage of this feature,

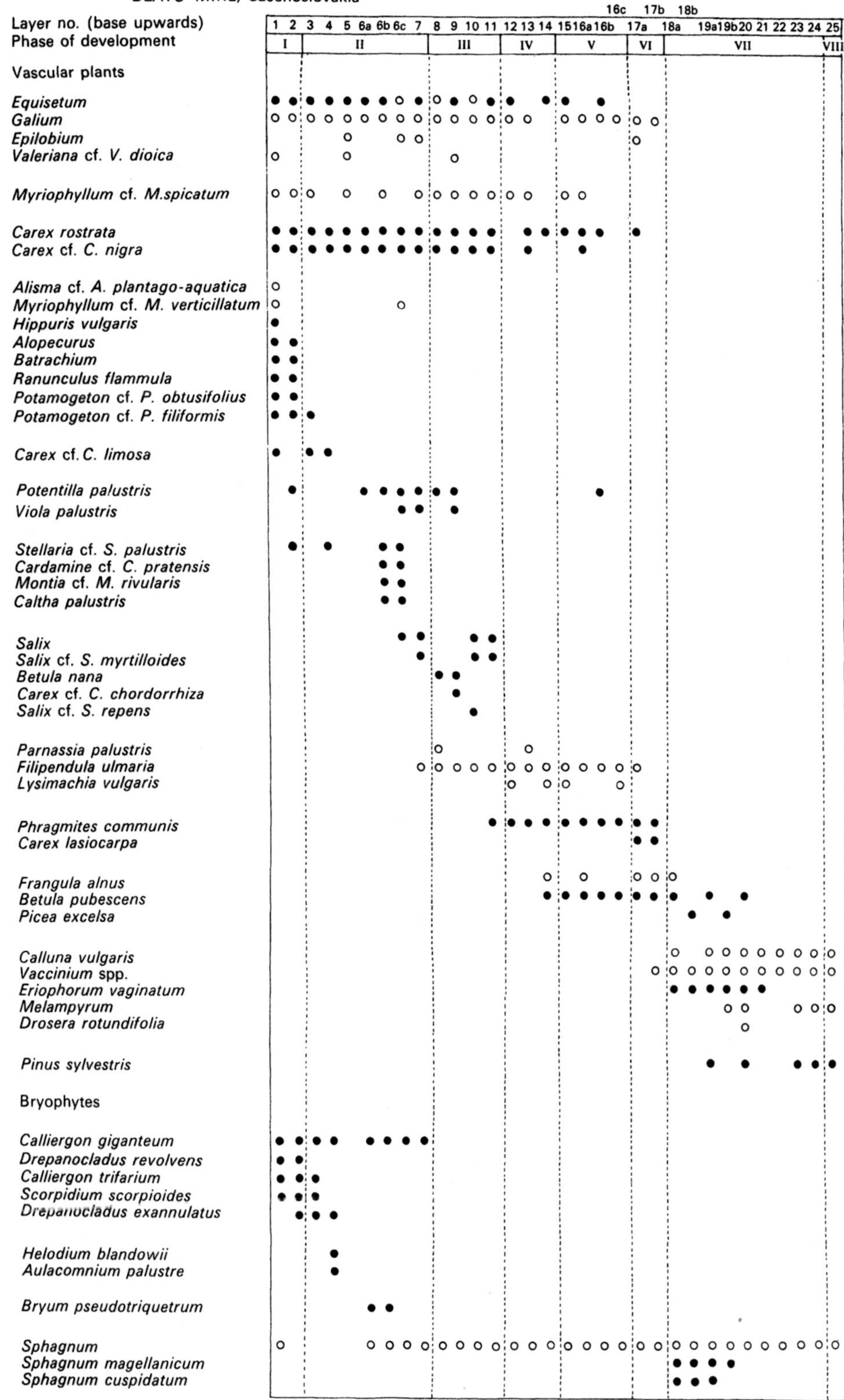

Fig. 5.1 The distribution of plant macrofossils and pollen in the profile from Bláto mire, Czechoslovakia, with the taxa arranged in the order of similar occurrence. For further explanation, see text. (After Rybníček and Rybníčková, 1968.)

and have attempted to use macrofossils to reconstruct past plant communities, by analysing autochthonous peats. Examples of such research are discussed below.

1. *Bláto mire, Czechoslovakia*

Rybníček and Rybníčková (1968) analysed a peat profile from Bláto mire in the mountains in southeastern Czechoslovakia for both pollen and macrofossils. Because of the local representation of macrofossils in peats, they attempted to reconstruct past communities, using the present behaviour and associations of the species as a guide. The approach could be called ***palaeophytosociology.***

Figure 5.1 shows the presence of each fossil taxon in each level of the profile. Pollen finds are also indicated, but they are not given as much weight as the macrofossils in community reconstruction because of the greater dispersal possibilities of pollen. The fossil taxa were arranged so that taxa of similar occurrences occurred next to each other.

As a result, they identified eight phases of mire succession, which are indicated on Fig. 5.1.

I. Aquatic plants, such as *Potamogeton filiformis* and *Ranunculus* subgenus *Batrachium* (aquatic buttercups), *Myriophyllum spicatum*, and *M. verticillatum*. In addition, emergent aquatic plants occurred, such as *Hippuris vulgaris* and *Alisma plantago-aquatica*. The lake environment allowed the transport of macrofossils from nearby fringing sedge swamp communities, including *Carex rostrata*, *C. nigra* (*C. fusca*), and *Ranunculus flammula*.

II. Sedge swamp, where the open water was invaded by *Carex* species which were later accompanied by other marsh species, such as *Potentilla palustris*, *Viola palustris*, *Stellaria* cf. *S. palustris* and *Cardamine* cf. *C. pratensis*.

III. Shrubby swamp. First *Betula nana* colonized the fen, closely followed by *Salix* cf. *S. myrtilloides* and *S. repens*.

IV. Open fen. As a result of a fire, the shrubs were destroyed and their place taken by species of fen, such as *Filipendula ulmaria*, *Lysimachia vulgaris*, and *Phragmites communis*.

V. *Phragmites*-dominated fen with a *Sphagnum*-rich understory. This suggests that conditions were becoming more acid as peat accumulated above the general water table.

VI. An acid sedge fen. As a result of increased water level due to increased precipitation, the fen became wetter and acidic. *Phragmites* was joined by *Carex lasiocarpa* associated with *Vaccinium oxycoccos* and *Sphagnum*, and the fen species declined.

VII. Bog vegetation. There is a sharp cut-off of all the fen species as the community became ombrotrophic. *Eriophorum vaginatum* became dominant, accompanied by *Vaccinium oxycoccos* and *Calluna vulgaris*. *Sphagna* typical of bogs such as *Sphagnum magellanicum* and *S.* sect. *Cuspidata* formed the ground layer.

VIII. Drier bog vegetation with *Pinus sylvestris*. Disturbance of the surface and drainage by human activity caused the peat to dry out, allowing pine to colonize, and dwarf-shrubs such as *Calluna* and *Vaccinium* spp. to increase in abundance. The bog is now too dry for peat growth.

Rybníček and Rybníčková interpreted the assemblages of plants from Bláto mire in terms of modern plant communities. Those which did not resemble modern communities, such as IV, they suggest were transitory phases after the drastic disturbance of the fire, which were not fully stabilized communities in equilibrium with the environment.

A similar approach was taken by Rybníček and Rybníčková (1974) in a study of the development of waterlogged meadow communities in Czechoslovakia.

2. *Numerical analysis of macrofossil data from peat deposits*

The study on Bláto mire outlined above arranged macrofossil taxa into groups of similar occurrence. In numerical terms, these are called ***recurrent groups.*** Recurrent groups can be detected by numerical classification techniques. Rybníček and Rybníčková delimited recurrent groups by eye. Their classification can be compared with a purely numerical classification, such as the method of minimum-variance cluster analysis. In essence, the numerical similarity between each possible pair of taxa is calculated. The most similar taxa are grouped together initially, and then the next most similar taxa are added to the groups, and so on, until all the taxa have been clustered (see Robichaux and Taylor, 1977). The relationship and level of similarity of all the taxa is shown diagrammatically by the dendrogram in Fig. 5.2 (see also Chapter 2). In this case, the groups of taxa can be interpreted as fossil assemblages.

Eleven groups (A–K) of taxa can be delimited in the dendrogram.

A. These are sedge, reedswamp, and fen plants today, comparable to Bláto mire phase II.

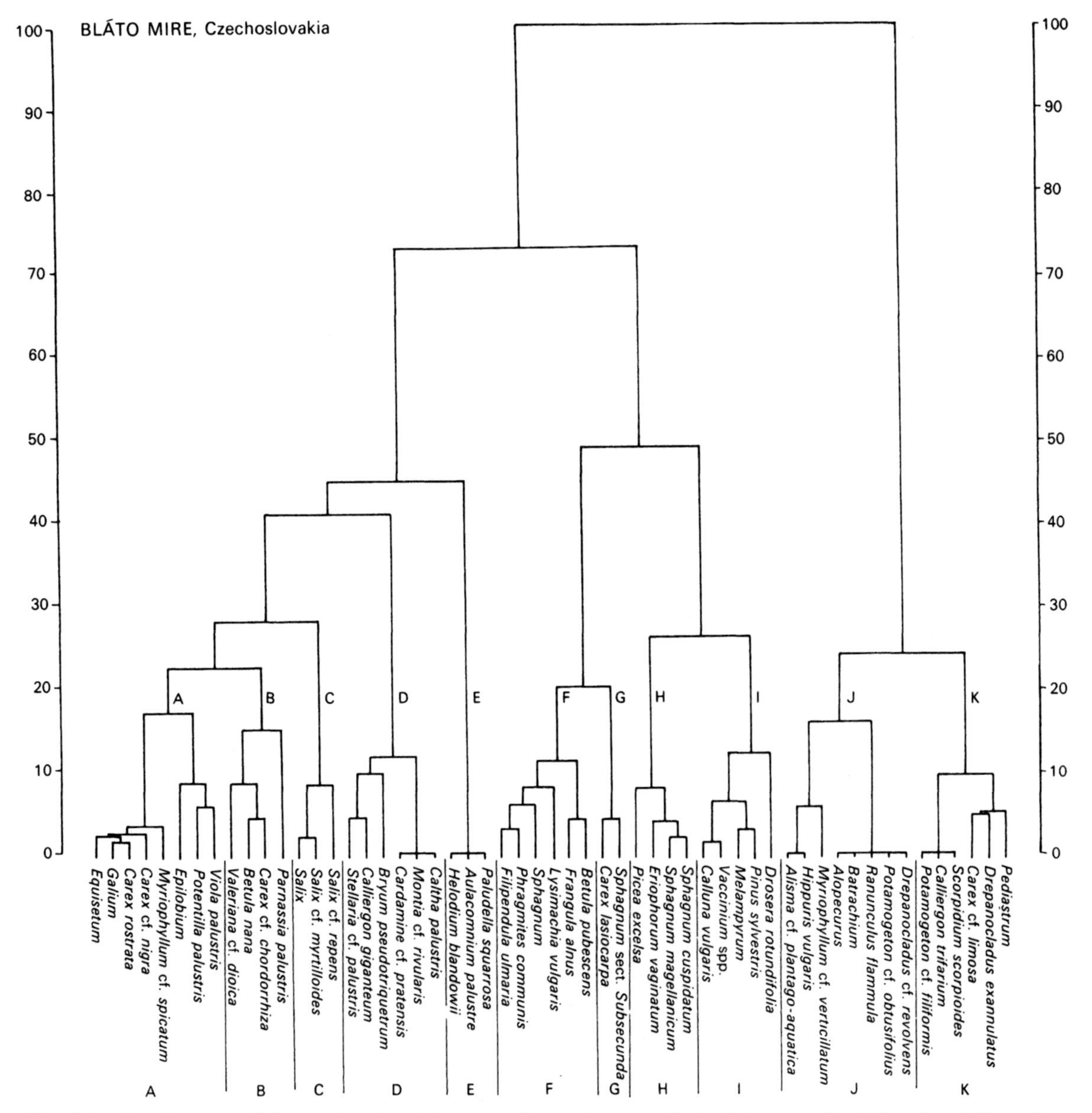

Fig. 5.2 Dendrogram of the results of a minimum-variance cluster analysis of the macrofossil data from Bláto mire. The clusters are denoted by A to K at the base, and they are discussed in the text.

B. These are species of fen and shrubby swamp, comparable to Bláto mire phase III.

C. These are the willows characteristic of Bláto mire phase III, suggesting that perhaps some division within phase III may be possible.

D. These are another group of fen species, which occur in the upper part of Bláto mire phase II.

E. Mosses of intermediate-fen occurring together at the beginning of Bláto mire phase II.

F. Fen plants and associated woody species of Bláto mire phase IV.

G. Acidic sedgeswamp species of Bláto mire phase VI.

H. The species of the actively growing bog phase

PAST ASSEMBLAGES										MODERN ASSEMBLAGES															
										Westhoff and den Held							Oberdorfer								
Lithology	Detritus lake mud		*Eriophorum-Carex* peat	*Carex-Phragmites* peat	*Sphagnum-Eriophorum* peat																				
Macrofossil zones	A	B	C	D	E	F_1	F_2	G	H	Reed swamps	Sedge fens	Brook vegetation	Rich meadow	Poor meadow	Intermediate fen	Poor fen and bog	Reed swamps	Sedge fens	Brook vegetation	Rich meadow	Poor meadow	Intermediate fen (sedge)	Intermediate fen (cotton grass)	Poor fen	Bog
Filipendula ulmaria	○	○	·	·	·	·	·	·	·	·	·	x	·	·	·	·	·	·	x	x	x	·	·	·	·
Salix	○	○	·	·	·	·	·	·	·	·	·	·	·	·	·	·	·	·	·	·	·	·	·	·	·
Scirpus sylvaticus	·	●	●	·	·	·	·	·	·	·	·	·	x	·	·	·	·	·	x	x	·	·	·	·	·
Potentilla palustris	●	○	●	●	·	·	·	·	·	·	·	·	·	·	x	·	·	·	·	x	·	x	x	x	·
Carex nigra	●	·	●	●	·	·	·	·	·	·	·	·	·	·	x	·	·	·	·	x	·	x	·	·	·
Carex undiff.	●	●	·	●	●	·	·	·	●	·	x	·	x	x	x	·	·	x	x	x	x	x	x	x	·
Carex pseudocyperus	·	●	●	●	·	·	·	·	·	·	x	·	·	·	·	·	·	x	·	·	·	·	·	·	·
Phragmites communis	·	●	●	●	●	·	·	·	·	x	·	·	·	·	·	·	x	x	·	x	x	·	·	·	·
Succisa pratensis	·	·	○	○	·	·	·	·	·	·	·	·	·	x	x	·	·	·	x	x	x	·	·	·	·
Sanguisorba officinalis	·	·	○	○	·	·	·	·	·	·	·	·	x	x	·	·	·	·	x	x	x	·	·	·	·
Cirsium palustre	·	●	·	●	·	·	·	·	·	·	·	·	·	x	x	·	·	·	x	x	x	x	x	·	·
Lychnis flos-cuculi	·	●	·	●	·	·	·	·	·	·	·	·	x	x	·	·	·	·	x	x	x	·	·	·	·
Valeriana dioica	·	·	·	○	·	·	·	·	·	·	·	·	·	x	x	·	·	·	x	x	x	x	x	·	·
Parnassia palustris	·	·	·	○	·	·	·	·	·	·	·	·	·	·	x	·	·	·	·	·	·	x	x	x	·
Caltha palustris	·	·	·	○	·	·	·	·	·	·	·	·	x	·	·	·	·	·	x	x	x	x	x	·	·
Sphagnum teres	·	·	·	●	·	·	·	·	·	·	·	·	·	·	x	x	·	·	·	·	·	x	x	x	·
Potentilla erecta	·	·	·	●	●	·	·	·	·	·	·	·	·	x	·	·	·	·	·	x	x	x	x	·	·
Sphagnum recurvum	·	·	·	●	●	·	·	·	·	·	·	·	·	·	x	x	·	·	·	·	·	x	·	x	·
Sphagnum squarrosum	·	·	·	●	●	·	·	·	·	·	·	·	·	·	x	·	·	·	·	·	·	·	·	·	·
Vaccinium undiff.	·	·	·	·	○	○	·	·	·	·	·	·	·	·	·	x	·	·	·	·	·	·	·	·	x
Andromeda polifolia	●	●	●	●	·	·	○	○	○	·	·	·	·	·	·	x	·	·	·	·	·	·	x	x	x
Eriophorum	·	·	●	●	●	●	●	●	●	·	·	·	·	·	x	x	·	·	·	·	·	x	x	x	x
Sphagnum magellanicum	·	·	·	·	●	●	●	●	●	·	·	·	·	·	·	x	·	·	·	·	·	·	·	·	x
Sphagnum rubellum	·	·	·	·	·	●	●	●	●	·	·	·	·	·	·	x	·	·	·	·	·	·	·	·	x
Sphagnum undiff.	·	·	·	·	○	·	·	○	○	·	·	·	·	·	·	x	·	·	·	·	·	·	x	x	x
Drosera	·	·	·	·	·	○	·	·	○	·	·	·	·	·	·	x	·	·	·	·	·	·	x	x	x
Pleurozium schreberi	·	·	·	·	·	·	●	·	●	·	·	·	·	·	·	x	·	·	·	·	·	·	·	·	·
Calluna vulgaris	·	·	·	·	·	○	○	●	●	·	·	·	·	·	·	x	·	·	·	·	·	·	·	·	x
Melampyrum	·	·	·	·	○	·	·	○	○	·	·	·	·	·	·	x	·	·	·	·	·	·	·	·	x
Empetrum	·	·	·	·	·	○	·	○	○	·	·	·	·	·	·	x	·	·	·	·	·	·	·	·	x
Vaccinium uliginosum	·	·	·	·	·	·	·	●	●	·	·	·	·	·	·	x	·	·	·	·	·	·	·	·	x
Dicranum	·	·	·	·	·	·	·	·	●	·	·	·	·	·	·	x	·	·	·	·	·	·	·	·	x
Aulacomnium palustre	·	·	·	·	·	·	·	·	●	·	·	·	·	·	·	x	·	·	·	·	·	·	·	·	x

Fig. 5.3 The distribution of plant fossils in the profile from Feigne d'Artimont, France, grouped into past assemblages on the left, and their comparison with recent assemblages of the same plants placed into phytosociological associations using the system of Westhoff and Den Held, centre, and Oberdorfer, right. ●, macrofossil; ○, pollen; ×, occurrence of plant in modern association. (After Janssen *et al.*, 1975.)

VII, dominated by *Eriophorum vaginatum* and *Sphagnum* spp.
I. The species characteristic of Bláto mire phase VIII, the upper drier bog phase with *Pinus*.
J. Aquatic plants with their earliest occurrence in Bláto mire phase I.
K. Aquatic plants whose occurrence extends from Bláto mire phase I through to Phase II. J and K do not correspond precisely to Rybníček and Rybníčková's grouping, although they distinguish a similar trend.

The numerically delimited groups correspond remarkably closely with Rybníček and Rybníčková's groups, confirming that there are indeed distinct groups within the data. These groups generally make good ecological and phytosociological sense, and can be readily related to modern analogues. In this example, the use of numerical methods has confirmed the non-numerical classification, and emphasized the potential for reconstructing past plant communities from peat deposits. We shall compare a similar exercise on macrofossil analyses from a lake environment later in this chapter (p. 76).

3. *Feigne D'Artimont, Vosges, France*

Janssen *et al.* (1975) reconstructed the past vegetational development of this fen/bog complex in the Vosges mountains, France, in a way similar to Rybníček and Rybníčková. They compared directly their groups of fossil taxa with the occurrence of the species in modern communities defined on two phytosociological systems, as shown in Fig. 5.3. Both comparisons revealed the same sequence of communities: reedswamp, sedge fen, brook vegetation, rich meadow, poor meadow, intermediate fen, poor fen, bog.

4. *Synthesis of past mire communities and their distribution in time*

From macrofossil and pollen studies in central Europe, Rybníček (1973) reconstructed 20 fossil mire communities, which could be recognized in at least two sequences. They can mostly be related to modern communities, some at a detailed association level, because of the specific ecological relationships of the components (indicator species), and some at a broader ecological level (e.g. rich fen). Rybníček arranged the occurrence of each community according to the time intervals in which it was recorded, and obtained the pattern shown in Fig. 5.4.

In the Weichselian late-glacial (II, III) period, the principal communities recorded are aquatic, particularly *Potamogeton-Batrachium* communities and rich fens. In the early post-glacial or Holocene (IV–V) these were succeeded by a range of rich and intermediate fens, until mid post-glacial times (VI, VII) when the main process of lake infilling took place and mire communities spread generally. Many shallow-water aquatic communities have their main expansion at this time, together with reedswamps and fen carr. A little later (VII, VIII), usually in areas of lower nutrient supply, and where ground water ceases to influence the vegetation, bog communities developed, succeeding the intermediate- and poor-fen communities. It is interesting that there is no fossil evidence, even in previous interglacials for the development of raised bog communities before the end of the interglacial. Some communities are even more recent in origin, such as poor-fens and forested bogs (e.g. Ledo-Pinion in X), and some of these may be a result of, or have been influenced by human activities.

Rybníček's data cover Central Europe, but a similar sequence of plant communities can be traced in north-west Europe including the British Isles. In arctic regions, however, bog communities are rare, and the earlier communities still predominate. This reflects the more adverse conditions for growth in the arctic, and the lesser degree of oligotrophication in the cooler climate. The oldest communities can still be found throughout Europe in suitable habitats where mineral enrichment is maintained, and environmental factors inhibit bog development. Such communities can thus be regarded as relics of a former age. Attempts to conserve them may be impossible because of the inexorable advance of natural processes, and the interference of man in wetland habitats.

Rybníček's synthesis has shown that mire communities are not continuous in space or time. The plant communities change in response to environmental changes, which, in the case of peat-forming communities, are often a result of plant growth itself. Here palaeoecology has provided an important historical insight into the nature of some plant communities, and contributed important facts on which to base ecological theories.

Other studies on the palaeoecology of peat communities include those by Grosse-Brauckmann (1963, 1968, 1976) in Germany, by Birks and Ransom (1969) on an interglacial peat in the Shetlands, Scotland, and by Griffin (1977) on the boreal peatlands of Minnesota.

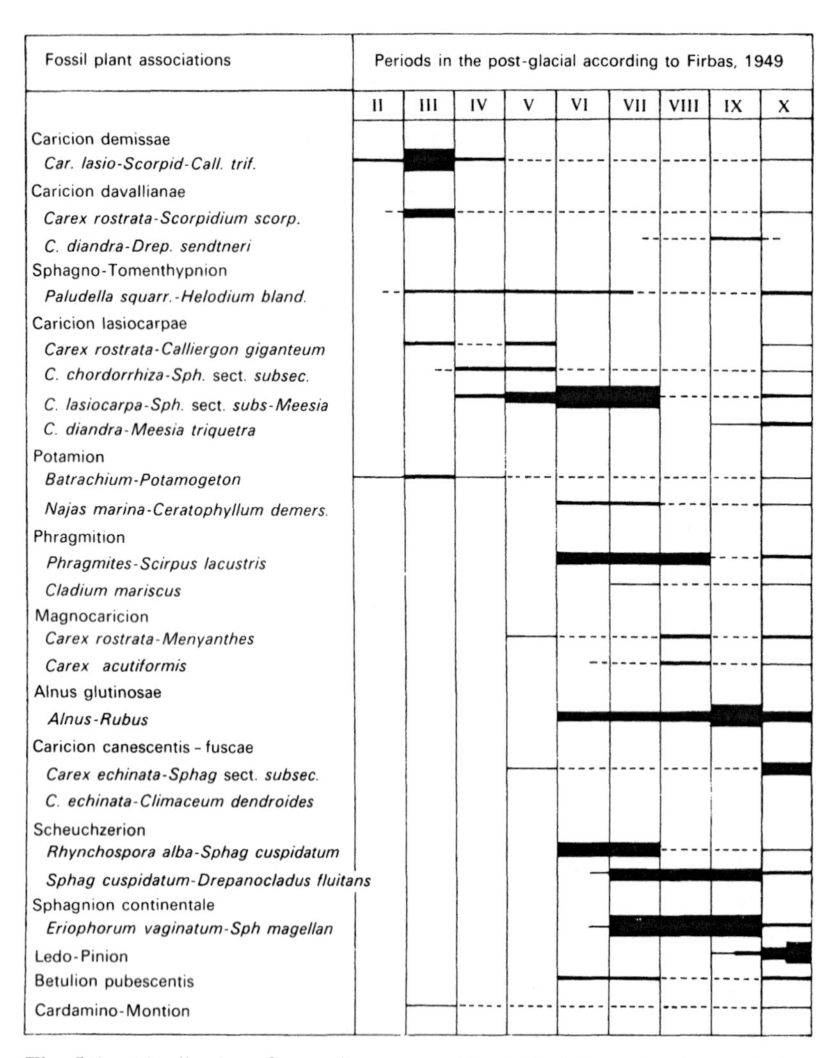

Fig. 5.4 Distribution of past mire communities in the Weichselian late-glacial and Holocene of central Europe. (After Rybníček, 1973.)

5. *Macrofossils in lake sediments: Kirchner Marsh, Minnesota, U.S.A.*

Several studies have now been published of macrofossil analyses from lake sediments (e.g. Watts and Winter, 1966; Watts, 1970; Birks, 1976; West, 1957; Wright and Watts, 1969; Watts and Bright, 1968; Birks and Mathewes, 1978). One of the first was that from Kirchner Marsh, in central Minnesota, U.S.A. by Watts and Winter (1966). If we recall briefly the time-space diagram of Chapter 2 and imagine that our sample is taken from a lake sediment, the basin of deposition will be the lake itself. In addition, the lake will receive material from its catchment area, and perhaps even from beyond that. Although macrofossils in general are locally distributed, some will reach the relatively deep water of the centre of a lake, where macrophyte growth is absent, by dispersal from the shallower water and the marginal swamp vegetation. A few macrofossils will be distributed from the upland vegetation of the catchment area, either by wind or by inflow streams if any are present. In the case of macrofossils, transport from beyond the catchment area is unlikely, although wind-dispersed fruits, such as those of *Betula*, may be transported great distances, and waterfowl may disperse seeds of aquatic plants over long distances.

Kirchner Marsh is a small cattail (*Typha*) marsh

which has overgrown a former lake. The pollen analyses by Winter (Wright *et al.*, 1963) and the macrofossil analyses by Watts (Watts and Winter, 1966) were from the same 12 m long core. Therefore the two sets of data can be directly compared.

A simplified version of the macrofossil diagram is shown in Fig. 5.5. For simplicity Watts called all reproductive bodies (seeds, fruits, oospores, etc.) 'seeds', and we will follow his example. Watts analysed samples large enough to provide more than 100 seeds if possible, and presented the data as percentages of the seed sum. The diagram was then divided horizontally with the pollen zones defined from the pollen diagram. Watts subsequently recorded similar macrofossil assemblages from other sites in the Mid-West of America (e.g. Wright and Watts, 1969), so the assemblages appear to possess some regional vegetational characteristics. Because we shall refer to Kirchner Marsh several times during this book, we shall present a fairly detailed account of the vegetational history it demonstrates.

Zones K and A-a. These are interpreted by Watts and Winter as representing a *Picea* forest dominating the upland, and with an understory containing *Rubus pubescens*, *Cornus canadensis*, *Arctostaphylos uva-ursi*, and *Fragaria virginica*. The lake shore was probably fringed by *Typha*, accompanied by other marsh plants such as *Hypericum virginicum*, *Epilobium*, *Lactuca biennis*, *Cirsium muticum*, and species of *Carex*. The lake contained a restricted flora of *Hippuris vulgaris*, *Heteranthera dubia*, and *Potamogeton praelongus*. Evidence of the upland vegetation derived from the pollen diagram shows that a spruce parkland during Zone K developed into spruce forest containing some thermophilous plants during Zone A-a.

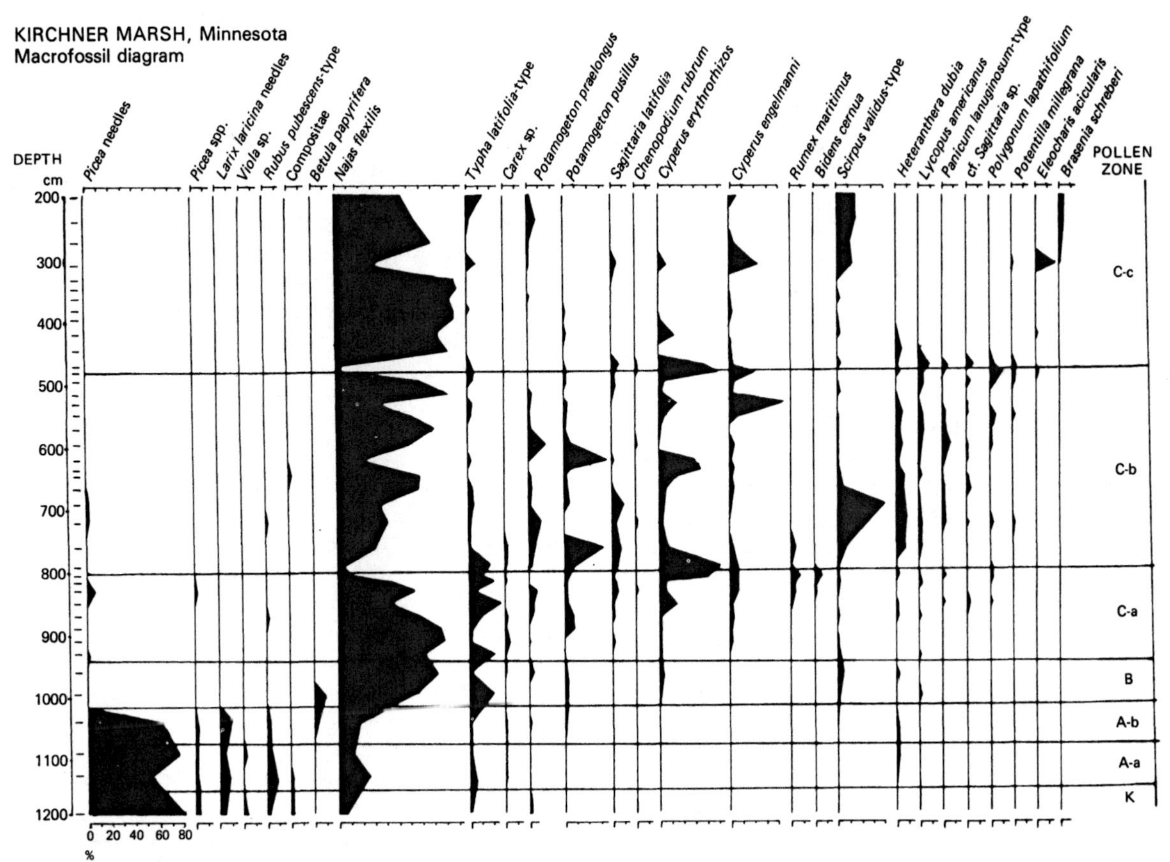

Fig. 5.5 Plant macrofossil diagram from Kirchner Marsh, Minnesota, showing selected macrofossils only. All values are expressed as percentages of the total macrofossils shown, and all the curves are drawn to the same scale. The pollen zones are shown. (After Watts and Winter, 1966.)

Zone A-b. Stabilization of the landscape proceeded, leading to the deposition of more organic lake mud, with decreasing amounts of silt and clay. A *Scirpus validus* and *Carex lasiocarpa* var. *americana* reed-swamp fringed the shores, beyond which *Sparganium* spp., and waterlilies, *Nymphaea* and *Nuphar lutea* grew in the shallower water. Spruce forest still covered the uplands.

Zone B. The aquatic vegetation resembled that of Zone A-b, but with increased abundance of the submerged aquatic *Najas flexilis*. However, fairly rapid changes in the upland forests were occurring. In the pollen diagram, curves of *Betula*, *Alnus*, *Fraxinus*, *Abies*, and *Pinus* rise sharply, and then drop again as curves for *Ulmus*, *Quercus*, and *Ostrya/Carpinus* rise, representing the successional development of the mixed hardwood forest. Of these trees, macrofossils of *Betula* and *Abies* were found, corresponding with peaks in their pollen curves.

Zone C-a. During this zone the aquatic vegetation changes from that of a relatively deep lake to that of shallower water and exposed shores. Damp ground perennials such as *Sagittaria latifolia*, *Lycopus americanus*, and *Eupatorium perfoliatum* appear, together with annuals characteristic of intermittently dried up shores, such as *Cyperus erythrorhizos*, *C. engelmanni*, *Chenopodium rubrum*, *Rumex maritimus* var. *fueginus*, and *Bidens cernua*. On the upland, more mesic trees declined, leaving *Quercus* as the principal tree pollen, accompanied by increasing amounts of herb pollen suggesting the development of an oak-savanna.

Zone C-b. During this zone, *Quercus* pollen values are very low, being replaced by pollen of herbaceous plants typical of prairie. The vegetational changes inferred from zones C-a and C-b indicate a climate change to lower summer rainfall conditions. This hypothesis is confirmed by the macrofossil data, which show steeply fluctuating curves of the muddy shore annuals and low values of submerged aquatics such as *Najas flexilis*, which alternate with higher values of aquatics and reedswamp taxa. Watts and Winter estimate from radiocarbon dates that this period of low, fluctuating water levels lasted about 200 years at Kirchner Marsh. Evidence for this eastward extension of prairie between about 8000 and 7000 years ago has also been found in other sites in Minnesota and S. Dakota, using evidence from pollen, macrofossils, and sediments (Wright, 1966).

Zone C-c. *Quercus* pollen values increase again at the expense of the herb pollen types, suggesting that conditions became less xeric and oak forest developed once more. The macrofossils show that the water level increased substantially, eliminating the muddy shore plants and allowing a luxurious development of aquatics, first by *Najas flexilis*, and then by waterlilies, *Nuphar*, *Nymphaea*, and *Brasenia*. A reedswamp containing *Scirpus validus*, *Carex comosa*, and *Dulichium arundinaceum* extended from the shores. Eventually it overgrew the sampling site, and then the whole lake. The overlying sedge peat contained too few macrofossils to make further analyses worthwhile.

At Kirchner Marsh, the macrofossils played a very important role in reconstructing the vegetational and also climatic history of central Minnesota. They provided an ecological insight into the changes within and just around the lake itself which would have been impossible from a study of pollen alone. Their local distribution and specific identification were both important characteristics which made the interpretation possible.

We saw on page 69 how numerical analysis of macrofossils from peats provided a useful classification of the data, from which valuable ecological interpretations could be made. If a similar minimum-variance cluster analysis is performed on the Kirchner Marsh macrofossil data using 76 taxa defined on a presence/absence basis in the 41 samples the dendrogram shown in Fig. 5.6 is obtained. The dendrogram can be divided into ten clusters, with cluster A being further subdivided into two. By comparing the content of the clusters with the macrofossil diagram, the clusters can be interpreted.

Cluster A1. *Najas flexilis*-group. This includes taxa which are frequent throughout the sequence, occurring in 75% or more of the samples.

Cluster A2. *Ceratophyllum demersum*-group. These are taxa which occur in Zones C-a and C-b, or at the C-a/C-b transition.

Cluster B. *Utricularia vulgaris*-group. These all occur in a single sample in Zone C-b.

Cluster C. *Juncus tenuis*-group. This group includes prairie species, centred on Zone C-b.

Cluster D. *Urtica gracilis*-group. This is a very heterogeneous group.

Cluster E. *Nuphar lutea*-group. This group contains many of the aquatic and marsh species characteristic of Zone B.

Cluster F. *Brasenia schreberi*-group. This includes

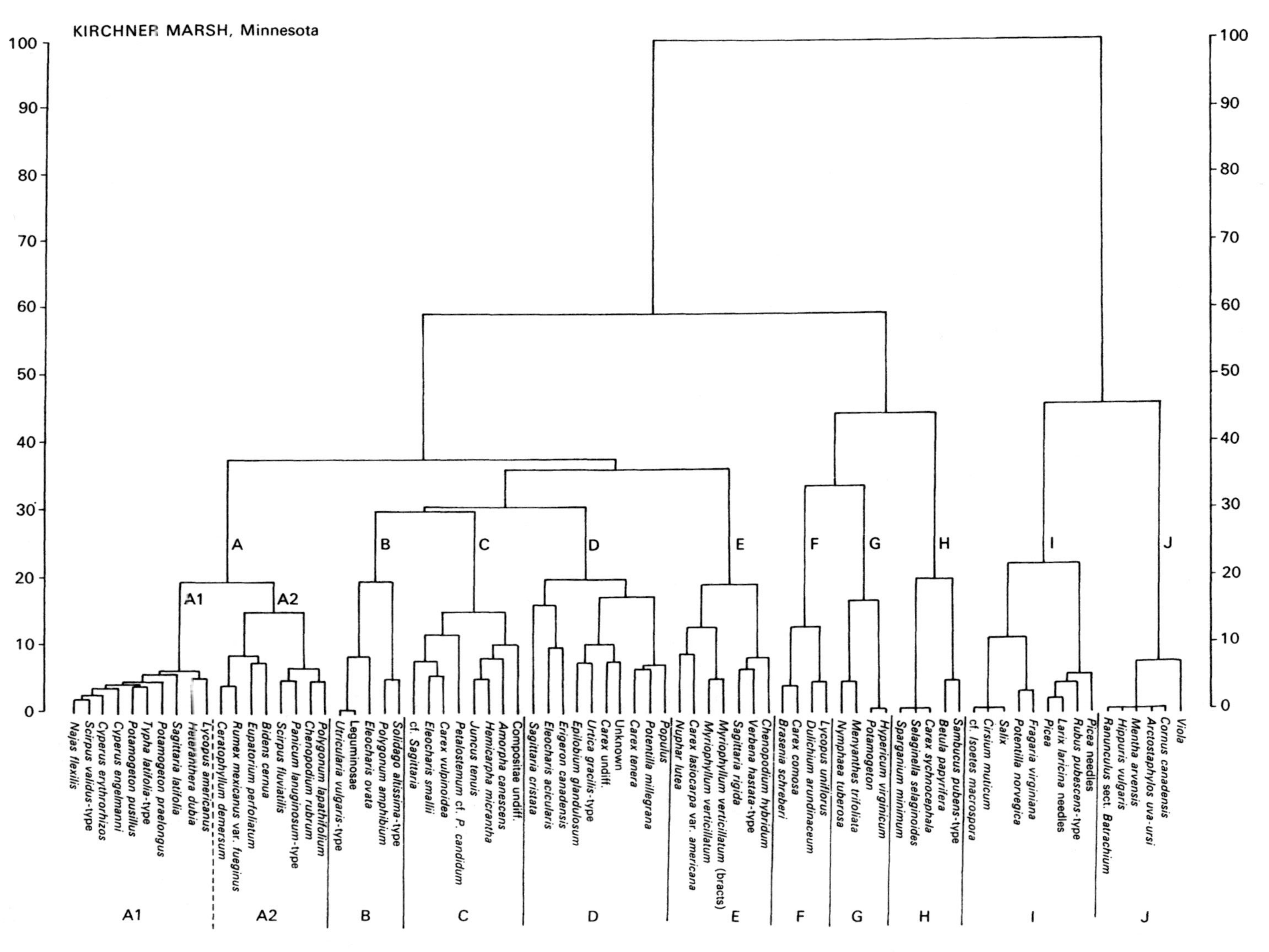

Fig. 5.6 Dendrogram of the results of a minimum-variance cluster analysis of the macrofossil data from Kirchner Marsh. The clusters are denoted by A1 to J at the base, and they are discussed in the text.

the aquatic and reedswamp taxa of the deepened lake at Zone C-c.
Cluster G. *Nymphaea tuberosa*-group. These taxa occur mainly at the Zone B/C-a transition.
Cluster H. *Sparganium minimum*-group. These taxa occur mainly in Zone A-b.
Cluster I. *Picea*-group. This includes the taxa of the boreal forest Zones A-a and A-b.
Cluster J. *Arctostaphylos uva-ursi*-group. These taxa occur in Zone A-a only.

Most of these clusters contain taxa of diverse ecology; aquatic, marsh, and upland. The clusters are clearly not ecological groups or past communities, but represent groups of taxa which occur together in time, and thus have similar patterns of occurrence throughout the sequence. Cluster analyses of other macrofossil data from lake sediments at Jacobson Lake (Wright and Watts, 1969) and Wolf Creek, Minnesota (Birks, 1976) give similar results, with clusters of species associated in time rather than in space as ecological groups. It would seem that cluster analysis of macrofossils from lake sediments does not provide, to any great extent, insight into past ecological conditions, in contrast to the analysis of macrofossil data from autochthonous peats. However, in macrofossil diagrams with very large numbers of taxa, cluster analysis is useful for sorting the data into groups of similar occurrence, which then enables the palaeoecologist to tackle ecological reconstructions more easily.

Other studies on macrofossils in lake sediments include those by Baker (1976), Matthews (1974, 1975), Watts (1970), Watts and Bright (1968), Wright and Watts (1969), and West (1957).

Modern representation and dispersal of macrofossils in lake sediments

We have seen how assemblages of macrofossils from lake sediments are derived from a variety of plant communities. We can now ask the question, how can we reconstruct plant communities from macrofossil assemblages in lake sediments?

Because macrofossils can often be identified to species level the ***indicator species*** approach to reconstructing plant communities is usually profitable. However, this approach relies basically on the presence of a taxon, and does not use the numerical information available. Apart from knowing a taxon was present, it is also important to know how abundant it was in the area. To provide this type of information, we need to know many more features of the relationships between the parent plants and their macrofossils. Such features include:

a) the macrofossil production of the taxon, and how this is affected by environmental conditions
b) the mode of dispersal of the macrofossils, whether by wind, water, or animal dispersal, or in the case of some seeds, self propulsion by explosive mechanisms (e.g. *Impatiens*)
c) the germination rate of seeds
d) the preservability of seeds in the environment of deposition
e) the palatability of macrofossils by, for example, ducks, which can feed almost exclusively on seeds or fruits of aquatic plants such as *Potamogeton*
f) the distance the macrofossil is likely to travel from its source. Heavy seeds will sink or fall relatively quickly compared with those adapted for wind dispersal, or those with good floating properties adapted for water dispersal
g) the relationship between the abundance of the plant and the abundance of its macrofossils. In theory, one would suppose that these abundances would be linearly related, but are they in nature?

Similar questions have been asked about the nature of pollen analysis (see Chapter 8) and Chapters 8, 9, 10, and 11 discuss some of the investigations prompted by these questions. Relatively few investigations have been made into the modern behaviour of macrofossils. For example, Chaney (1924) attempted to relate modern leaf assemblages in riverine environments to the interpretation of Tertiary leaf floras. McQueen (1969) studied the dispersal of macrofossils into a lake in New Zealand to try to determine the amount of transport from higher altitudes *via* the inflow streams. Ryvarden (1971, 1975) studied macrofossil dispersal onto recently deglaciated terrain in Norway using traps, so that he could measure the annual influx of macrofossils, and the relative efficiency of wind and water transport for each species in that environment. Birks (1973) studied the representation of macrofossils in modern surface samples from lake and swamp sediments in Minnesota, U.S.A. This project was designed specifically to provide basic information on macrofossil representation, and thus to be of use in the detailed interpretation of macrofossil diagrams.

Birks (1973) collected modern surface samples from surface muds and peat-forming communities from a range of small lakes in Minnesota, U.S.A.

She recorded the vegetation from which each surface sample was collected, and also the other types of vegetation and their extent in and around the lake. She demonstrated that lakes in the prairie, deciduous forest, and coniferous forest regions of Minnesota could be characterized by their modern floras, and that the macrofossil floras, not surprisingly, also followed this pattern. The composition of the flora was strongly related to the water chemistry of the lake. Lakes in the coniferous forest region have soft, often acid or brown water; those in the deciduous forest region have hard water rich in calcium carbonates; and lakes in the prairie have very hard, tending to saline water rich in carbonates and sulphates. It is therefore possible to interpret some features of the lake environment from the macrofossil assemblage, if the tolerances of the modern plants are known.

Apart from a purely floristic analysis, Birks also acquired information about the representation of macrofossils in different vegetational and sedimentary environments. A typical example was the results from Missouri Pond, shown in Fig. 5.7. A diagrammatic cross-section of the lake is shown at the top, with the positions of the surface samples indicated. Below, the cover values of the species in the vegetation are shown by vertical bars, and the values of macrofossils by open silhouettes. The taxa fall into five groups. (i) Of the obligate aquatic taxa, no macrofossils were recovered of the *Potamogeton* species, *Utricularia vulgaris*, *Sparganium americanum*, and *Nuphar variegatum*. (ii) Macrofossils of *Scirpus subterminalis*, *Brasenia schreberi*, and *Nymphaea odorata* were rare in comparison with their vegetational cover. (iii) In contrast macrofossils of *Chara globularis* and *Najas flexilis* were abundant in comparison with their vegetation occurrence, illustrating their high macrofossil production and relatively good dispersal. *Najas gracillima* seeds were found but no plants were recorded. This *Najas* is also an annual, and may have not been present at the sampling time or confused in the field with *N. flexilis*. (iv) Sedge-mat species macrofossils are found most abundantly in samples near the shore, showing a sharp fall-off with distance from the source. (v) Macrofossils of upland taxa again are commonest near the shore. The only macrofossils present in the deep water in the centre of the lake were *Larix* needles. The needles of this deciduous conifer are possibly blown across the ice during winter, and then deposited after the spring thaw.

Similar studies of other lakes confirmed that macrofossil dispersal is poor, except for taxa adapted for water or animal dispersal. Birks' results showed that, in general, the number of macrofossils in a sample increased with the cover value of the taxon in the vegetation. However, some taxa, for example *Najas flexilis*, produced very high numbers of seeds, which were often recorded when the plant was not present at the plot but growing some distance away, or even recorded as absent from the lake. *Najas* is an annual and depends on high seed production for survival. In contrast other perennial taxa, particularly waterlilies such as *Brasenia schreberi* and *Nuphar variegatum*, were often present in the vegetation but no macrofossils were recovered. Either they produce low numbers of seeds, or else the seeds are poorly preserved, either because they all germinate or are eaten, or decay rapidly. *Potamogeton* species are even more poorly represented, and their seeds are well known to be a source of food for wildfowl.

An analysis of the dispersal of macrofossils of swamp or marsh taxa into Minnesotan lakes demonstrated in general that taxa with no specific dispersal adaptations such as *Viola* showed a sharp decrease in numbers near the shore. Fruits adapted for floating, such as *Ranunculus sceleratus* could be found up to 150 m from the shore. Similarly, taxa with adaptations for animal dispersal, such as *Bidens cernua* and *Dulichium arundinaceum*, and taxa with adaptations for wind dispersal such as *Typha*, were found, sometimes in abundance, up to 100 m or 150 m from the shore.

These results emphasize the type of information that is necessary for the interpretation of fossil samples of macroscopic plant remains. An abundance of *Najas flexilis* seeds does not impart as much palaeoecological information as an abundance of *Brasenia schreberi* seeds, for example. In addition, we must be careful in interpreting changes in numbers of macrofossils. They may be a result of changing plant communities, or else a change in environmental conditions which cause the sampling site to be further away from the communities represented; for example, a higher water level would result in the marginal communities being further from the sampling site, and lead to the representation of only those taxa with some adaptations for dispersal.

Birks' results on macrofossil representation need considerably enlarging in order to provide data for other species and other geographical areas, particularly in Europe, where the number of macrofossil

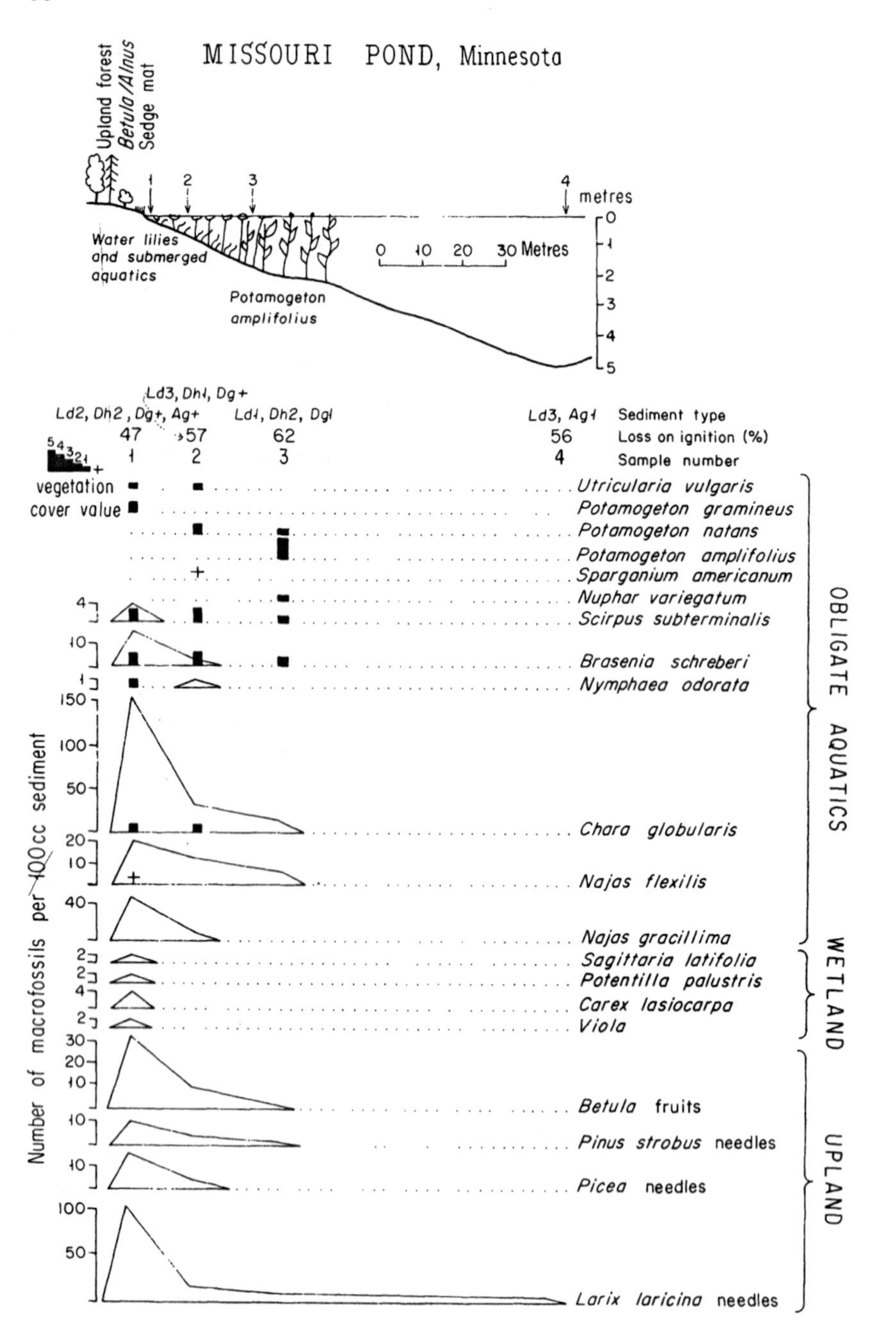

Fig. 5.7 The distribution of macrofossils in the surface sediments of Missouri Pond, Minnesota. The vegetation along the transect is indicated at the top, the cover values of species in the vegetation are indicated by upright bars, and the numbers of macrofossils per 100 cm^3 sediment are shown by open silhouettes. (From Birks, 1973.)

diagrams is slowly increasing (e.g. West, 1957; Wasylikowa, 1964; Birks and Ransom, 1969; Birks and Mathewes, 1978; Ralska-Jasiewiczowa, 1980).

Comparison of macrofossils and pollen in the same cores

Following our discussion of macrofossil representation in sediment samples, it is of interest to compare directly the macrofossils and pollen produced by the same vegetation in and around a lake. Both pollen and macrofossil analyses were performed by Birks and Mathewes (1978) on a sediment core from a site in Abernethy Forest, in the pinewoods of the Spey Valley, north of the Cairngorm mountains in Scotland (Fig. 5.8).

The macrofossil analysis recorded several taxa not found during the pollen analysis. Of these, *Chara*, *Nitella*, and all mosses except *Sphagnum* do not produce recognizable pollen or spores, and pollen of *Najas flexilis*, *Juncus* spp., and *Luzula* are very rarely preserved. Macrofossils are therefore the major source of information about their Quaternary history and palaeoecology. Other taxa such as Cruciferae, *Cirsium*, *Drosera rotundifolia*, *Narthecium ossifragum*, and *Typha* all produce recognizable pollen, which did not happen to be found in the Abernethy Forest sediments.

The pollen analyses recorded several taxa not found as macrofossils. Of these, the pteridophytes very rarely produce identifiable macrofossils. Several very important pollen types such as those of the trees, *Alnus glutinosa*, *Corylus avellana* (or *Myrica gale*), *Fraxinus excelsior*, *Quercus*, and *Ulmus*, and the shrubs *Hedera helix* and *Juniperus communis* rely on pollen records for reconstructing vegetational history. Several herbs recorded by pollen are seldom recorded as macrofossils, and some, such as *Filipendula* and *Artemisia* can be abundant and thus important in vegetational and environmental reconstruction. Many of the pollen types not recorded as macrofossils are plants of dry, upland habitats, and this makes their representation by macrofossils in lake sediments rather low. The pollen rain is tending to represent a very much wider area than the macrofossils.

The remainder of the taxa at Abernethy Forest were recorded as both pollen and macrofossils, and their curves are directly compared in Fig. 5.8. The pollen and spore taxa are plotted as annual influx. Several macrofossils may be equivalent to one pollen or spore taxon because macrofossils can usually be identified to a lower taxonomic level than pollen or spores. This is particularly so in the Cyperaceae, where the pollen is very difficult to identify beyond family level, and where many taxa may be involved, with diverse ecological requirements.

The pollen representation of several taxa, particularly Cyperaceae and *Calluna vulgaris* is low compared with the abundance of the equivalent macrofossils. In the case of *Salix*, pollen influx is very low when *Salix herbacea* macrofossils are abundant in zone AFM-3. However, the prostrate-growing *S. herbacea* probably has a very low pollen production, whereas the bud scales and leaves would probably be readily blown and dispersed by wind and water in the open unstable environment in which it lives.

In contrast, the pollen and seed curves of *Nymphaea* coincide well, and such high values of both fossils implies the considerable former abundance of *Nymphaea* at the Abernethy Forest site.

The pollen curves tend to be more continuous than the macrofossil curves. This is because pollen is produced and fossilized in much greater quantities than seeds or vegetative parts. The scale of pollen influx can be compared to that of the macrofossil influx in Fig. 5.8. The pollen influx is number cm^{-2} $year^{-1}$, whereas macrofossil influx is number 1000 cm^{-2} $year^{-1}$. Therefore the chance of recording a pollen grain is much greater than that of recording a macrofossil of any particular taxon.

Another detailed study involving both pollen and macrofossils on the same core was done by H. J. B. Birks and H. H. Birks at Wolf Creek, in Minnesota (Birks, 1976). Although they did not make a direct comparison as at Abernethy Forest, the macrofossils proved to be very useful in the reconstruction of the vegetational and environmental history at that site.

The comparison between macrofossils and pollen in lake sediments and their contribution to the reconstruction of vegetational history is summarized in Table 5.1. It is obvious that each complements the other, and that singly, each would provide a rather biased vegetational reconstruction.

Other plant macrofossils

In addition to studying seeds, fruits, needles, wood, and other parts which can be classed as macrofossils, plant tissue fragments can be extracted from sediments, especially peats. Frequently plant cuticles can be identified, along with other charac-

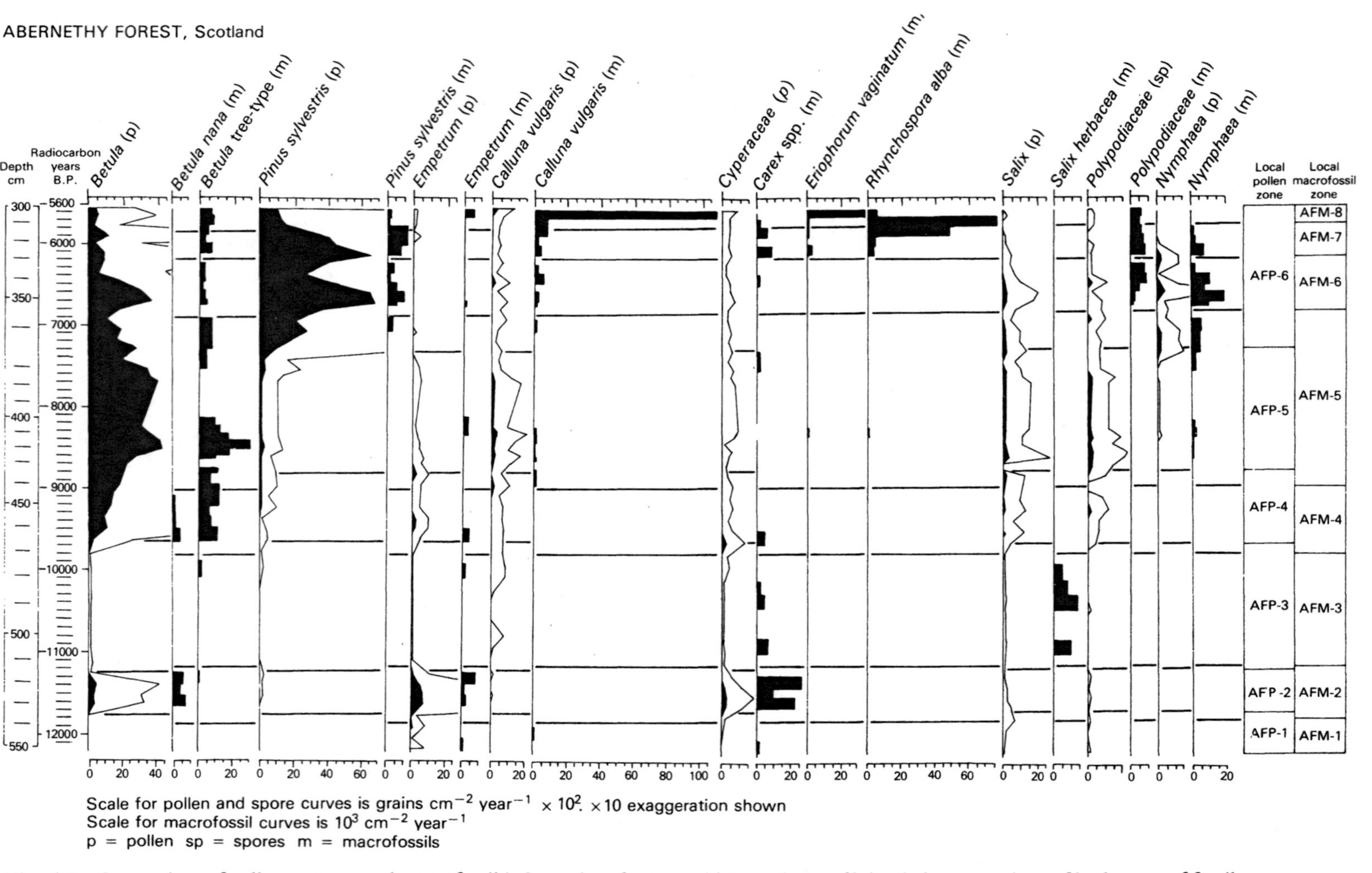

Fig. 5.8 Comparison of pollen or spore and macrofossil influx values for taxa which produce sufficiently large numbers of both types of fossil at Abernethy Forest, Scotland. The pollen or spore taxon (taxa) are presented followed by the equivalent macrofossil taxon (taxa). The values are plotted against radiocarbon age of the sediment. The pollen zones are drawn across the pollen curves, and the macrofossil zones across the macrofossil curves. (After Birks and Mathewes, 1978.)

Table 5.1 General features of macrofossils and pollen from lake sediments.

Macrofossils	Pollen
Derived from small area, mainly local wetland habitats	Derived from large area, including dry upland habitats
Can often be identified to low taxonomic level	Can rarely be identified to low taxonomic level
Relatively low annual influx	Relatively high annual influx
Some taxa only recorded as macrofossils	Some taxa only recorded as pollen
Can reconstruct local vegetation in floristic and even phytosociological detail	Can reconstruct general character of vegetation in the catchment and some local detail

teristic fragments. The monographs of Gross-Brauckmann (1972, 1974) illustrate and describe some of the commoner plant remains found in peats. Recently, Palmer (1976) developed techniques of studying cuticles and opal phytoliths (silica bodies found in cells of grass leaves; see also Moore, 1978) using both light microscopy and the scanning electron microscope. Palmer was able to identify several African grasses to subfamily, tribe, or even genus. This technique is important in studying grass-dominated vegetation types such as the African savanna, because all grass pollen grains are remarkably similar, and grass fruits are seldom preserved.

Studies on plant tissues such as cuticles have potential in the reconstruction of past vegetation, in both the Quaternary and the pre-Quaternary, where other fossils may be rare. Such work is discussed by Dilcher (1974).

References

Techniques and identification manuals

BEIJERINCK, W. (1947) *Zadenatlas der Nederlandsche Flora.* Veenman.

BERTSCH, K. (1941). *Fruchte und Samen. Handbücher der praktischen Vorgeschichtsforschung.* Ferdinand Enke.

CONOLLY, A. P. (1976). Use of the scanning electron microscope for the identification of seeds, with special reference to *Saxifraga* and *Papaver*. *Folia Quaternaria*, **47**, 29–32.

DICKSON, C. A. (1970). The study of plant macrofossils in British Quaternary deposits. In *Studies in the Vegetational History of the British Isles* (eds. D. Walker and R. G. West). Cambridge.

DILCHER, D. L. (1974). Approaches to the identification of angiosperm leaf remains. *Bot. Gaz.*, **40**, 1–157.

HUCKERBY, E., MARCHANT, R. AND OLDFIELD, F. (1972). Identification of fossil seeds of *Erica* and *Calluna* by scanning electron microscopy. *New Phytol.*, **71**, 387–92.

KATZ, N. J., KATZ, S. V. AND KIPIANI, M. G. (1965). *Atlas and Keys of Fruits and Seeds Occurring in the Quaternary Deposits of the U.S.S.R.* (in Russian). Nauka.

MARTIN, A. C. AND BARKLEY, W. D. (1961). *Seed Identification Manual.* University of California Press.

MOORE, P. D. (1978). Botanical fingerprints. *Nature*, **274**, 18.

PALMER, P. G. (1976). Grass cuticles: a new paleo-ecological tool for East African lake sediments. *Can. J. Bot.*, **54**, 1725–34.

ROBICHAUX, R. H. AND TAYLOR, D. W. (1977). Vegetation analysis techniques applied to late Tertiary fossil floras from the Western United States. *J. Ecol.*, **65**, 643–60.

Bryophyte macrofossils and identification manuals

BOROS, A. AND JARAI-KOMLODI, M. (1975). *An Atlas of Recent European Moss Spores.* Akademiai Kiado.

DICKSON, J. H. (1973). *Bryophytes of the Pleistocene.* Cambridge University Press.

DIXON, H. N. (1924). *The Student's Handbook of British Mosses* (3rd edition). Sumfield and Day.

NYHOLM, E. (1954–69). *Illustrated Moss Flora of Fennoscandia* 1–6. Gleerup and Swedish Natural Science Research Council.

SMITH, A. J. E. (1978). *The Moss Flora of Britain and Ireland.* Cambridge University Press.

Macrofossil studies of peat deposits

BIRKS, H. J. B. AND RANSOM, M. E. (1969). An interglacial peat at Fugla Ness, Shetland. *New Phytol.*, **68**, 777–96.

GRIFFIN, K. O. (1977). Paleoecological aspects of the Red Lake Peatland, northern Minnesota. *Can. J. Bot.*, **55**, 172–92.

GROSSE-BRAUCKMANN, G. (1963). Über die Artenzusammen setzung von Torfen aus dem Nordwestdeutschen Marshen-Randgebiet. *Vegetatio*, **11**, 325–41.

GROSS-BRAUCKMANN, G. (1968). Einige Ergebnisse einer Vegetationskundlichen Auswertung Botanischer Torfuntersuchungen, besonders im Hinblick auf Sukzessionsfragen. *Acta Bot. Neerl.*, **17**, 59–69.

GROSSE-BRAUCKMANN, G. (1972). Über pflanzliche Makrofossilen mitteleuropäischer Torfe. 1. Gewebereste Krautiger Pflanzen und ihre Merkmale. *Telma*, **2**, 19–55.

GROSSE-BRAUCKMANN, G. (1974). Über pflanzliche Makrofossilen mitteleuropäischer Torfe. 2. Weitere Reste (Früchte und Samen, Moose u.a.) und ihre Bestimmungsmöglichkeiten. *Telma*, **4**, 51–117.

GROSSE-BRAUCKMANN, G. (1976). Zum Verlauf der Verlandung bei einem eutrophen Flachsee (nach quartärbotanischen Untersuchungen am Steinhuder Meer). II. Die Sukzessionen, ihr Ablauf und ihre Bedingungen. *Flora*, **165**, 415–55.

JANSSEN, C. R. *et al.* (1975). Ecologic and paleoecologic studies in the Feigne d'Artimont (Vosges, France). *Vegetatio*, **30**, 165–78.

RYBNÍČEK, K. (1973). A comparison of the present and past mire communities of central Europe. In *Quaternary Plant Ecology* (eds. H. J. B. Birks and R. G. West). Blackwell.

RYBNÍČEK, K. AND RYBNÍČKOVÁ, E. (1968). The history of flora and vegetation on the Bláto mire in southeastern Bohemia, Czechoslovakia. *Folia Geobot. Phytotax.*, **3**, 117–42.

RYBNÍČEK, K. AND RYBNÍČKOVÁ, E. (1974). The origin and development of waterlogged meadows in the central part of the Šumava Foothills. *Folia Geobot. Phytotax.*, **9**, 45–70.

Macrofossil studies of lake deposits

BAKER, R. G. (1976). Late Quaternary vegetation history of the Yellowstone Lake Basin, Wyoming. *U.S. Geol. Survey Prof. Paper*, **729-E**, 48 pp.

BIRKS, H. H. AND MATHEWES, R. W. (1978). Studies in the vegetational history of Scotland. V. Late Devensian and early Flandrian pollen and macrofossil stratigraphy at Abernethy Forest, Inverness-shire. *New Phytol.*, **80**, 455–84.

BIRKS, H. J. B. (1976). Late-Wisconsinan vegetational history at Wolf Creek, Central Minnesota. *Ecol. Monogr.*, **46**, 395–429.

MATTHEWS, J. V. (1974). Quaternary environments at Cape Deceit (Seward Peninsula, Alaska): evolution of a tundra ecosystem. *Bull. Geol. Soc. Amer.*, **85**, 1353–84.

MATTHEWS, J. V. (1975). Incongruence of macrofossil and pollen evidence: a case from the Late Pleistocene of the Northern Yukon coast. *Geol. Survey Canada Paper*, **75-1**, B, 139–46.

RALSKA-JASIEWICZOWA, M. (1980). *The Late Glacial and Holocene Vegetation History of the Biesczady Mountains, Poland* (in press).

WASYLIKOWA, K. (1964). Pollen analysis of the Late-Glacial sediments in Witow near Leczyca, Middle Poland. *Report of the VIth International Congress on Quaternary, Warsaw*, 1961, Vol. **II**, 497–502.

WATTS, W. A. (1970). The full-glacial vegetation of northwestern Georgia. *Ecology*, **51**, 17–33.

WATTS, W. A. AND BRIGHT, R. C. (1968). Pollen, seed, and mollusk analysis of a sediment core from Pickerel Lake, northeastern South Dakota. *Bull. Geol. Soc. Amer.*, **79**, 855–76.

WATTS, W. A. AND WINTER, T. C. (1966). Plant macrofossils from Kirchner Marsh, Minnesota – a paleoecology study. *Bull. Geol. Soc. Amer.*, **77**, 1339–60.

WEST, R. G. (1957). Interglacial deposits at Bobbitshole, Ipswich. *Phil. Trans. R. Soc.* B, **241**, 1–31.

WRIGHT, H. E. (1966). Stratigraphy of lake sediments and the precision of the palaeoclimatic record. In *World Climate from 8000 to 0 B.C.* (ed. J. J. Sawyer). Royal Meteorol. Society, London.

WRIGHT, H. E. AND WATTS, W. A. (1969). Glacial and vegetational history of Northeastern Minnesota. *Minn. Geol. Survey Spec. Publication*, **11**, 59 pp.

WRIGHT, H. E., WINTER, T. C., AND PATTEN, H. L. (1963). Two pollen diagrams from southeastern Minnesota: problems in the regional late-glacial and postglacial vegetational history. *Bull. Geol. Soc. Amer.*, **74**, 1371–96.

Modern macrofossil assemblages

BIRKS, H. H. (1973). Modern macrofossil assemblages in lake sediments in Minnesota. In *Quaternary Plant Ecology* (eds H. J. B. Birks and R. G. West). Blackwell.

CHANEY, R. W. (1924). Quantitative studies of the Bridge Creek Flora. *Am. J. Sci.*, **8**, 127–44.

MCQUEEN, D. R. (1969). Macroscopic plant remains in recent lake sediments. *Tuatara*, **17**, 13–19.

RYVARDEN, L. (1971). Studies in seed dispersal. I. Trapping of diaspores in the alpine zone at Finse, Norway. *Norw. J. Bot.*, **18**, 215–26.

RYVARDEN, L. (1975). Studies in seed dispersal. I. Winter-dispersed species at Finse, Norway. *Norw. J. Bot.*, **22**, 21–4.

6
Palaeolimnology

Introduction

Limnology is the study of lakes, including their animal and plant life, the physical and chemical features of their water, characteristics of the bottom sediment, and the relationship of these to the physical, chemical, and biological features of the catchment area. Palaeolimnology is the study of lake environments in the past; 'the understanding of the sequential changes and their causal relationships that bodies of water have experienced during their history' (Frey, 1964).

Limnology and palaeolimnology are both huge subjects about which whole books can be written. Here we must restrict ourselves to a single chapter, and so our account has to be selective and condensed. However, we will mention several references of general importance which will provide a lead into the vast palaeolimnological literature.

First, we must briefly describe the features of a lake which are an essential background for any limnological discussions.

In autumn, the surface water of a lake loses heat and cools down. Water is most dense at 4°C, so at this temperature it sinks to the bottom of the lake and colder water floats to the surface. Eventually all the water will reach 4°C, and there will be no temperature gradient from surface to bottom. Any further cooling affects the upper layers, perhaps leading to ice formation. In spring, the heat from the sun warms the upper water, which, being less dense, floats on the surface, and remains there as an insulating layer. The deeper water heats up very slowly, if at all. The upper layer is called the ***epilimnion***, the lower layer the ***hypolimnion***, and the zone of sharp temperature change, the ***thermocline***. This situation is schematically illustrated in Fig. 6.1. In autumn, the epilimnion loses more heat than it receives, and cools down. The natural mixing which results is often accentuated by autumn gales, and a lake may cease to be stratified quite suddenly. The process is called ***overturn***.

The summer stratification of lakes and the autumn overturn are important in that they cause mixing and redistribution of the bottom sediments. Therefore, the results obtained from a palaeoecological investigation of lake sediments are not necessarily representative of the conditions at the particular part of the lake that was sampled.

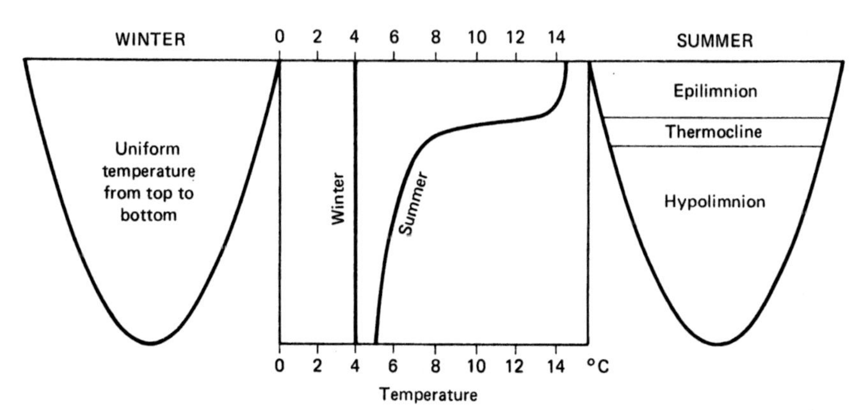

Fig. 6.1 Diagrammatic representation of temperature conditions in the water of a lake in winter and summer.

Figure 6.2 summarizes the formation of lake sediment. The sediment is the net result of input from the watershed and from lake productivity. Any organic matter is decomposed to some degree by bacteria and bottom-dwelling animals (benthos). The organic fraction of the sediment therefore consists largely of resistant substances such as lignin, chitin, and a certain amount of cellulose. There are biologically manufactured inorganic components such as diatom frustules and sponge spicules. There is also a wide variety of simpler organic compounds resulting from decomposition processes, such as sugars, amino acids, photosynthetic pigment derivatives, and many others (see Vallentyne, 1957). Oxygen depletion by the processes of decomposition in the more or less stagnant hypolimnion of eutrophic lakes can lead to anaerobic conditions. Anoxia inhibits further breakdown and the result is better preservation of organic materials. Once these are buried below the active surface layers, they remain more or less unchanged. If there is very little disturbance by currents and benthic organisms the sediments may be visibly laminated. The laminae result from different types of deposition at different times of the year. If such laminae can be proved to be annual, they form a reliable basis for estimating the chronology of the lake sediments. Laminated sediments are relatively rare, however, because the sediment surface is usually disturbed by burrowing animals and by the turbulence of overturn.

There may be considerable redistribution of fine sediments from the littoral to the profundal regions. This process can be strong enough to redistribute smaller fossils, such as pollen grains, diatoms, and cladocera. Because shallow water and deep water biotas in a lake are substantially different, the fossil assemblages from shallow and deep water sediments are also different. If we want to interpret the past conditions of the whole lake from one or a few offshore cores, we must have some appreciation of the extent to which littoral and offshore sediments become mixed. In no case so far investigated has the mixing been sufficient to produce sediments with a uniform composition over the whole lake.

Apart from reflecting their internal biota and environment, lakes are very sensitive to changes in their catchment area. Such changes may be caused by climate, or by human interference with the landscape. We shall be discussing some of the ways

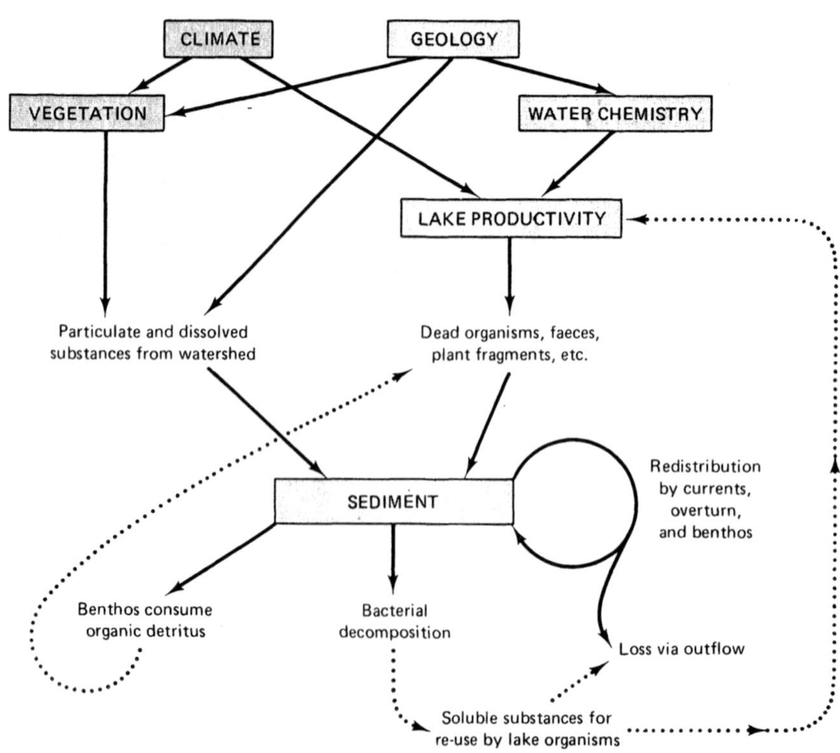

Fig. 6.2 Summary of the factors affecting the formation of lake sediment.

terrestrial events are registered by lake sediments later. Lake sediments are therefore particularly valuable to the palaeoecologist in that they hold both the palaeolimnological record, and a reflection of terrestrial events; indeed the former is partially controlled by the latter.

Good reviews on the subject of palaeolimnology are by Frey (1969), Livingstone (1968), and Oldfield (1977). The review by Frey (1964) also discusses palaeolimnology, and is an extensive account of animal fossils found in bog and lake sediments.

A wide variety of organisms can be preserved in lake sediments. Plant remains include pollen grains, fruits, seeds, and other macrofossils of higher plants, microspores and macrofossils of pteridophytes and bryophytes, oospores of some algae, fungal hyphae and spores, and frustules of a major group of algae, the diatoms. Among the animal fossils are rhizopods and some other protozoans, sponges, preserved as spicules, turbellarians (flatworms), bryozoans, particularly *Cristatella* and *Plumatella*, crustaceans, particularly cladocerans (water fleas) and ostracods, insects (particularly the head capsules of midge larvae (chironomids) and beetles), mites (particularly *Hydrozetes*), molluscs, and vertebrates (most usually fishes, but sometimes small mammals and birds) (see Frey, 1964). Here we shall be mainly concerned with the most common and most widely studied groups, the Cladocera, the insects, and the molluscs. Amongst the plant remains in lake sediments, we have already discussed plant macrofossils (Chapter 5) and we will later turn our attention to pollen grains and spores. Here we shall discuss the other major plant group which has been studied, the diatoms.

Diatoms

Diatoms are unicellular algae with a unique silica shell, or frustule, which is readily preserved in lake sediments. The structure of two 'typical' diatoms is shown in Fig. 6.3.

Fig. 6.3 Two 'typical' diatoms (**a**) Pennales, or long diatoms (the raphe is absent in Araphidineae). (**b**) Centrales, or disc-shaped diatoms.

The two valves of the frustule fit together like a date box. When the diatom reproduces asexually, two new valves are made to fit inside the original two, and the living part of the diatom simply divides, each half to a new frustule.

Diatoms come in two basic shapes, circular (Centrales) and elongated (Pennales). The Centrales are mostly planktonic and the Pennales are benthic, apart from the planktonic Araphidineae. Modern diatom taxonomy is based on frustule characteristics, and therefore, with good preservation, a fossil diatom can be identified as accurately as a living one. The frustules are intricately sculptured, and the amazing and beautiful structures have been strikingly illustrated by scanning electron microscope photographs. Such photographs have led to further advances in diatom taxonomy (see, for example, Miller, 1969; Florin, 1970; Haworth, 1975), using both modern and fossil frustules.

Diatom frustules are often abundant and well preserved in lake sediments. However, they are not indestructible, and they may be physically or chemically eroded. Long thin diatoms, such as *Fragilaria crotonensis* are rarely found intact, and diatoms brought into lakes by streams are often physically eroded. Chemical erosion occurs under certain conditions of water chemistry by solution of the silica. The frustules are most frequently dissolved under very alkaline conditions, especially if these are reducing, but they can also be dissolved in very acid situations. Round (1964) discusses this problem, but comes to the general conclusion that frustule solution is rarely a major factor affecting fossil diatom assemblages in freshwater sediments, in contrast to marine conditions, where up to 75% of the frustules may be affected.

Diatoms are useful palaeoecological indicators. Many have distinct ecological tolerances, and a substantial amount of autecological information is available. There are several diatom habitats in a lake: pelagic or planktonic species float in the water, usually in the epilimnion, during the summer; epiphytic species grow attached to higher aquatic plants; epipelic species live on the surface of the mud; epilithic species live on stones. In addition, lake sediments may contain representatives of two other habitats, streams and soil, which have been washed in by stream-flow or soil disturbance. In addition, the diatoms from each habitat are sensitive to conditions of pH, water chemistry (Rawson, 1956), and salinity (e.g. Alhonen, 1971), to climate, and to light intensity, which is itself related to water depth, turbulence, presence of other planktonic organisms, and currents. A diatom assemblage can therefore be fairly easily divided into several ecological groups.

Diatoms are usually so abundant in lake sediments that they can be counted. The data are often presented as percentages. An absolute method, such as that proposed by Battarbee (1973), may be used to obtain concentration figures, and, if the sediment accumulation rate is known, the annual diatom influx can be calculated, in much the same way as for pollen grains (see Chapter 10).

We may briefly summarize the features which make diatoms suitable tools for the palaeoecologist.

1. The fossil frustules can be readily identified.
2. Fossils are frequently well preserved in lake sediments.
3. Many diatoms have distinct ecological characteristics.
4. Fossils are abundant and hence can be counted.

Good reviews of the use of diatoms in palaeoecology, including the difficulties encountered in their use are given by Round (1964) and Bradbury (1975).

Modern diatom assemblages

Even though fossil diatoms represent the whole organism, it is necessary to know to what degree a modern diatom assemblage represents the contemporary diatom population before making any reconstructions from fossil assemblages. As we have commented before, the surface of lake sediments is often disturbed before being finally buried, and most lakes also receive material from the catchment area *via* inflow streams, so that many different habitats in and around the lake are represented in any fossil assemblage.

At Blelham Tarn in the English Lake District, Haworth (1976a) took monthly samples of (a) sediments caught in a trap suspended 2 m above the bottom, and of (b) the uppermost 5 cm of sediment. She monitored the representation of a bloom of *Stephanodiscus astraea* var. *minutula** which had newly arrived in the lake plankton. The analyses showed that the bloom was registered in the sediments for two months, but subsequently it became diluted by some downward movement of the small round frustules in the loose upper sediment, and by the great abundance of species such as *Asterionella formosa* which flourished in the plankton after *Stephanodiscus* had declined. In this lake at least it would appear that the planktonic diatoms are deposited in orderly succession with little disturbance until the autumnal overturn which causes resuspension and mixing of the upper 5 cm of sediment. It would appear that *Stephanodiscus* will not be prominent in the lake sediments unless it persists for several years. A single year's bloom becomes too diluted by the other abundant member of the phytoplankton to become a marker horizon.

Meriläinen (1969, 1971) studied surface sediment samples from four lakes, to investigate the distribution of diatoms over the lake. He chose ***meromictic*** lakes, which are sufficiently deep and sheltered from strong winds that the very bottom layer of water is never disturbed. It is therefore devoid of oxygen, and rich in dissolved substances, particularly, in Meriläinen's lakes, iron compounds. He discovered that diatoms tended to be dissolved in the bottom layer. Otherwise, the distribution of the different species corresponded in general with their habitats. Planktonic species were more abundant near the centre of the lakes, and epiphytic and benthonic taxa tended to be most abundant in shallow waters. This work suggests that a single core from a lake will not be able to provide an integrated diatom assemblage from the whole lake, but will record the diatoms growing locally as being more abundant. However, the position is not so extreme as with larger fossils such as plant macrofossils, and some diatoms from all parts of the lake are likely to be found at any one point on the bottom, because of internal sediment redistribution. Very similar conclusions about the distribution of diatoms over Sallie Lake, Minnesota were reached by Bradbury and Winter (1976).

Koivo and Ritchie (1978) analysed modern diatom samples from 20 lakes along a transect from northern boreal forest to tundra near the Mackenzie Delta in Canada. Most of the diatom assemblages were characteristic of oligotrophic lakes, but different assemblages were found in saline, meromictic, hardwater, and N-enriched lakes, confirming that water chemistry is an important factor controlling diatom occurrence. Similar conclusions were reached by Bright (1968) in a survey of lakes with varied water chemistry in Minnesota, U.S.A.

* The taxonomy of small *Stephanodiscus* species is controversial at present (Bradbury, 1975). Because we cannot re-examine material, we have had to use the author's original nomenclature, which is not necessarily correct in light of modern electron microscope studies.

Diatoms in palaeolimnological studies

There are now numerous stratigraphic studies of diatoms in lake sediments. We shall describe a few varied studies which have been used to reconstruct successfully the palaeoecology of the lakes concerned and their surrounds, and even climatic and geomorphological changes.

1. *English Lake District*

Pennington (1943) presented the first diatom analyses of the sediments of Lake Windermere. Subsequently, analyses have been made on several Lake District lakes by Round (1957, 1961), Evans (1970), and Haworth (1969). Here we will discuss the studies at Esthwaite Water (Round, 1961) and Blea Tarn (Haworth, 1969), which form an interesting contrast.

Round (1961) distinguished five phases in the diatom stratigraphy of Esthwaite Water, which are summarized in Table 6.1.

These changes in Esthwaite Water can be matched in cores from Windermere (Pennington, 1943) and Kentmere (Round, 1957). The main features of all are a sparse diatom flora in Devensian late-glacial sediments; a sparse but increasing flora in the early post-glacial (Flandrian) sediments reflecting alkaline conditions; an abundant flora in mid- and late post-glacial sediments reflecting a change to acidic water; a recent change to eutrophic conditions as man's activities in the catchment led to an increased input of nutrients. This eutrophic phase is most marked at Esthwaite Water. This lake is in subdued, low-lying cultivated terrain, which contains several villages. Eutrophication is enhanced by the addition of sewage waste into the lake, and blooms of blue-green algae may suppress summer diatom production.

The upland Blea Tarn, Langdale, described by Haworth (1969) contrasts with Esthwaite Water. Like Esthwaite Water, diatoms were sparse and there were no planktonic species in the late-glacial sediments. The early post-glacial flora suggested that the lake was moderately base-rich and contained enough nutrients to support *Asterionella formosa* for a short period. At 7000 B.P. the flora became characteristic of acid water, and the only changes from then until the present day indicate a gradual depletion of nutrients. There is no recent eutrophic development, because insufficient nutrients have been released from the catchment by human activity. The catchment of Blea Tarn contains only one house and the relatively recently cleared hillsides are used for grazing rather than cultivation.

2. *North-west Scotland*

Haworth has also studied the late- and post-glacial diatom history of the large Loch Sionascaig

Table 6.1 Phases in the diatom stratigraphy of Esthwaite Water. (From Round, 1961.)

Phase	Time period	Diatoms
1	Devensian late-glacial (12 000–10 000 B.P.)	Low diatom productivity and few species. Plankton absent. Water alkaline
2	Early post-glacial (10 000–7000 B.P.)	Low but gradually increasing productivity. Epiphytic, epipelic, and planktonic forms. Water alkaline
3	Mid-post-glacial (7000–5000 B.P.)	High productivity, including large increase in epipelic forms. Base-rich indicators decrease, acid indicators increase, i.e. water becoming increasingly acid
4	Late post-glacial (5000–1000 B.P.)	Optimum growth of planktonic and epiphytic forms. No base-rich indicators. Water acid (soil leaching and bog development in catchment)
5	Recent (1000 B.P.–present)	*Asterionella formosa* appears together with other eutrophic indicators, responding to increased nutrients washed into the lake as a result of forest clearance of the catchment

in the Lewisian Gneiss country of north-west Scotland (Pennington *et al.*, 1972). This lake showed a similar succession to Blea Tarn, except that the acidification in the mid-post-glacial was more marked. During the interstadial of the Devensian late-glacial (the Allerød) the diatom flora contained numerous planktonic species. Investigations of Cam Loch and Loch Borralan in the same region of Scotland (Haworth, 1976b) did not show extensive planktonic development in the Allerød. So far Loch Sionascaig is unique in this respect. Figure 6.4 is an example of the way in which Haworth reconstructed the late-glacial environment from the diatom assemblages at Loch Borralan. Before the Allerød, diatoms were sparse probably due to turbidity of the water and a small amount of organic matter. During the Allerød (zone 2), planktonic species increased, although benthic forms remained overwhelmingly predominant. During the cold stadial after the Allerød (zone 3) there is a peak of *aerophilous* taxa (characteristic of terrestrial habitats), which were washed in during the soil disturbances of the cold period. At the opening of the post-glacial, aerophilous species decline. The apparent indicators of cold conditions change very little and do not respond to the supposed climatic changes of the late-glacial period. This demonstrates the relative insensitivity of diatoms to temperature, compared with their response to the nutrient and pH status of the lake.

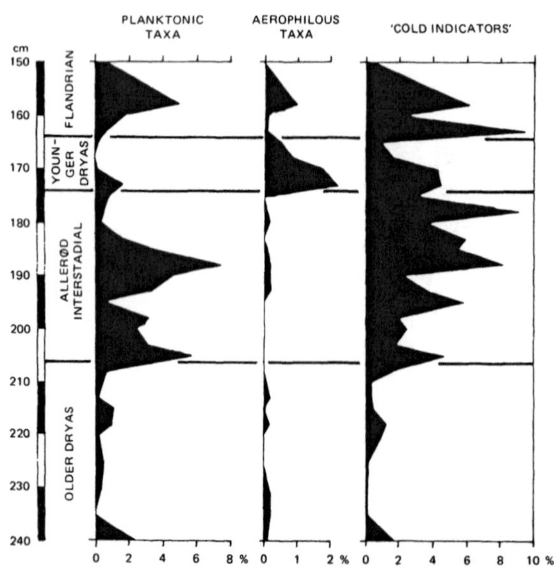

Fig. 6.4 Ecological assemblages of diatoms from Devensian late-glacial sediments in Loch Borralan, Scotland. (After Haworth, 1976.)

3. *Kirchner Marsh, Minnesota, U.S.A.*

When Kirchner Marsh was investigated using pollen and plant macrofossil stratigraphy, Watts and Winter (1966) (see Chapter 5) described a layer of plant 'trash' at the base of the core, which appeared to be of terrestrial origin (it contained many macrofossils characteristic of spruce forests), yet it was overlain by limnic sediments. It was suggested that this layer originated from the forest growing on a buried ice block. When it melted, the hollow containing Kirchner Marsh was formed, with the forest layer at the bottom. Florin (1970) analysed the diatoms from this layer. Figure 6.5 summarizes her results. Her interpretation is illustrated in Fig. 6.6. The results can be applied to buried ice block melting in two general situations; dead ice in an outwash plain, and in an ice-cored moraine (Florin and Wright, 1969). In diatom zone 1, aquatic species are absent, and the flora is typical of a mossy forest floor which grew on the ice block. In zone 2, these species are joined by diatoms which can either live in moist terrestrial habitats, or on lake shores (facultative terrestrial). The ice surface began to melt to form small pools. Zone 3 contains aquatic diatoms only, typical of benthic littoral habitats. The pools had increased in size, reducing the terrestrial surface. In zone 4, planktonic diatoms appear, suggesting that the lake was fully developed.

4. *Mexico City*

Mexico City is built on the lacustrine sediments of former Lake Texcoco. Bradbury (1971) investigated the history of this huge former lake by means of diatom analysis. His coring site was near the shore of the lake, and he discovered alternating phases of deep-water planktonic species, and littoral and marsh diatoms. About 100 000 years ago Lake Texcoco was cool and deep, but it slowly dried up to become a series of saline pools. Freshwater marsh diatoms were maintained by springs around the lake margin. A subsequent rise in water level flooded the springs with brackish water, shown first by the appearance of benthic taxa and, as the lake became deeper, by planktonic species characteristic of brackish water.

The lake never became freshwater again, but remained brackish. The water level continued to fluctuate, brackish water and marsh floras alternating, throughout the last glaciation. During the

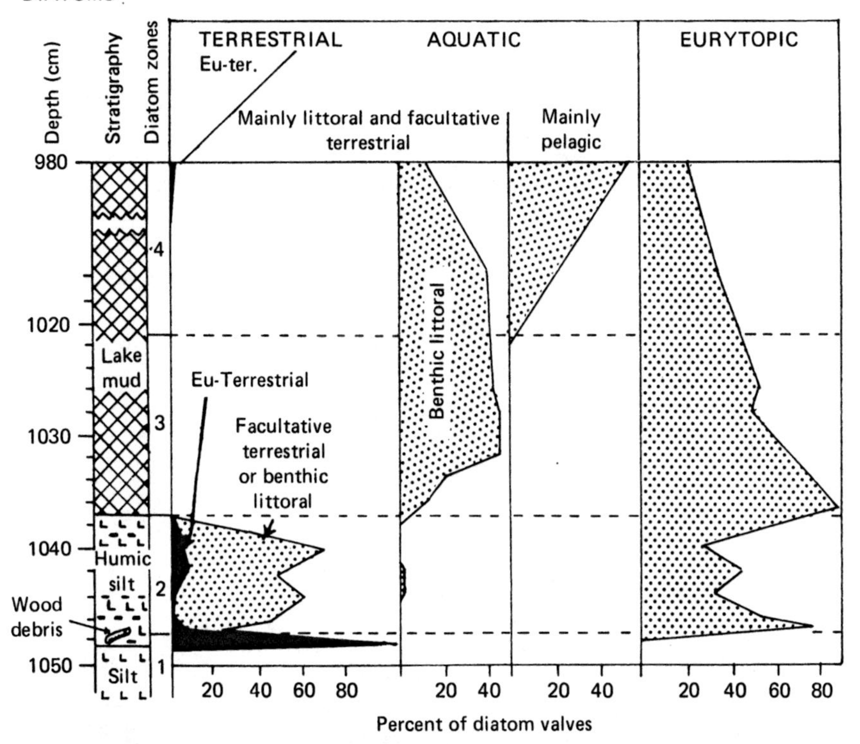

Fig. 6.5 Summary diagram of the diatom analyses from the basal sediments of Kirchner Marsh, Minnesota, showing how terrestrial and benthic littoral taxa are associated with the lowest sediments of terrestrial origin. (After Florin and Wright, 1969.)

last 10 000 years the marsh flora predominated, suggesting a drier warmer climate during the Holocene (post-glacial). Eventually the lake became reduced to a few saline pools, and periodic floods have been controlled by a system of dykes and drains. This study used the sensitivity of diatoms to water chemistry to deduce the history of changing water levels in Lake Texcoco, which could be correlated with changes in climate.

5. *Diatoms in saline African lakes*

Bradbury (1971) was able to deduce brackish or saline conditions in ancient Lake Texcoco, because certain diatoms are characteristic of these conditions. The modern diatom floras of lakes in the semi-arid climate of East Africa were examined in relation to water chemistry by Hecky and Kilham (1973). As we have seen, diatoms are useful indicators of changing water-levels, which are very important in the palaeoecology of East Africa, especially where the environment of early man is being worked out. Hecky and Kilham found a distinct alkalinity series, each diatom having its own range of tolerance in which it dominated the plankton. Thus, an assemblage of fossil diatoms could be interpreted with some confidence in terms of water chemistry. They also found that the diatoms were well preserved in these alkaline and saline lake sediments, perhaps due to the presence of humic acids preventing solution of the silica frustules.

6. *Pickerel Lake, Dakota, U.S.A.*

The study of Pickerel Lake, South Dakota, U.S.A. by Haworth (1972a) used the chemical tolerances of diatoms to interpret a phase of saline water in that lake during the mid-Holocene. The lake developed from a relatively dilute alkaline lake into a saline lake, which then became progressively more dilute once more. These phases of lake development corresponded to the vegetational changes around the lake

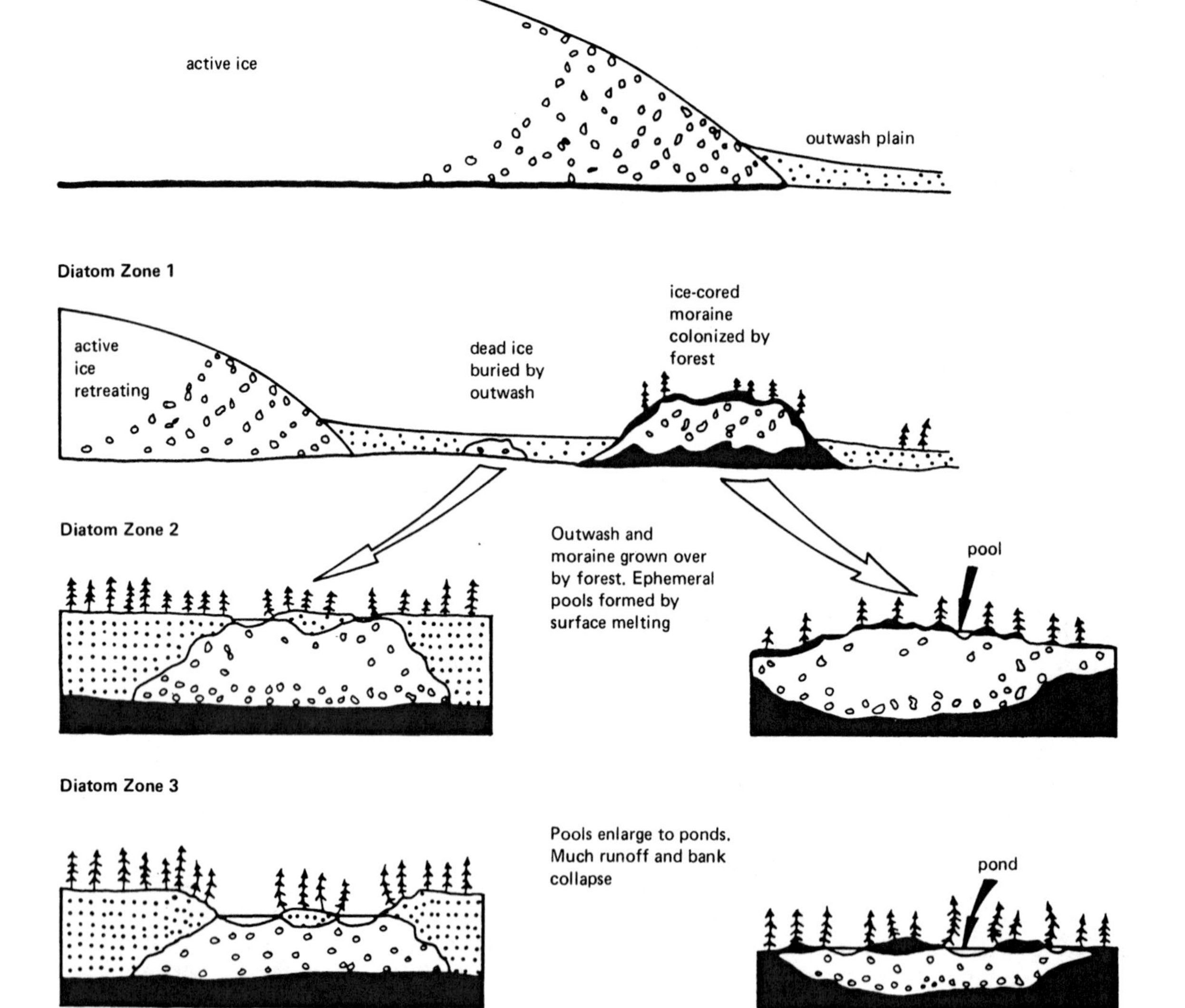

Fig. 6.6 Illustration of the events interpreted from Fig. 6.5 at Kirchner Marsh. See text for explanation. (After Florin and Wright, 1969.)

deduced from pollen and macrofossil analysis by Watts and Bright (1968). The alkaline phase corresponded with the presence of spruce forest, the increasing concentration of salts with the development of deciduous forest, and the saline phase with the development of open prairie. The dry prairie climate caused salts to be concentrated in the soils and lake by evaporation. Subsequent decreasing salinity corresponded with the time when woodland again spread round the site in response to increased rainfall.

7. *Diatoms and recent lake eutrophication*

We have already mentioned how the diatom assemblage at Esthwaite Water changed after the Norse clearances of about 1000 A.D., indicating an increased input of nutrients into the water, resulting at first from forest clearance and agriculture, and later also from input of sewage. Today the lake supports blooms of the blue-green alga *Oscillatoria*, which indicates a substantial amount of enrichment, or so-called eutrophication.

The flora of many lakes has changed in this direction in modern times, due to the input of sewage or agricultural fertilizers. Even thoroughly treated sewage effluent is still a rich solution of nitrates and phosphates which encourage plant growth. The blue-green blooms of *Oscillatoria* create a 'pea-soup' effect which is generally regarded as unpleasant pollution if the lake is used for recreation.

A study of the course of eutrophication in Lake Washington, adjacent to Seattle, Washington, U.S.A. was made by Stockner and Benson (1967). They argued that, of the planktonic diatoms, the Centrales are usually characteristic of oligotrophic lakes, and the Araphidineae of eutrophic lakes. Therefore, the ratio of Araphidineae to Centrales (A/C) in an assemblage gives an indication of the trophic status of the lake. Stockner and Benson's study of this ratio in Lake Washington (Fig. 6.7) showed increases in eutrophication associated with the early development of Seattle since about 1850, a period of raw sewage input between 1910 and 1930, peaking at about 1925, and a period of sewage diversion until 1941, followed by a period of treated sewage input until, in 1967, all sewage was diverted from the lake.

The A/C ratio was subsequently applied to several lakes in North America and the English Lake District by Stockner (1972). The ratio remained low in recent sediment cores from oligotrophic lakes such as Wastwater and Ennerdale in the English Lake District, but it rose to a level indicating the development of eutrophy in Windermere (South Basin), Blelham Tarn, and Esthwaite Water. The ratio indicated that Esthwaite Water was the most eutrophic lake studied, which accords with its status assessed by many other methods. In all cases, the course of the ratio could be correlated with historical changes in the catchment areas, in terms of agriculture and human settlement.

An exception to the use of the A/C ratio has been described by Haworth (1972b) from Loch Leven, south Scotland, where increasing eutrophication and partial drainage of the lake resulted in an increase in the centric diatoms *Stephanodiscus* spp. and *Cyclotella*, rather than in the Araphidinae (*Asterionella* and *Fragilaria*). Similar responses have also been recorded in North American lakes, for

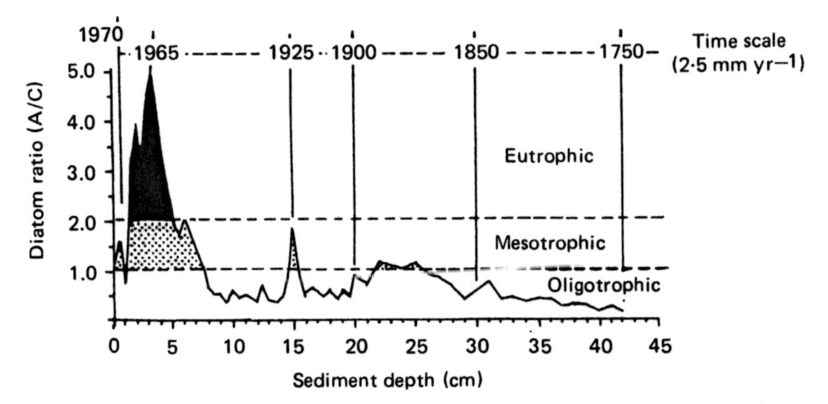

Fig. 6.7 The Araphidinae/Centrales diatom ratio in the upper sediments of Lake Washington. If a constant rate of sediment accumulation of 2.5 mm year^{-1} is assumed, the profile can be related to calendar years as shown.
(After Stockner and Benson, 1967.)

Table 6.2 Summary of the general changes in the diatom floras of enriched lakes. (From Bradbury, 1975.)

	Predisturbance assemblage	Postdisturbance assemblage
Diatom diversity	Tends to be higher in regard to planktonic and littoral diatoms	Tends to be lower, with only a few species dominating
Planktonic	More or less even representation of spring, summer, and fall species	Spring and sometimes late fall-blooming species, avoiding summer blooms of blue-green algae
	Usually several eurytopic species are present	Few eurytopic species present. *Stephanodiscus hantzschii* found in the most eutrophic lakes; *S. minutus* in less enriched lakes
Benthic and epiphytic species	Greater diversity of types, and greater number of individuals	Benthic flora reduced after disturbance; epiphytic types may increase on increased macrophytes, but they are often swamped by massive blooms of planktonic algae
Diatom productivity	Depends upon trophic state of lake	Increases with enrichment of lake
Total flora	Distinctive floras in different lake types and habitats	Similar floras tend to be found in highly enriched lakes. The species are not unique to polluted lakes

example, by Bradbury and Megard (1972), Birks *et al.* (1976), and Koivo and Ritchie (1978), and in Northern Ireland by Battarbee (1978). Obviously, the A/C ratio is only useful in those lakes where *Stephanodiscus* does not become prominent in eutrophic conditions. An inspection of the contents of the A and C categories is necessary for detailed palaeolimnological reconstructions, and indicator species should be employed.

Similar reservations were made by Bradbury (1975) in a review of recent diatom stratigraphy of several lakes in Minnesota and South Dakota, U.S.A. He conveniently summarizes the changes likely to occur when a lake is affected by human activity in the watershed in a table which we reproduce here as Table 6.2.

8. *Annual diatom sequence preserved in varves*

In any lake there is a succession of diatoms throughout the year. This succession is occasionally preserved when there is no disturbance of the sediment. In Lower Saxony, Germany, Benda (1974) described the annual diatom succession in varved sediments of Eemian and Holstenian interglacial age. In the spring, a bloom of planktonic *Stephanodiscus astraea* and *S. astraea* var. *minutula* laid down a light coloured band. During the summer, *Melosira granulata* predominated, and laid down dark coloured sediment, which resulted in annual light-dark pairs of laminations.

Simola (1977) studied annual diatom laminations from recent sediments in Lake Lovojärvi, Finland. He peeled off very thin layers by laying transparent adhesive tape on the slightly dried frozen sediment, and he was able to reconstruct the annual diatom succession in great detail (Fig. 6.8). He could identify the 1974 layer by comparison with phytoplanktonic observations on the lake made in that year.

Other algae in lake sediments

Fossils of algae other than diatoms have rarely been recorded from lake sediments. Colonies of

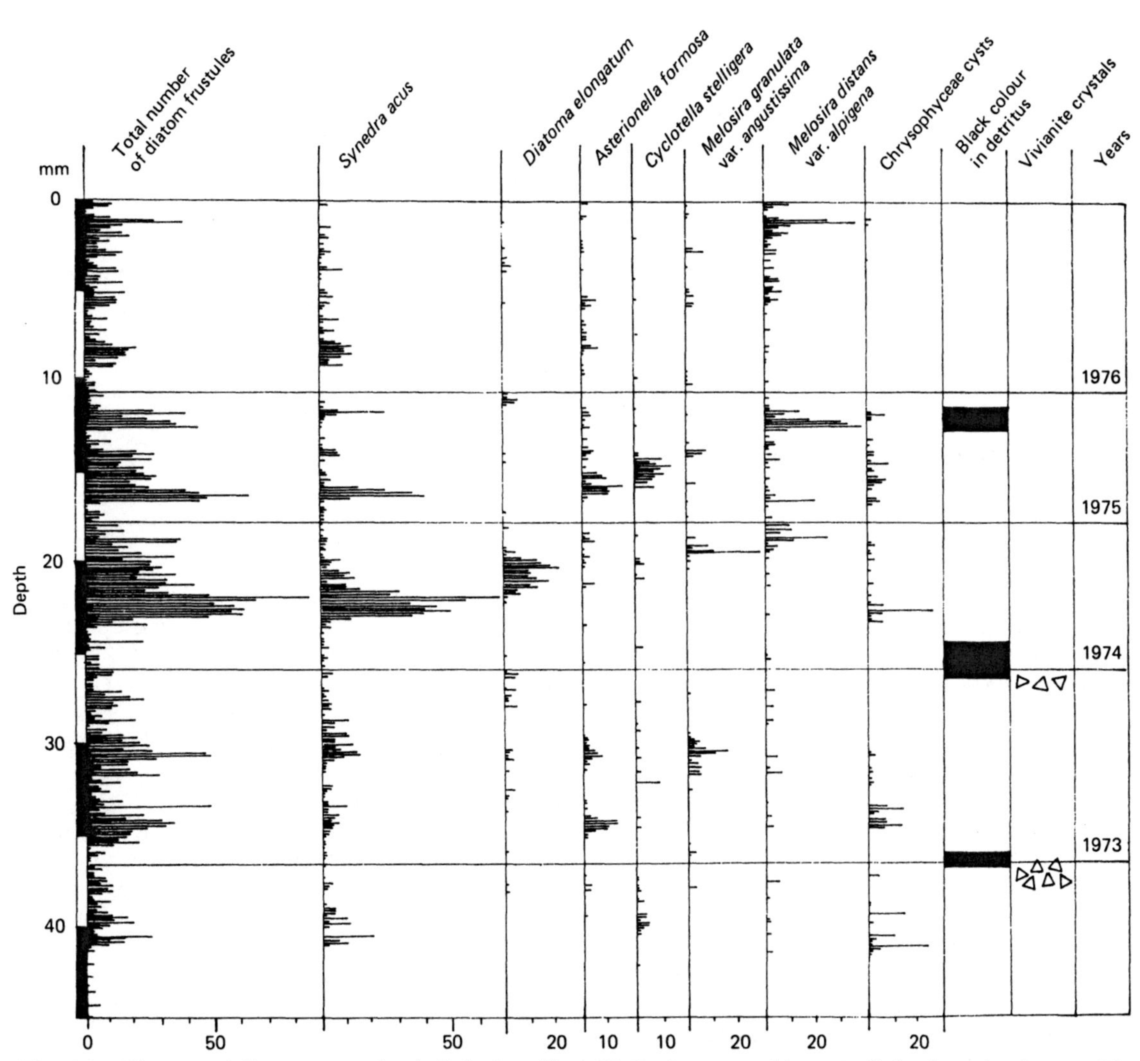

Fig. 6.8 The annual diatom succession in Lake Lovojärvi, Finland, preserved in annually laminated sediments. The values are numbers of objects in successive horizontal microscope fields of view 150 μm wide across the adhesive tape. (After Simola, 1977.)

Pediastrum spp. and *Botryococcus* are sometimes recorded by pollen analysts (e.g. Birks, 1976; Birks *et al.*, 1976). The presence of certain algae has also been detected by analysis for pigments and other organic compounds in lake sediments, which we will discuss later in this chapter (p. 107). Leventhal (1970) analysed the chrysophycean cysts from the sediments of Lago di Monterosa in Italy. However, so little is known of chrysophycean ecology, that the results were difficult to interpret. Other unusual fossils were analysed in this study, such as sponge spicules (Hutchinson, 1970). Some dinoflagellate cysts were identified by Brugam (1978b).

Cladocera

Cladocera are the most abundant of the crustaceans preserved in lake sediments, most particularly the families Bosminidae and Chydoridae. The Daphniidae are usually only represented by their large resistant egg cases (ephippia), or by their minute mandibles. A typical chydorid is illustrated in Fig. 6.9.

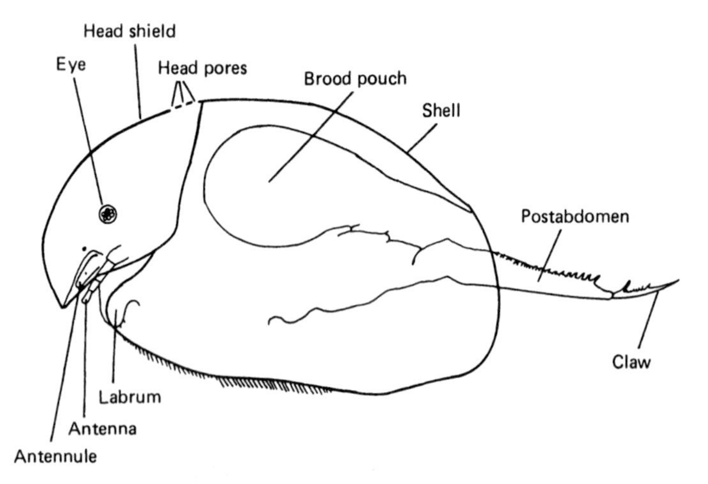

Fig. 6.9 Outline drawing of *Camptocerus rectirostris*, a typical chydorid cladoceran, showing the relation of the various exoskeletal fragments to the whole animal. (After Frey, 1962.)

In general, *Bosmina* and *Daphnia* are planktonic, and the chydorids are littoral. However, *Chydorus sphaericus* can become planktonic, especially when blooms of blue-green algae provide a 'raft' in eutrophic conditions.

Modern cladoceran assemblages

Because of seasonal variations in the populations of each species, the simplest way to sample the total population of Cladocera in a lake is to take a sample of the surface sediment, in the same way as for other limnic fossils.

Frey (1960), investigating five lakes near Madison, Wisconsin, U.S.A. showed that the live cladoceran population was closely related to the abundance of shells in the sediments, and that the different species could be quantitatively assessed (see also De Costa (1964) and Goulden (1969)). Frey (1960) discovered that Cladocera were more abundant with increasing lake productivity, caused either by shallowing or by a greater input of nutrients.

This approach was developed further by Whiteside (1970), who analysed surface mud samples from 77 Danish lakes. On the basis of their physical characteristics, the lakes could be grouped into (1) clear water oligotrophic lakes, (2) ponds and bogs or lakes with brown peaty water, and (3) eutrophic or 'polluted' clear-water lakes. Each chydorid species had its own range of ecological preference between the three lake types, some being more widespread than others. Oligotrophic waters supported the greatest number of species, probably because the habitat was more diverse, with abundant rooted aquatic plants, in comparison to his eutrophic lakes which generally lacked macrophytes. *Chydorus sphaericus* and *Alona rectangula* were the two most common species which preferred eutrophic lakes.

Chydorids live either in aquatic vegetation or on the bottom mud. Quade (1969) has shown that individual species tend to prefer different plant species. For palaeolimnological studies, a small shallow lake will be ideal for studying changes in the chydorid populations, whereas a large deep lake will buffer any environmental changes, and the relatively small proportion of shallow to deep waters will dilute the chydorid fossils. Keen (1973), Goulden (1971), and Whiteside (1974) have followed changes in chydorid populations through a year, and provided some insight into what controls their abundance.

The most important planktonic cladocerans for palaeolimnologists are in the genus *Bosmina*. *Daphnia* is poorly preserved and the ephippial egg cases are hard to identify to species. The replacement of *B. coregoni* by *B. longirostris* during a change to eutrophic conditions was first recorded by Minder (1938) in Zürichsee after 50 years of sewage input from Zürich, Switzerland. Such a replacement was also recorded by Deevey (1942) in the sediments of Linsley Pond, Connecticut, U.S.A. on the change to eutrophic conditions and it has subsequently been found by several workers. Brooks (1968), Brooks and Dodson (1965), Kerfoot (1974a,b; 1975), Stenson (1976), and Zaret and Kerfoot (1975) have demonstrated that the change may be actually controlled by fish predation. The fish first select the largest prey, the *Daphnia* spp., and then the somewhat smaller *Bosmina coregoni*. With these competitors removed, *B. longirostris* increases in abundance, but it is too small to be valuable prey for most fish species. Small size is selected for, because the fish eat the larger *Bosmina* first, and in lakes with large fish populations, *B. longirostris* reaches maturity at a smaller size than in lakes with less fish. As a lake becomes more eutrophic, its plankton productivity increases, and thus the fish population increases, perhaps also with a change in species as the hypolimnion becomes deoxygenated, and as a result, the larger cladocerans suffer.

Some fossil evidence of the correlation with eutrophication was obtained by Deevey (1969) from Rogers Lake, Connecticut. Large *Bosmina* species were replaced by *B. longirostris* at the same time as remains of the larvae of the planktonic predator *Chaoborus* were found in the sediments. *Chaoborus* is the midge larva most tolerant of anaerobic eutrophic conditions. At Linsley Pond, Connecticut, Brugam (1978b) proposed that fish predation decreased with increasing eutrophication, and daphniids became common.

Cladocera in palaeolimnological studies

1. *Esthwaite Water, English Lake District*

Goulden (1964a,b) analysed cladocerans from a core of sediment from Esthwaite Water. He discovered four main periods of abundance, each associated with climatic or cultural changes in the drainage basin. (1) In the Allerød interstadial of the Devensian late-glacial, chydorids were comparatively abundant, and *Bosmina coregoni* was the dominant planktonic species. (2) In the mid-post-glacial (Flandrian) time (pollen zones VI and VII) cladoceran fossils were more abundant than at any other time, reflecting a large population. These two peaks probably represent climatic changes to warmer conditions allowing an increase of lake productivity. (3) The onset of Neolithic occupation of the catchment at about 5000 B.P. is reflected by another peak of abundance, resulting from an influx of ions from soil disturbance as forests were cut down. (4) The Norse invasion at about 1000 A.D. resulted in the last rise in the cladoceran population. Increasing amounts of nutrients were supplied from the catchment following deforestation, the establishment of sheep walks, agriculture, and later from sewage input from the increasing human population.

During the fourth increase in the cladoceran population, a change in dominance occurred. In the plankton, *Bosmina coregoni* was replaced by *B. longirostris*, a certain species of *Daphnia* ('B') became abundant, and *Ceriodaphnia* became important. Simultaneously, a change in the midge larva fauna occurred in the bottom muds; *Sergentia* and *Tanytarsus* were replaced by *Chironomus*, indicating deoxygenation of the hypolimnion due to enrichment of the lake. Goulden's results confirm closely the reconstruction of the lake's history using diatoms, described earlier (p. 89). Goulden concluded that chydorid cladocerans are useful indicators of climatic change, whereas planktonic Cladocera may be used for deducing past limnological changes.

A very similar series of changes in the Cladocera in the sediments of Blelham Tarn, about 6 km east of Esthwaite Water was described by Harmsworth (1968). Here also, the replacement of *Bosmina coregoni* by *B. longirostris* was accompanied by a change from *Tanytarsus* and *Sergentia* to *Chironomus* midge larvae.

2. *Other studies*

Frey (1962) illustrates cladoceran fossils from Eemian (last) interglacial deposits in Denmark and shows that there has been negligible morphological evolution since that time. In addition, palaeolimnological interpretations made on the basis of their modern ecology are consistent with other lines of evidence, suggesting that cladoceran ecology also has not changed since the Eemian.

Megard (1964) analysed the Cladocera in the sediments of Dead Man Lake in the Chuska Mountains, south-west U.S.A. During the Wisconsinan glaciation, only seven species were recorded, mostly littoral chydorids, reflecting alpine conditions in the mountains during the glaciation. Cladocera were more abundant during the Holocene. Megard used the ecology of the Cladocera and the midge larvae to deduce the limnological history of the lake, and showed that it had always been shallow even though it now contains some 16 m of sediment.

The Cladocera of Lake Zeribar in Iran were also studied by Megard (1967) as part of the Iranian Prehistoric Project. The aim was to provide environmental information for the time when men first domesticated plants and animals about 11 000 years ago (see p. 109). Megard demonstrated a cool dry climate between 23 000 and 11 000 years ago, a slow warming to present temperatures up to 5000 years ago, and relative stability since then. The expansion of *Bosmina longirostris* at about 5000 B.P. indicated that the lake became deeper. This was probably caused by an increase in rainfall, and blockage and raising of the outlet level by alluvial fans.

Crisman and Whitehead (1978) and Whitehead and Crisman (1978) detected changes in the productivity of two small lakes in Massachusetts, U.S.A. using cladocerans, diatoms, pollen, sediment chemistry, sedimentary pigments, and changes in lake bathymetry. They inferred that some changes were caused by climatic and vegetational changes, but that others reflected short-term changes in the watershed, namely the decline of *Tsuga* in the surrounding forest about 4800 B.P., and shallowing in the lake leading to continuous water circulation. Changes in the cladoceran communities during these changes in productivity (Crisman and Whitehead, 1978) reflected the gradual colonization and population development associated with an increase in diversity through the late-glacial and the early Holocene, and several other processes, such as the relatively greater increase in planktonic *Bosmina* than in littoral cladocerans following an increase in nutrients, changes in the pattern of organic influx, such as leaves, leading to an increase in epipelic forms, gradual or sudden changes in the water level affecting the abundance of macrophytes, and changes in predation pressures, particularly by *Chaoborus* and fish. A replacement of *Bosmina coregoni* by *B. longirostris* could not be fully explained in terms of eutrophication. They suggest an increased littoral zone may have reduced the totally planktonic *B. coregoni*, and encouraged the facultative planktonic-littoral *B. longirostris*. In addition, predation by *Chaoborus* may have lead to selection for reduced size, or different sized particles may have been selectively sedimented at the sampling point.

Other organisms in lake sediments

Molluscs and beetles have been studied from lake sediments, but because many of the species live in terrestrial or marsh habitats they will be discussed separately in Chapter 7. Other purely aquatic animals which have been studied include fish, chironomid larvae, and ostracods.

Fish

Fish fossils are rarely recorded. Some fish vertebrae and scales were recovered from sediments of Devensian late-glacial age from Esthwaite Water and identified as either trout (*Salmo trutto*) or char (*Salvelinus* sp.) by Pennington and Frost (1961). Only trout occurs in the lake today.

Chironomid (midge) larvae

We have already mentioned a few studies on chironomid larvae in conjunction with cladoceran studies. Frey (1964) and Stahl (1959) thoroughly review the literature on chironomids and their palaeoecology is also reviewed by Stahl (1969). Midge larvae live in the bottom mud of lakes. Different species are sensitive to different levels of oxygen availability. In oligotrophic lakes with an oxygenated hypolimnion, *Tanytarsus* predominates, in mesotrophic lakes with a poorly oxygenated hypolimnion, *Stichtochironomus* and *Sergentia* predominate, and *Chironomus* predominates in eutrophic lakes with an anaerobic hypolimnion. *Chaoborus* (the phantom midge) larvae can withstand the greatest degree of oxygen depletion, and this species is therefore a very useful palaeoecological indicator. Alhonen and Haavisto (1969) have used the remains of midge larvae to reconstruct the history of a small lake in southern Finland, and they suggest a change from eutrophic to oligotrophic conditions during the Flandrian.

Ostracods

Ostracods are minute crustaceans (Delorme, 1967), but in contrast to the Cladocera, their shells are calcified. Therefore they are best preserved under calcareous or saline conditions where the action of carbonic acid is minimal. Consequently, they are rare in typical organic lake mud, but they can be abundant in marls, and they can sometimes comprise almost the whole of the sediment.

Ostracods have been relatively little studied (Frey, 1964) until recently, when Delorme (1969) by analysing the environmental conditions of modern Canadian ostracod assemblages has demonstrated their usefulness as ecological indicators. They are particularly sensitive to the concentration of dissolved solids in the lake water, each species having a distinct range of tolerance, from very dilute 'soft water' lakes to extremely concentrated 'saline' lakes.

Indeed, many ostracods are marine. Ostracods are also sensitive to a lesser degree to the depth and permanence of the lake or stream, and to climate, in as much that it determines the type of vegetation (e.g. forest or prairie) round the lake, and the temperature and evaporation of the water. Delorme (1971a,b) and Delorme *et al.* (1977) have used ostracods to reconstruct past palaeoecological and climatic conditions from fossil assemblages.

Trace fossils

Gibbard and Stuart (1974) demonstrated traces of animals in Middle Pleistocene pro-glacial lake sediments north of London. They consisted of various types of tracks and burrows preserved in laminated silty lake muds. Gibbard and Stuart tentatively assigned them to various invertebrates, such as an isopod crustacean, a gastropod mollusc, chironomid midge larva, and other worms and arthropods. The muds are otherwise sterile, and this is the only evidence for a fairly abundant fauna in proglacial lakes. Gibbard (1977) has subsequently reported fossil tracks, possibly formed by benthonic insect larvae, in pro-glacial silts and clays in southern Finland.

Lake sediments in palaeolimnology

The animal and plant fossils found in lake sediments are the main source of information on the environment and the ecology of a lake in the past. However, the sediments themselves are also an important source of information. We have already discussed the importance of the composition of sediments in Chapters 3 and 4. However, sediments can be subjected to much more refined methods of study, many of which lead to important conclusions about the origin of the sediments, the environment in which they were deposited, and events occurring in the catchment area.

Sedimentary structures

1. *Laminations*

Lake sediments are frequently laminated. The laminae may be merely colour variations, or they may be due to structural features. The annual laminations due to cyclical diatom deposition described by Benda (1974) and Simola (1977) have already been mentioned on p. 94. Many lakes in glacial environments deposit laminated sediments. In response to the melting of ice and more active inflow of meltwater in spring and summer, relatively course sediments are deposited, followed by fine sediments during winter when the lake is ice covered. These annual structural laminations are called varves. They have been particularly studied in southern Scandinavia, where they were laid down in lakes during the ice retreat at the end of the last glaciation. A long varve chronology has been built up. Cross correlations between lakes were made by matching the relative thickness of the varves in the sequences, which corresponded to annual variations in temperature.

Sometimes very deep (>30 m), steep-sided meromictic (p. 88) lakes in sheltered situations deposit annually laminated sediments, due to the seasonal differences in the input to the sediments being preserved in the undisturbed conditions, even during overturns. These laminations in organic sediments provide an absolute chronology independent of radiocarbon dating, and hence they provide a useful check on the latter. Stuiver (1971) radiocarbon-dated the annually laminated sediments of Lake of the Clouds in Minnesota, U.S.A., and found close agreement between radiocarbon ages and varve ages for the last 2500 years. For earlier times, the radiocarbon ages are consistently younger by as much as 700 years. This departure of the ^{14}C time scale from the lamination count closely resembles its departure from absolute years as measured by tree-ring counts as far back as tree rings can go, about 7400 years, suggesting a systematic deviation in the $^{12}C:^{14}C$ ratio in the atmosphere before 2500 years ago.

There have been several interesting studies on laminated lake sediments. In many cases the laminations have been proved to be annual. For example, Tippett (1964) described laminations in lakes in Ontario, Canada, where the light parts were deposited in the summer, as proved by the types of tree pollen and diatoms they contained, and the dark parts were deposited in the autumn and winter. The light summer colour was produced by relatively large deposition of calcium carbonate, resulting from photosynthesis of aquatic plants in high pH conditions (see also Brunskill, 1969). Another cause of laminations may be eutrophication; the resulting anoxic conditions leading to meromixis, and the preservation of large seasonal diatom blooms which form a relatively thick pale layer of frustules (e.g. Ludlam, 1976; Vuorinen, 1976). Decreased organic deposition during winter months can lead to the formation of clay-rich layers (e.g. Digerfeldt *et al.*,

1975). At Lake of the Clouds, Minnesota, the light bands contain iron oxides formed during increased oxygenation of the water during spring and autumn overturns, whereas the dark layers contain more organic matter produced in the summer (Anthony, 1977).

In several cases, the annual nature of the laminations has been confirmed by correlations of stratigraphically distinct layers, such as clay layers, with events in the watershed, for example bridge or road building and lake drainage (Digerfeldt *et al.*, 1975; Saarnisto *et al.*, 1977). However, at Lake of the Clouds, the annual laminations have been used to date events in the catchment which have no historical record. Craig (1972) studied changes in the vegetation as revealed by pollen analysis over the last 10 000 years, and Swain (1973) studied the fire history of the area over the last 1000 years by means of pollen analysis and charcoal quantities in the sediments (see Chapter 10). Swain (1978) and Cwynar (1978) also used laminated sediments to study fire and climatic history in northern Wisconsin and Ontario, respectively.

In Hoxnian interglacial sediments at Marks Tey in Essex, England, Turner (1970) used presumed annual diatom-rich laminations to determine the relative times of some of the vegetational changes, including a short phase of reduced forest cover possibly brought about by Acheulian man.

In Finland, Saarnisto *et al.* (1977) measured the sedimentation rate in Lake Lovojärvi using the laminations, and found that periods of slow sediment accumulation corresponded quite closely in age with periods of glacial advance in Scandinavia, back to 1590 A.D. They suggest that relative lamination thickness can be an indication of climatic conditions, narrower layers being laid down in colder conditions, especially during the summer.

2. *Other structures*

Other structures may be present in sediments apart from obvious laminations. Calvert and Veevers (1962) first demonstrated that objects such as shells can be detected beneath the surface of the sediment core by X-radiography, and that structures poorly visible or undetectable in ordinary light can be revealed. Subsequently X-radiography has been a useful rapid technique for detecting sedimentary horizons in cores which can then be correlated over a lake basin. For example, Karlén (1976) used X-radiography to detect silty and more organic layers in sediments in a lake receiving outflow from a nearby glacier in northern Sweden. Edmondson and Allison (1970) demonstrated characteristic laminations not visible to the naked eye in recent sediments of Lake Washington, U.S.A. The strongest lamination could be assigned to the effect of lowering the lake level in 1916. An ingenious method of X-raying cores of soft sediment before they are distorted by extrusion from the core tube involving the use of a rectangular core tube is described by Axelsson and Händel (1972).

The chemical composition of lake sediments

The chemical composition of lake sediments is a reflection of the substances that have accumulated, their transformation by biological and purely chemical processes at the sediment surface, and the exchange of substances between the sediment and the water.

Inorganic chemistry of lake sediments

Evidence is accumulating that in most lakes with active inflow and outflow streams, the sediment is largely derived from the soils of the catchment area, with a greater or lesser amount being contributed by internal lake productivity. Therefore, changes in the chemistry of the lake sediments will be a reflection of changes in the catchment area related to differing rates of soil erosion and leaching.

Basic work on lake sediment chemistry was done by Mackereth (1965, 1966) using sediments from the English Lake District. Original work by Pearsall *et al.* (1960) had suggested that lake muds were derived from the fine fraction of soils eroded from the catchment. Mackereth confirmed this view. He argued that sodium and potassium were very soluble, and therefore that any sodium or potassium in the ***mineral*** fraction of lake sediment would be derived from mineral soil erosion, as any soluble sodium or potassium would have been leached from the soils and be associated with the organic and water component of the sediment. He found that the amount of sodium and potassium plotted against depth corresponded to periods of erosion and soil stability as interpreted from pollen analysis (Fig. 6.10). In addition, there was an excellent relationship between the amount of mineral matter in the sediment and the amount of sodium and potassium in the mineral matter component. The fact that the carbon curve with

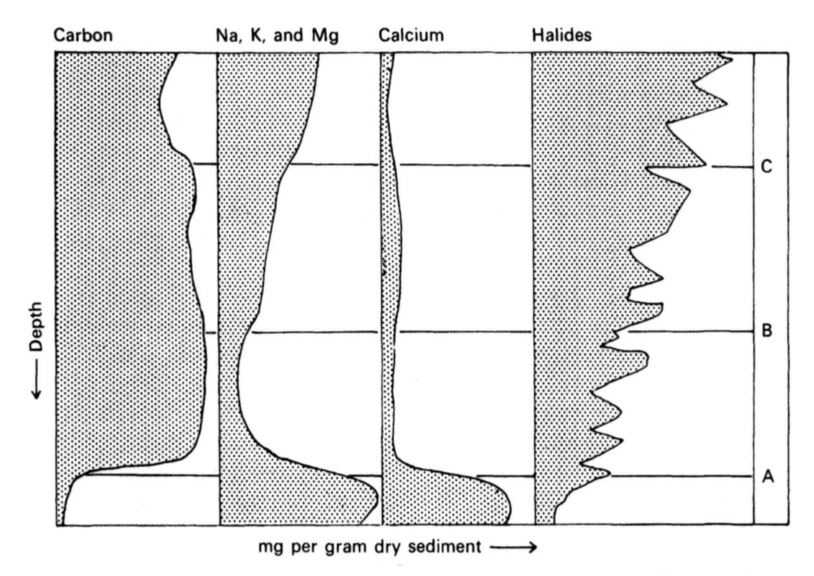

Fig. 6.10 Generalized curves for carbon, sodium, potassium and magnesium, calcium, and halides in sediment cores from Lake District Lakes. See discussion in text. **A** is the end of Devensian late-glacial; **B** is the onset of Neolithic forest clearance; **C** is increased landscape disturbance since the Norse invasions. (After Mackereth, 1965; 1966.)

depth was a mirror image of the mineral matter curve, and also the sodium and potassium, and, to a certain extent, magnesium curve, indicated therefore that the absolute amounts of carbon were varying little, and that its proportion was being controlled by the amount of mineral matter being added to the sediment.

Mackereth discovered that calcium behaved somewhat differently. It appeared to be more easily dissolved from mineral matter. It was only present in high concentration in sediments deposited during periods of rather intense erosion, particularly at the end of the glaciation before forest was established in areas of drift containing a high proportion of calcium. Otherwise, calcium appears to be steadily leached from the soils in relatively low quantities.

From his data, Mackereth could interpret events in the catchment areas of his lakes during the Flandrian. During the Devensian late-glacial, before about 10 000 B.P. the landscape was incompletely vegetated, and the soils were undeveloped and contained little organic matter. As forest development proceeded after the Flandrian amelioration of climate, the soils became stabilized, and leaching was the predominant source of ions and humus compounds in lake sediments. The change from erosion to leaching processes accounts for the rise in carbon content of the sediments at A in Fig. 6.10. At about 5000 B.P. Neolithic people started to clear the forests. This is reflected by the fall in carbon and rise in sodium and potassium at B in Fig. 6.10. These trends continued as men continued to disturb the landscape, and were particularly intensified at about 1000 B.P., at level C in Fig. 6.10, corresponding to the Norse settlement of the area.

Mackereth argued that the only source of halides in lake sediments was from sea spray carried by clouds and deposited by rain. Therefore, an increase in halide should reflect an increase in rainfall. However, he showed that such a simple explanation is not sufficient. Halides, particularly iodine, become bound in the organic fraction of the soil, particularly in acid peaty soils and bogs. Iodine reaches the lake sediments by inwash of these soils, reflecting first the development of acid and peaty soils in the catchment, and later the erosion and inwashing of them. Pennington and Lishman (1971) compared iodine profiles from the English Lake District and Scotland in conjunction with pollen analyses and found that the iodine content and the iodine : carbon ratio did not correlate with differences in rainfall between the sites, but rather with the type of soil and changes it had undergone, such as deforestation and peat development.

Ericsson (1973) attempted to recognize former saline conditions by analysing the exchangeable cations in sediment profiles. During sea-level changes in the Baltic Sea, inlets and low-lying lakes have been periodically cut off and become freshwater, or flooded by sea-water during a marine transgression phase. Salinity changes are readily detectable by diatom analysis. This, however, is a laborious process requiring a great deal of skill in the identification of the diatom frustules. It would be simpler and quicker to perform chemical analyses. Unfortunately, sodium and potassium behaved very irregularly, potassium particularly showing an association with the mineral fraction of the sediment rather than with salinity changes. Magnesium showed good correlation with salinity changes, but calcium was not so well correlated. Therefore an analysis of magnesium through a profile gives a reasonable indication of salinity changes, which can then be confirmed and characterized by diatom analyses at selected levels in circumstances where diatoms are present (e.g. Alhonen, 1971).

In lake basins with no inflow or outflow, the sediment chemistry reflects the events occurring *in* the lake to a greater degree than in the types of lakes we have already discussed, with active inflows and outflows. For example, high amount of carbonate and chloride or sulphate may be present in sediments deposited during arid conditions, when evaporation exceeded precipitation. In Linsley Pond, Connecticut, U.S.A., Livingstone (1957) and Livingstone and Boykin (1962) analysed the phosphorus content of the sediments, and showed that high quantities were present in the lower third of the sequence. Earlier work had shown that the productivity of Linsley Pond was low when the pond was young, some 10 000 years ago, rose slowly to a high level, and remained high ever since. The earlier workers supposed that the changes in productivity were controlled internally by reactions within the living ecosystem, the biomass. However, Livingstone and Boykin demonstrated that phosphorus, an element essential for the growth of organisms, is probably adsorbed and bound to mineral material. In the early Holocene, much mineral material was being washed into the lake as the landscape gradually became stabilized. Phosphorus would be quickly buried before it could equilibrate with the lake water and the large amount of mineral material would ensure that it was tightly adsorbed. It would be trapped in the mud, which would then contain relatively high values of phosphorus. With the stabilization of the surrounding soil, the sediment became more organic, thus allowing more phosphorus to be released into the lake water, which could then support a greater biomass. With the decrease in mineral input to the sediment the phosphorus would be readily recycled and it would be able to support large populations of organisms in the lake.

Another attempt to study past lake productivity and its relation to sedimentation and events in the catchment area was made by Likens and Davis (1975) at Mirror Lake, New Hampshire, U.S.A. They had a good radiocarbon chronology for the sediments back to 11 000 B.P. and calculated the absolute influx of various substances through time. The highest influx rates were found at the base of the profile, associated with mineral inwash at the end of the Wisconsinan glacial stage. Low but increasing influxes of organic matter and pigments suggested that lake productivity was low but increasing. In conjunction with evidence of the vegetation of the catchment from pollen analysis and from previous work upon the effects of experimentally deforesting the catchment (Likens *et al.*, 1977), Likens and Davis were able to deduce changes in the productivity of Mirror Lake through the Holocene. However, they emphasized that it is very important to consider as many lines of evidence as possible, as one or a few parameters may give a distorted impression. Because forest largely controls the input of nutrients into a lake, they hoped that differences would be detectable during the major changes of forest composition. However, they were not, and they recommend that evidence from biological fossils, such as diatoms, may give a much more sensitive indication of lake conditions in the past than chemical analyses of the sediments.

Chemical analyses of recent lake sediments

As man has modified his environment at an ever increasing rate, so lakes have registered changes in their sediments. Chemical analyses can detect the changes and estimate their magnitude. Some of the changes can be dated by radionuclides or by widespread pollen stratigraphical changes. Other changes in the lake sediments may then be dated, and perhaps assigned to a specific cause.

1. *The dating of recent lake sediments*

i) The use of ^{14}C is difficult in recent sediments because of the relatively large error in the measurements (standard deviation of about 100 years), and because large amounts of ^{14}C-deficient carbon have been released into the atmosphere as a result of the burning of fossil fuels since the Industrial Revolution (the so-called Suess effect).

ii) Pollen stratigraphical horizons are useful markers. In north-west Europe, there is a widespread decline of *Ulmus* pollen dated consistently to about 5000 B.P. In North America, the colonization by European man has been recent, and it is marked by an increase in *Ambrosia* pollen. The *Ambrosia* rise can be dated by historical records in different areas as colonization proceeded westwards. In eastern North America, a well-documented widespread decline of *Castanea* pollen resulted from a fatal epidemic disease caused by the blight *Endotheca parasitica* (Anderson, 1974). Other pollen horizons have been correlated with historical events, but they tend to be of more local significance.

iii) Caesium-137. This isotope was first introduced into the atmosphere as a result of nuclear testing in 1954. Since then, the amount of ^{137}Cs in undisturbed sediments has followed the course shown in Fig. 6.11. Caesium-137 reached maximum levels in 1963, since when there has been a decline. Caesium-137 is washed out of the atmosphere by rain, and it becomes incorporated into lake sediments within 6–12 months of its formation. Caesium-137 has become a useful method of dating recent sediments (e.g. Ritchie *et al.*, 1973; Battarbee, 1978).

iv) Lead-210 can be used to date sediments up to 150 years old. The method is well explained by Brugam (1978a). Briefly, radium-226 decays in soils to radon-222 which escapes to the atmosphere. Here it decays to lead-210, which is washed out by precipitation, and eventually incorporated into lake sediments. This is known as unsupported ^{210}Pb. Supported ^{210}Pb is produced within the sediments, and it is assessed by measuring its ^{226}Ra grandparent. The excess ^{210}Pb over the expected ^{210}Pb gives the amount of unsupported ^{210}Pb. The half-life of ^{210}Pb is 22.26 years. By assuming either a constant initial concentration in sediments which have accumulated at a constant rate, or by assuming a constant rate of ^{210}Pb supply to the sediments, the age of the sediment can be calculated. Oldfield (1977) and Oldfield *et al.* (1978) have demonstrated by using independant dating horizons, that the second model gives more accurate dates when the rate of sedimentation has changed, which is frequently the case in lakes which have been affected by eutrophication.

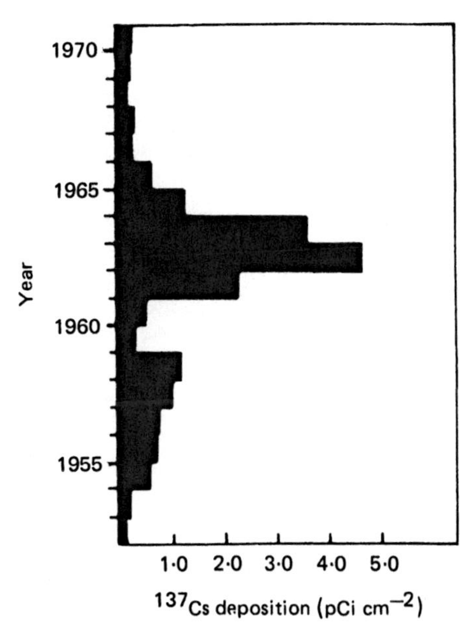

Fig. 6.11 Idealized annual deposition of ^{137}Cs in sediments of Windermere, English Lake District. (After Pennington *et al.*, 1973.)

v) Thorium-228 : Thorium-232. This isotope ratio can be used to date sediments up to 10 years old. Bruland *et al.* (1975) used it to determine whether modern sediment was present in cores from Lake Superior, U.S.A. They could then profitably date the sediment using ^{210}Pb and by the *Ambrosia* pollen rise.

vi) Palaeomagnetism. Mackereth (1971) measured the intensity and direction (declination) of the ***remanent magnetism*** in sediment cores from Windermere. The horizontal component of the magnetic field has oscillated about a mean direction with an amplitude of about 20° and a frequency of 2700 years. The time scale was measured by a series of radiocarbon dates. Mackereth showed that palaeomagnetism could be used to correlate and date other cores. Creer *et al.* (1972) showed that most of the remanent magnetism was in haematite produced by chemical changes as the sediment was deposited. They confirmed the potential of palaeomagnetism for dating by showing that the movements of the geomagnetic north pole coincided with its movements recorded in London since 1580.

Thompson (1973) confirmed Mackereth's and Creer *et al.*'s results from Lough Neagh, N. Ireland. Subsequently Tolonen *et al.* (1975) showed that the palaeomagnetic curve from Lake Lojärvi in Finland could be correlated with the English results, and they were able to correct radiocarbon dates which were suspected of being too old due to inwash of old carbon in soils during an Iron Age clearance. The palaeomagnetic dating corresponded well with archaeological and pollen dates, and the date of land uplift which separated the lake from the Baltic Sea.

Thompson *et al.* (1975) measured the ***magnetic susceptibility*** of sediments from Lough Neagh. This is the extent to which the sediment can be magnetized. The most susceptible sediments contain the most allochthonous inorganic material, washed in from soils in the catchment as a result of forest clearance and cultivation. Cores from one lake could be correlated easily and quickly, even before extrusion from the core tubes. This method does not, of course, date sediments.

2. *Mercury in recent lake sediments*

About half the mercury in the present environment has been released by man's activities, including denudation of land surfaces, heavy industry, mining and quarrying, burning of fossil fuels, and sewage disposal. Mercury is naturally present in the environment, derived from rocks. Fredriksson and Qvarfort (1973) showed that mercury levels in lake sediments in an area of Sweden unaffected by industrial pollution had remained constant through the Holocene. The actual amounts varied between lakes due to differences in the bedrock. Aston *et al.* (1973) analysed mercury from the upper sediments of Windermere, England, and discovered sharp increases over the last 150 years. The sediments were dated using the ^{210}Pb technique, and the results are shown in Fig. 6.12. Up to about 1300 A.D., there was a low background level of mercury. The small increase up to 1800 probably resulted from increasing farming activity in the catchment. As a result of the Industrial Revolution, the mercury level doubled in less than a century, and it has doubled again in the last 50 years. Windermere is in a rural setting, and the mercury has probably been supplied in rainfall and to a greater degree by the increasing input of sewage and of engine use on the lake.

If levels of mercury are sufficiently high in lake waters, it is concentrated by the lake organisms, and it can reach sufficient levels in fish to be poisonous to human beings who eat them.

3. *Studies on recent eutrophication*

Many lakes have suffered in the last century from alterations in their water quality as a result of increased human activity in the watershed. Generally, there is an increased inwash of soil, and of sewage if there are habitations around the shore, and perhaps also of industrial effluents. The water becomes richer in nutrients such as nitrate and phosphate, plant and animal growth accelerates, and as organisms die, oxygen is increasingly removed from the water by the bacteria responsible for decomposition. In serious cases, the hypolimnion becomes anaerobic, resulting in the death of fish, and the epilimnion supports great blooms of algae, mainly diatoms and blue-green algae, which make the water murky and smelly.

Bortelson and Lee (1972) studied the sedimentary record of the progressive eutrophication of Lake Mendota, Wisconsin, U.S.A. A buff marl recorded stable conditions before the arrival of European settlers in 1820. This horizon is adequately dated by the rise of *Ambrosia* pollen. Until 1880, a grey mud was formed containing increased amounts of organic-C, Fe, Mn, K, Al, and P. As Madison became urbanized and other smaller towns sprang up, the nutrient levels were considerably raised, leading to the deposition of black mud at a greatly increased rate which contained high levels of Fe, Mn, Al, and P.

There has been much concern over the deteriorating quality of Lake Erie in the Great Lakes System

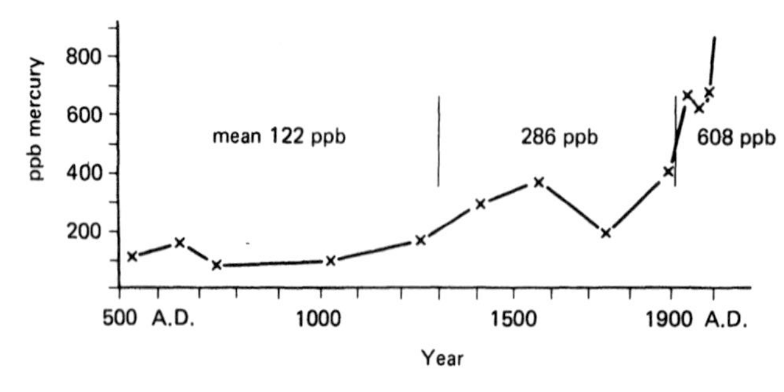

Fig. 6.12 Mercury in recent sediments of Windermere. (After Aston *et al.*, 1973.)

of North America. Kemp *et al.* (1974) analysed Lake Erie sediments and compared them with sediments from the less polluted Lake Ontario, and the scarcely polluted Lake Huron. They dated the sediments by the *Ambrosia* pollen rise (1850 in this region) and the decline of *Castanea* pollen (between 1930 and 1935 in this region). They confirmed their time scale by ^{137}Cs measurements. They discovered that there had been a three-fold increase in sedimentation rate in Lake Erie since 1935, and in the relatively small Kingston Bay of Lake Ontario, sedimentation rate had also increased, particularly since 1950. However, rates in Lake Huron had remained more or less constant. Their results show how eutrophication and pollution can effect even the largest lakes to such an extent as to give rise to public concern. Beeton (1969) also discusses the effects of the pollution of the Great Lakes.

Pennington *et al.* (1973) used ^{137}Cs to date recent sediments from five lakes in the English Lake District. Table 6.3 shows that, although all the lakes had comparable sediment accumulation rates during the Flandrian, the recent rates had increased greatly in the two eutrophic lakes, Esthwaite Water and Blelham Tarn. There was less of an increase in Windermere, and virtually none at all in the oligotrophic lakes Ennerdale Water and Wastwater, when factors such as the consolidation of the sediments are taken into account (Pennington, 1974). They concluded that the increased sedimentation must result from the addition of organic material to the sediment from internal lake productivity in eutrophic conditions.

Table 6.3 Past and present sediment accumulation rates (cm $year^{-1}$) in five Lake District lakes. (From Pennington, 1974.)

	Mean for last 10 000 years	Mean since 1954*	Mean since 1963*	Annual sediment trapped†
Esthwaite Water	0.048	0.85	0.83	2.16
Blelham Tarn	0.045	0.95	0.64	1.67–1.93
Windermere	0.032	0.48	0.36	0.26
Ennerdale Water	0.058	0.43	0.22	0.41
Wastwater	0.030	0.25	0.16	0.25–0.33

* Based on ^{137}Cs levels in sediment cores (see Pennington *et al.*, 1973).

† Based on sediment trapping for periods of at least one year (see Pennington, 1974).

The recent increase in sediment accumulation rate in Windermere was further studied by Pennington (1973). She measured the rate by four different methods: (i) Collection of annual sediment increment in sediment traps. (ii) The amount of sediment accumulated above a distinct stratigraphic horizon between brown mud and black ooze. (iii) The position of the westernmost magnetic deflection in 1820. (iv) The amount of ^{137}Cs. All four methods gave similar results, which were further confirmed by ^{210}Pb determinations. The mean Flandrian rate was 7–10 times less than the recent rate. The increase coincided with a rise in the diatom *Asterionella formosa*, suggesting that the cause lay in increased lake productivity following eutrophication.

Pennington (1974) followed up these results by measuring the seston (living, dead, and inorganic particles suspended in lake water) in five Lake District lakes. From seston trapped throughout the year, she demonstrated considerable resuspension of fine sediment at the autumn overturn. When the recent sediment accumulation rate was calculated from the annual seston catch, it mostly exceeded the Flandrian rate. The greatest difference was found in the eutrophic Esthwaite Water and Blelham Tarn, where the hypolimnion becomes deoxygenated, in contrast to the other lakes (see Table 6.3). As a result, less of the algal matter is oxidized, and it becomes incorporated into the sediments.

In a detailed study on recent sedimentation in Blelham Tarn, Pennington *et al.* (1976) calculated sediment accumulation rates in five cores from different parts of the lake basin, using ^{210}Pb, ^{137}Cs, and ^{14}C, in conjunction with two stratigraphic markers which are present throughout the lake, but at different depths. A change from brown to black mud was consistently dated at 1930, and a paler layer in the brown mud at 1860–70. The ^{210}Pb dating showed that this layer had been sedimented very rapidly, as a result of inwash of material. The most likely cause was the diversion of one of the inflow streams at that time. An important limnological conclusion was that sediment is not deposited evenly over the lake bottom, but is differentially transported to deeper areas. Thus, in one year, sediment of uniform composition is deposited in differing thicknesses depending upon the morphometry of the lake basin. Similar conclusions were reached in America by Davis (1976) at Frains Lake, Michigan, and at Mirror Lake, New Hampshire by Likens and Davis (1975).

Table 6.4 Soil erosion rates before and after deforestation.

Frains Lake, Michigan (Davis, 1976)	Pre-settlement (10 000 B.P.–1830 A.D.) 9 tonnes km^{-2} $year^{-1}$	Post-settlement (1830–1976 A.D.) 90 tonnes km^{-2} $year^{-1}$
Hubbard Brook, New Hampshire (from Bormann *et al.*, 1974)	Forested landscape 2.5 tonnes km^{-2} $year^{-1}$	Deforested landscape 38 tonnes km^{-2} $year^{-1}$

Lake sediments and the history of the watershed

The majority of inorganic material in lake sediments originates in soils in the watershed. Davis (1976) used this fact to estimate the rate of soil erosion in the watershed of Frains Lake, Michigan, U.S.A. under different kinds of land use.

The area was settled in 1830, and this is reflected by a well marked rise in *Ambrosia* pollen. This consistently occurs at a prominent sedimentary change from lake mud to clay-rich mud. Frains Lake has no inflows or outflows, and is therefore an effective sediment trap. Davis measured the amount of sediment deposited in the last 146 years over the lake by identifying the settlement horizon of 1830 from the sediment change, and by pollen analysis. The horizon was identified in cores from a neighbouring lake by X-ray analysis. The rate of deposition of the pre-settlement sediment was measured by means of radiocarbon dates.

The sedimentation in Frains Lake was complicated by the fact that it was some ten times greater in a small central depression than in the rest of the lake (compare results from Blelham Tarn, p. 105). However, the sediment composition in the depression appeared to be comparable to the thinner layers elsewhere.

Davis estimated the volume of sediment deposited since 1830, measured its ash content, and hence estimated the annual influx of inorganic material. By knowing the area of the watershed, she could calculate the rate of erosion, which was about 90 tonnes km^{-2} $year^{-1}$ (see Table 6.4). Similarly, she estimated that the pre-settlement erosion rate was about 9 tonnes km^{-2} $year^{-1}$. Comparison with the Hubbard Brook Forest, New Hampshire, which was experimentally deforested, is made in Table 6.4. Here there was about 15-fold increase in erosion immediately after felling.

Davis continued her study by investigating changes in the erosion rate since 1830. She calculated a time scale by assuming (i) a constant rate of sediment accumulation, and (ii) a constant pollen influx. Both methods gave similar results, shown in Fig. 6.13. After the initial forest clearance and ploughing, soil erosion increased to 30 times the pre-settlement rate. However, as the land use became stabilized, the erosion rate become stabilized at about ten times the pre-settlement rate.

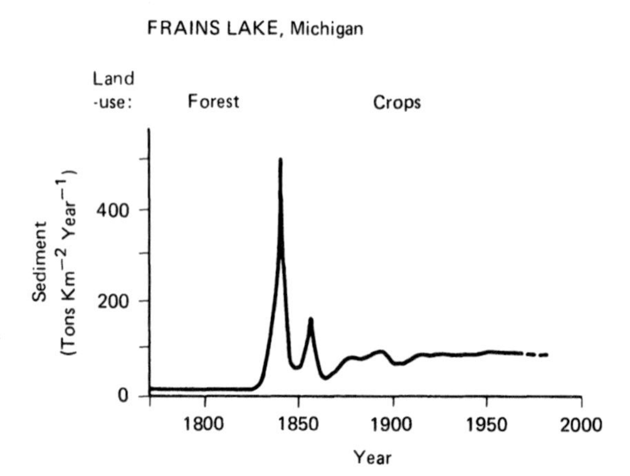

Fig. 6.13 Generalized annual sediment yield to the sediments of Frains Lake, Michigan, since 1800. (After Davis, 1976.)

Organic components of lake sediments

1. *Pigments*

The main organic molecules which have been studied in lake sediments are plant pigments or their derivatives. Pigments may persist in the mud long after any morphologically recognizable remains of the organisms are detectable, and in some cases, they are the only evidence of the previous existence of organisms which are almost never preserved at all, such as most algae and bacteria. For example, Brown (1968) used carotenoid pigment analysis to identify the probable abundance of the bacterium *Rhodopseudomonas spheroides* from Little Round Lake, Ontario.

Large numbers of pigments are extractable and identifiable from lake mud (see Vallentyne, 1960) and several appear to be specific to certain organisms. For example, Züllig (1960) showed that myxoxanthophyll was specific to blue-green algae, and used it to demonstrate the course of eutrophication in three Swiss lakes. Blue-green algae are not preserved as morphological fossils, and they can be very important components of the biota of eutrophic lakes. Similarly, Brown and Coleman (1963) showed that oscillaxanthin was produced only by the blue-green algae family Oscillatoriaceae, and Griffiths *et al.* (1969) examined the vertical distribution of oscillaxanthin in recent Lake Washington sediments, to illustrate the changes in abundance of *Oscillatoria* since sewage enrichment of the lake in the 1950s.

Sedimentary plant pigments are derived either from allochthonous sources in the watershed (green or decaying leaves or soil humus) or from autochthonous sources within the lake (algae, aquatic macrophytes, bacteria). Ideally, pigments can be used to demonstrate:

i) the floristic changes within a lake. Some pigments are specific to different plant groups, but their identification depends on the degree of their preservation and their taxonomic specificity

ii) any changes in abundance of algae and macrophytes in the past, assuming this to be proportional to the amount of pigment preserved

iii) any changes in the physical or chemical environment, which result in changes in the algal population and the proportion of autochthonous and allochthonous input to the sediment.

The last point was investigated by Sanger and Gorham (1970) who analysed the pigments in surface muds, soils, decaying leaves, green leaves, etc. They concluded that pigments in sediments are much more abundant both quantitatively and qualitatively than pigments in material of terrestrial origin, and also than pigments in living aquatic macrophytes and algae. They concluded that decomposing algae on the surface of the lake mud were the most likely source for the diversity of pigments in the sediment. This work was followed up by Gorham *et al.* (1974) who measured algal productivity in a range of oligotrophic, mesotrophic, and eutrophic lakes in the English Lake District. They found that algal productivity was proportional to the trophic status of the lake. The high correlations between algal standing crop, epilimnetic chlorophyll, and the concentration of pigments in the sediment lend support to the theory that sedimentary pigments are good indices of lake productivity. In addition, their data indicated that much of the organic matter in the sediments of productive lakes, especially those such as Blelham Tarn and Esthwaite Water with anoxic hypolimnions, was derived from autochthonous production rather than from the catchment area, and that even in poor oligotrophic lakes with oxygenated hypolimnions, a certain amount of organic matter was derived from within the lake. In general the proportion of chlorophyll derivatives to carotenoid derivatives in sediment is indicative of its place of origin. Terrestrial material is rich in chlorophyll derivatives whereas lake organisms produce a higher proportion of carotenoids, and a much greater diversity of pigments.

Stratigraphic studies using pigments in lake sediments. Fogg and Belcher (1961) extracted and measured the chlorophylls and carotenoids in a core from Esthwaite Water. They showed a maximum of lake productivity in the early post-glacial, followed by a general decline in pigment concentration indicating a more or less stable lake productivity. After disturbance of the landscape by Neolithic man, and his successors, pigment concentration fell, indicating a greater allochthonous input to the sediments. Recently, sedimentary pigment concentration rose as a result of lake eutrophication, and the excellent preservation of the pigments indicated deoxygenation of the hypolimnion.

More recently Sanger and Gorham (1972) measured the concentration and diversity of pigments in a core from Kirchner Marsh, Minnesota, in order to test the use of pigments as palaeoecological indicators in a situation where a considerable amount was already known (e.g. Watts and Winter, 1966; p. 74). They deduced that lake productivity reached a peak in the early Holocene about 10 000 years ago. Productivity then declined to a minimum and the proportion of chlorophyll to carotenoid increased about 6000 years ago during zone C-b, when the lake shallowed and become surrounded by annual communities during the warm dry 'prairie' period. During the succeeding zone C-c aquatic conditions were gradually restored, but large fluctuations in the concentration of pigments and the chlorophyll : carotenoid ratio indicate an unsteady rise, and perhaps a considerable inwash of allochthonous material as the lake level rose. Gradually, reedswamp encroached, and at about 1700 B.P. a sharp decrease in concentration and numbers of pigments associated with a sharp rise in the chlorophyll : carotenoid ratio indicated final overgrowth of the coring site.

2. *Other organic molecules*

i) *n*-alkanes

Cranwell (1973) demonstrated that *n*-alkanes derived from the lipid fraction of higher plants consist of molecules of carbon chain lengths C_{23}–C_{35}, with a predominance of odd numbers of carbon atoms. In contrast, lower plants produced *n*-alkanes with no preference towards odd numbers of carbon atoms. Chains of C_{31} were dominant in peat and mor humus formed in acidic environments, and chains of C_{27} and C_{29} predominated in mull humus characteristic of deciduous forest, and formed at neutral pH. The two types of humus can also be characterized by their electron spin resonance (ESR) (Atherton *et al.*, 1967).

Cranwell found that sediments could also be categorized by the chain lengths of *n*-alkanes, depending upon the type of humus from which they were derived. Sediments containing mor humus washed in from the catchment contained *n*-alkanes of predominantly C_{31} chain length and those containing inwashed mull humus had *n*-alkanes dominated by C_{27} and C_{29} chain lengths. These results were confirmed by measuring the ESR of the sediments. The results compared perfectly with deductions made about the soils of the catchment at the time the sediment was deposited by means of pollen analysis. Therefore an analysis of the carbon chain length of *n*-alkanes in lake sediments is useful in determining the type of humus in the soils of the catchment area.

ii) *n*-alkanoic acids, branched/cyclic alkanoic acids, and alkanoic acids.

Cranwell (1974) isolated these three types of unsaturated fatty acid from the monocarboxylic acid fraction from soils, peat, and lake sediments. He found that autochthonous material is the main source of C_{12}–C_{18} *n*-alkanoic acids, whereas C_{22}–C_{28} *n*-alkanoic acids are derived from terrestrial sources. Therefore the relative amounts of these two groups indicates the proportion of autochthonous and allochthonous input to the sediments. Not surprisingly, he discovered that there was a greater proportion of autochthonous material in sediments of productive lakes, which both produced more algal material and possessed reducing conditions at the sediment surface which favoured preservation.

Cranwell also demonstrated a high correlation between large amounts of autochthonous *n*-alkanoic acids and branched/cyclic alkanoic acids. The latter are produced mainly by bacteria, which are more abundant in eutrophic lakes. Also correlated were high amounts of alkanoic acids, which are best preserved in reducing conditions associated with productive, eutrophic lakes.

Cranwell used these results to interpret the late-glacial environment of Cam Loch, Scotland, from the alkanoic acids in the sediments. He concluded that Cam Loch was a productive lake with periods of deoxygenation in the hypolimnion. His conclusions are supported by Haworth's (1976a) diatom analyses. This contrasts to the oligotrophic status of the lake at present, associated with a deforested catchment largely covered in blanket peat.

The occurrence of other organic compounds in lake sediments is reviewed by Cranwell (1976), who shows that several may be used as indices of trophic status and of the place of origin of the organic matter.

Coordinated palaeolimnological studies

There have been a number of projects where analyses of sediments and several groups of organisms have been performed on sediments from a single lake. Here we will mention the study on Lake Zeribar, Iran, which was designed to elucidate the environment in which early man first cultivated cereal crops at about 11 000 years ago, and several studies on the effects of eutrophication on lake ecosystems.

Lake Zeribar, Iran

Lake Zeribar occupies a valley in the Zagros Mountains in western Iran. The valley is dammed by alluvial fans, and at present the lake is about 4–5 m deep, and surrounded by a sedge mat, which floats at periods of high water level, when the lake flows over the fans. The sediments are at least 20 m deep, and have accumulated as the lake level gradually rose in response to the build up of the alluvial fans. Figure 6.14 illustrates stratigraphic curves for selected chemical characteristics, and plant and animal fossils.

The pollen diagram (van Zeist and Wright, 1963) shows a change at the end of the Pleistocene at about 11 000 years ago. Using a survey of modern pollen spectra from different vegetation types in Iran, Wright, McAndrews, and van Zeist (1967) (see p. 246) concluded that open treeless vegetation

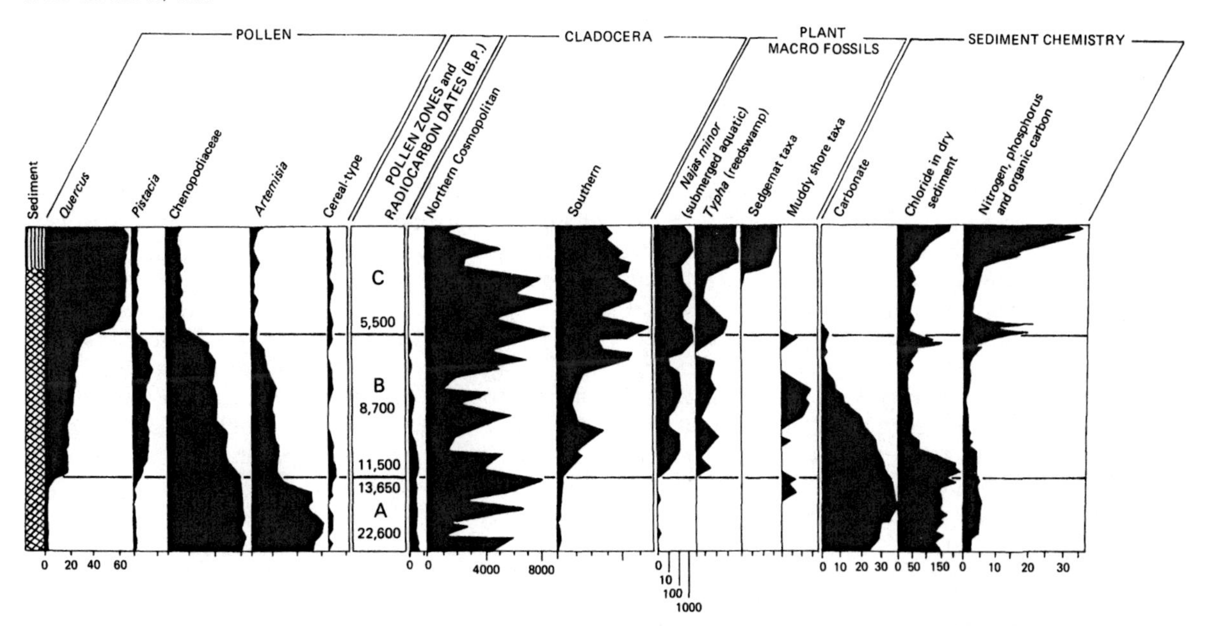

Fig. 6.14 The palaeolimnology of Lake Zeribar, western Iran. Selected generalized curves for plant and animal fossils and some chemical characteristics are shown. Scale for pollen is percentages of *Quercus*, *Pistacia*, Chenopodiaceae, *Artemisia*, and *Plantago*. Scale for cladocera is number of fragments per cm^3 sediment. Scale for plant macrofossils is log number in 100 cm^3 sediment. Scale for chloride is ppm. Scale for other chemical components is percentage. Radiocarbon dates are in radiocarbon years B.P. (After the references listed in text.)

gradually changed to fully developed oak forest between 11 000 and 5000 years ago, as a result of gradually increasing rainfall. This is represented by the *Quercus* pollen curve in Fig. 6.14.

Plant macrofossil analyses (Wasylikowa, 1967) showed that the aquatic macrophyte flora became much more diverse at 11 000 B.P. in response to climatic warming. However, by 5500 B.P. the lake became surrounded by marsh vegetation, and shortly afterwards the floating sedge mat developed. These events were a result of increased rainfall leading to higher water levels.

The cladoceran analyses by Megard (1967) have already been mentioned on p. 98. The Cladocera record the warming at the opening of the post-glacial, with conditions becoming stabilized at about 5000 B.P. The expansion of *Bosmina longirostris* suggests the development of deep open water at this time.

Chemical analyses of the sediments by Hutchinson and Cowgill (1963) showed high concentrations of chloride and carbonate in the Pleistocene sediments, which gradually decreased through the first half of the post-glacial. They concluded that the lake had a greater evaporative capacity during the Pleistocene, resulting in the concentration of ions. This was gradually reduced with increasing rainfall during the post-glacial. This corresponds well with the previously recorded Pleistocene aridity in the Near East and elsewhere. The later post-glacial sediments are littoral, and the chemical analyses are consistent with the formation of the sedge mat and extensive beds of aquatic macrophytes.

Wright (1968, 1976) has summarized the palaeoenvironmental evidence from Lake Zeribar and other sites in the Near East. The climatic amelioration about 11 000 years ago is associated in time with the move of palaeolithic men from cave dwellings to open village sites, where grain was cultivated and animals were domesticated. He proposed that this climatic change also allowed the immigration of wild grain plants, and thus the environmental stage was set for the domestication of plants and animals.

Eutrophication studies

Faegri (1954), Likens (1972), and Hutchinson (1973) have discussed the meaning of eutrophy in lake systems, and agree that the essential feature is a high concentration of nutrients and high pH in the lake water. Likens (1972) defines eutrophication as 'nutrient or organic matter enrichment, or both, that results in high biological productivity and a reduced volume within an ecosystem'. Eutrophication has long been realized to be a natural process, resulting in the 'ageing' of a lake, and its gradual infilling by sediment. When this process is accelerated by the effects of man, it may be called 'cultural eutrophication'. When the results are offensive to man, eutrophication is considered to be a form of pollution. Likens (1972) considers models which may be useful in assessing the processes leading to eutrophy. Faegri (1954) is concerned with the definition of eutrophy and its distinction from other lake types in a classificatory sense. Hutchinson (1973) briefly reviews the evolution of the term eutrophy, and reviews limnological work on ecosystem processes in eutrophic situations, and also palaeolimnological work on eutrophication in the past.

In recent years many lakes have suffered cultural eutrophication as a result of the effects of forest clearance, agriculture, sewage, and industrial effluents. Several integrated palaeolimnological studies have been carried out to determine the course of eutrophication and the effects it has had on the lake organisms. We shall describe four such studies here.

1. *Shagawa Lake, north-eastern Minnesota, U.S.A.*

Changes in the pollen, diatom, Cladocera, and pigment content of the upper sediments of Shagawa Lake have been studied by Bradbury and Megard (1972), Bradbury and Waddington (1973), and Gorham and Sanger (1976). Their results are summarized in Fig. 6.15.

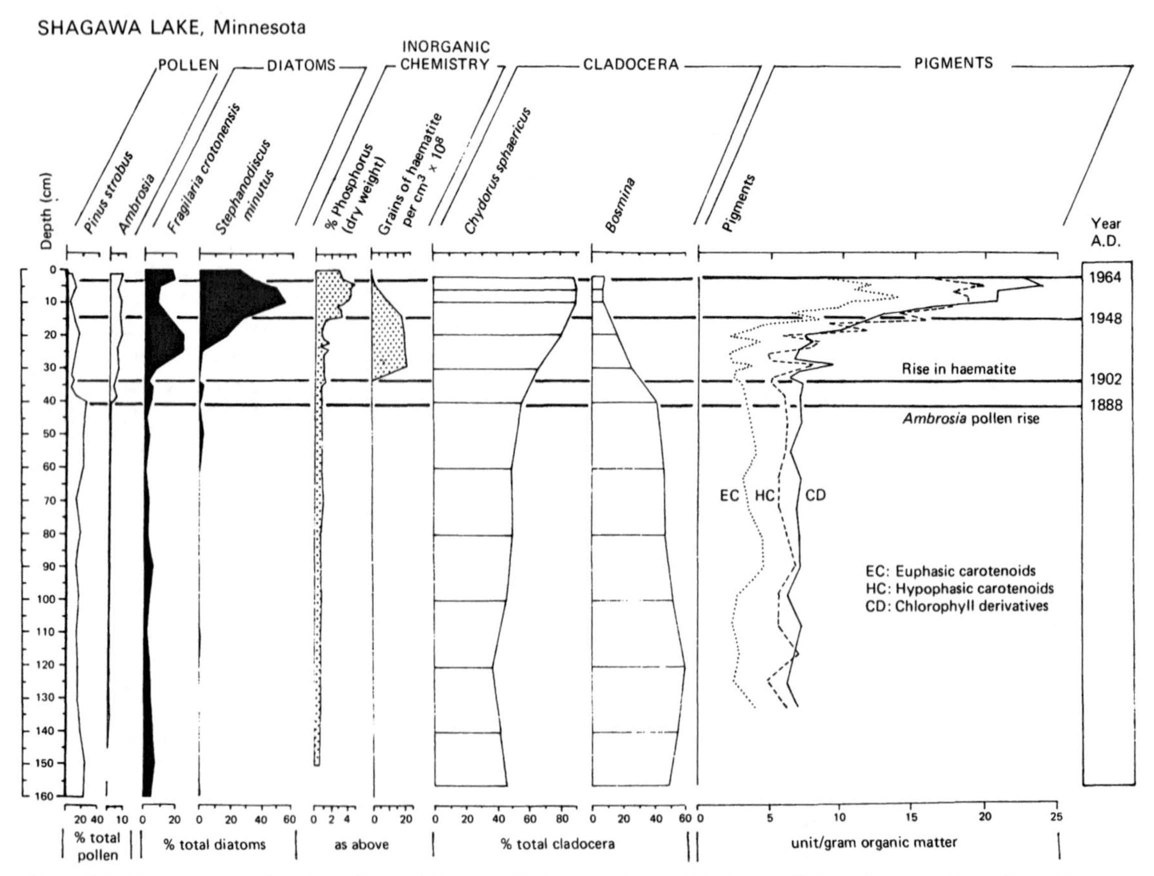

Fig. 6.15 The recent palaeolimnology of Shagawa Lake, north-east Minnesota. Selected curves for pollen, diatoms, cladocera, inorganic components and pigments. CD, chlorophyll derivatives; EC, epiphasic carotenoids; HC, hypophasic carotenoids. (Compiled from Bradbury and Waddington, 1973; and Gorham and Sanger, 1976.)

At 40 cm, *Ambrosia* pollen rises and *Pinus strobus* (white pine) pollen falls, reflecting the first European settlement in the area, with the establishment of Ely on the lake shore, and the logging of white pine in about 1850. At 34 cm, grains of haematite (ferric oxide) begin to be deposited in large numbers, reflecting the opening of iron mines along the southern shore. Algae and *Chydorus sphaericus* begin to rise at this level as a result of nutrient enrichment due to logging and other landscape disturbance.

Between 34 and 15 cm there is a steady rise of *Chydorus sphaericus*, *Fragilaria crotonensis*, *Stephanodiscus minutus*, and the sedimentary pigments, associated with maximal input of haematite. These sediments were laid down between 1900 and 1948 when mining was extensive, and the population of Ely reached its maximum. At 15 cm there is a sharp rise in sedimentary phosphorus, probably originating from synthetic detergents. *Chydorus sphaericus* replaces the planktonic *Bosmina*, and the pigments show sharp increases associated with blooms of blue-green algae. The blue-green algae replaced *Fragilaria*, discouraged the planktonic *Bosmina*, and formed a raft which allowed *Chydorus sphaericus* to become effectively planktonic. *Stephanodiscus* becomes abundant, because it blooms very early in spring, when it can utilize available nutrients before the blue-green algae develop. Also at 15 cm, the amounts of haematite grains decline, reflecting the decrease in mining as the mines gradually became less profitable and were closed. However, developing tourism kept the summer population of Ely at a high level.

The lake remained in this polluted state until 1964, when a secondary sewage treatment plant was constructed. The sediments record a fall in phosphorus, *Stephanodiscus*, and pigments, and a return of *Fragilaria crotonensis*, suggesting some slight improvement in the lake's condition. However, blue-green algal blooms still occur, rendering the lake water offensive in summer.

2. *Three lakes in north-western Minnesota, U.S.A.*

Birks *et al.* (1976) used a combination of pollen, plant macrofossils, diatoms, some other algae, cladocerans, including *Daphnia* ephippia and chydorids, molluscs, and features of lake sediments, in a palaeolimnological study of three lakes in north-western Minnesota over the last 100 years. The rise of *Ambrosia* pollen reflected the onset of European settlement in all the lakes. This date was used to estimate sediment accumulation rates.

Elk Lake was considered as a control: although its catchment has been logged and its level raised by a dam, there has been no settlement on its shore. Sallie Lake and St. Clair Lake are both in the same catchment which was logged and then farmed, and in which the town of Detroit Lakes was established. The shallow St. Clair was partially drained but the larger, deeper Sallie Lake has high amenity value, although at present it is considered to be polluted because of summer blooms of blue-green algae. Both lakes receive agricultural run-off and sewage effluent from Detroit Lakes and shore settlements.

The catchment of Elk Lake was logged in 1890, after which there was regrowth of secondary forest. The lake organisms were affected only slightly with some increase in aquatic macrophytes and reedswamp plants and their associated fauna. The dam built in 1935 prevented fluctuating water levels from adding sand and silt to the sediments, which then became more organic. The aquatic plants near the sampling site decreased in density as a result of the deeper water, and the reedswamp was also reduced.

After the *Ambrosia* rise at 1870, Sallie Lake showed an increase in planktonic diatoms, and an expansion of *Chydorus sphaericus* as a result of an increase in its water-weed habitat. The addition of sewage effluent in 1900 caused greater changes. Aquatic macrophytes became abundant in the shallow water, together with the algae and animals associated with them, particularly diatoms and cladocerans tolerant of eutrophic conditions, such as *Fragilaria* spp., *Stephanodiscus astraea* var. *minutula*, and *Chydorus sphaericus*. The construction of a secondary sewage treatment plant in 1929 reduced the input of organic carbon, but not the input of phosphates and nitrates. Aquatic macrophyte growth was greatly stimulated, and blooms of blue-green algae appeared. As a result the hypolimnion became anoxic and the sediments are black, and rich in sulphides rather than sulphates. Pigments were not studied, but the blooms of blue-green algae are historically documented, and they resulted in the reduction of summer-blooming diatoms. As at Shagawa Lake, spring blooming *Stephanodiscus* increased. In spite of the mechanical removal recently of waterweeds, the lake remains highly productive and offensive in summer, and only a reduction in nutrient input will result in any improvement.

St. Clair Lake is smaller and shallower than Sallie Lake and it was already eutrophic at the *Ambrosia* rise. Forest clearance and farming had only a little effect on its biota. However, its partial drainage in 1915 resulted in very large changes. *Ceratophyllum demersum* became the dominant aquatic macrophyte, and planktonic diatoms greatly exceeded benthic taxa. Pulmonate snails replaced prosobranchs and fingernail clams, indicating a reduction in oxygen and higher water temperatures as a result of the drainage. Amongst the Cladocera, the usual increase of *Chydorus sphaericus* occurred. The variety of organisms studied by Birks *et al.* enabled the changes deduced from the response of a certain set of organisms to be used to interpret the behaviour of another set whose ecology is not so well known. The inclusion of aquatic macrophytes was important, because these provide a habitat for so many other organisms.

3. *Lake Washington, Washington State, U.S.A.*

We have already outlined Stockner and Benson's (1967) work on the diatoms of Lake Washington in relation to its history of eutrophication. In this case, the proportion of araphidinate to centric (A/C) diatoms was a good reflection of the degree of eutrophication.

Various physical, chemical, and biological features have been recorded in Lake Washington since 1930 (Edmonson, 1969a). Subsequent examination of the organisms and the chemistry of the recent sediments have shown how such changes are reflected in the palaeolimnological record. These changes are summarized in a schematic way in Fig. 6.16. Before 1910 the lake was sufficiently eutrophic to support *Bosmina longirostris* and *Asterionella formosa* and the sediment was organic. A sand layer deposited as a result of construction work on the inflow in 1916 caused a sharp reduction in organic and phosphorus content of the sediment, but the biota were unaffected. The increasing input of raw sewage until 1930 is reflected in a rise of all the curves in Fig. 6.16. The response to the diversion of sewage between 1930 and 1941 was a marked reversal in the eutrophication process. However, the increasing input of treated sewage effluent after 1941 led to renewed eutrophication, resulting in the first bloom of *Oscillatoria rubescens* in 1955, recorded by the rise in the curve for chlorophyll-derived sedimentary pigments.

All sewage effluent has been diverted from the lake since 1967, and there has been a rapid improvement in its condition. Shapiro *et al.* (1971) compared the chemical analyses of cores taken in 1958 and 1959 with analyses of new cores obtained in 1968 and 1970. Although the surface layer was

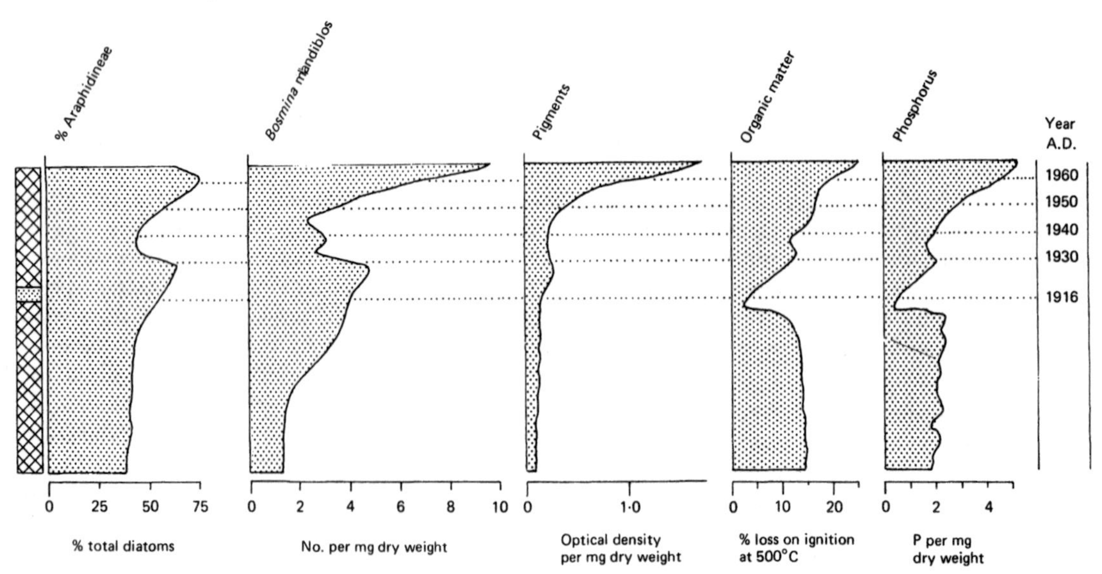

Fig. 6.16 Palaeolimnology of Lake Washington, U.S.A. Generalized curves for the percentage of Araphidinate diatoms, *Bosmina* mandibles, pigments, organic matter (loss on ignition at 500°C) and phosphorus. (Compiled from Edmonson, 1969; and Shapiro *et al.*, 1971.)

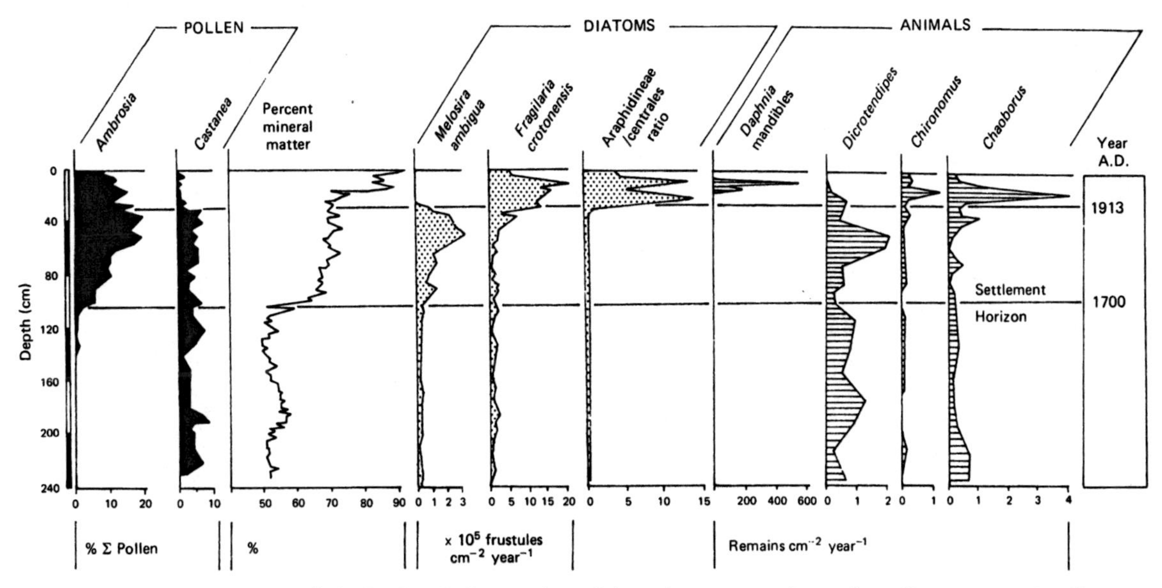

Fig. 6.17 Palaeolimnology of Linsley Pond, Connecticut. Selected curves are shown for pollen types, major diatoms, the Araphidinae/Centrales diatom ratio, *Daphnia* mandibles, and chironomid remains. The sediments have been dated by radiocarbon, and the known times of the *Ambrosia* pollen rise and the *Castanea* decline. (After Brugam, 1978a,b.)

still the richest in phosphorus and nitrogen, the more recent values were less than previously. The peak of phosphorus at the time of maximum enrichment was preserved, but at a lower concentration below the 1970 surface, but the peak of nitrogen was not detectable at all due to diagenesis within the sediment. The Lake Washington study is important in demonstrating how documented changes in lake characteristics are registered in the sediments, thereby validating many of the techniques used in palaeolimnology.

4. *Linsley Pond, Connecticut, U.S.A.*

Brugam (1978a,b) undertook a detailed survey of the recent palaeolimnology of Linsley Pond, using pollen, diatoms, some other algae, chironomids, cladocerans, rotifers, and chemical analysis of the sediments. He related the changes in the lake to documented changes in the watershed. The sediments were independently dated using ^{210}Pb and ^{14}C. The *Ambrosia* pollen rise occurred at 1700 A.D. when the first farm was established in the watershed by the Linsley family. Figure 6.17 shows the changes in some of the fossils. An increased influx of mineral matter resulted from soil disturbance, but there was little effect on the lake organisms. Similarly little change was registered after the establishment of a second farm beside the lake in about 1800. There was a gradual increase in diatoms, principally *Melosira ambigua*, and in the midge *Dicrotendipes*. This may have been because the lake was already moderately eutrophic, and no major thresholds were passed, in contrast to Sallie Lake and Shagawa Lake in Minnesota.

In 1915, as a result of changing farming practices near the lake, the diatom and midge assemblages became dominated by types more tolerant of eutrophic conditions. *Fragilaria crotonensis* and *Asterionella formosa* replaced *Melosira*, and the araphidinate/centric ratio increased (Fig. 6.17), and *Chironomus* began to replace *Dicrotendipes* as the hypolimnion became less oxygenated. However, the zooplankton, principally *Bosmina longirostris*, showed little response.

In 1960, houses were built along the shore, and a major change in the zooplankton ensued. Daphniids became abundant, and *Bosmina* carapaces increased in size. Brugam suggests that this may be a result of the detrimental effects of eutrophication on the alewife fish population, which led to a relaxation of the predation pressure on the Cladocera. The midge larvae also respond at this time, with a decline in *Dicrotendipes* and great increases in *Chironomus* and *Chaoborus*. This 1960 increase in eutrophication was a result of increased soil erosion reflected by the increased mineral matter reaching the sediment, and also of increased organic matter and nutrient influx from sewage effluent.

Chemical analyses of the sediment showed changes in the major elements, but these were difficult to interpret in terms of events in the catchment. The zooplankton showed little response to changes which caused large responses in the diatom flora. Zooplankton appear to be rather insensitive directly to changes in nutrient levels, but they are very sensitive to changes in predation pressure, which may be brought about by changes in nutrient level, as in 1960 in Linsley Pond.

Linsley Pond is small, and its biota responded rapidly to changes in the catchment. This contrasts with a large lake such as Lake Washington, which showed more gradual changes in response to eutrophication. Consequently, it should be relatively easy to restore small lakes to an acceptable condition by removing the nutrient source.

References

General principles of limnology

HUTCHINSON, G. E. (1957, 1967, 1975). *A Treatise on Limnology I, II, III*. J. Wiley and Sons.

MACAN, T. T. AND WORTHINGTON, E. B. (1951). *Life in Lakes and Rivers*. Collins.

RUTTNER, F. (1952). *Fundamentals of Limnology*. University of Toronto Press.

WELCH, P. S. (1952). *Limnology*. McGraw-Hill.

General principles of palaeolimnology

DEEVEY, E. S. (1955a). Some biogeographic implications of paleolimnology. *Proc. Int. Ass. Theor. Appl. Limnology*, **12**, 654–9.

DEEVEY, E. S. (1955b). The obliteration of the hypolimnion. *Mem. Ist. Ital. Idrobiol.*, suppl. **8**, 9–38.

FREY, D. G. (1964). Remains of animals in Quaternary lake and bog sediments and their interpretation. *Arch. Hydrobiol. Beih.*, **2**, 1–114.

FREY, D. G. (1969). The rationale of paleolimnology. *Mitt. Int. Verein. Limnol.*, **17**, 7–18.

LIVINGSTONE, D. A. (1968). Paleolimnology. *AAAS Symposium on Reconstruction of Past Biological Environments*, 1–4.

OLDFIELD, F. (1977). Lakes and their drainage basins as units of sediment-based ecological study. *Progress Physical Geogr.*, **1**, 460–504.

VALLENTYNE, J. R. (1957). The molecular nature of organic matter in lakes and streams with lesser reference to sewage and terrestrial soils. *J. Fish. Res. Bd. Canada*, **14**, 33–82.

WINTER, T. C. AND WRIGHT, H. E. (1977). Paleohydrologic phenomena recorded by lake sediments. *Eos*, **58**, 188–96.

WRIGHT, H. E. (1966). Stratigraphy of lake sediments and the precision of the paleoclimatic record. In *World Climate from 8000 to 0 B.C.* (ed. J. J. Sawyer). Royal Meteorol. Society, London.

Modern diatom assemblages, diatom ecology, and techniques

BATTARBEE, R. W. (1973). A new method for the estimation of absolute numbers, with reference especially to diatoms. *Limnol. Ocean.*, **18**, 647–53.

BRADBURY, J. P. (1975). Diatom stratigraphy and human settlement in Minnesota. *Geol. Soc. Amer. Spec. Paper*, **171**, 74 pp.

BRADBURY, J. P. AND WINTER, T. C. (1976). Areal distribution and stratigraphy of diatoms in the sediments of Lake Sallie, Minnesota. *Ecology*, **57**, 1005–14.

BRIGHT, R. C. (1968). Surface-water chemistry of some Minnesota lakes, with preliminary notes on diatoms. *Limnological Research Center, Univ. of Minnesota Interim Report*, **3**, 58 pp.

DUTHIE, H. C. AND SREENIVASA, M. R. (1972). The distribution of diatoms on the superficial sediments of Lake Ontario. *Proc. 15th Conf. Great Lakes Res.*, 45–52.

FLORIN, M.-B. (1970). The fine structure of some pelagic freshwater diatom species under the scanning electron microscope I. *Svensk Bot. Tidskr.*, **64**, 51–64.

HAWORTH, E. Y. (1975). A scanning electron microscope study of some different frustule forms of the genus *Fragilaria* found in Scottish late-glacial sediments. *Br. phycol. J.*, **10**, 73–80.

HAWORTH, E. Y. (1976a). The changes in the composition of the diatom assemblages found in the surface sediments of Blelham Tarn in the English Lake District during 1973. *Ann. Bot.*, **40**, 1195–205.

KILHAM, P. (1971). A hypothesis concerning silica and the freshwater planktonic diatoms. *Limnol. Ocean.*, **16**, 10–18.

KOIVO, L. K. AND RITCHIE, J. C. (1978). Modern diatom assemblages from lake sediments in the boreal-arctic transition region near the Mackenzie Delta, N.W.T., Canada. *Can. J. Bot.*, **56**, 1010–20.

MERILÄINEN, J. (1967). The diatom flora and the hydrogen-ion concentration of the water. *Annls. Bot. Fenn.*, **4**, 51–8.

MERILÄINEN, J. (1969). Distribution of diatom frustules in recent sediments of some meromictic lakes. *Mitt. Int. Verein. Limnol.*, **17**, 186–92.

MERILÄINEN, J. (1971). The recent sedimentation of diatom frustules in four meromictic lakes. *Annls. Bot. Fenn.*, **8**, 160–76.

MILLER, U. (1969). Fossil diatoms under the scanning electron microscope. A preliminary report. *Sver. Geol. Unders.*, C, **63**(5), 65 pp.

RAWSON, D. S. (1956). Algal indicators of trophic lake types. *Limnol. Ocean.*, **1**, 18–25.
ROUND, F. E. (1964). The diatom sequence in lake deposits: some problems of interpretation. *Verh. Int. Verein. Limnol.*, **15**, 1012–20.
ROUND, F. E. (1971). The growth and succession of algal populations in freshwaters. *Mitt. Int. Verein. Limnol.*, **19**, 70–99.

Diatom palaeoecological studies

ALHONEN, P. (1971). The stages of the Baltic Sea as indicated by the diatom stratigraphy. *Acta Bot. Fennica*, **92**, 18 pp.
BENDA, L. (1974). Diatomeen der niedersächsischen Kieselgur-Vorkommen, palökologische Befunde und Nackweis einer Jahresschichtung. *Geol. Jb.*, **A21**, 171–97.
BIRKS, H. H., WHITESIDE, M. C., STARK, D. M. AND BRIGHT, R. C. (1976). Recent paleolimnology of three lakes in northwestern Minnesota. *Quat. Res.*, **6**, 249–72.
BRADBURY, J. P. (1971). Paleolimnology of Lake Texcoco, Mexico. Evidence from diatoms. *Limnol. Ocean.*, **16**, 180–200.
BRADBURY, J. P. AND MEGARD, R. O. (1972). Stratigraphic record of pollution in Shagawa Lake, Northeastern Minnesota. *Bull. Geol. Soc. Amer.*, **83**, 2639–48.
DAVIS, R. B. AND NORTON, S. A. (1978). Paleolimnologic studies of human impact on lakes in the United States, with emphasis on recent research in New England. *Pol. Arch. Hydrobiol.*, **25**, 99–115.
DUTHIE, H. C. AND SREENIVASA, M. R. (1971). Evidence for the eutrophication of Lake Ontario from the sedimentary diatom succession. *Proc. 14th Conf. Great Lakes Res.*, 1–13.
EVANS, G. H. (1970). Pollen and diatom analysis of late-Quaternary deposits in the Blelham Basin, North Lancashire. *New Phytol.*, **69**, 821–74.
FLORIN, M.-B. (1970). Late-glacial diatoms of Kirchner Marsh, southeastern Minnesota. *Nova Hedwigia*, **31**, 667–756.
FLORIN, M.-B. AND WRIGHT, H. E. (1969). Diatom evidence for the persistence of stagnant glacial ice in Minnesota. *Bull. Geol. Soc. Amer.*, **80**, 695–704.
HAWORTH, E. Y. (1969). The diatoms of a sediment core from Blea Tarn, Langdale. *J. Ecol.*, **57**, 429–39.
HAWORTH, E. Y. (1972a). Diatom succession in a core from Pickerel Lake, northeastern South Dakota. *Bull. Geol. Soc. Amer.*, **83**, 157–72.
HAWORTH, E. Y. (1972b). The recent diatom history of Loch Leven, Kinross. *Freshwater Biol.*, **2**, 131–41.
HAWORTH, E. Y. (1976b). Two late-glacial (Late Devensian) diatom assemblage profiles from northern Scotland. *New Phytol.*, **77**, 227–56.
HECKY, R. E. AND KILHAM, P. (1973). Diatoms in alkaline, saline lakes: ecology and geochemical implications. *Limnol. Ocean.*, **18**, 53–71.
KOIVO, L. K. (1976). Species diversity in post-glacial diatom lake communities of Finland. *Palaeogeography, Palaeoclimatology, Palaeoecology*, **19**, 165–90.
PENNINGTON, W. (1943). Lake sediments: The bottom deposits of the North Basin of Windermere; with special reference to the diatom succession. *New Phytol.*, **42**, 1–27.
PENNINGTON, W., HAWORTH, E. Y., BONNY, A. P. AND LISHMAN, J. P. (1972). Lake sediments in northern Scotland. *Phil. Trans. R. Soc.* B, **264**, 191–294.
ROUND, F. E. (1957). The late-glacial and post-glacial diatom succession in the Kentmere Valley deposit. *New Phytol.*, **56**, 98–126.
ROUND, F. E. (1961). The diatoms of a core from Esthwaite Water. *New Phytol.*, **60**, 43–59.
SIMOLA, H. (1977). Diatom succession in the formation of annually laminated sediment in Lovojärvi, a small eutrophicated lake. *Ann. Bot. Fennica*, **14**, 143–8.
SREENIVASA, M. R. AND DUTHIE, H. C. (1973). The post-glacial diatom history of Sunfish Lake, southwestern Ontario. *Can. J. Bot.*, **51**, 1599–609.
STOCKNER, J. G. (1972). Paleolimnology as a means of assessing eutrophication. *Verh. Int. Verein. Limnol.*, **18**, 1018–30.
STOCKNER, J. G. AND BENSON, W. W. (1967). The succession of diatom assemblages in the recent sediments of Lake Washington. *Limnol. Ocean.*, **12**, 513–32.
WATTS, W. A. AND BRIGHT, R. C. (1968). Pollen, seed, and mollusk analysis of a sediment core from Pickerel Lake, northeastern South Dakota. *Bull. Geol. Soc. Amer.*, **79**, 855–76.
WATTS, W. A. AND WINTER, T. C. (1976). Plant macrofossils from Kirchner Marsh, Minnesota – a palaeoecological study. *Bull. Geol. Soc. Amer.*, **77**, 1339–60.

Palaeoecological studies of algae other than diatoms

BIRKS, H. H., WHITESIDE, M. C., STARK, D. M. AND BRIGHT, R. C. (1976). Recent paleolimnology of three lakes in northwestern Minnesota. *Quat. Res.*, **6**, 249–72.
BIRKS, H. J. B. (1976). Late-Wisconsinan vegetational history at Wolf Creek, Central Minnesota. *Ecol. Monogr.*, **46**, 395–429.
HUTCHINSON, G. E. (1970). Ianula: an account of the history and development of the Lago di Monterosi, Latium, Italy. *Trans. Amer. Phil. Soc.*, **60**, 178 pp.
LEVENTHAL, E. A. (1970). The Chrysomandina. In Hutchinson (1970). *Trans. Amer. Phil. Soc.*, **60**, 123–42.
WHITESIDE, M. C. (1965). On the occurrence of *Pediastrum* in lake sediments. *J. Ariz. Acad. Sci.*, **3**, 144–6.

Modern cladoceran assemblages and cladoceran ecology

BROOKS, J. L. (1968). The effects of prey size selection by lake planktivores. *Syst. Zool.*, **17**, 272–91.

BROOKS, J. L. AND DODSON, S. I. (1965). Predation, body size, and composition of plankton. *Science*, **150**, 28–35.

DEEVEY, E. S. (1942). Studies on Connecticut lake sediments. III. The biostratonomy of Linsley Pond. *Am. J. Sci.*, **240**, 233–64, 313–38.

DEEVEY, E. S. (1964). Preliminary account of fossilization of zooplankton in Rogers Lake. *Verh. Int. Verein. Limnol.*, **15**, 981–92.

DEEVEY, E. S. (1969). Cladoceran populations of Rogers Lake, Connecticut, during late- and postglacial time. *Mitt. Int. Verein. Limnol.*, **17**, 56–63.

DE COSTA, J. J. (1964). Latitudinal distribution of chydorid cladocera in the Mississippi Valley, based on their remains in surficial lake sediments. *Invest. Ind. Lakes Streams*, **6**, 65–101.

FREY, D. G. (1960). The ecological significance of cladoceran remains in lake sediments. *Ecology*, **41**, 785–90.

FREY, D. G. (1967). Cladocera in space and time. *Proc. Symp. Crustacea*, **1**, 1–9.

GOULDEN, C. E. (1969). Interpretative studies of cladoceran microfossils in lake sediments. *Mitt. Int. Verein. Limnol.*, **17**, 43–55.

GOULDEN, C. E. (1971). Environmental control of the abundance and distribution of the chydorid cladocera. *Limnol. Ocean.*, **16**, 320–31.

HARMSWORTH, R. V. AND WHITESIDE, M. C. (1968). Relation of cladoceran remains in lake sediments to primary productivity of lakes. *Ecology*, **49**, 998–1000.

KEEN, R. (1973). A probabilistic approach to the dynamics of natural populations of the Chydoridae (Cladocera, Crustacea). *Ecology*, **54**, 524–34.

KEERFOOT, W. C. (1974a). Net accumulation rates and the history of cladoceran communities. *Ecology*, **55**, 51–61.

KEERFOOT, W. C. (1974b). Egg-size cycle of a cladoceran. *Ecology*, **55**, 1259–70.

KEERFOOT, W. C. (1975). The divergence of adjacent populations. *Ecology*, **56**, 1298–313.

KEERFOOT, W. C. (1977). Implications of copepod predation. *Limnol. Ocean.*, **22**, 316–25.

MINDER, L. (1938). Der Zürichsee als Eutrophierungsphänomen. Geol. Meere Binnengewaesser, **2**, 284–99.

MUELLER, W. P. (1964). The distribution of cladoceran remains in surficial sediments from three northern Indiana lakes. *Invest. Ind. Lakes Streams*, **6**, 1–63.

QUADE, H. W. (1969). Cladoceran faunas associated with aquatic macrophytes in some lakes in northwestern Minnesota. *Ecology*, **50**, 170–9.

STENSON, J. A. E. (1976). Significance of predator influence on composition of *Bosmina* spp. populations. *Limnol. Ocean.*, **21**, 814–22.

WHITESIDE, M. C. (1970). Danish chydorid cladocera: modern ecology and core studies. *Ecol. Monogr.*, **40**, 79–118.

WHITESIDE, M. C. (1974). Chydorid (Cladocera) ecology: seasonal patterns and abundance of populations in Elk Lake, Minnesota. *Ecology*, **55**, 538–50.

WHITESIDE, M. C. AND HARMSWORTH, R. V. (1967). Species diversity in chydorid (Cladocera) communities. *Ecology*, **48**, 664–7.

ZARET, T. M. AND KEERFOOT, W. C. (1975). Fish predation on *Bosmina longirostris:* body-size selection *versus* visibility selection. *Ecology*, **56**, 232–7.

Cladoceran palaeoecological studies

ALHONEN, P. (1971). The Flandrian development of the pond Hyrynlampi, southern Finland, with special reference to the pollen and cladoceran stratigraphy. *Acta Bot. Fennica*, **95**, 19 pp.

ALHONEN, P. (1972). Gallträsket: The geological development and palaeolimnology of a small polluted lake in southern Finland. *Comm. Biologicae*, **57**, 34 pp.

CRISMAN, T. L. AND WHITEHEAD, D. R. (1978). Paleolimnological studies on small New England (U.S.A.) ponds. II. Cladoceran community responses to trophic conditions. *Pol. Arch. Hydrobiol.*, **25**, 75–86.

DE COSTA, J. (1968). The history of the Chydorid (Cladocera) community of a small lake in the Wind River Mountains, Wyoming, U.S.A. *Arch. Hydrobiol.*, **64**, 400–25.

FREY, D. G. (1961). Developmental history of Schleinsee. *Verh. Int. Verein. Limnol.*, **14**, 271–8.

FREY, D. G. (1962). Cladocera from the Eemian interglacial of Denmark. *J. Paleont.*, **36**, 1133–54.

GOULDEN, C. E. (1964a). Progressive changes in the cladoceran and midge fauna during the ontogeny of Esthwaite Water. *Verh. Int. Verein. Limnol.*, **15**, 1000–5.

GOULDEN, C. E. (1964b). The history of the Cladoceran fauna of Esthwaite Water (England) and its limnological significance. *Arch. Hydrobiol.*, **60**, 1–52.

GOULDEN, C. E. (1966). La Aguada de Santa Ana Vieja: an interpretative study of the cladoceran microfossils. *Arch. Hydrobiol.*, **62**, 373–404.

HARMSWORTH, R. V. (1968). The developmental history of Blelham Tarn (England) as shown by animal microfossils, with special reference to the cladocera. *Ecol. Monogr.*, **36**, 223–41.

MEGARD, R. O. (1964). Biostratigraphic history of Dead Man Lake, Chuska Mountains, New Mexico. *Ecology*, **45**, 529–46.

MEGARD, R. O. (1967). Late-Quaternary cladocera of Lake Zeribar, western Iran. *Ecology*, **48**, 179–89.

TSUKADA, M. (1972). The history of Lake Nojiri, Japan. *Trans. Conn. Acad. Arts & Sci.*, **44**, 334–65.

WHITEHEAD, D. R. AND CRISMAN, T. L. (1978). Paleolimnological studies of small New England (U.S.A.) ponds. I. Late-glacial and postglacial trophic oscillations. *Pol. Arch. Hydrobiol.*, **25**, 471–81.

Fish remains in lake sediments

PENNINGTON, W. AND FROST, W. E. (1961). Fish vertebrae and scales in a sediment core from Esthwaite Water (English Lake District). *Hydrobiologia*, **17**, 183–90.

Chironomid remains in lake sediments

ALHONEN, P. AND HAAVISTO, M. L. (1969). The biostratigraphical history of Lake Otalampi in southern Finland, with special reference to the remains of subfossil midge fauna. *Bull. Geol. Soc. Finland*, **41**, 157–64.

FREY, D. G. (1964). Remains of animals in Quaternary lake and bog sediments and their interpretation. *Arch. Hydrobiol. Beih.*, **2**, 1–114.

STAHL, J. B. (1959). The developmental history of the chironomid and *Chaoborus* faunas of Myers Lake. *Invest. Ind. Lakes Streams*, **5**, 47–102.

STAHL, J. B. (1966). The ecology of *Chaoborus* in Myers Lake, Indiana. *Limnol. Ocean.*, **11**, 177–83.

STAHL, J. B. (1969). The uses of chironomids and other midges in interpreting lake histories. *Mitt. Int. Verein. Limnol.*, **17**, 111–25.

Ostracod remains in lake sediments

DELORME, L. D. (1967). Field key and methods of collecting freshwater ostracodes in Canada. *Can. J. Zool.*, **45**, 1275–81.

DELORME, L. D. (1969). Ostracodes as Quaternary paleoecological indicators. *Can. J. Earth Sci.*, **6**, 1471–6.

DELORME, L. D. (1971a). Paleoecological determinations using Pleistocene freshwater ostracodes. *Bull. Centre Rech. Pan-SNPA*, **5** suppl., 341–7.

DELORME, L. D. (1971b). Paleoecology of Holocene sediments from Manitoba using freshwater ostracodes. *Geol. Assoc. Canada Spec. Paper*, **9**, 301–4.

DELORME, L. D., ZOLTAI, S. C. AND KALAS, L. L. (1977). Freshwater shelled invertebrate indicators of paleoclimate in northwestern Canada during late glacial times. *Can. J. Earth Sci.*, **14**, 2029–46.

Trace fossils

GIBBARD, P. L. (1977). Fossil tracks from varved sediments near Lammi, South Finland. *Bull. Geol. Soc. Finland*, **49**, 53–7.

GIBBARD, P. L. AND STUART, A. J. (1974). Trace fossils from proglacial lake sediments. *Boreas*, **3**, 69–74.

Sedimentary structures and laminated sediments

ANTHONY, R. S. (1977). Iron-rich rhythmically laminated sediments in Lake of the Clouds, northeastern Minnesota. *Limnol. Ocean.*, **22**, 45–54.

AXELSSON, V. AND HÄNDEL, S. K. (1972). X-radiography of unextruded sediment cores. *Geogr. Annls.*, **54A**, 34–7.

BRUNSKILL, G. J. (1969). Fayetteville Green Lake, New York. II. Precipitation and sedimentation of calcite in a meromictic lake with laminated sediments. *Limnol. Ocean.*, **14**, 830–47.

CALVERT, S. E. (1964). Factors affecting distribution of laminated diatomaceous sediments in Gulf of California. *Marine Geology of the Gulf of California, Amer. Assoc. Petrol. Geol. Memoir*, **3**, 311–30.

CALVERT, S. E. (1966). Origin of diatom-rich, varved sediments from the Gulf of California. *J. Geol.*, **74**, 546–65.

CALVERT, S. E. AND VEEVERS, J. J. (1962). Minor structures of unconsolidated marine sediments revealed by X-radiography. *Sedimentology*, **1**, 287–95.

CRAIG, A. J. (1972). Pollen influx to laminated sediments: a pollen diagram from northeastern Minnesota. *Ecology*, **53**, 46–57.

CWYNAR, L. (1978). Recent history of fire and vegetation from laminated sediment of Greenleaf Lake, Algonquin Park, Ontario. *Can. J. Bot.*, **56**, 10–21.

DEEVEY, E. S. (1946). An absolute pollen chronology in Switzerland. *Am. J. Sci.*, **244**, 442–7.

DIGERFELDT, G., BATTARBEE, R. W. AND BENGTSSON, L. (1975). Report on annually laminated sediment in lake Järlasjön, Necka, Stockholm. *Geol. Fören. Förhandl. Stock.*, **97**, 29–40.

EDMONDSON, W. T. AND ALLISON, D. E. (1970). Recording densitometry of X-radiographs for the study of cryptic laminations in the sediment of Lake Washington. *Limnol. Ocean.*, **15**, 138–44.

HOLMES, P. W. (1968). Sedimentary studies of late Quaternary material in Windermere Lake (Great Britain). *Sediment. Geol.*, **2**, 201–24.

KARLÉN, W. (1976). Lacustrine sediments and tree-line variations as indicators of Holocene climatic fluctuations in Lappland, Northern Sweden. *Geogr. Annls.*, **58A**, 1–34.

LUDLAM, S. D. (1976). Laminated sediments in holomictic Berkshire Lakes. *Limnol. Ocean.*, **21**, 743–6.

RENBERG, I. (1976). Annually laminated sediments in Lake Rudetjärn, Medelpad province, northern Sweden. *Geol. Fören. Förhandl. Stock.*, **98**, 355–60.

SAARNISTO, M., HUTTUNEN, P. AND TOLONEN, K. (1977). Annual lamination of sediments in Lake Lovojärvi, southern Finland, during the past 600 years. *Ann. Bot. Fenn.*, **14**, 35–45.

SIMOLA, H. (1977). Diatom succession in the formation of annually laminated sediments in Lovojärvi, a small eutrophicated lake. *Ann. Bot. Fenn.*, **14**, 143–8.

SMITH, A. J. (1959). Structures in the stratified late-glacial clays of Windermere, England. *J. Sed. Petrol.*, **29**, 447–53.

STUIVER, M. (1971). Evidence for the variation of atmospheric C^{14} content in the Late Quaternary. In *The Late Cenozoic Glacial Ages* (ed. K. K. Turekian). Yale University Press.

SWAIN, A. M. (1973). A history of fire and vegetation in northeastern Minnesota as recorded in lake sediments. *Quat. Res.*, **3**, 383–96.

SWAIN, A. M. (1978). Environmental changes during the past 2000 years in north-central Wisconsin: analysis of pollen, charcoal, and seeds from varved lake sediments. *Quat. Res.*, **10**, 55–68.

TIPPETT, R. (1964). An investigation into the nature of the layering of deep-water sediments in two eastern Ontario lakes. *Can. J. Bot.*, **42**, 1693–709.

TURNER, C. (1970). The Middle Pleistocene deposits at Marks Tey, Essex. *Phil. Trans. R. Soc.* B, **257**, 373–440.

VUORINEN, J. (1976). The influence of past land use on the sediments of a small lake. *Abstracts 2nd Int. Symp. Paleolimnol. Mikolajki, Poland.*

Inorganic chemistry of lake sediments

ANDERSON, T. W. (1974). The chestnut pollen decline as a time horizon in lake sediments in Eastern North America. *Can. J. Earth Sci.*, **11**, 678–85.

APPLEBY, P. G. AND OLDFIELD, F. (1978). The calculation of lead-210 dates assuming a constant rate of supply of unsupported ^{210}Pb to the sediment. *Catena*, **5**, 1–8.

ASTON, S. R., BRUTY, D., CHESTER, R. AND PADGHAM, R. C. (1973). Mercury in lake sediments: a possible indicator of technological growth. *Nature*, **241**, 450–1.

BEETON, A. M. (1969). Changes in the environment and biota of the Great Lakes. In *Eutrophication: Causes, Consequences, Correctives.* National Academy of Sciences, Washington, D.C.

BORMANN, F. H., LIKENS, G. E., SICCAMA, T. G., PIERCE, R. S. AND EATON, J. S. (1974). The export of nutrients and recovery of stable conditions following deforestation at Hubbard Brook. *Ecol. Monogr.*, **44**, 255–77.

BORTELSON, G. C. AND LEE, G. F. (1972). Recent sedimentary history of Lake Mendota, Wis. *Env. Science & Tech.*, **6**, 799–808.

BRUGAM, R. B. (1978a). Pollen indicators of land-use change in southern Connecticut. *Quat. Res.*, **9**, 349–62.

BRULAND, K. W., KOIDE, M., BOWSER, C., MAHER, L. J. AND GOLDBERG, E. D. (1975). Lead-210 and pollen geochronologies on Lake Superior sediments. *Quat. Res.*, **5**, 89–98.

DAVIS, M. B. (1976). Erosion rates and land-use history in southern Michigan. *Envir. Conservation*, **3**, 139–48.

ERICSSON, B. (1973). The cation content of Swedish post-glacial sediments as a criterion of palaeosalinity. *Geol. Fören. Förhandl. Stock.*, **95**, 1–40.

FREDRIKSSON, I. AND QVARFORT, U. (1973). The mercury content of sediments from two lakes in Dalarna, Sweden. *Geol. Fören. Förhandl. Stock.*, **95**, 237–42.

KEMP, A. L. W., ANDERSON, T. W., THOMAS, R. L. AND MUDROCHOVA, A. (1974). Sedimentation rates and recent sediment history of Lakes Ontario, Erie and Huron. *J. Sed. Petrol.*, **44**, 207–18.

KRISHNASWAMY, S., LAL, D., MARTIN, J. M. AND MEYBECK, M. (1971). Geochronology of lake sediments. *Earth Planet. Sci. Letters*, **11**, 407–14.

LIKENS, G. E., BORMANN, F. H., PIERCE, R. S., EATON, J. S. AND JOHNSON, N. M. (1977). *Biogeochemistry of a Forested Ecosystem.* Springer-Verlag.

LIKENS, G. E. AND DAVIS, M. B. (1975). Post-glacial history of Mirror Lake and its watershed in New Hampshire, U.S.A.: an initial report. *Verh. Int. Verein. Limnol.*, **19**, 982–93.

LIVINGSTONE, D. A. (1957). On the sigmoid growth phase in the history of Linsley Pond. *Am. J. Sci.*, **255**, 364–73.

LIVINGSTONE, D. A. AND BOYKIN, J. C. (1962). Vertical distribution of phosphorus in Linsley Pond mud. *Limnol. Ocean.*, **7**, 57–62.

MACKERETH, F. J. H. (1965). Chemical investigations of lake sediments and their interpretation. *Proc. R. Soc.* B, **161**, 295–309.

MACKERETH, F. J. H. (1966). Some chemical observations on post-glacial lake sediments. *Phil. Trans. R. Soc.* B, **250**, 165–213.

OLDFIELD, F., APPLEBY, P. G. AND BATTARBEE, R. W. (1978). Alternative ^{210}Pb dating: results from the New Guinea Highlands and Lough Erne. *Nature*, **271**, 339–42.

PEARSALL, W. H., GAY, J. AND NEWBOULD, J. (1960). Post-glacial sediments as a record of regional soil drifts. *J. Soil Sci.*, **11**, 68–76.

PENNINGTON, W. (1973). The recent sediments of Windermere. *Freshwater Biol.*, **3**, 363–82.

PENNINGTON, W. (1974). Seston and sediment formation in five Lake District lakes. *J. Ecol.*, **62**, 215–51.

PENNINGTON, W., CAMBRAY, R. S. AND FISHER, E. M. (1973). Observations on lake sediments using fallout ^{137}Cs as a tracer. *Nature*, **242**, 324–6.

PENNINGTON, W., CAMBRAY, R. S., EAKINS, J. D. AND HARKNESS, D. D. (1976). Radionuclide dating of the recent sediments at Blelham Tarn. *Freshwater Biol.*, **6**, 317–31.

PENNINGTON, W. AND LISHMAN, J. P. (1971). Iodine in lake sediments in Northern England and Scotland. *Biol. Rev.*, **46**, 279–313.

RITCHIE, J. C., MCHENRY, J. R. AND GILL, A. C. (1973). Dating recent reservoir sediments. *Limnol. Ocean.*, **18**, 254–63.

SASSEVILLE, D. R. AND NORTON, S. A. (1975). Present and historic geochemical relationships in four Maine lakes. *Limnol. Ocean.*, **20**, 699–714.

Palaeomagnetism of lake sediments

CREER, K. M., THOMPSON, R., MOLYNEUX, L. AND MACKERETH, F. J. H. (1972). Geomagnetic secular variation recorded in the stable magnetic remanence of recent sediments. *Earth Planet. Sci. Letters*, **14**, 115–27.

MACKERETH, F. J. H. (1971). On the variation in direction of the horizontal component of remanent magnetisation in lake sediments. *Earth Planet. Sci. Letters*, **12**, 332–8.

MOLYNEUX, I., THOMPSON, R., OLDFIELD, F. AND MCCALLAN, M. E. (1972). Rapid measurement of the remanent magnetization of long cores of sediment. *Nature*, **237**, 42–3.

STOBER, J. C. AND THOMPSON, R. (1977). Palaeomagnetic secular variation studies of Finnish lake sediments and the carriers of remanence. *Earth Planet. Sci. Letters*, **37**, 139–49.

THOMPSON, R. (1973). Palaeolimnology and palaeomagnetism. *Nature*, **242**, 182–4.

THOMPSON, R. (1974). Palaeomagnetism. *Sci. Prog.*, **61**, 349–73.

THOMPSON, R. (1975). Long period European geomagnetic secular variation confirmed. *Geophys. J. R. astr. Soc.*, **43**, 847–59.

THOMPSON, R., BATTARBEE, R. W., O'SULLIVAN, P. E. AND OLDFIELD, F. (1975). Magnetic susceptibility of lake sediments. *Limnol. Ocean.*, **20**, 687–98.

TOLONEN, K., SIIRIÄNEN, A. AND THOMPSON, R. (1975). Prehistoric field erosion sediment in Lake Lojärvi, S. Finland and its palaeomagnetic dating. *Ann. Bot. Fenn.*, **12**, 161–4.

Organic chemistry of lake sediments

ATHERTON, N. M., CRANWELL, P. A., FLOYD, A. J., AND HAWORTH, R. D. (1967). Humic acid – I. E.S.R. spectra of humic acids. *Tetrahedron*, **23**, 1653–67.

BROWN, S. R. (1968). Bacterial carotenoids from freshwater sediments. *Limnol. Ocean.*, **13**, 233–41.

BROWN, S. R. (1969). Paleolimnological evidence from fossil pigments. *Mitt. Int. Verein. Limnol.*, **17**, 95–103.

BROWN, S. R. AND COLMAN, B. (1963). Oscillaxanthin in lake sediments. *Limnol. Ocean.*, **8**, 352–3.

CRANWELL, P. A. (1973). Chain-length distribution of n-alkanes from lake sediments in relation to post-glacial environmental change. *Freshwater Biol.*, **3**, 259–65.

CRANWELL, P. A. (1974). Monocarboxylic acids in lake sediments: indicators derived from terrestrial and aquatic biota, of paleoenvironmental trophic levels. *Chem. Geology*, **14**, 1–14.

CRANWELL, P. A. (1976). Organic geochemistry of lake sediments. In *Environmental Biogeochemistry* 1 (ed. J. O. Nriagu). Michigan.

CRANWELL, P. A. (1977). Organic geochemistry of Cam Loch (Sutherland) sediments. *Chem. Geology*, **20**, 205–21.

FOGG, G. E. AND BELCHER, J. H. (1961). Pigments from the bottom deposits of an English lake. *New Phytol.*, **60**, 129–42.

GORHAM, E., LUND, J. W. G., SANGER, J. E. AND DEAN, W. E. (1974). Some relationships between algal standing crop, water chemistry, and sediment chemistry in the English Lakes. *Limnol. Ocean.*, **19**, 601–17.

GRIFFITHS, M., PERROTT, P. S. AND EDMONDSON, W. T. (1969). Oscillaxanthin in the sediments of Lake Washington. *Limnol. Ocean.*, **14**, 317–26.

ROGERS, M. A. (1965). Carbohydrates in aquatic plants and associated sediments from two Minnesota lakes. *Geochem. Cosmochim. Acta*, **29**, 183–200.

SANGER, J. E. AND GORHAM, E. (1970). The diversity of pigments in lake sediments and its ecological significance. *Limnol. Ocean.*, **15**, 59–69.

SANGER, J. E. AND GORHAM, E. (1972). Stratigraphy of fossil pigments as a guide to the postglacial history of Kirchner Marsh, Minnesota. *Limnol. Ocean.*, **17**, 840–54.

SANGER, J. E. AND GORHAM, E. (1973). A comparison of the abundance and diversity of fossil pigments in wetland peats and woodland humus layers. *Ecology*, **54**, 605–11.

SWAIN, F. M. (1970). *Non-marine Organic Geochemistry*. Cambridge University Press.

VALLENTYNE, J. R. (1956). Epiphasic carotenoids in post-glacial lake sediments. *Limnol. Ocean.*, **1**, 252–62.

VALLENTYNE, J. R. (1960). Fossil pigments. In *Comparative Biochemistry of Photoreactive Systems* (ed. M. B. Allen). Academic Press.

ZÜLLIG, H. (1960). Die Bestimmung von Myxoxanthophyll in Bohrprofilen zum Nachweis vergangener Blaulgen-entfaltungen. *Verh. Int. Verein. Limnol.*, **14**, 263–70.

Integrated palaeolimnological studies

Lake Zeribar, Iran

HUTCHINSON, G. E. AND COWGILL, U. M. (1963). Chemical examination of a core from Lake Zeribar, Iran. *Science*, **140**, 67–9.

MEGARD, R. O. (1967). Late-Quaternary cladocera of Lake Zeribar, Western Iran. *Ecology*, **48**, 179–89.

VAN ZEIST, W. (1967). Late Quaternary vegetation history of Western Iran. *Rev. Palaeobot. Palynol.*, **2**, 301–11.

VAN ZEIST, W. AND WRIGHT, H. E. (1963). Preliminary pollen studies at Lake Zeribar, Zagros Mountains, Southwestern Iran. *Science*, **140**, 65–7.

WASYLIKOWA, K. (1967). Late Quaternary plant macrofossils from Lake Zeribar, Western Iran. *Rev. Palaeobot. Palynol.*, **2**, 313–18.

WRIGHT, H. E. (1966). Stratigraphy of lake sediments and the precision of the paleoclimatic record. In *World Climate from 8000 to 0 B.C.* (ed. J. J. Sawyer). Royal Meteorol. Society, London.

WRIGHT, H. E. (1968). Natural environment of early food production north of Mesopotamia. *Science*, **161**, 334–9.

WRIGHT, H. E. (1976). The environmental setting for plant domestication in the Near East. *Science*, **194**, 385–9.

WRIGHT, H. E., MCANDREWS, J. H. AND VAN ZEIST, W. (1967). Modern pollen rain in western Iran, and its relation to plant geography and Quaternary vegetational history. *J. Ecol.*, **55**, 415–43.

Eutrophication studies

BATTARBEE, R. W. (1978). Observations on the recent history of Lough Neagh and its drainage basin. *Phil. Trans. R. Soc.* B, **281**, 303–45.

BIRKS, H. H., WHITESIDE, M. C., STARK, D. M. AND BRIGHT, R. C. (1976). Recent paleolimnology of three lakes in northwestern Minnesota. *Quat. Res.*, **6**, 249–72.

BRADBURY, J. P. (1975). Diatom stratigraphy and human settlement in Minnesota. *Geol. Soc. Amer. Spec. Paper*, **171**, 74 pp.

BRADBURY, J. P. AND MEGARD, R. O. (1972). Stratigraphic record of pollution in Shagawa Lake, northeastern Minnesota. *Bull. Geol. Soc. Amer.*, **83**, 2639–48.

BRADBURY, J. P. AND WADDINGTON, J. C. B. (1973). The impact of European settlement on Shagawa Lake, Northeastern Minnesota, U.S.A. In *Quaternary Plant Ecology* (eds H. J. B. Birks and R. G. West). Blackwell.

BRUGAM, R. B. (1978b). Human disturbance and the historical development of Linsley Pond. *Ecology*, **59**, 19–36.

EDMONDSON, W. T. (1969a). Eutrophication in North America. In *Eutrophication: Causes, Consequences, Correctives*. National Academy of Sciences, Washington, D.C.

EDMONDSON, W. T. (1969b). Cultural eutrophication with special reference to Lake Washington. *Mitt. Int. Verein. Limnol.*, **17**, 19–32.

FAEGRI, K. (1954). Some reflections on the trophic system in limnology. *Nytt. Mag. Bot.*, **3**, 43–9.

GORHAM, E. AND SANGER, J. E. (1976). Fossilized pigments as stratigraphic indicators of cultural eutrophication in Shagawa Lake, northeastern Minnesota. *Bull. Geol. Soc. Amer.*, **87**, 1638–42.

HUTCHINSON, G. E. (1970). Ianula: an account of the history and development of the Lago di Monterosi, Latium, Italy. *Trans. Amer. Phil. Soc.*, **60**, 178 pp.

HUTCHINSON, G. E. (1973). Eutrophication. *Amer. Sci.*, **61**, 269–79.

LIKENS, G. E. (1972). Eutrophication and aquatic ecosystems. In *Nutrients and Eutrophication*. Special Symposia Vol. 1, Amer. Soc. Limnology and Oceanography, 3–13.

LUND, J. W. G. (1972). Eutrophication. *Proc. R. Soc.* B, **180**, 371–82.

SHAPIRO, J., EDMONDSON, W. T. AND ALLISON, D. E. (1971). Changes in the chemical composition of sediments of Lake Washington, 1958–70. *Limnol. Ocean.*, **16**, 437–52.

7

Molluscs, insects, and vertebrates in Quaternary palaeoecology

Introduction

Although molluscan and insect fossils are found in lake sediments, they are not exclusive to them, unlike most of the organisms that we discussed in Chapter 6. Insects, in particular beetles (Coleoptera), are frequently terrestrial organisms, whereas molluscs range in habitat from lakes and streams through terrestrial habitats to the driest grassland and rock habitats. Therefore, although molluscs and insects can play an important role in palaeolimnology, they also have wider applications in the interpretation of Quaternary palaeoecological events on dry land.

Molluscan ecology and zoogeography

Introduction

Molluscs occur in marine, freshwater, and terrestrial habitats, but here we shall only be concerned with the non-marine molluscs. Marine species are sometimes recovered from terrestrial deposits, and are valuable in determining the occurrence of a marine transgression, for example the highest sea-level reached during interglacial periods in East Anglia (see West, 1977).

Non-marine molluscs are divided into two main groups; a) the bivalves, lamellibranchs, or pelecypods (freshwater mussels and clams) and b) the univalves, or gastropods (snails). The snails may breathe either by gills (prosobranchs), in which case they have an operculum to shut the mouth of the shell, or they may breathe by a 'lung' (pulmonates). Prosobranchs are mostly aquatic, with some forms able to live in damp places on land. Pulmonates can be aquatic, but are primarily land snails. Pulmonates with reduced internal shells are the slugs.

Freshwater molluscs thrive best in alkaline situations where there is a readily available supply of calcium carbonate for their shells. Their shells are also preserved best in such conditions.

Mollusc shells are fairly large objects, and they therefore tend to be deposited and fossilized near to their source of origin, in a way similar to plant macrofossils (see Chapter 5). However, shells from different habitats are often found mixed together; for example, shells from all the different habitats within a lake may occur together in one sediment sample, usually with some lake-shore and marsh species. Frequently a predominantly moving-water fauna is associated with xerophilous land species. This is because moderately fast-moving streams lack fringing marshes, and thus may erode material from the banks which contains land shells. These are often blocked with dirt, and float readily. Sparks (1964) discusses the composition of mollusc assemblages, and critically examines the processes of deposition, fossilization, and preservation, the effects of contamination by older or younger shells, and the problems of sampling and identification (see Chapter 2) relevant to the study of fossil molluscs.

Molluscs are reliable indicators of local habitat conditions near the site of their deposition. However, they are not such reliable indicators of regional conditions or climate, because, being small, they can occupy locally favourable habitats under a generally unfavourable climate. An example of this is seen towards the end of the Ipswichian interglacial in Britain, where species characteristic of the warmest middle part of the interglacial regularly persist in reduced numbers at a time when the botanical evidence indicates climatic deterioration prior to the onset of Devensian glacial stage (Sparks, 1963). Similarly, on a smaller time scale, Kerney (1963) demonstrated that relatively thermophilous species which spread during the warm Allerød phase of the Devensian late-glacial persisted through the following renewed cold conditions of the Younger Dryas phase, into early Flandrian time. References are given at the end of the chapter to mollusc identification manuals and to relevant works on molluscan ecology and distribution.

Ecological groups of molluscs

Because of their ability to indicate local habitat conditions, Sparks (1961) was able to divide molluscs into ecological groups, and then to use changes in the proportions of the groups to deduce local environmental changes.

After the obvious division into freshwater and land species Sparks subdivided each of the groups.

1. *Freshwater molluscs* can be divided into four major groups.

a) A 'Slum' group. These species tolerate poor water conditions, such as small water bodies subject to drying up, stagnation, and considerable temperature changes. Typical British members are the snails *Lymnaea truncatula*, *Aplexa hypnorum*, and *Planorbis leucostoma*, and the fingernail clams *Sphaerium lacustre*, *Pisidium casertanum*, *P. personatum*, and *P. obtusale*.

b) A 'Catholic' group. These species tolerate a very wide range of freshwater conditions apart from the worst 'slums'. Typical British members are *Lymnaea palustris*, *L. peregra*, *Planorbis crista*, *P. contortus*, *Segmentina complanata*, *Sphaerium corneum*, *Pisidium milium*, *P. subtruncatulum*, and *P. nitidum*.

c) A Ditch group. These species occur in ditches or streams with clear, slow-moving water and an abundant growth of aquatic plants. Typical British members include *Valvata cristata*, *Planorbis planorbis*, *Segmentina nitida*, *Acroloxus lacustris*, and *Pisidium pulchellum*.

d) A Moving-water group. These species occur in faster streams (not mountain torrents) and larger water bodies, such as lakes where the water is disturbed by wind and currents. Typical British members include *Valvata piscinalis*, *Bithynia* spp., *Lymnaea stagnalis*, *Physa fontinalis*, *Pisidium amnicum*, *P. henslowanum*, *P. moitessierianum*, and the larger freshwater mussels.

2. *Land molluscs* can also be subdivided. These groups are less distinct ecologically than the freshwater molluscan groups.

a) Marsh and associated habitats group.

b) Dry-land group of open country, including, for example, *Vallonia costata*, *V. excentrica*, *Pupilla muscorum*, and *Helicella* spp.

Other groups may also be delimited when required, such as woodland species, or species associated with man (anthropophiles). We shall discuss how molluscs have been used in palaeoecological interpretations after a consideration of their geographical distribution.

Geographical groups of molluscs

Throughout Europe and beyond, molluscs have specific distribution patterns, many of which are, as yet, little known. To be of use as climatic indicators, their climatic tolerances and limits must be known. It is often the case that the limits of distribution of a species are controlled by climatic conditions. As a rough guide to temperature tolerance of molluscs in fossil assemblages, Sparks (1964) divided them into geographical groups on the basis of their northern limits in Fennoscandia.

A. Species reaching to or almost to the Arctic Circle.

B. Species reaching 63°N.

C. Species reaching 60–61°N (approximately the present northern limit of oak).

D. Species reaching only southernmost Sweden, or confined to the European mainland.

Sparks used these groups to indicate the general climatic trends at the end of the Ipswichian (Eemian) interglacial at Histon Road, Cambridge (Sparks and West, 1959) and at Stutton, Suffolk (Sparks and West, 1963). The behaviour of the groups at Stutton is illustrated in Fig. 7.1. The sequence covers the last part of the interglacial, and three sections have been combined to show the complete succession. Group A, the most cold-

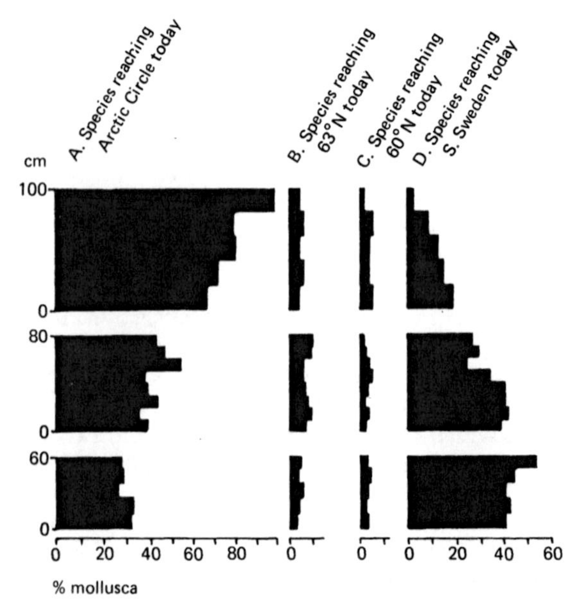

Fig. 7.1 The climatic deterioration at the end of the Ipswichian (last) interglacial illustrated at Stutton, England, by changes in groups of mollusca tolerant of different temperature conditions. (After Sparks, 1964.)

tolerant, increases gradually at the expense of Group D, the most thermophilous, whilst the intermediate groups show little response to the gradually deteriorating climatic conditions.

Kerney (1968) demonstrated the use of individual species in determining climatic changes during the Flandrian (post-glacial). The mid-Flandrian, between about 8000 and 2500 B.P., had a somewhat warmer climate than at present, deduced from finds of the European pond tortoise (*Emys orbicularis*), now extinct in Britain, and changes in the abundance of *Viscum*, *Hedera*, and *Ilex* pollen (Iversen, 1944; Godwin, 1975). No snails have become extinct due to the temperature decline, but three in particular have retreated from their previous northern limits. Kerney presents European maps of their distribution, and describes their restriction in Britain. He suggests that *Pomatius elegans* and *Lauria cylindracea* are oceanic species unable to withstand winter cold, whereas *Ena montana* is a continental species which requires summer warmth. It would appear that both winter and summer temperatures have decreased since the Flandrian climatic optimum in Britain.

The climate of the Ipswichian interglacial can be seen to contrast with that of the Flandrian, using the evidence of these three molluscs. *Ena montana* is abundant, *Lauria cylindracea* is very rare, and *Pomatius elegans* is absent. The conclusion that the Ipswichian climate was therefore warmer and more continental than that of the Flandrian is supported by evidence from plants and other animals (see West, 1977).

Kerney (1976a) has illustrated the value of detailed mapping of mollusc species in determining factors which influence their distribution. Besides its sensitivity to winter cold, *Pomatius elegans* is restricted to calcicolous habitats on chalk or limestone. *Discus ruderatus* is a widespread continental species which is presently absent from western Norway, and western Europe including Britain. Since it has been recorded in sediments of early Flandrian age in Britain, its subsequent extinction is probably due to an increased oceanicity of climate. It probably requires low winter temperatures for survival. The British distribution of non-marine molluscs is now sufficiently well known that an Atlas has recently been produced (Kerney, 1976b). Such maps are of great value in assessing the modern ecology and geography of a species and in indicating its potential value as an indicator of past conditions.

Molluscan palaeoecology

Molluscs have been studied in a wide range of Quaternary deposits in Britain, both glacial and interglacial. However, information about deposits older than the last interglacial (Ipswichian) is still sparse, and as yet there is no comprehensive picture of the fauna during these older glacial and interglacial phases (Kerney, 1977a). The molluscs of the Ipswichian interglacial, the last glaciation (Devensian), and the Flandrian (post-glacial) are, in contrast, relatively well known, although there is scope for a lot more basic exploratory work, particularly outside southern and eastern England.

Ipswichian (last) interglacial

The molluscan fauna of the last interglacial is very well known, largely through the work of Sparks. Over 125 species are known from last interglacial deposits in south-east England, where fossiliferous deposits are concentrated. Sparks (1963) systematically reviews the occurrence and ecology of most of these species. The greatest abundance of species is in the middle of the interglacial. Many of them are thermophilous and presently distributed in southern and eastern Europe, suggesting a warmer, more continental climate during the Ipswichian interglacial. Many of these species survive into the 'pine-woodland zone' and 'steppe-tundra zone' of the end of the interglacial, perhaps persisting in favourable microhabitats.

The local palaeoecology of several Ipswichian deposits has been worked out by Sparks using his ecological groups (see p. 122). As one example, we shall refer to his work at Histon Road, Cambridge (Sparks and West, 1959; Sparks, 1961).

At the base, A, of Fig. 7.2 marshy conditions predominate, the slum group being most common. Between levels A and B, the ditch and catholic groups replace the slum group, culminating at level B with a peak in the moving-water group. Stream conditions persist until level C. These conditions led to a larger proportion of land snails, mostly from dry-land habitats, being represented in the sediments. At level C, there is a reversal to marsh conditions, the slum group becomes dominant, and marsh-habitat land snails outnumber the dry-land snails. At level D, there is an abrupt return to moving-water conditions, but with too few land snails for them to be graphed. Studies of pollen and plant macrofossils on the same samples support the interpretation based on the mollusc stratigraphy.

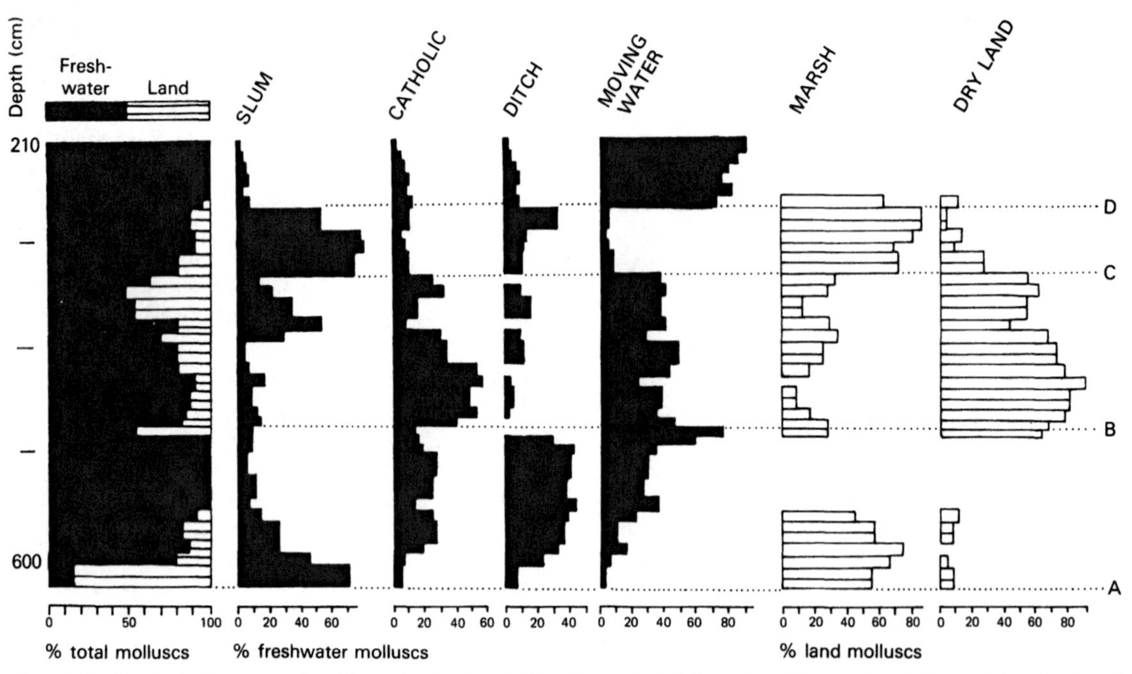

Fig. 7.2 Ecological groups of mollusca in the Ipswichian interglacial deposit at Histon Road, Cambridge, England. The freshwater and land groups are shown as percentages of the total freshwater and land molluscs respectively. (After Sparks, 1961.)

The changes may be explained by a meandering stream shifting across its floodplain.

Devensian late-glacial

Kerney (1963) studied the molluscs in several fossil soils of Late Devensian age on the chalk downs of south-east England. His results from Dover Hill, Folkestone (Fig. 7.3) are representative of the area as a whole.

The molluscan fauna is restricted to fourteen species, all characteristic of open-ground habitats. The five species in molluscan Zone I are all widespread in northern Europe well beyond the Arctic Circle. They all increase in absolute abundance in Subzone Ib, which has led Kerney to infer a small climatic amelioration in the periglacial conditions, which he correlates with the Bølling interstadial of Denmark. At the beginning of Zone II, the same species increase once more, and other species appear, such as the relatively thermophilous *Hygromia hispida*, *Abida secale*, and *Helicella itala*, suggesting a greater rise in temperature than in Subzone Ib. At the top of Zone II, soil developed over the chalk rubble and chalk mud, suggesting a fairly closed plant community. Fossil charcoal fragments were identified as juniper and possibly birch. Kerney therefore correlated Zone II with the Allerød interstadial, which was confirmed by a radiocarbon date on the charcoal (see Fig. 7.3). The beginning of Zone III is marked by a decline in all species except *Hygromia hispida*, and it is correlated with the Younger Dryas stadial. However, there are no extinctions, suggesting that the Zone III climate differed from that of Zone I, probably being more humid and somewhat warmer, thus allowing the thermophiles to persist, and even flourish. Other investigations of Devensian faunas are reviewed by Kerney (1977a).

Flandrian or Holocene (post-glacial)

Several studies have been made on Flandrian molluscs. Those by Sparks (1962) at Haweswater on the Carboniferous limestone of the southern Lake District produced a very poor fauna with virtually

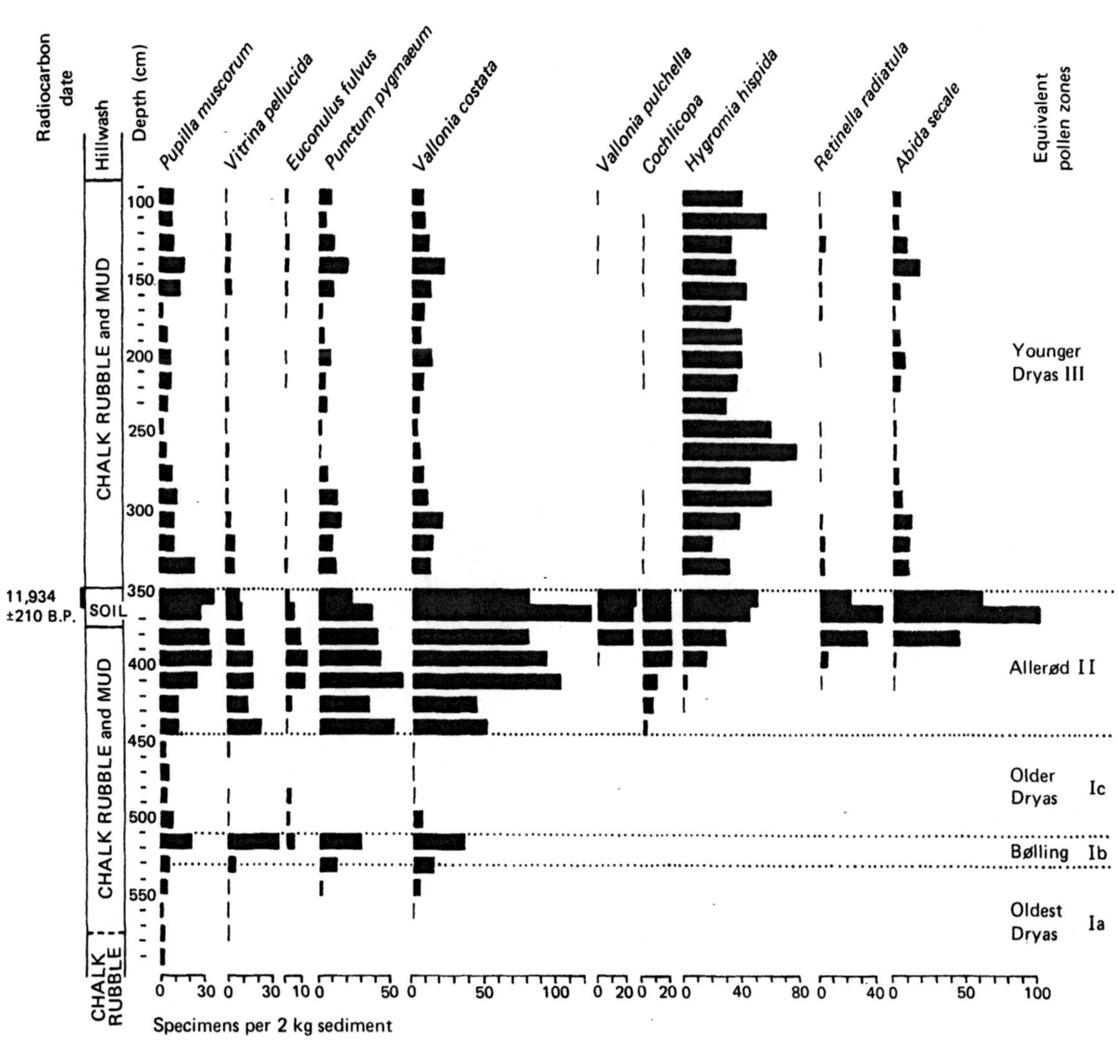

Fig. 7.3 Mollusca from Devensian late-glacial soil at Dover Hill, Folkestone, Kent, England, showing number of specimens in a 2 kg sample. (After Kerney, 1963.)

no change throughout the deposit. However, the site at Apethorpe, Northamptonshire (Sparks and Lambert, 1961) proved more interesting. The molluscs and plant fossils combined to reveal a situation similar to that at Histon Road in the Ipswichian interglacial, with a stream moving across its valley, occasionally flooding the whole area. A core from the edge of the valley deposits contained a greater representation of xerophilous and woodland species.

A more general picture of Flandrian molluscan changes was obtained by Kerney, Brown, and Chandler (1964) from Brook in Kent (Fig. 7.4). Kerney (1977b) has since formalized these changes in a series of molluscan assemblage zones for the Flandrian in Kent. At the end of the last glaciation (pollen Zones I–III; molluscan Zones y and z of Kerney, 1977b), the fauna was composed of northern taxa of palaeoarctic, holarctic, or boreal-alpine distribution. Most of them subsequently became extinct in south-east England. Open-ground species predominated at the opening of the Flandrian (molluscan Zone a), but were soon replaced by

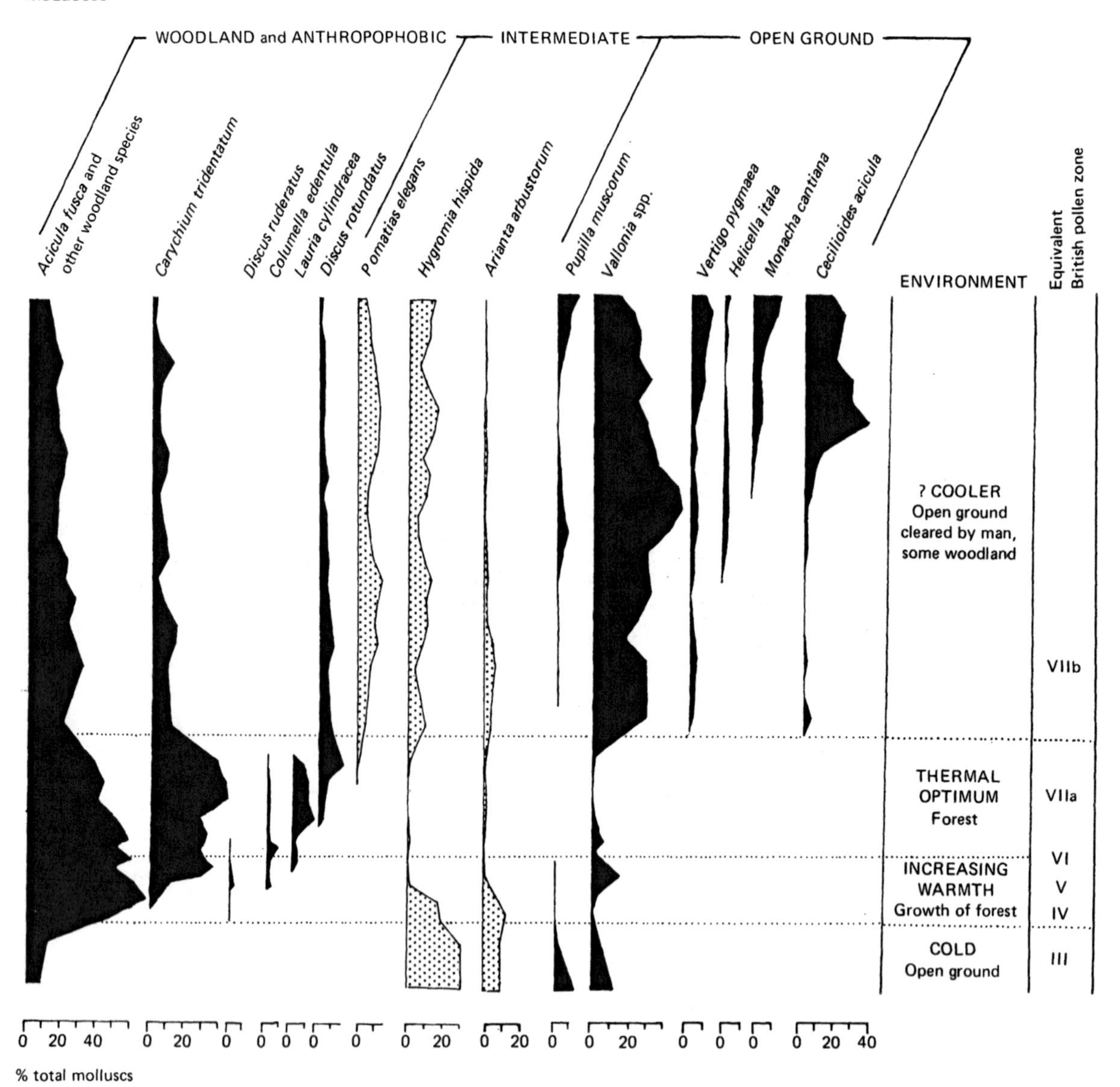

Fig. 7.4 The Flandrian (post-glacial) molluscan succession at Brook, Kent, England. (After Kerney, 1968.)

woodland and anthropophobic (adversely affected by human influences) species as deciduous forest developed and clothed the landscape (pollen Zones IV–VIIa; molluscan Zones b–d). There is a dramatic change at the beginning of pollen Zone VIIb (molluscan Zone e). The dominance of the grassland genus *Vallonia*, the return of some late-glacial species such as *Pupilla muscorum*, and the reduction in the woodland group are reflections of human disturbance of the landscape, which resulted in the destruction of forest, and the creation of more open, xeric habitats.

Kerney (1966) has discussed in detail the effects of man on the British molluscan fauna. Man has produced an over-all drying of the landscape with the reduction of woodlands and the drainage of wetlands. Man has thus favoured xerophilous molluscs at the expense of hygrophiles. The net result, however, is a diversification of habitat, with small relict woods, scrub, hedgebanks, ditches, grassy margins of fields and woods, stone walls, and grassland. Besides a reduction of woodland and marsh species and the expansion of open-ground species, some snails thrive in association with man. In addition, man has introduced several alien species into Britain, particularly since Roman times. *Helix aspersa*, the garden snail, is the most notable introduction and its occurrence defines molluscan Zone f. On balance, man's influence has been favourable to the molluscan fauna, although in recent times with the development of heavy industry, the use of insecticides, and widespread pollution, many molluscs have declined in their abundance.

Evans (1969, 1972) has discussed the use of molluscs as chronological and palaeoenvironmental indicators in archaeology, particularly in connection with archaeological sites on the Chalk of south-east England where reliable pollen analytical data are lacking (Dimbleby and Evans, 1974). Molluscs are, as we have seen, useful indicators of the destruction of woodland and the creation of grassland habitats. Their chronological use is rather limited, however, to the relatively few species which have probably been introduced by man into Britain, such as *Helix aspersa*, the Roman snail *H. pomatia*, and various species of the Mediterranean genus *Helicella* in Medieval times.

The palaeoecology of Pickerel Lake, U.S.A.

So far we have only discussed British molluscan studies. Relatively few palaeoecological studies involving molluscs have been done in other countries, but the investigation of Pickerel Lake in South Dakota by Watts and Bright (1968) presented a rather different approach which combined mollusc and plant evidence to reconstruct the palaeolimnology and palaeoecology of the lake and its catchment. We have already mentioned the study of the diatoms from Pickerel Lake by Haworth (1972) (p. 91). Her data generally confirmed the results of Watts and Bright. The molluscs and aquatic plants were used to deduce conditions of water depth and chemical quality, which are summarized in Table 7.1.

At the time of its formation as a hollow produced by the melting of a buried ice block at the end of the last (Wisconsinan) glaciation, Pickerel Lake had alkaline but not very hard water, and was surrounded by spruce forest. As the climate warmed up, prairie gradually replaced the forest. At the maximum temperature and dryness, between 8000 and 4000 years ago, the water-level of Pickerel Lake fluctuated and fell very low. However, unlike many prairie lakes, it never became sufficiently concentrated by evaporation to produce conditions of high sulphate ions, which would have been reflected in its fauna and flora. During the last 4000 years, some deciduous forest recolonized the lake shores, and the water level stabilized.

In order to produce such an interpretation Watts and Bright sampled the present-day mollusc and plant communities. The ecology of American molluscs is not well known, and more work needs to be done. There is some indication that particular mollusc assemblages occur under certain conditions of lake water chemistry and aquatic or semi-aquatic vegetation. These conditions are correlated with the over-all vegetation type (prairie, deciduous, or coniferous forest) which is largely controlled by differences in regional climate (R. C. Bright and H. H. Birks, unpublished data).

Coleoptera as fossils

The majority of insect fossils identified from Quaternary sediments belong to the Coleoptera, or beetles. This is partly because beetles have robust chitinous exoskeletons, which are often well preserved, and partly because the order Coleoptera is taxonomically the best documented insect group, and therefore is one of the least difficult to identify. Other insect fossils may be very abundant in sediments, such as Diptera remains, but these are often broken or distorted. The heads of Chironomidae larvae which are discussed in Chapter 6 are an exception.

During the last 20 years our knowledge of Quaternary beetles has developed rapidly, due principally to the work of G. R. Coope and his colleagues. They have shown that it is possible to make many specific identifications from a careful study of the fossil fragments, and that whole beetles are not required. Recent work has involved the use of high magnification under the scanning electron microscope to show minute sculpturing details for comparison with modern reference material (e.g. Morgan, 1969, 1973). As in diatoms, the modern taxonomy of the group is based on exoskeleton

Table 7.1 Summary of the inferred changes in vegetation and ecological conditions in and near Pickerel Lake during the past 11 000+ years*. (From Watts and Bright, 1968.)

Pollen zones	Date (years B.P.)	Terrestrial plant communities	Aquatic and semi-aquatic plant communities	Climate	Conditions of Pickerel Lake				
					Depth at core site (m)	pH	Alkalinity	$SO_4^{=}$	Hardness
4		Prairie still dominant in the region. Deciduous forest common around lakes and in gullies again; *Quercus* and *Fraxinus* abundant, with *Tilia*, *Acer*, *Ostrya*, *Salix*, and *Ulmus*	Aquatic vegetation essentially as in zone 3, but with more *Myriophyllum exalbescens*-type. Reed marshes common around the lake margin	Warm, but with more summer precipitation than during zone 3 time	(stable level) <3	Now 8.5 8±	Now 162 ppm High	Now 38 ppm Low	Very hard
	4000 (estimated)								
3		Prairie on the upland, with more herbs than for zone 2; similar to modern blue-stem prairie. Perhaps a few groves of deciduous trees around lakes and in gullies. Many 'drawdown' herbs on lake bottom exposed by desiccation	*Najas*, *Zannichellia*, *Potamogeton*, and *Ceratophyllum* the most common aquatics. Reed marshes around the lake margin more common than previously	Warm, with recurring summer drought	<3 (periods of very low water level) <2 <3	>7	High, with high carbonate	Low	Very hard
	8000 (estimated)								
2		Mixed mesophytic deciduous forest on north-facing slopes on the upland and around lakes with *Ulmus* the most common tree, and with *Acer*, *Ostrya*, *Carya*, *Fraxinus*, *Betula*, *Alnus*, and *Abies*. Oak savanna on drier sites on the upland with prairie of grass and perhaps *Pteridium* in the openings	*Najas flexilis* the most common aquatic. Marshes with *Typha* and *Scirpus* more common around the lake than previously	Warming	3.5 (stable level) <3 3.5	>7	Slightly higher	Low	Not very hard
	10 670 ± 140 (Y-1361)								
1		Boreal forest with shrubby areas dominant. *Picea glauca* on well-drained sites. *P. mariana* (wet soils) and *Larix laricina* and *Fraxinus nigra* (boggy areas) mostly in depressions and around incipient lakes	Pioneer species of *Carex* and *Sparganium*, with *Najas flexilis* *Sagittaria latifolia*, *Eleocharis smallii*, and *Typha latifolia*. Marshes around the margin of the lake	Cool and moist	(stable level) <3	>7	Low to moderate	Low	Not very hard

* Read upwards from base.

features, thus enabling fossils to be specifically identified from exoskeleton fragments.

Coleoptera appear to have undergone little morphological evolution since the beginning of the Quaternary. The oldest Quaternary fossils are frequently recognizable as modern species. Similarly, little physiological evolution is thought to have occurred, because groups of beetles which are associated today can often be found together in sediments, suggesting that their ecology has probably not changed. There may be a few exceptions, but these are generally explained by a change or restriction in the species ecological tolerance perhaps at the edge of its geographical range. The stability of very closely related species of *Pterostichus* subgenus *Cryobius* was illustrated by Matthews (1974) from a remarkable long profile at Cape Deceit on the Seward Peninsula of Alaska. It had been supposed that all these closely related species of *Pterostichus* had evolved fairly recently during the last (Wisconsinan) glaciation. However, many of the species were determined from fossils of about 400 000 to 900 000 years old, from the Middle Pleistocene. It is now thought that during the Pliocene, the tundra habitat was disjunct, and that the widespread boreal *Cryobius* may have evolved independently in each tundra area in a slightly different way. This initial radiation probably happened rapidly, until all the suitable habitats had been filled, when evolution again may have become minimal. On the coalescence of the different tundra areas during the Pleistocene, the various *Cryobius* races intermingled and became widespread, but remained genetically and taxonomically distinct.

An example of morphological evolution during the Pleistocene is also described by Matthews (1974) from Alaska. In the Middle Pleistocene, *Tachinus apterus* (a ground beetle) had fairly large wings, and correspondingly long elytra. Modern species are brachypterous, or short winged, and flightless. Matthews found all intermediate stages of this development. Coope (1977) suggests that these differences reported by Matthews (1974) are of the same order of magnitude as would be expected *today* between different geographical races of the same species. It is necessary to extend back to the late Miocene to see unambiguous signs of evolution at the species level (Matthews, 1976). Because of this almost total lack of evolution during the Pleistocene, it is felt justified to reconstruct the environment of the past from the known environmental preferences of the modern species using the principle of uniformitarianism. Coope (1970, 1977) discusses this topic in detail.

Few fossils have been definitely determined as extinct species. If they are unknown, it is usually the case that a living example is eventually found, perhaps from some obscure part of the world. A spectacular example was described by Coope (1973). An unknown species of *Aphodius* (dung beetle) was frequently found in deposits of Devensian (last glaciation) age from the English Midlands. It was eventually identified as *Aphodius holdereri* which now lives on the northern slopes of the Himalayas, Tibet, between 3000 and 5000 m altitude. Several other unknown fossil beetles have been tracked down to arctic species living in Russia and Siberia today.

Beetle fossils may be extracted from sediments by splitting the sediment along bedding planes and by hand picking the fossils. This is laborious, and leads to the over-representation of the more conspicuous fossils. Alternatively, the sediment may be washed through sieves with water, and the fossils picked from the residue under a binocular microscope, as in the extraction of plant macrofossils (see Chapter 5). In certain cases, the chitinous beetle remains may be separated by flotation from the matrix.

Frequently the fossils are brightly iridescent, as in living beetles, but often a different colour. Repeated wetting and drying destroys the colours, because the interstitial waxes are lost. After extraction the fossils are kept for further study either gummed on cards or in tubes of alcohol, or, in special cases, embedded in balsam or resin.

The fossils are examined under a binocular microscope, identified, and counted. If necessary the scanning electron microscope can be used to reveal the smallest structures.

When the numbers of separate exoskeletal parts have been counted, the minimum number of individuals can be inferred from the most abundant part. As yet there has been little attempt to reconstruct past beetle populations, perhaps because their modern ecology is rather poorly known. Pearson (1962) addressed the question 'To what extent does an assemblage of beetle fossils from Devensian late-glacial deposits at St. Bees, Cumbria, represent a natural population?' He used the fact that, in a natural population, doubling the sampling area increases the number of species recorded by a constant amount. Species-area graphs were linear if plotted on logarithmic scales. Pearson's results are

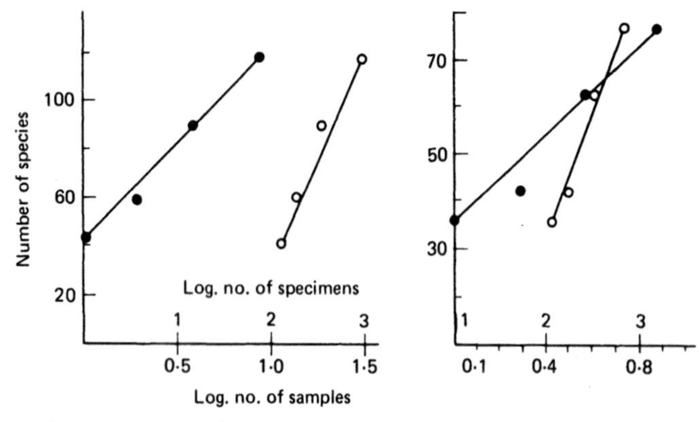

Fig. 7.5. The number of species of beetle obtained from the samples plotted against the logarithm of the number of specimens, ○, and the logarithm of the number of samples, ●. (After Pearson, 1963.)

shown in Fig. 7.5. Because the graphs were linear, he concluded that his assemblage did represent a natural population preserved in the Devensian late-glacial pond, and that his sampling procedure was reliable.

Coleopteran ecology and zoogeography

Ecology

Some beetles have a wide range of habitats (***eurytopic***) but many are very particular, and have a precise habitat (***stenotopic***). Coope (1967) surveys the factors which influence the local distribution of species.

1. *Thermal factors.* The correct range of temperature is often one of the most important environmental factors limiting the distribution of a species. Different thermal conditions may be necessary at different stages of the life cycle. In temperate climates, most beetles complete their life cycle in the summer and hibernate during the winter. The success of a species may depend upon the number of day degrees to complete its life history or to reach a suitable condition for hibernation, and different species have different temperature thresholds at which they become inactive. Some beetles tolerate a wide range of temperatures (***eurytherms***), but others tolerate only a narrow range (***stenotherms***). The thermal environment varies greatly with local conditions within a habitat. For example, south-facing slopes or bare soil exposed to sunshine are warmer, and the temperature varies through the day and night, whereas some places are buffered against large temperature changes, such as water bodies, piles of vegetation refuse or dung which maintain warmth during decay, and underground burrows.

2. *Soil conditions.* The type of soil is important for ground beetles, such as the Carabidae and Staphylinidae. Particle size, clay, sand, chalk, gravel, and organic content may all be ecologically significant.

3. *Water conditions.* Many beetles live in still or running water during their life histories, and because Quaternary sediments frequently accumulate in water, these beetles are commonly found as fossils. Water beetles are influenced by the degree of eutrophy of the water, its pH, and temperature. Other hygrophilous species live in marshy places beside water, and some ground beetles hunt for stranded insects along the water's edge.

4. *Chemical factors.* As already mentioned, water conditions are important for water beetles. Ground beetles may be influenced by the pH of the soil directly, whereas others are influenced indirectly because their food plants are restricted by soil conditions. Some beetles prefer salt (NaCl) perhaps in very small quantities, and are thus coastal in distribution.

5. *Botanical factors.* Many beetles eat plants (they are phytophages), and some are restricted to one plant species or genus. Phytophagous beetles include two large suborders, the Phytophaga and the Rhynchophora. Some species (Byrrhidae) eat moss, and several ground beetles live in moss, although they do not eat it. Leaf litter is the food of several species, often under specific trees, such as willows.

Besides providing food, forest trees also provide habitats for bark-boring beetles. Decaying wood is a favourite habitat for other beetles. Some beetles feed on specific herbs, for example *Adoxus obscurus* eats willowherb (*Epilobium*), and many beetles of open sun-exposed habitats (heliophiles) feed on plants typical of these habitats.

6. *Other factors.* Many beetles are carnivorous, and hunt a certain type of prey. The large *Dytiscus* water beetles eat tadpoles and sticklebacks, and some ground beetles eat snails and caterpillars. Other beetles feed on dead animals, and are often found fossilized in association with the bones of the animal upon whose carcass they were feeding. Yet other beetles live in dung of large herbivorous mammals. Their remains are often abundant in pool sediments, suggesting that the pools were used for drinking by large herbivores.

Some beetles are intimately associated with man, living in his food stores, the timber of his houses, and in domestic debris. Little is known of their ecology, as shown by Kenward (1975a), who tried to unravel the history of *Aglenus brunneus*, a blind flightless beetle found in organic debris associated with man, after it had been discovered in the debris on the floor of an Anglo-Scandinavian tannery excavated in York, England.

Zoogeography and climate

Beetles are often very good indicators of climate, particularly of temperature regimes. This is because stenothermic beetles only live under certain narrow temperature conditions. Beetles have been particularly important in elucidating the climatic fluctuations in Britain during the Devensian glaciation (Coope, 1977).

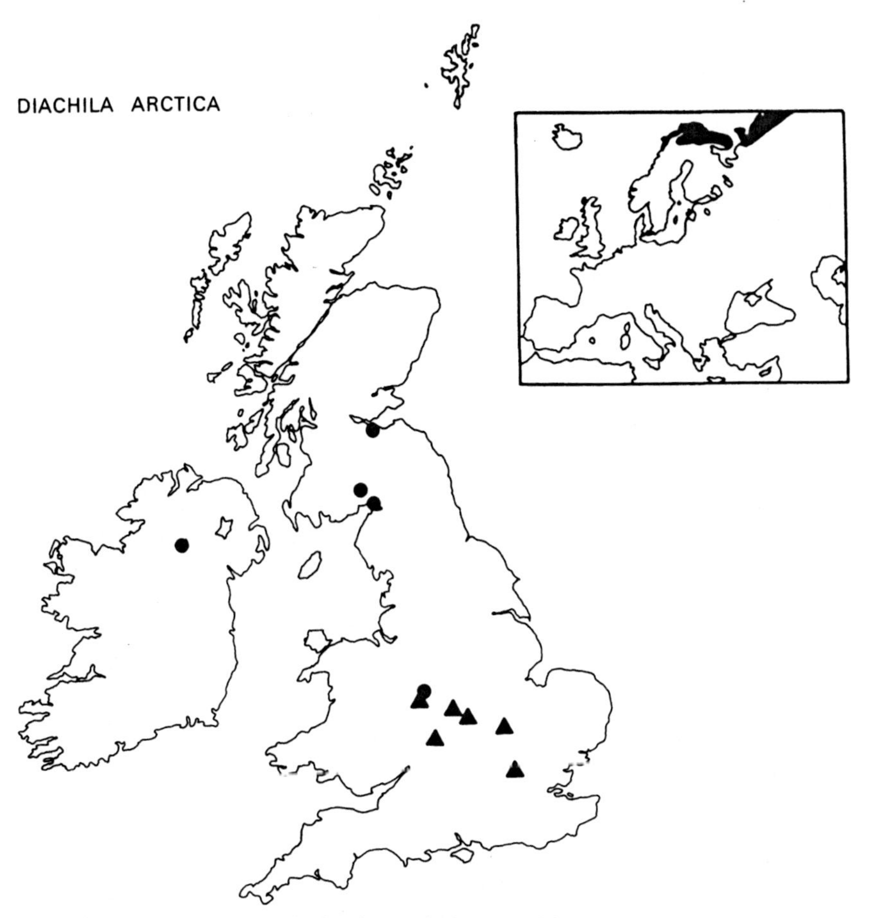

Fig. 7.6 Fossil localities of *Diachila arctica* in the British Isles. ▲, Mid-Devensian sites. ●, Devensian late-glacial sites. Its present European distribution is also shown. (After Coope, 1969.)

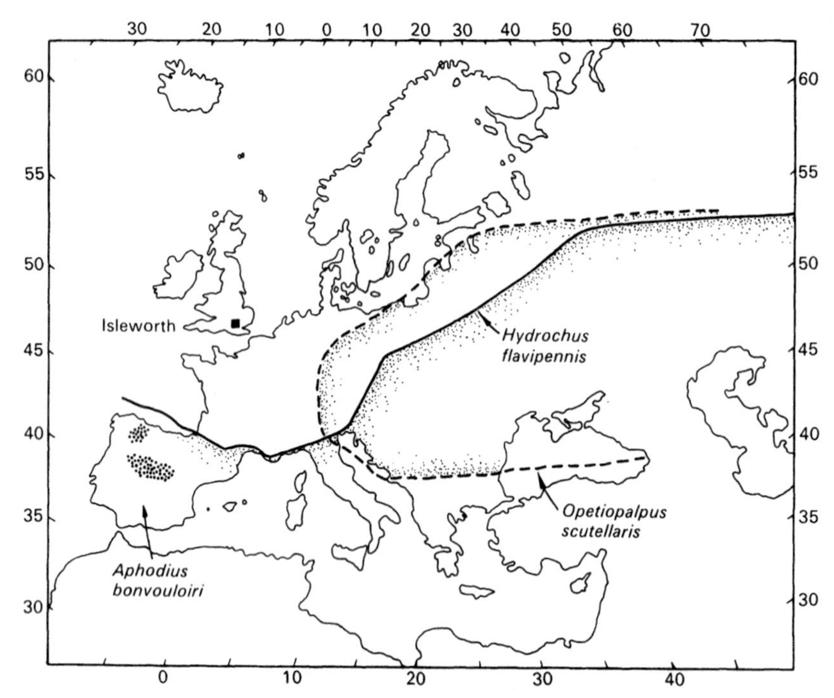

Fig. 7.7 Present day European distributions of *Hydrochus flavipennis*, *Aphodius bonvoulieri*, and *Opetiopalpus scutellaris*, all of which have been found fossil in the deposits of mid-Devensian age at Isleworth, England. (After Coope and Angus, 1975.)

Most beetles can fly, and they also have a high rate of reproduction. These features enable them to colonize rapidly new areas as they become available. Hence beetles have a short response time to climatic changes, much faster than most flowering plants.

To be able to use beetles as climatic indicators we must understand their present-day tolerances, and then we can extrapolate these tolerances back into the past, assuming meanwhile that the species concerned have not changed their ecological tolerances since the period of time being investigated. By using modern beetle data to reconstruct past environments, a consistent and ecologically sensible picture usually emerges.

At present, there is very little direct information available on the environmental tolerances of beetles. Most of it is inferred indirectly from the geographical distribution of species. In many cases, the distribution of a beetle species is limited by environmental, particularly thermal factors, and the limits of its tolerance can be estimated by comparing its distribution with the distribution of climatic parameters, particularly isotherms. Summer-temperature isotherms are most valuable, because hibernating beetles are largely indifferent to winter temperatures.

Figure 7.6 shows the present-day European distribution of *Diachila arctica* together with the localities in Britain where it has been recovered from Devensian sites. Its previous occurrence indicates the occurrence of a much colder climate in Britain. Figure 7.7 shows the modern distributions of three species found in Devensian deposits at Isleworth in southern England (Coope and Angus, 1975) which are no longer present in Britain. They all have more southerly ranges today, suggesting that some periods during the Devensian were warmer than the present British climate. We shall discuss the climatic reconstruction of the Devensian stage below.

Most beetles have widespread distributions, and are thus assumed to be eurytherms. The most important indicators are the stenotherms with narrow tolerances, and particularly those species which have unspecialized diets, and are not dependent, for example, on particular plants. When a beetle assemblage is analysed for climatic information, the modern distributions of all the indicator species are examined, and where they all coincide is taken as the place with the equivalent modern climate. By this method, the climate under which the Early Devensian deposits at Chelford, England were laid down was deduced to be similar to the cool temperate, moderately continental climate of southern Finland or Russian Karelia today (Coope, 1959). This evidence accorded well with evidence from the plant remains (Simpson and West, 1958).

Preservation and representation of Coleoptera in fossil assemblages

The state of preservation of beetle remains may indicate whether the fossils are autochthonous, if delicate or ill-developed species are found (Kenward, 1978). Similarly, two levels of preservation in an assemblage may indicate that some specimens have been secondarily redeposited.

The interpretation of beetle assemblages has been discussed extensively by Kenward (1978). The results of his analyses of modern beetle assemblages from various habitats have revealed that many species may originate well beyond the deposition site, either by flying from their normal habitat, by transport by predators, or in bird droppings. Kenward called this long-distance transported component the background fauna. Its potential importance in influencing the interpretation of an assemblage was demonstrated from a sample taken from a drain in a barren, built-up area of the city of York, England, by Kenward (1975b). An ecological interpretation of the assemblage suggested the presence of an open pool surrounded by short vegetation containing some weeds and nettles, an abundance of dead wood, and heaps of decaying vegetation and herbivore dung. Human proximity was strongly suggested by beetles which characteristically occupy buildings used for living and storage. The lack of local fossils in this situation emphasizes the importance of the background fauna. Kenward (1978) discusses the representation and separation of the background fauna for a series of modern and fossil samples.

The death assemblage may be further modified from the life assemblage by redeposition, either by natural causes, such as along a stream bank, or by human activity, especially in archaeological sites. Beetles themselves may also penetrate older deposits by burrowing or by creeping into crevices.

Kenward (1978) emphasizes the lack of ecological information about individual species, and about communities and populations of beetles. He advocates an approach to interpretation using the whole assemblage, in addition to the indicator value of individual species approach preferred by Coope, Osborne, and others. Statistical and numerical methods have potential value in the assemblage approach, and there is great scope for future work on the interpretation of beetle assemblages.

Coleopteran palaeoecology

Most palaeoecological studies using Coleoptera have been made in Britain, and they have concentrated on Devensian (last glaciation) deposits. There are relatively few studies on interglacials in Britain. Palaeoecological studies involving Coleoptera have been initiated recently in North America by Matthews (1968, 1974), Ashworth (1977), Ashworth *et al.* (1972), and Ashworth and Brophy (1972).

Hoxnian interglacial

Shotton and Osborne (1965) studied the varied fauna of the Hoxnian interglacial deposits at Nechells, near Birmingham, England, to complement the palaeobotanical investigations made there by Kelly (1964). They were able to make an extensive palaeoecological reconstruction of the first half of the interglacial, using fossils of fish, Cladocerans, ostracods, mites, molluscs, polyzoans, and insects. Beetles were prominent amongst the insects, the majority of the rest being chironomid larvae. Their results are summarized in Fig. 7.8. Geomorphological evidence led Kelly (1964) to propose that the lower sediments were deposited in a deep glacially dammed lake, and this suggestion is supported by the types of fish, ostracod, chironomid, and mollusc fossils found in them.

At the Zone IINa/b transition (Fig. 7.8) the carbonate-based fossils of molluscs, ostracods, *Chara*, and fish-teeth all disappear, indicating that the water and sediments became acidic. The lake became overgrown by the end of Zone IINc and further deposits were laid down in temporary pools and floods in the river floodplain.

Fossil beetles were abundant in the deposit, and they suggested that the climate was cold but not arctic at the base. The number of insect fossils was taken as a rough guide to the insect population. This increased during the warmer lacustrine phase. However, insect remains were sparse in the two

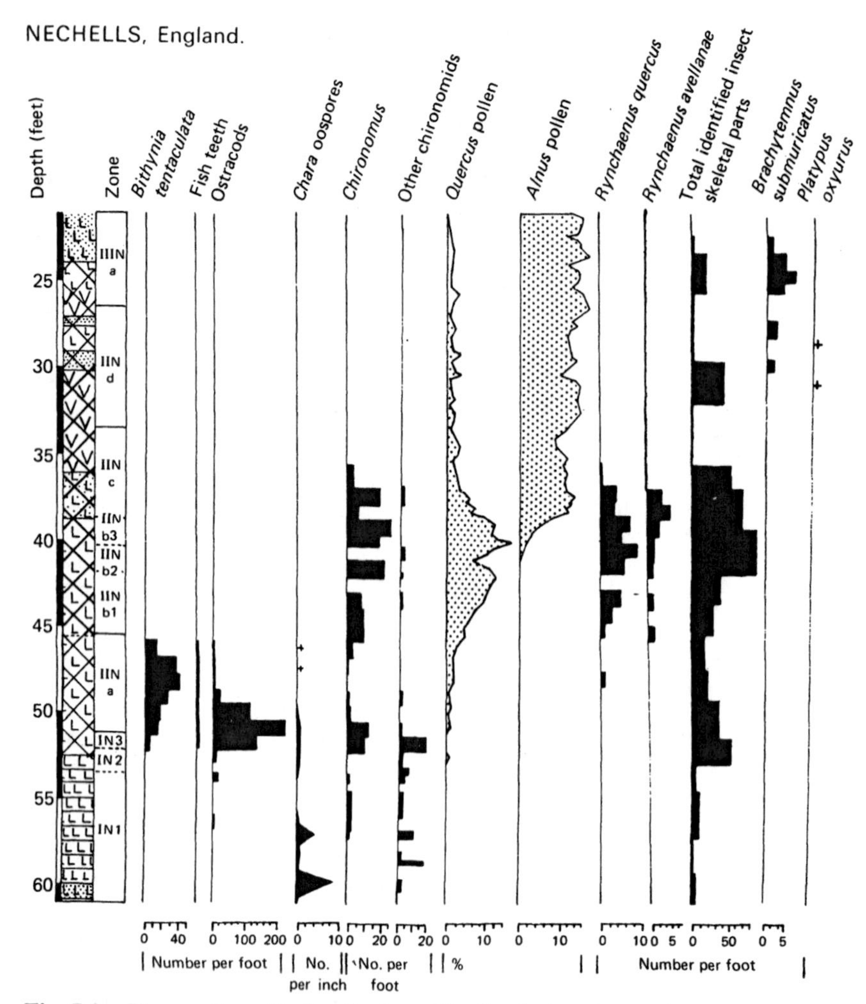

Fig. 7.8 The stratigraphic distribution of fossils of selected organisms in the deposits of Hoxnian interglacial age at Nechells, Birmingham, England. (After Shotton and Osborne, 1965.)

phases of woodpeat deposition. Most of the species occur in Britain today, suggesting that the climate was similar to the modern climate. Many of them were phytophages, and the numbers of the weevils *Rhynchaenus avellanae* and *R. quercus* are illustrated in Fig. 7.8 in relation to the pollen curve of their food plant *Quercus*. *R. avellanae* also eats *Corylus avellana*, but the latter was rare at Nechells at this time. The *Rhynchaenus* remains follow the oak pollen closely until they suddenly decline, even though oak pollen, although reduced, does not disappear entirely. Shotton and Osborne supposed that the rise in *Alnus* pollen at this time meant that alder occupied the site of deposition, which reduced the amount of *Quercus* pollen and prevented *Rhynchaenus* remains from reaching the site.

At the top of the deposit, two beetle species occur which today are absent from Britain and occur further south. *Platypus oxyurus* feeds on *Abies* bark, and is presumably restricted by the occurrence of *Abies alba*. *Abies* grows in montane conditions today in central Europe, but not necessarily in a warmer climate than Britain's. *Brachytemnus submuricatus* feeds on the bark of *Populus* and *Salix*, and this beetle may indicate that conditions were somewhat warmer in the Hoxnian. During the decline of the water beetles in Zone IINc, there is an increase in species characteristic of running water, indicative of the developing riverine environment.

We have discussed the Nechells investigation in some detail because it is an integrated study of many organisms, including plants and their associated beetles.

Ipswichian (last) interglacial

Studies on beetles from this interglacial are sparse. Franks *et al.* (1958) briefly described an assemblage from Trafalgar Square, London, which contained many southern molluscs, vertebrates, and plants, together with many thermophilous insects, several of which are now absent from Britain and occur today in southern and central Europe. Many of these were southern scarabaeine dung beetles, presumably associated with the large herbivorous mammals. A similar fauna was found at Bobbitshole, Ipswich (Coope, 1974). The distribution of *Oodes gracilis* is shown in Fig. 7.9. It appears (Lindroth, 1943) to require a summer temperature about 2.0°C higher than that at Ipswich today. Pearson (1963) describes two small beetle faunas from deposits at Selsey Bill and Sidgwick Avenue, Cambridge. They have a southern aspect, but all the species occur in Britain today.

Devensian glacial stage

The majority of coleopteran studies in the British Isles have been concentrated on Devensian deposits, particularly in the Midlands where such deposits are well represented. The main interest has been in the reconstruction of climatic changes during the Devensian, and a clear picture has now emerged, which is most recently reviewed by Coope (1975, 1977).

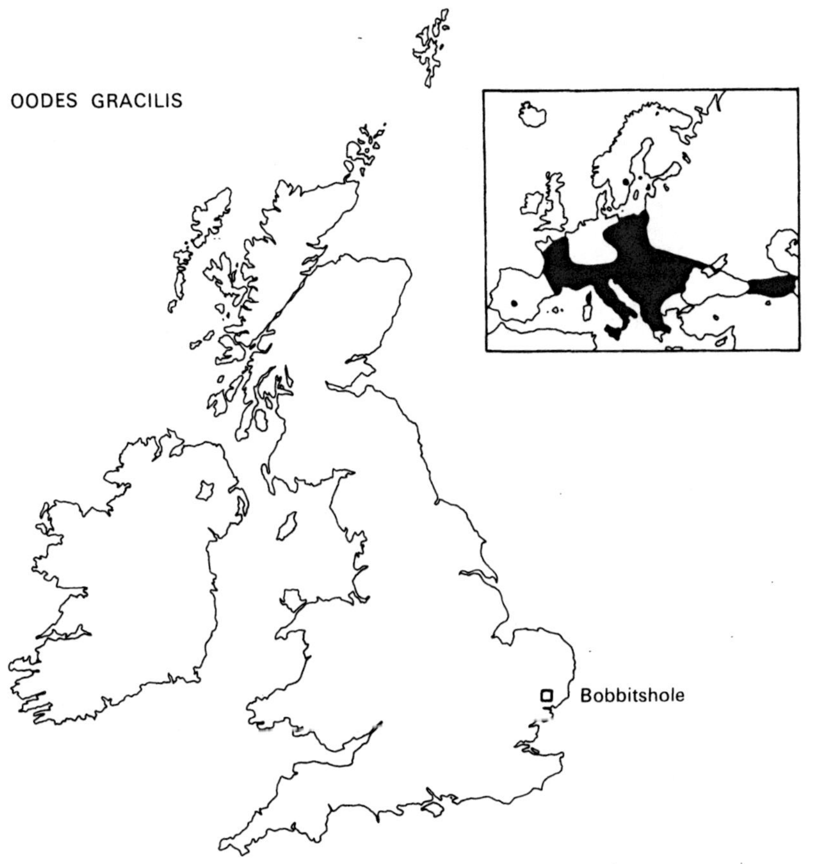

Fig. 7.9 The fossil occurrence of *Oodes gracilis* in the Ipswichian interglacial site at Bobbitshole is shown by a square. Its present European distribution is shown. (After Coope, 1974.)

1. *Chelford interstadial*

Tree stumps in the deposits at Chelford have been radiocarbon-dated to 60 800 ± 1500 B.P. Both the fauna and the flora indicate a July temperature of about 15 °C, some 2 °C lower and somewhat more continental than in Cheshire today. A very similar flora and beetle fauna was described from Four Ashes in the English Midlands by Morgan (1973), and correlated with the Chelford interstadial. At Wretton, Norfolk (West *et al.*, 1974), deposit WG was assigned on pollen evidence to the Chelford interstadial, but the beetle fauna had high arctic affinities, and indicated a barren tundra landscape. Coope (1975) suggests that the tree pollen may be derived from long distance transport, and be over-represented because of the extremely low local pollen production.

2. *Upton Warren interstadial complex of the Middle Devensian*

This term was proposed by Coope and Sands (1966) to cover the period between 50 000 and 25 000 years ago, before the major Devensian ice-advance. Assemblages of coleoptera have now been described from more than twenty sites in the English Midlands. The data can be summarized by a curve of average July temperature for the period reconstructed by Coope (1975, 1977) (see Fig. 7.10). Earlier reviews include those by Coope (1969) and Coope, Morgan and Osborne (1971), and the study of the complex Early and Middle Devensian sequences at Four Ashes by Morgan (1973).

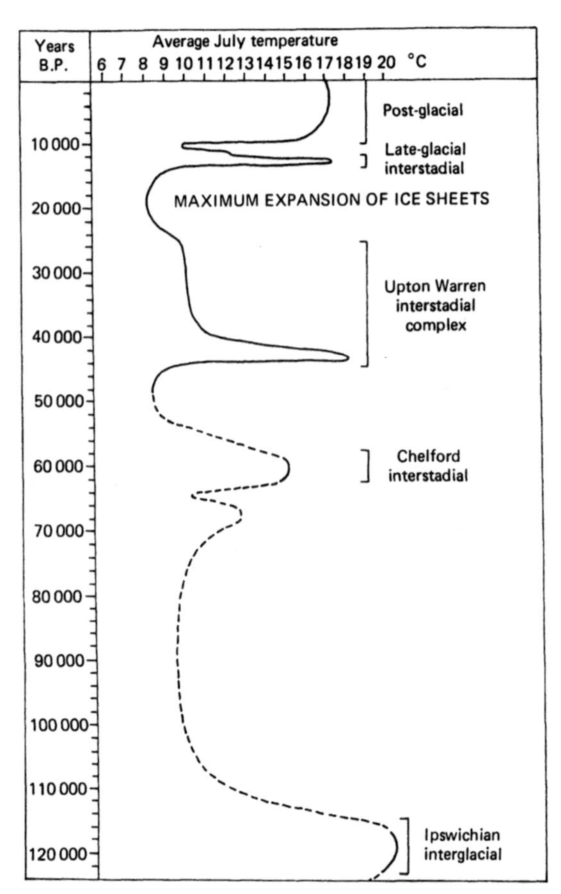

Fig. 7.10 Fluctuations in the average July temperature in lowland Britain since the last (Ipswichian) interglacial, as deduced from finds of fossil beetles. (After Coope, 1975.)

After the temperate Chelford interstadial, the climate became cold for about 20 000 years, when cold stenotherms, restricted today to arctic areas, replaced the temperate fauna. Subsequent deposits at Upton Warren (Coope, Shotton and Strachan, 1961), Four Ashes (Morgan, 1973), and Isleworth (Coope and Angus, 1975) have been radiocarbon-dated to between about 43 000 and 38 000 B.P. They all contain thermophilous Coleoptera, some species of which occur only in southern and central Europe today (see Fig. 7.7 and Coope and Angus, 1975), and there are no arctic stenotherms. Coope (1975) estimates that, at its thermal maximum, this interstadial had an average July temperature of about 18 °C, which is 1 °C or 2 °C higher than that in southern and central England today. However, plant remains from these sites consistently show an absence of trees and a predominance of grassland vegetation. There is also a lack of any beetles associated with trees. This anomaly is readily explicable in terms of the different migration rates of beetles and trees. As we mentioned before, beetles can rapidly occupy new territory as it becomes available. Trees, however, have a long generation time and relatively slow dispersal. During the pre-Upton Warren cold phase, trees and beetles alike probably migrated far to the south to suitable climates. All the available evidence indicates that the climatic amelioration was rapid and short-lived, perhaps leaving insufficient time for trees to recolonize Britain, but sufficient time for the mobile beetles to invade. An additional possibility is that large herbivorous mammals retarded tree regeneration and spread by extensive grazing.

After the thermal maximum (Fig. 7.10) the July temperature declined slowly and steadily, and faunas from this period show steadily decreasing

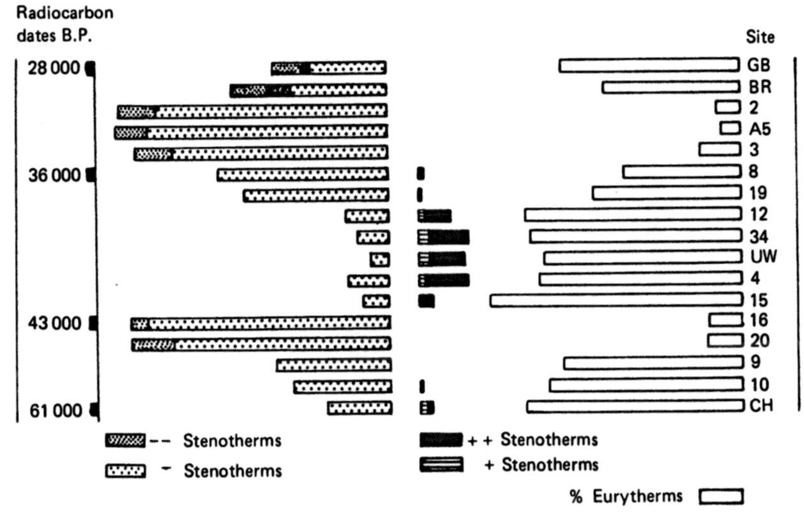

Fig. 7.11 Suggested chronology for the Four Ashes beetle faunas (numbered samples) and some other sites (CH, Chelford; UW, Upton Warren; BR, Brandon; GB, Great Billing). The proportions of cold (– and – –) stenotherms and warm (+ and + +) stenotherms are indicated on the left. The percentage of eurytherms is shown by open bars on the right. (After Morgan, 1973.)

numbers of thermophilous beetles and a corresponding increase in cold and arctic species. This succession is illustrated in Fig. 7.11 from the changes observed at Four Ashes (Morgan, 1973). The July temperature probably stayed at about 10°C, although the modern distributions of the beetles suggest that the climate was much more continental. Several species are now restricted to eastern Europe and Asia, including *Aphodius holdereri* from Tibet. This cold continental phase lasted until about 25 000 B.P. when the major Devensian ice-advance began. The few sparse coleopteran faunas from between 25 000 and 13 000 B.P. indicate extremely severe conditions, although perhaps not quite so continental. A decrease in continentality may have led to an increase in precipitation, which may have resulted in the ice advance. During the retreat phase of the ice, temperatures were again around 10°C in July, typical of tundra environments today, and the beetle fauna has a high proportion of arctic stenotherms with eastern distributions. For example, *Helophorus splendidus* was found in deposits about 13 000 years old at Glan Llynnau, North Wales (Coope and Brophy, 1972). Figure 7.12 shows its present known distribution.

3. *Devensian late-glacial*

After the main ice retreat, completed by about 13 000 B.P., there was a climatic amelioration, the so-called Allerød interstadial. This was followed by a deterioration lasting about 500–1000 years, during which small glaciers redeveloped and treeless conditions prevailed in Britain. This phase is called the Younger Dryas stadial by Mangerud *et al.* (1974). There is still considerable argument as to the reality of the Bølling interstadial and the Older Dryas stadial before the Allerød interstadial. The coleopteran evidence has added to the controversy because it apparently conflicts with the classical interpretation based on plant evidence. However, coleopteran evidence from an increasing number of sites appears to be consistent, ranging from Hawks Tor, Cornwall, the London area, the Midlands, North Wales, the Isle of Man, St. Bees, Cumbria, and southern Scotland (see Coope, 1975, 1977; Bishop and Coope, 1977).

Both coleopteran and plant evidence agree about the existence of the cold Younger Dryas stadial, in which arctic conditions returned and glaciers redeveloped in mountainous areas. By contrast, the beetle evidence indicated that the warmest time of the late-glacial interstadial occurred shortly after

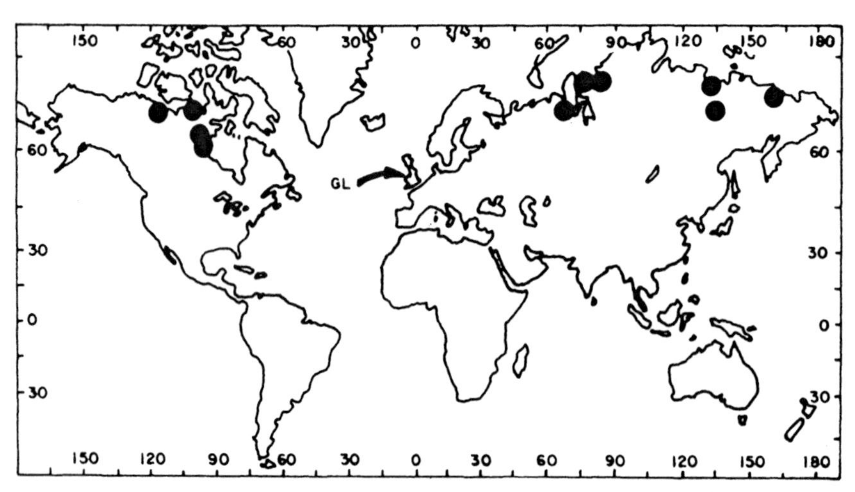

Fig. 7.12 The present world distribution of the arctic stenotherm *Helophorus splendidus*, which was found in Devensian late-glacial deposits at Glan Llynnau (GL), North Wales. (After Coope and Brophy, 1972.)

13 000 B.P. and that temperatures declined throughout Allerød time, finally dropping during the Younger Dryas stadial. The suggested temperature curve is illustrated in Fig. 7.13, along with the period(s) of tree development, pollen zones, and geological time units. This apparent anomaly can again perhaps be explained in terms of the greater mobility of insects. Although the temperature was declining at the opening of Allerød time, it had not crossed the threshold for tree growth, so birch woodland could develop in Britain as birch reached the country from its glacial refuges. Although trees are poor climatic indicators because of their slow response time, other plants respond more quickly, such as *Juniperus communis*. Although a general increase in pollen influx at about 13 000 B.P. indicating more favourable flowering conditions was recorded by Pennington and Bonny (1970) from Blelham Bog in the English Lake District, juniper pollen in Devensian late-glacial diagrams frequently only expands just before the birch pollen increase. It is difficult to reconcile this late expansion of a plant already present in the landscape with over a thousand years previously of summer temperatures as warm as those of today.

According to the coleopteran evidence temperatures began to fall during the time of the Older Dryas. However, there is no strong evidence for a decidedly colder period. Indeed unambiguous plant and geological evidence for this short-lived cold period is scanty, both in Britain and in Europe.

Flandrian or Holocene (post-glacial)

After the arctic conditions of Younger Dryas time, the temperature appears to have ameliorated rapidly at or just before 10 000 B.P. July temperatures appear to have risen rapidly to those of today or even slightly warmer during about 500 years.

Little detailed work has been done on coleopteran faunas from the early Flandrian. In general the terrestrial fauna consists primarily of woodland species (Osborne, 1965). Several of these, such as *Ernopocerus caucasicus*, which feeds on lime trees, are now extinct in Britain. However, this is probably not as a result of climatic change, but because the beetles' habitat has been almost completely destroyed by man. As with the molluscs, many beetles of open ground and grassland habitats expand with increasing forest clearance, and species dependent on trees or rotten wood for food or habitat decline. The faunas at Church Stretton, Shropshire (Osborne, 1972) cover the Neolithic period when forest clearance was patchy, and neighbouring woodland faunas and grassland faunas of the same age are recorded. A later Bronze Age fauna obtained from a well shaft at Wilsford, Wiltshire (Osborne, 1969) contains species of grassland and many dung beetles reflecting the abundance of grazing animals. Such species must have been confined to small areas where trees could not grow during the time of forest dominance, and subsequently they have been able to expand their range.

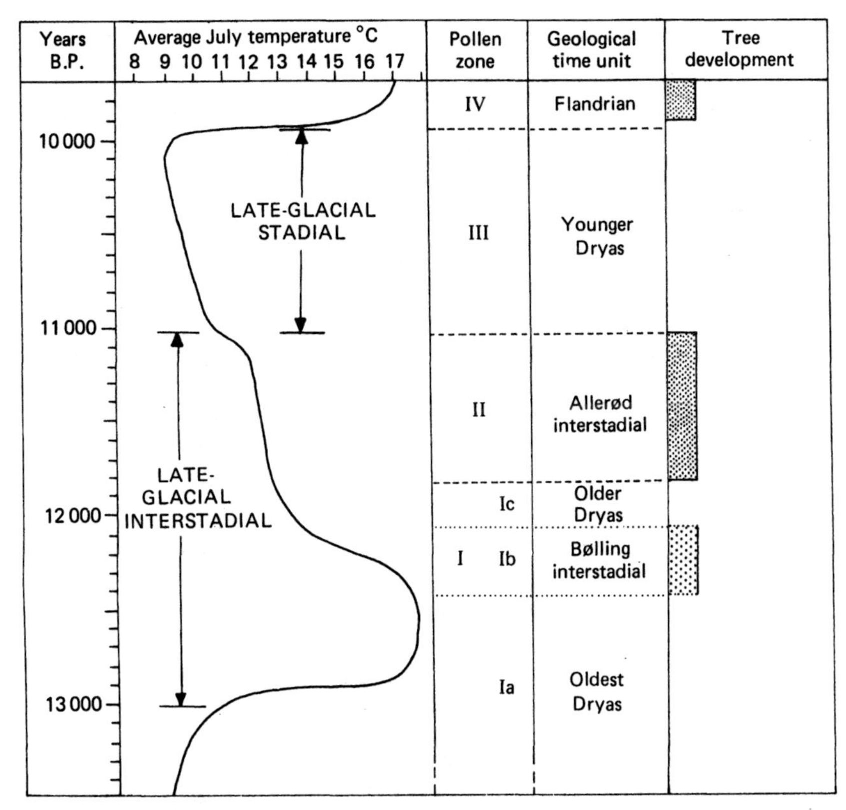

Fig. 7.13 Fluctuations in average July temperatures in lowland Britain during the Devensian late-glacial, as deduced from finds of fossil beetles. Periods of tree development (birch) are indicated. There is no unambiguous evidence of tree growth in Britain during Bølling Interstadial time. (After Coope, 1975.)

Osborne (1976) reviewed Flandrian coleopteran assemblages and he attempted to construct a temperature curve based on the evidence of present beetle distributions. Cooler periods could be correlated with periods of glacial advance in Europe during the Flandrian (Denton and Karlén, 1973).

Habitat reconstructions using fossil Coleoptera

Besides being useful climatic indicators, beetles can be used to reconstruct other features of past environments. Some interpretations have already been mentioned. Many beetles have rather specific habitat requirements (see p. 130) and if their fossils are found, it can be inferred that these conditions prevailed in that place in the past, assuming that the fossils originally lived at the site of deposition (see p. 129).

Beetle exoskeletons tend to be fairly locally deposited, in a similar way to plant macrofossils and molluscs. They are rather fragile, and cannot withstand substantial transport or repeated wetting and drying, which cause the chitin to deteriorate, warp, and break (Morgan, 1973). Therefore beetle remains provide a local picture of habitat conditions, primarily of the wetland and of the upland immediately surrounding it. We have already mentioned that the decline of *Rhynchaenus quercus* at Nechells was interpreted as an increase in the distance of oak trees from the site of deposition, as a result of the occupation of the damp ground by alders.

Most studies on beetles involve some reconstruction of local habitats to some degree. We shall discuss two studies here, but other good examples of habitat reconstructions are presented by Osborne (1972) for Late Devensian and Flandrian deposits at Church Stretton, Shropshire, by Coope and Brophy

(1972) for Late Devensian deposits at Glan Llynnau, N. Wales, by Morgan (1973) for Early and Middle Devensian deposits at Four Ashes, Warwickshire, and by Bishop and Coope (1977) for Late Devensian deposits in south-west Scotland.

1. *Early Devensian deposits at Isleworth, Middlesex (Greater London)*

Coope and Angus (1975) describe and discuss this deposit, which they correlate with the warm phase of the Upton Warren Interstadial Complex (p. 136). The organic layers were deposited in a silty pool with neutral or slightly calcareous, eutrophic water, which supported both submerged and emergent aquatic vegetation. Nearby, wet silt or clay soils supported luxuriant growths of plants nowadays considered to be weeds of cultivation such as *Artemisia* and thistles. The open vegetation was probably maintained by the disturbance of abundant large herbivorous mammals. Many bones of reindeer and bison were discovered, together with a large range of dung-inhabiting beetles.

Drier, more sandy soils seemed to have carried sparse vegetation characteristic of recently exposed ground with some acid open heathland and open calcicolous communities. There is no evidence from the plant remains or from the beetles for any tree growth in the landscape of this time, even though the climate, as reconstructed from the beetle assemblage, was warm enough for tree growth, perhaps even a little warmer than that of modern Isleworth. The likely explanation for this anomaly involving the relatively slow migration rate of trees is discussed on p. 136.

2. *Bronze Age deposits at Wilsford, Wiltshire*

Osborne (1969) describes the deposits recovered from a Bronze Age well on the chalk of southern England which are radiocarbon-dated to 3330 $\pm$ 90 B.P. By far the most abundant beetle remains were those of dung beetles, suggesting an abundance of livestock, and therefore of grassland. The diets of the phytophagous species confirmed this interpretation. There were also species which ate nettles and thistles, and these presumably grew round the well on soils enriched by the animal dung. Most of the beetles were heliophiles (living in open sunny habitats) and no tree-dependent species were recorded, suggesting extensive deforestation of the landscape, at least locally.

Osborne suggests that cattle were kept penned near the well for easy watering, thus leading to large amounts of dung, and also vegetable refuse resulting from the bringing in of taller meadow vegetation for fodder. The presence of considerable numbers of *Anobium punctatum* (the furniture beetle) suggests that perhaps some wooden structures were built above and around the well shaft.

The Scandinavian beetle fauna

Certain plants and beetles have restricted and disjunct distributions in Norway, and arguments have been put forward that they survived the last glaciation in local glacial refugia (e.g. Lindroth, 1949), particularly on the west coast.

Coope (1969) argues that these species could have recolonized Scandinavia from Britain near the end of the last glaciation. He argues that many of the species were present in Britain during the Devensian, although they have subsequently become extinct there. He argues that beetles could have flown, drifted on ice flows or flood refuse, or even walked across the then dry bed of the southern North Sea to the newly deglaciated areas of western Norway. In addition to beetles of cold climates, thermophilous species colonized Britain during the Devensian late-glacial interstadial, and these could have supplied the relatively thermophilous element amongst the so-called 'glacial relics'. During the Devensian late-glacial ice re-advance of Younger Dryas times, the fauna would have become restricted, and their disjunct present distributions may be due to this event. Further recolonization became impossible as the depth of the North Sea increased during the early Flandrian.

Vertebrates as fossils

Fossils of vertebrates are usually bones and teeth. Under exceptional circumstances, soft parts, such as skin and hair, may be preserved, the most remarkable being the finds of woolly mammoths and woolly rhinoceroses frozen in the Siberian permafrost, and the woolly rhinoceros 'pickled' in salt in Poland. Fortunately, a great deal about the mode of life of a vertebrate may be deduced from features of its skeleton and teeth.

Naturalists have always been interested in vertebrate fossils, and as a result there are many fine specimens in museums. However, these are usually accompanied by little or no stratigraphic information, and are therefore of limited scientific value. As in other aspects of palaeoecology, it is necessary to take carefully documented samples at the site. Vertebrate fossils vary greatly in size. The larger ones can be hand picked, but the sediment should also be sieved carefully in order to extract the smaller bones and teeth of animals such as voles, frogs, and fish.

Although the study of vertebrates has been actively undertaken on the continent during the twentieth century, comparatively little work has been done in Britain until recent years. The most extensive and synthetic work has been done by A. J. Stuart on British faunas, particularly those from the glacial and interglacial deposits of East Anglia. B. Kurtén has synthesized the European data in two books (Kurtén, 1968, 1972), the latter being a beautifully illustrated account for the general reader.

Just as in other palaeoecological studies, it is important to realize the mode of origin of the fossil assemblage (see Fig. 1.5) if meaningful palaeoecological communities are to be recognized. There are several situations which are good collectors of vertebrate remains (Stuart, 1974).

1. *Cave deposits*

Many animals shelter or make their dens in caves. Cave earth accumulates as the result of deposition of material such as dust, rock falls from the roof, substances dissolved in water, faeces, and animal carcasses. The occupants of caves are generally predators and scavengers, such as hyenas, wolves, and owls, which bring their prey into the cave to devour it there. Cave earths thus contain remains of both predators and their prey. Other animals may use the caves primarily for shelter, such as bats and, in particular, bears. In Europe during the Pleistocene, the now extinct cave bear (*Ursus spelaeus*) hibernated in caves. In Britain, it was geographically replaced by brown bears (*U. arctos*) which sheltered in caves here, but not in the regions occupied by the cave bear. Huge accumulations of cave bear bones have been found representing the bear population over thousands of years (see p. 151). Kurtén (1958, 1972) gives an excellent account of bears and other animals in caves.

The most interesting inhabitant of Pleistocene caves was early man. Much of our knowledge of his tools and way of life comes from the study of his remains and artifacts found in caves. Cave deposits are often stratigraphically disturbed, and fossils are often difficult to date accurately. In addition the fossils may not be representative of the contemporary fauna, due to the selectivity of the predators for different prey, and the occupation of the cave by a few species only, which may be mutually exclusive.

In addition to the cave dwellers, cave earths may also contain remains of animals which became accidentally trapped, for example, by falling down a hidden fissure. In this category the remains of mammoths and rhinoceroses preserved in ice or salt could be regarded as a result of accidentally falling into the depositional environment.

2. *Fluviatile deposits*

These include river terraces and alluvial floodplain deposits. Fluviatile deposits contain fossils of animals which lived near or in rivers, such as fish, amphibians, water birds, and water voles. Remains of terrestrial animals can be washed in by floods or by erosion of the banks as we have already noted for the Mollusca (p. 121).

Occasionally, a dramatic flood may drown a whole population of animals, or changes in the course of the river may isolate an aquatic community which dies when the pool dries up. Redeposition may often occur as the river changes its course and erodes old banks. The fossils are carried downstream until the current becomes too slow to carry them any further. As a result, large heavy fossils will be deposited first and the lighter ones will be carried longer. Considerable sorting may result.

3. *Lake sediments*

These contain fossils of the inhabitants of the lake and its shores. However, remains of large terrestrial animals may also be found in lake sediments. They may have drowned after breaking through thin ice, or come down to drink and become stuck in the mud. Many fossils of the giant Irish deer (*Megaloceros giganteus*) have been found in Irish lake sediments, and other large herbivores found in lake deposits include elk (*Alces alces*) and elephants.

Men often chose to live on lake shores. The Mesolithic site at Star Carr, Yorkshire, England (Clark, 1954) contains bones of many animals that were hunted, including birds.

4. *Tar pits*

The best known tar pits are in the south-western United States, the most famous of which is Rancho La Brea in Los Angeles (Stock, 1965). Animals became trapped in the sticky tar as they ventured

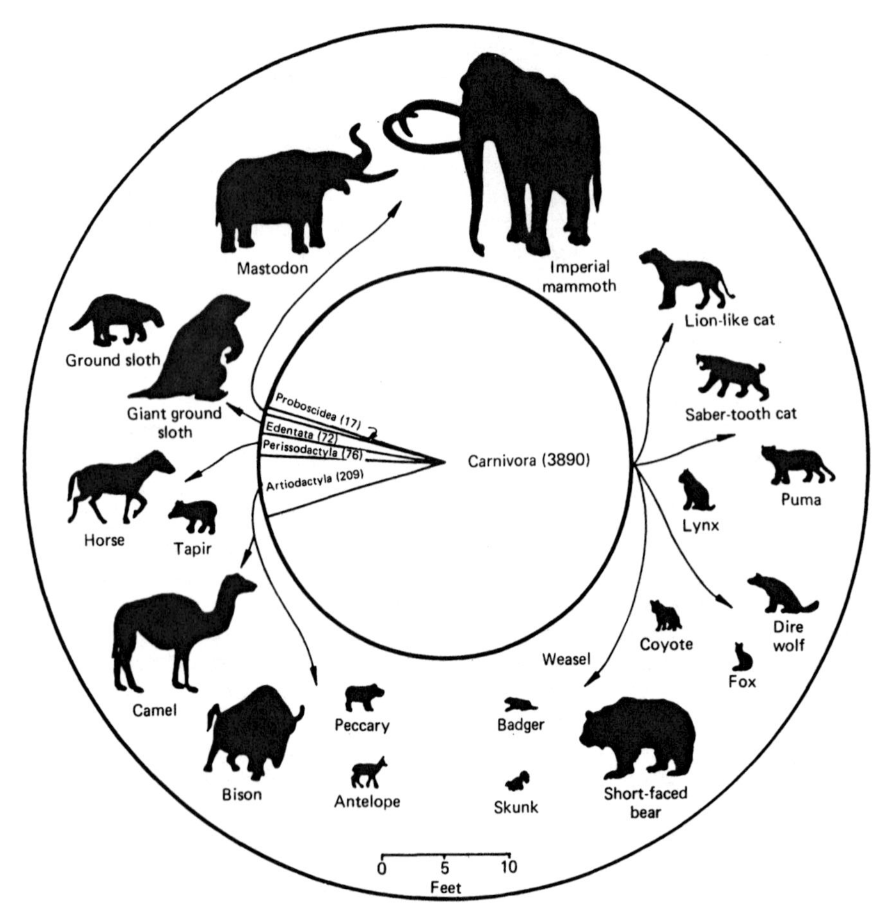

Fig. 7.14 The relative numbers of individuals in the mammalian orders (except rodents, lagomorphs, insectivores and bats) occurring in the Rancho La Brea Pleistocene tar pit fauna. The Carnivora predominate. (After Stock (1965) by courtesy of the Natural History Museum of Los Angeles County, California.)

down to drink, and scavengers and carnivores also became trapped as they sought an easy meal. As a result, the overwhelming majority of mammal and bird remains are of carnivores and scavengers, and birds of prey (Figs 7.14 and 7.15), which is a complete reversal of the usual distribution of animal types.

Vertebrate evolution in the Quaternary

The Mollusca and the Coleoptera have shown little evidence of morphological evolution during the Quaternary, even in response to the great climatic changes. Similarly, vertebrates apart from mammals have undergone only minor evolutionary change. By contrast, the mammals have shown enormous evolutionary changes throughout the Tertiary and Quaternary. They are a relatively young group of vertebrates, and became dominant after the 'Age of Reptiles' in the Mesozoic Era.

The rate of mammal evolution has greatly increased in response to the repeated glaciations of the later Pleistocene. It was relatively slow during the Tertiary and early Pleistocene. During the Pleistocene the rate became about four times greater (Kurtén, 1972), no doubt in response to the oscillations in the climate and the corresponding changes in the environment. The names and approximate ages of corresponding glacial and inter-

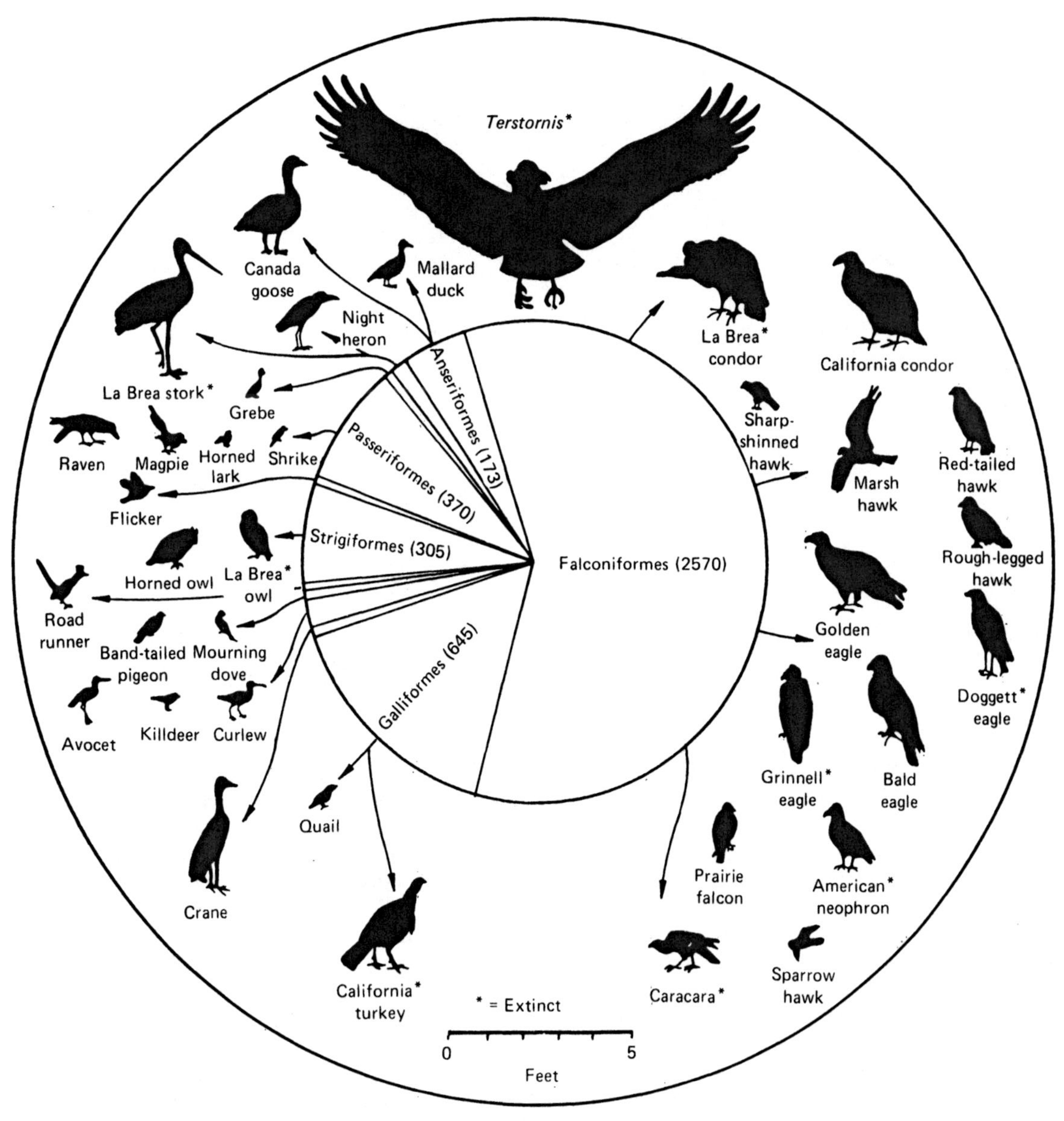

Fig. 7.15 The relative numbers of individuals in the avian orders occurring in the Rancho La Brea Pleistocene tar pit fauna. (After Stock (1965) by courtesy of the Natural History Museum of Los Angeles County, California.)

glacial stages and ages in Britain, Europe, and North America are shown in Table 7.2. Opinions on these correlations are constantly changing, and more evidence is necessary, particularly from the Alps and North America.

Throughout the Pleistocene, evolution can be traced in all the mammalian groups. Very few species have survived unchanged since the Tertiary. Particularly marked changes have occurred in the elephants, aurochs and bison, deer, horses, rhinoceroses, beavers, small rodents such as voles, and the primates. These latter include the evolution of hominids with the eventual emergence of modern man. We cannot discuss detailed evolutionary changes here, but we will briefly illustrate them with two groups, the elephants and man. These

Table 7.2 Nomenclature of episodes of the Pleistocene in Britain, north-west Europe, and North America. The names for the glaciations in the Alps are shown in brackets. An approximate time scale is shown on the right.

Britain	Europe	N. America	Approximate time scale (years B.P.)
Flandrian	Flandrian	Holocene	0–10 000
Devensian	Weichselian (Würm)	Wisconsinan	10 000–80 000
Ipswichian	Eemian	Sangamon	100 000
Wolstonian	Saalian (Riss)	Illinoian	200 000
Hoxnian	Holsteinian	Yarmouth	250 000
Anglian	Elsterian (Mindel)	Kansan	350 000
Cromerian	Cromerian	Aftonian	400 000
Beestonian	Menapian (Günz)	Nebraskan	
			Glacial Pleistocene
			Preglacial Pleistocene
	Early Pleistocene (Villafranchian)		1 000 000
			3 000 000
			Pleistocene
	Reuverian		Pliocene

accounts are largely based on information from Kurtén (1972).

1. *Elephant evolution*

In the early Pleistocene, often called the Villafranchian, there were true elephants and mastodonts. The mastodonts' teeth were adapted for eating soft leaves, and an enormous straight-tusked form developed in the early Pleistocene. The true elephant (*Elephas planifrons*) appears to have evolved in Africa during the Pliocene, and it spread rapidly through Eurasia. Its teeth were adapted for eating tough abrasive grasses. It gave rise to the elephants and mammoths. In southern Europe, the southern elephant (*Mammuthus meridionalis*) evolved and eventually replaced the mastodonts there. It gave rise to the mammoth lineage, progressing through the steppe mammoth (*M. trongontherii*) during the first glaciation to the woolly mammoth (*M. primigenius*) of the later glaciations. All mammoths became extinct at the end of the last glaciation, probably due to hunting by man. Meanwhile, the early Pleistocene elephant probably developed into the straight-tusked *Elephas namadicus* (synonymous with *E. antiquus*), which was adapted to living in open woodland and browsing on leaves. It survived in Europe until the middle of the last glaciation. Elsewhere, the elephants of Africa became adapted to a savanna environment, and the elephants of India lived in dense jungle. Mastodonts survived in North America until the end of the last glaciation.

2. *Evolution of man*

This topic is of essential interest to human beings, but it has generated an enormous amount of controversy. The broad outline of human evolution is accepted but the details are still contested.

The first true apes were found in the Oligocene of Egypt. The split into apes and hominids is recorded in the Miocene of Kenya. By the end of the Miocene, hominids had spread to Europe and Asia as far as China. The Late Tertiary *Ramapithecus* possibly used sticks and stones as tools, and his remains are associated with those of forest animals.

Remains of *Australopithecus* have been found in Pliocene and early and middle Pleistocene deposits in Africa but not in Europe. *A. robustus* was larger and more heavily built than *A. africanus*. He had huge jaw muscles which were inserted on a crest on top of his skull. He probably could not move very fast, and scavenged on tough abrasive food, probably in a humid forested environment. *A. africanus*, by contrast, probably lived in a drier savanna environment. His jaw muscles were less developed, and he probably actively hunted other animals. Both types of *Australopithecus* had a small brain capacity of about 500 ml, and used roughly shaped stone tools.

Homo erectus appears in the middle Pleistocene in Africa and Asia. Figure 7.16 is a reconstruction of Peking man from the Chouketian cave in China. He had a brain capacity of about 1100 ml and a higher forehead than his contemporary, Java man, whose brain capacity was only about 850 ml. *Homo erectus* used Abbeville type stone tools, which he gradually developed into the Acheulian type. He also used fire, as shown by the dating of hearths in China (about 300 000 years old), and in Hungary (about 400 000 years old). Fire was most important to early man's ecology. It enabled him to keep other animals out of caves, to cook his food and to keep warm.

Fig. 7.16 The reconstruction of Peking man (*Homo erectus*) from China.

Deposits of Holsteinian (Hoxnian) interglacial age in Germany and S. England have yielded fossils intermediate between *Homo sapiens* and *H. neanderthalensis*. Presumably this type gave rise to Neanderthal man in Europe and modern man elsewhere. Neanderthal men gradually developed more and more extreme characteristics, with heavy eyebrow ridges, thick, rather bent limb bones, and powerful necks. Their brain size exceeded that of modern man, but as a result of development of the hindbrain rather than the forebrain. Their tools gradually became more sophisticated, developing through several cultural types up to the last glaciation.

There is no firm evidence as to where *Homo sapiens* evolved, but it may have been in the Far East, around 40 000 years ago. Different races soon evolved, such as the mongoloid type in China and Asia, and the aboriginal type in Australia. Africa was occupied by the caucasian type. The earliest fossil negroid type surprisingly is only some 6000 years old. Modern man replaced Neanderthal man in Europe during the last glaciation. He had more sophisticated stone and bone tools, similar to those used by modern Eskimos, and was responsible for the beautiful cave paintings found particularly in France.

In North America evidence is increasing that man crossed the Bering Land Bridge from northern Asia about 13 000 years ago, near the end of the last (Wisconsinan) cold stage. We have already discussed Martin and Mosimann's model of his rapid spread, which resulted in the mass extinction of large mammals (p. 3). The first records of man in South America date from about 11 000 B.P. implying that men migrated the length of the two continents in about 2000 years. More diagnostic data are required to amplify these events.

Quaternary vertebrate palaeoecology and palaeogeography

The Pleistocene is characterized by great climatic changes, which, in Europe, resulted in the alternation of steppe and tundra in glacial periods with temperate forests in the interglacials. As a consequence, the animals had to migrate with their habitat, or else adapt to changing conditions. Vertebrates adopted both solutions; most animals migrated, but several appear to have had a broad enough ecological tolerance to adapt to climatic and vegetational changes. Mammals are warm-blooded, and therefore tolerant of a rather wide range of temperature conditions, unlike the beetles. Their over-all environment is generally more important than the temperature regime. Many herbivores need open steppe and grassland to roam in herds, and during the later Pleistocene, they did so in temperature conditions ranging from interglacial to full-glacial. Similarly, their predators also tolerated the temperature changes. However, the development of forest changes the character of the environment, and consequently the fauna is different in character.

The past environment is best deduced from an assemblage of organisms, because individual species, especially vertebrates, may well have changed their ecological preferences during the processes of evolution. Such changes have probably occurred in the mammals (Stuart, 1974). For example, the hamster (*Cricetus cricetus*) is today an animal of the Asian steppe. During the Cromerian interglacial, its remains were recovered from sediments deposited in a forested environment. Similarly, horses (*Equus*) are typically animals of grasslands and open forests, but their remains have been recovered from sediments registering a forested environment in the Cromerian and Hoxnian interglacials, as well as from glacial steppe environments. An example of an animal which could not adapt to changing conditions is the giant Irish deer (*Megaloceros giganteus*). It had been very successful during the late Pleistocene, and it was abundant in Ireland during the Allerød interstadial some 12 000 years ago, when it grazed the lush meadow vegetation (Mitchell and Parkes, 1949). The climate quickly and severely deteriorated during the Younger Dryas and food became scarce. *Megaloceros* became extinct because it could not migrate from Ireland during this climatic change (Gould, 1974), because it was virtually surrounded by sea by that time. In addition, giant deer may well have suffered from the Palaeolithic hunters in other parts of Europe.

During the Pleistocene, mammals changed their geographical ranges considerably, in a way comparable to the Coleoptera. Some of the restrictions of range today are just as remarkable, although there is no doubt that man has played a large part in determining the distributions of the larger mammals. Stuart (1977) discusses the palaeozoogeography of various mammals whose remains have been found in British deposits of the Devensian cold stage. As with the beetles and molluscs, a diverse assemblage of mammals has been found consisting of several elements which would be regarded as incompatible today.

1. *Arctic and boreal animals*

These are well represented in Devensian deposits, as might be expected. Many of them do not occur in Britain today. Arctic fox (*Alopex lagopus*), reindeer (*Rangifer tarandus*), and elk (*Alces alces*) still live in Scandinavia, tending to migrate southwards in the winter (Fig. 7.17). Lemmings do not now occur in Britain. The Norway lemming (*Lemmus lemmus*) and several voles (*Microtus* spp.) still live in Scandinavia, but the arctic lemming (*Dicrostonyx torquatus*) is confined to tundra, and is nowadays totally absent from Scandinavia. This distribution is reminiscent

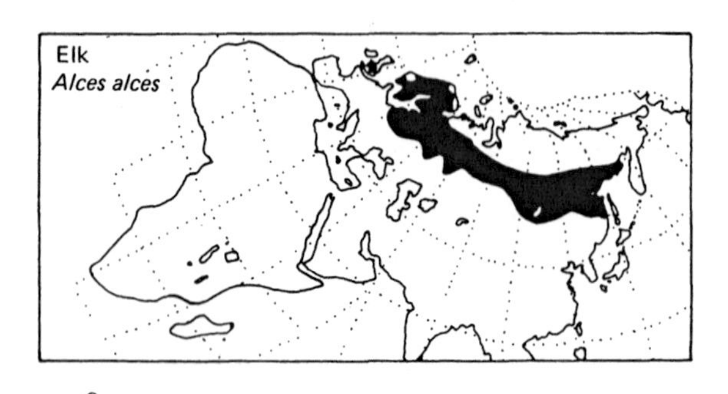

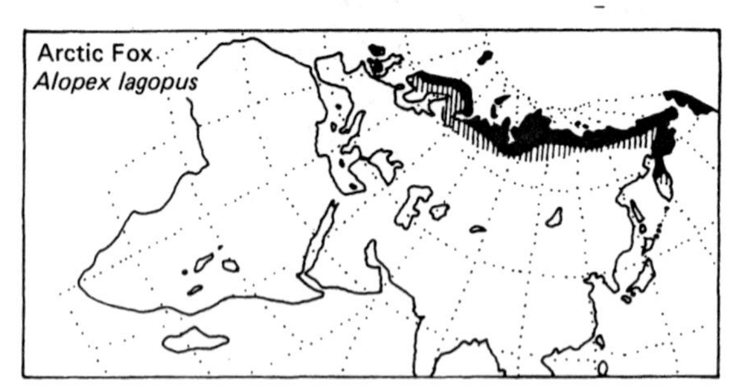

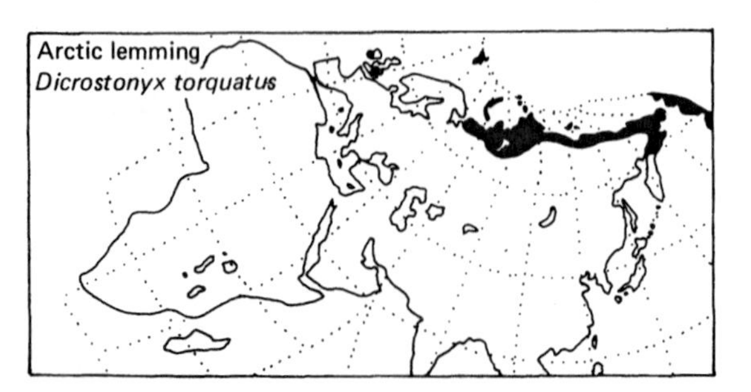

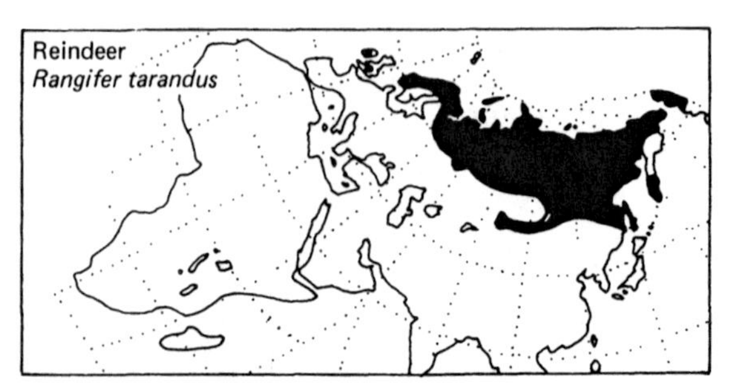

Fig. 7.17 Present Old World distributions of arctic fox, reindeer, elk, and arctic lemming. The summer range of the arctic fox is shown in solid black, with its winter extension hatched. Devensian fossil occurrences in the British Isles are indicated by dots. (After Stuart, 1977.)

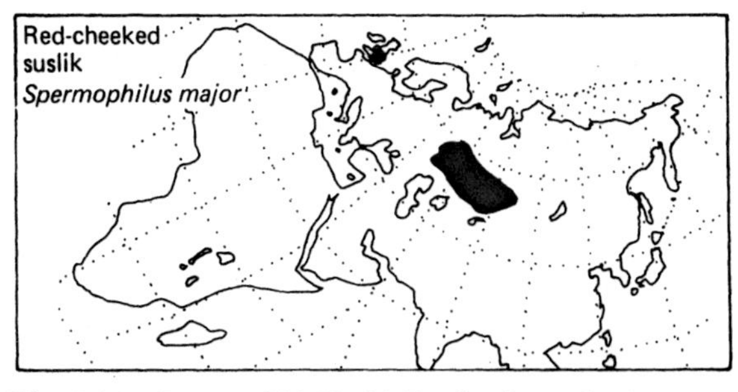

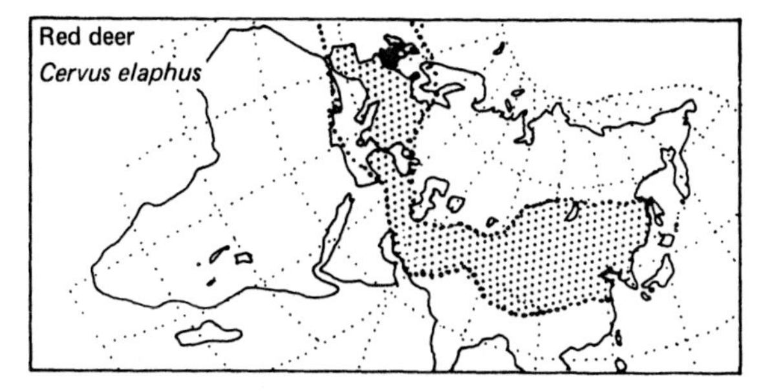

Fig. 7.18 Present Old World distributions of red-checked suslik and red deer. Devensian fossil occurrences in England are indicated by dots. (After Stuart, 1977.)

of many of the arctic coleopteran distributions (see p. 132).

The musk ox (*Ovibos moschatus*) and polar bear (*Ursus maritimus*) have also been found in British Devensian deposits. Polar bears are now confined to the arctic sea and adjacent coasts, whereas the musk ox is now restricted to the tundra of North America and Greenland. Besides glacial deposits, musk ox has also been recorded from warm steppe environments at the end of the Ipswichian interglacial. The polar bear has probably shown a genuine spectacular restriction of range, but the musk ox in Europe was probably exterminated by Neanderthal and early modern man during the Devensian cold stage.

2. *Steppe animals*

The red-cheeked suslik (*Spermophilus major*) has shown an amazing restriction of its range to the central Asian steppe (Fig. 7.18). The saiga antelope (*Saiga tartarica*) has shown a similar restriction, having been widespread through Europe to England, through Siberia and into north-west America during the Devensian. Horses (*Equus caballus*) are primarily animals of steppe or open forest, and they were one of man's staple prey during the Devensian. Today, wild horses are virtually extinct, apart from zebras, and man has taken advantage of their great evolutionary potential to breed all the different modern strains.

3. *Temperate animals*

The red deer (*Cervus elaphus*) is nowadays typically an animal of deciduous forests (Fig. 7.18), although its range has been considerably distorted by man. It is naturally absent from boreal forests or tundra. However, it was widespread during the Devensian, and was probably able to adapt to the treeless habitat, in a similar way to which red deer live on the Scottish hills today.

4. *Southern animals*

The Devensian fauna included lions (*Panthera leo*) and hyenas (*Crocuta crocuta*), which are both currently restricted to Africa south of the Sahara (Fig. 7.19). Both were widespread in Europe during the Pleistocene. In fairly recent times, lions occupied India and south-west Asia, and according to Herodotus, lions lived in northern Greece about 2500 years ago (Stuart, 1977).

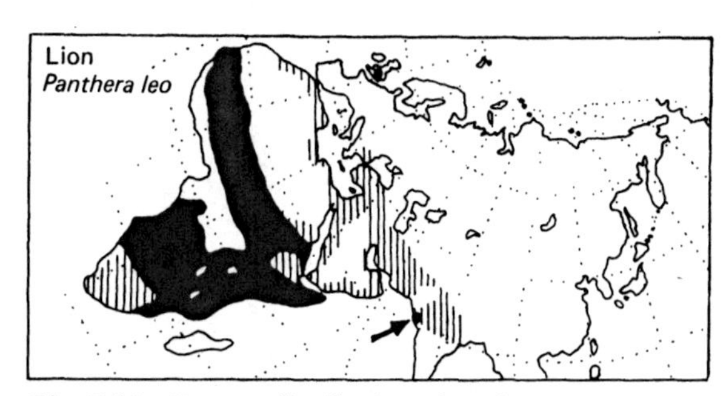

Fig. 7.19 Present distribution of the lion, shown in solid black. The isolated population in India is arrowed. The hatched area indicates approximately the areas in which the lion has become extinct in the last 2500 years. The Devensian fossil occurrences in England are indicated by dots. (After Stuart, 1977.)

5. *Extinct animals*

Several large mammals characteristic of the European Pleistocene are now extinct. They include woolly rhinoceros, mammoth, aurochs, bison, giant Irish deer, cave bears, and scimitar-toothed cats (*Homotherium*). In America, such animals as glyptodonts (huge relatives of armadillos), giant ground sloths (e.g. *Eremotherium*, *Nothrotherium*), camels, and mastodonts also became extinct at the end of the Wisconsinan, leaving no descendants. Most of these animals have long Pleistocene histories, and their evolution can be traced. Woolly rhinoceros and mammoths were adapted to very cold environments, but also occurred in warmer interstadial and interglacial conditions. For example, the deforested floodplain interglacial conditions at Barrington (Cambridge, England) supported a wide variety of large and small herbivores, including straight-tusked elephant and hippopotamuses (Gibbard and Stuart, 1975). Many people consider that man was largely responsible for the extinction of the large late-Pleistocene mammals although there is no definite evidence for this (see Martin and Wright (1967) for a detailed review of this question). Previously, faunas had adapted successfully to changes from glacial to interglacial and back to glacial conditions. Although specific forms may have become extinct, their modified descendants can be traced as evolution progressed. Man was a major new factor in the Devensian ecosystem and thus he could have been responsible for the mass extinction. Alternatively, it has been argued that a specially unfavourable spell of climate could have contributed to animal extinction, as has been shown for the giant Irish deer in Ireland. Further evidence pointing to man's involvement is the asynchroneity of extinction in different places, which is represented in Fig. 7.20, which is therefore independent of the world-wide climatic change between 13–10 000 years ago at the end of the Devensian. Large animals survived in uninhabited parts of the world, such as Madagascar, Mauritius, and New Zealand until man's arrival in relatively recent historical times led to their rapid extinction.

Palaeoecology of British Quaternary vertebrates

We have discussed how the ranges of certain animals have changed since the last glaciation.

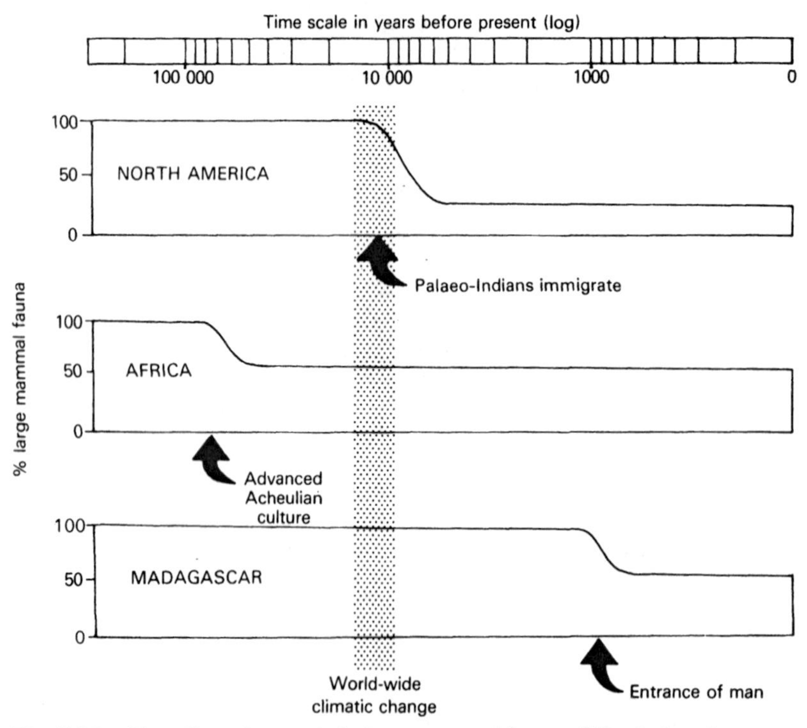

Fig. 7.20 The effect of man on the large mammal fauna of North America, Africa, and Madagascar. In Africa, the reduction occurs with the development of the advanced Acheulian culture associated with *Homo erectus*. In North America it occurs with the immigration of palaeoindians, during the period of world-wide climatic amelioration. In Madagascar, it occurs with the arrival of white men. (After Kurtén, 1972.)

These migrations uncomplicated by the activities of man have been occurring since the onset of climatic changes, particularly since the middle Pleistocene when cold periods culminated in extensive glaciation.

In the forest fauna of the Cromerian interglacial in Britain (Stuart, 1975), beavers (*Castor fiber*), wild boars (*Sus scofa*), roe deer (*Capreolus capreolus*) and desmans (*Desmana moschata*) occurred together with horses and hamsters. *D. moschata* is nowadays restricted to European Russia.

A similar assemblage occurred in Britain during the Hoxnian interglacial (Stuart, 1974). However, some of the Cromerian animals failed to reappear, most noticeably the hippopotamus. On the other hand, animals such as cave bears, giant Irish deer, and aurochs are found for the first time.

Most of the same animals or their evolved descendants reappeared in the Ipswichian interglacial (Stuart, 1974, 1976). Absentees included giant beaver (*Trogontherium cuvieri*) which had become extinct, and pine vole (*Pitymys*). The hippopotamus returned after its absence in the Hoxnian, perhaps reflecting the warmer climate of the Ipswichian interglacial.

Little is known of the intervening glacial faunas (Stuart, 1974) except for a growing amount of information from the Devensian (Stuart, 1977), some of which we have discussed above. There is no reason to suppose that earlier glaciations were much different. Herds of large herbivores probably roamed the treeless landscape, including horses, deer, reindeer, mammoths, bison, and woolly rhinoceroses. They were undoubtedly preyed upon by hunters such as lions, wolves, and dogs. Hyenas and scavenging birds would have made the most of abandoned carcasses. Man would have had relatively little impact before the Devensian stage, and he probably lived primarily as a scavenger on carcasses and on vegetable material.

The Holocene or Flandrian fauna was relatively impoverished compared with the previous interglacials due to the extinction of many of the larger animals (Stuart, 1974). With the spread of forest, the tundra and steppe species were forced northwards and eastwards. Several southern species failed to reach Britain before the English Channel was opened, and many others failed to reach Ireland across the early Flandrian landbridges. Similar distributions are also seen in plants. Some animals may have colonized southern England, and died out subsequently on the slight cooling of climate after the mid-Flandrian. An example is *Emys orbicularis*, the pond tortoise (Stuart, 1979). There is no doubt that the activities of man have been responsible for the introduction of several small mammals, such as mice and voles in Britain and Ireland, and he has also been responsible for the extermination of other species, such as beavers, wolves, and wild boar by hunting. Many other animals have been reduced as a result of the felling of forests and draining of wetlands. Ironically, in recent years some of the animals which would probably have been native to Europe in the Flandrian have been introduced in wild-life parks. Lions, zebras, and many others adapt successfully to the present-day British climate.

Ecological effects of vertebrates on Quaternary vegetation

The end of the Ipswichian interglacial is characterized by the development of open forest or steppe, whilst the climate, according to the evidence from plants, molluscs, and beetles, was still warm and continental. Stuart (1976) and Turner (1975) discuss the effect the fauna would have had on the vegetation. They conclude that along river floodplains in which many of the Ipswichian deposits are located, animals, particularly hippopotamuses, may have been responsible for deforestation. Today hippopotamuses in Africa graze a belt up to 1 km wide along the rivers, leaving it treeless and trampled (Lock, 1972). In addition, elephants habitually uproot small trees and eat the bark. Beavers gnaw trees for food and to build their dams. Once the forest was opened, other herbivores, both large and small, would enter and prevent forest regeneration by eating seedlings. Early man is unlikely to have had a great effect on the vegetation, although he may have used fire to drive game. In the Hoxnian interglacial deposits at Hoxne and Marks Tey, England, there is a small phase where some of the tree pollen is replaced by grass pollen. This may be due to deforestation by animals or to burning by man. However, the evidence is slender, and there may also be a climatic explanation, such as a period of drought which killed the trees.

Reconstruction of past populations of vertebrates

So far, we have been concerned with reconstructing the history of various animal and plant groups, and the environments in which they lived, and we have mentioned very little about the reconstruction of fossil populations. The study of the ecology and dynamics of living populations is a very large and complex subject, which we cannot hope to go into here. It is understandably difficult to reconstruct fossil populations because the material required to provide the data is seldom available. The material should ideally fulfil several conditions.

1. It should be large enough to allow statistically significant conclusions to be drawn.
2. The animals should (a) all have died at about the same time, being killed, for example, by some large disaster, such as flood or drought, so that a census of the population can be taken, or (b) the animal remains should have accumulated steadily over a long period of time, thus providing a regular sample of a standing population.
3. It must be possible to age the specimens. (a) This may be straightforward in animals with distinct growth phases, such as arthropods and crustaceans which undergo several moults before reaching maturity. (b) In continuously growing animals, annual growth rings may be present, such as in fish scales, bivalve shells, and the horny sheaths of some bovid horns. Additionally, the teeth of mammals can be used for aging the animal. The sequence of replacement of milk teeth occurs over the juvenile phase in most mammals, but in elephants and their relatives, the teeth are renewed from the back throughout life, up to six teeth. The relative amount of wear on the teeth may also be used to age the animal. If the animal has a seasonal parturition (time of giving birth) the amounts of tooth wear will tend to show annual discontinuities in a population sampled at one time. In his classic paper Kurtén (1953) has demonstrated the value of these teeth features in aging fossil populations. His results are summarized in a shorter paper (Kurtén, 1964).

The life table

Having obtained data on the fossil population a life table can be constructed. A life table for either living or fossil populations is a means of expressing facts of mortality in terms of probabilities (Reyment, 1971, page 26). Life tables were first used by ecologists (see Deevey, 1947), and adopted for fossil populations by Kurtén (1953).

A life table involves a series of calculations which we have no space to describe here, especially as there is an excellent account by Reyment (1971, Chapter 5), which clearly explains all the stages involved in the construction of a life table. A life table follows a theoretical cohort of animals which all began life at the same time and in which there is no immigration or emigration. Using the data on the numbers of individuals in different age classes, the life table displays, at every interval of age (standard time span) the number of deaths, the number of remaining survivors, the rate of mortality, and the further life expectation. The standard time span may be one year, or other intervals, such as the time between moults in a crustacean.

One of the most useful columns in the life table is the number of survivors in each span out of the original standard cohort. From it can be constructed a survivorship curve. Survivorship curves fall into three main shapes, illustrated in Fig. 7.21.

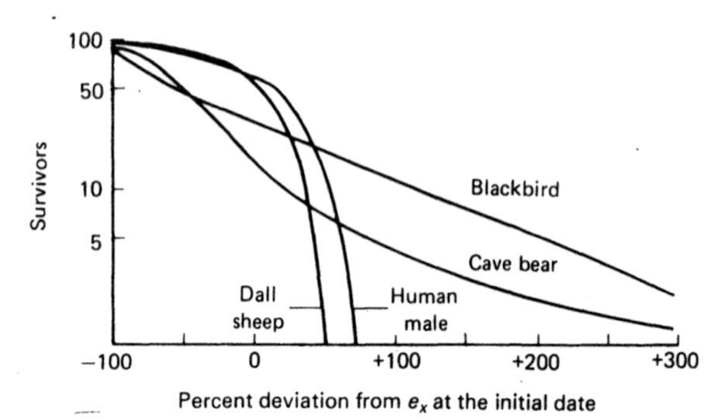

Fig. 7.21 Survivorship curves of human males, Dall sheep, cave bear, and blackbird, showing the three major types of curve. (After Kurtén, 1964.)

In many birds, typified by the blackbird (*Turdus merula merula*) the rate of mortality is almost constant through life, resulting in a straight diagonal survivorship curve (Deevey, 1947). The oldest blackbirds live more than six times as long as the mean life span. The convex curve, typified here by man, and the Dall sheep of Alaska shows that after an initially high juvenile mortality, there are very few deaths in the population until senescence is reached, when the rate of mortality rapidly increases. The third type of curve, the concave type, seems to be rather rare in nature, but is approached by the now extinct cave bear, *Ursus spelaeus* (Kurtén, 1964). An initially high juvenile mortality is gradually reduced, so that the oldest individuals may live more than four times as long as the mean life expectancy. Kurtén has worked extensively on cave bears (e.g. Kurtén, 1957, 1958, 1976).

He has shown that the bears had few natural predators, and that the greatest cause of mortality was insufficient food reserves to last the bears through their winter sleep. The bears died in their wintering caves, and great accumulations of skeletons have resulted. Different caves were prefered by male and female adults, but the juveniles were impartial. However, it is impossible that the total cave bear population of Europe could hibernate in the available number of caves, and it should be borne in mind that Kurtén's data may be atypical of the total population. In addition, only the weaker bears would have died in the caves, thus perhaps biasing the sample of the total population. Kurtén (1957, 1958) has used the life-table analyses of bear populations together with anatomical characteristics, particularly of the teeth, to demonstrate the action of selection and its rate in the populations. The older age groups tended to show less variation round the mean, because the more variable individuals had already died. However, the direction of selection varied in different localities.

Palaeoecological uses of life tables

Reyment (1971) briefly lists some of the uses of life tables in palaeoecology. We have already discussed the example of the cave bear.

1. Comparison of population structures between separate populations of the same species using the survivorship curve in particular.
2. Comparison of population structures between different organisms living at the same or different places and times.
3. Calculation of the population either in the past or future from a certain sample of the population.
4. Changes in the ecological situation may affect a population. This is particularly obvious for man during the enormous technological advances during the last 200 years. However, climatic changes greatly affected the mammalian populations of the Pleistocene and life-table analysis may indicate changes in the rates of mortality in different age groups.
5. The fertility of the fossil population may be

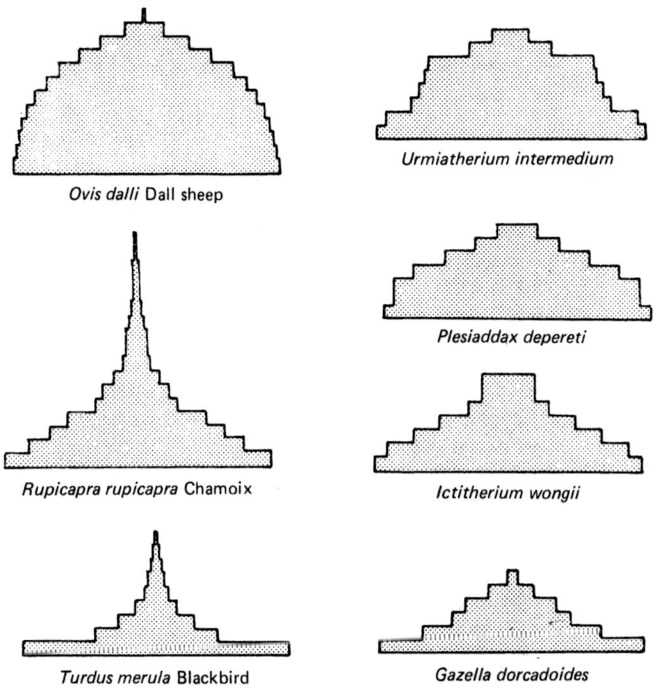

Fig. 7.22 Age pyramids of Dall sheep, Chamoix, and blackbird, and some extinct Pliocene Chinese mammals. The width of each bar represents the number of living individuals in the age class, with the youngest at the bottom and the oldest at the top. The survivorship curve of chamoix resembles that of blackbird (diagonal type). (After Kurtén, 1964.)

calculated by the ratio of adult females to juveniles, either by direct measurement or by assuming a replacement rate equal to the yearly mortality of the total population.

6. If the data are of the census type, for example, from a mass death of a complete population, age pyramids can be constructed which give an idea of the mortality of the species. Kurtén (1964) illustrates age pyramids of different types, which we reproduce here in Fig. 7.22. The lowest section represents the number of individuals in the youngest age class, and the uppermost section represents the oldest individuals at any one time.

The life-table approach to population studies only involves one species. When the actual causes of mortality are considered, there is surprisingly little information available, even for living species. Obviously disease and accidents will take their toll, and widespread or local ecological changes will have their effect, such as the onset of a glaciation, or the drainage of a marsh. However, one of the major causes of mortality is competition with other animals. The animals may be competing for food or living space, or they may be hunted by predators. There are some experimental studies on the competition between animals such as flour beetles (*Tribolium*) and cladocerans (*Daphnia* and *Simocephalus*) (Reyment, 1971) and between plant species (Harper, 1977). However, it is almost impossible to demonstrate the effects of competition in a palaeoecological situation apart from recording the successions of different animals and plants over long periods of time in response to changes in climate and environment. The mechanisms for these long term changes must be based on the competitive power of different organisms under changing ecosystem regimes. Competition between organisms under constant conditions in the past is very hard to demonstrate (see Reyment, 1971, for a discussion of this).

References

Molluscan ecology, distribution, and identification

BAKER, F. C. (1928). *The Freshwater Mollusca of Wisconsin. I. Gastropoda.* Trans. Wisconsin Academy of Science, Arts, and Letters, 494 pp.

BEEDHAM, G. E. (1972). *Identification of the British Mollusca.* Hulton.

BOYCOTT, A. E. (1934). The habitats of land mollusca in Britain. *J. Ecol.*, **22**, 1–38.

BOYCOTT, A. E. (1936). The habitats of freshwater mollusca in Britain. *J. Anim. Ecol.*, **5**, 116–86.

ELLIS, A. E. (1926). *British Snails.* Oxford University Press.

ELLIS, A. E. (1940). The mollusca of a Norfolk Broad. *J. Conch.*, **21**, 224–43.

KERNEY, M. P. (1976a). European distribution maps of *Pomatias elegans* (Müller), *Discus ruderatus* (Férrusac), *Eobania vermiculata* (Müller), and *Margaritifera margaritifera* (Linné). *Arch. Moll.*, **106**, 243–9.

KERNEY, M. P. (1976b). *Atlas of the Non-marine Mollusca of the British Isles.* Conchological Society of Great Britain and Ireland, and Natural Environment Research Council.

SPARKS, B. W. (1961). The ecological interpretation of Quaternary non-marine mollusca. *Proc. Linn. Soc., Lond.*, **172**, 71–80.

SPARKS, B. W. (1964). Non-marine mollusca and Quaternary ecology. *J. Anim. Ecol.*, **33** (Suppl.), 87–98.

WEST, R. G. (1977). *Pleistocene Geology and Biology.* Longman. (Chapters 7 and 13).

Molluscan palaeoecology

General

KERNEY, M. P. (1977a). British Quaternary non-marine mollusca: a brief review. In *British Quaternary Studies: recent advances* (ed. F. W. Shotton). Oxford University Press.

Hoxnian

KERNEY, M. P. (1971). Interglacial deposits in Barnfield Pit, Swanscombe, and their molluscan fauna. *Jl. geol. Soc. Lond.*, **127**, 69–93.

Ipswichian

SPARKS, B. W. (1957). The non-marine mollusca of the interglacial deposits at Bobbitshole, Ipswich. *Phil. Trans. R. Soc.* B, **241**, 33–44.

SPARKS, B. W. (1963). The distribution of non-marine mollusca in the last interglacial in south-east England. *Proc. Malac. Soc. Lond.*, **36**, 7–25.

SPARKS, B. W. AND WEST, R. G. (1959). The palaeoecology of the interglacial deposits at Histon Road, Cambridge. *Eiszeit. Gegenwart.*, **10**, 123–43.

SPARKS, B. W. AND WEST, R. G. (1963). The interglacial deposits at Stutton, Suffolk. *Proc. Geol. Assoc.*, **74**, 419–32.

SPARKS, B. W. AND WEST, R. G. (1970). Late Pleistocene deposits at Wretton, Norfolk. *Phil. Trans. R. Soc.* B, **258**, 1–30.

WEST, R. G., LAMBERT, C. A. AND SPARKS, B. W. (1964). Interglacial deposits at Ilford, Essex. *Phil. Trans. R. Soc.* B, **247**, 185–212.

WEST, R. G. AND SPARKS, B. W. (1960). Coastal interglacial deposits of the English Channel. *Phil. Trans. R. Soc.* B, **243**, 95–133.

Devensian

KERNEY, M. P. (1963). Late-glacial deposits on the Chalk of south-east England. *Phil. Trans. R. Soc.* B, **246**, 203–54.

WEST, R. G., DICKSON, C. A., CATT, J. A., WEIR, A. H. AND SPARKS, B. W. (1974). Late Pleistocene deposits at Wretton, Norfolk. II. Devensian deposits. *Phil. Trans. R. Soc.* B, **267**, 337–420.

Flandrian

DIMBLEBY, G. W. AND EVANS, J. G. (1974). Pollen and land-snail analysis of calcareous soils. *J. Arch. Sci.*, **1**, 117–33.

EVANS, J. G. (1969). Land and freshwater mollusca in archaeology: chronological aspects. *World Archaeology*, **1**, 170–85.

EVANS, J. G. (1972). *Land Snails in Archaeology*. Seminar Press.

GODWIN, H. (1975). *The History of the British Flora* (2nd Edition). Cambridge University Press.

HAWORTH, E. Y. (1972). Diatom succession in a core from Pickerel Lake, northeastern South Dakota. *Bull. Geol. Soc. Amer.*, **83**, 157–72.

IVERSEN, J. (1944). *Viscum*, *Hedera* and *Ilex* as climatic indicators. *Geol. Fören. Förhandl. Stock.*, **66**, 463–83.

KERNEY, M. P. (1966). Snails and man in Britain. *J. Conchol.*, **26**, 3–14.

KERNEY, M. P. (1968). Britain's fauna of land mollusca and its relation to the post-glacial thermal optimum. *Symp. zool. soc. Lond.*, **22**, 273–91.

KERNEY, M. P. (1977b). A proposed zonation scheme for late-glacial and postglacial deposits using land mollusca. *J. Arch. Sci.*, **4**, 387–90.

KERNEY, M. P., BROWN, E. H. AND CHANDLER, T. J. (1964). The late-glacial and post-glacial history of the Chalk escarpment near Brook, Kent. *Phil. Trans. R. Soc.* B, **248**, 135–204.

SPARKS, B. W. (1962). Post-glacial mollusca from Hawes Water, Lancashire illustrating some difficulties of interpretation. *J. Conchol.*, **25**, 78–82.

SPARKS, B. W. AND LAMBERT, C. A. (1961). The post-glacial deposits at Apethorpe, Northamptonshire. *Proc. Malacol. Soc. Lond.*, **34**, 302–15.

TAYLOR, D. W. (1965). The study of Pleistocene non-marine mollusks in North America. In *The Quaternary of the United States* (eds H. E. Wright and D. G. Frey). Princeton.

WATTS, W. A. AND BRIGHT, R. C. (1968). Pollen, seed, and mollusk analysis of a sediment core from Pickerel Lake, Northeastern South Dakota. *Bull. Geol. Soc. Amer.*, **79**, 855–76.

Coleopteran palaeoecology

General

COOPE, G. R. (1967). The value of Quaternary insect faunas in the interpretation of ancient ecology and climate. In *Quaternary Paleoecology* (eds E. J. Cushing and H. E. Wright). Yale University Press.

COOPE, G. R. (1970). Interpretations of Quaternary insect fossils. *Ann. Rev. Entomology*, **15**, 97–120.

COOPE, G. R. (1973). Tibetan species of dung beetle from Late Pleistocene deposits in England. *Nature*, **245**, 335–6.

COOPE, G. R. (1975). Climatic fluctuations in northwest Europe since the Last Interglacial, indicated by fossil assemblages of Coleoptera. *Geol. J. Spec. Issue*, **6**, 153–68.

COOPE, G. R. (1977). Fossil coleopteran assemblages as sensitive indicators of climatic changes during the Devensian (Last) cold stage. *Phil. Trans. R. Soc.* B, **280**, 313–48.

KENWARD, H. K. (1975a). The biological and archaeological implications of the beetle *Aglenus brunneus* (Gyllenhae) in ancient faunas. *J. Arch. Sci.*, **2**, 63–9.

KENWARD, H. K. (1975b). Pitfalls in the environmental interpretation of insect death assemblages. *J. Arch. Sci.*, **2**, 85–94.

KENWARD, H. K. (1978). The analysis of archaeological insect assemblages: a new approach. *The Archaeology of York*, **19**, 1–68.

MATTHEWS, J. V. (1974). Quaternary environments at Cape Deceit (Seward Peninsula, Alaska): evolution of a tundra ecosystem. *Bull. Geol. Soc. Amer.*, **85**, 1353–84.

MATTHEWS, J. V. (1976). Evolution of the subgenus *Cyphelophorus* (Genus *Helophorus*, Hydrophilidae, Coleoptera): description of two new fossil species and discussion of *Helophorus tuberculatus* Gyll. *Can. J. Zool.*, **54**, 652–73.

PEARSON, R. G. (1963). Coleopteran associations in the British Isles during the Late Quaternary period. *Biol. Rev.*, **38**, 334–63.

Hoxnian

KELLY, M. R. (1964). The Middle Pleistocene of North Birmingham. *Phil. Trans. R. Soc.* B, **247**, 533–92.

SHOTTON, F. W. AND OSBORNE, P. J. (1965). The fauna of the Hoxnian interglacial deposits at Nechells, Birmingham. *Phil. Trans. R. Soc.* B, **248**, 353–78.

Ipswichian

COOPE, G. R. (1974). Interglacial coleoptera from Bobbitshole, Ipswich, Suffolk. *Jl. geol. Soc. Lond.*, **130**, 333–40.

FRANKS, J. W., SUTCLIFFE, A. J., KERNEY, M. P. AND COOPE, G. R. (1958). Haunt of elephant and rhinoceros: the Trafalgar Square of 100 000 years ago – new discoveries. *Illustrated London News*, 1011–13.

LINDROTH, C. H. (1943). *Oodes gracilis* Villa. Eine thermophile Carabide Schweidens. *Notulae Entomol.*, **22**, 109–57.

Early and Middle Devensian

COOPE, G. R. (1959). A Late Pleistocene insect fauna from Chelford, Cheshire. *Proc. R. Soc.* B, **151**, 70–86.

COOPE, G. R. (1968). An insect fauna from Mid-Weichselian deposits at Brandon, Warwickshire. *Phil. Trans. R. Soc.* B, **254**, 425–56.

COOPE, G. R. (1969). The response of Coleoptera to gross thermal changes during the Mid-Weichselian interstadial. *Mitt. Int. Verein. Limnol.*, **17**, 173–83.

COOPE, G. R. AND ANGUS, R. B. (1975). An ecological study of a temperate interlude in the middle of the last glaciation, based on fossil Coleoptera from Isleworth, Middlesex. *J. Anim. Ecol.*, **44**, 365–91.

COOPE, G. R., MORGAN, A. AND OSBORNE, P. J. (1971). Fossil coleoptera as indicators of climatic fluctuations during the last glaciation in Britain. *Palaeogeography, Palaeoclimatology, Palaeoecology*, **10**, 87–101.

COOPE, G. R. AND SANDS, C. H. S. (1966). Insect faunas of the last glaciation from the Tame Valley, Warwickshire. *Proc. R. Soc.* B, **165**, 389–412.

COOPE, G. R., SHOTTON, F. W. AND STRACHAN, I. (1961). A late Pleistocene fauna and flora from Upton Warren, Worcestershire. *Phil. Trans. R. Soc.* B, **244**, 379–421.

MORGAN, ANNE (1969). A Pleistocene fauna and flora from Great Billing, Northamptonshire, England. *Opus. Entomol.*, **34**, 109–29.

MORGAN, ANNE (1973). Late Pleistocene environmental changes indicated by fossil insect faunas of the English Midlands. *Boreas*, **2**, 173–212.

SIMPSON, I. M. AND WEST, R. G. (1958). On the stratigraphy and palaeobotany of a Late-Pleistocene organic deposit at Chelford, Cheshire. *New Phytol.*, **57**, 239–50.

WEST, R. G., DICKSON, C. A., CATT, J. A., WEIR, A. H. AND SPARKS, B. W. (1974). Late Pleistocene deposits at Wretton, Norfolk. II. Devensian deposits. *Phil. Trans. R. Soc.* B, **267**, 337–420.

Late Devensian

ASHWORTH, A. C. (1972). A late-glacial insect fauna from Red Moss, Lancashire, England. *Ent. scand.*, **3**, 211–24.

ASHWORTH, A. C. (1973). The climatic significance of a Late Quaternary insect fauna from Rodbaston Hall, Staffordshire, England. *Ent. scand.*, **4**, 191–205.

BISHOP, W. W. AND COOPE, G. R. (1977). Stratigraphical and faunal evidence for Late glacial and early Flandrian environments in south-west Scotland. In *Studies in the Scottish Late Glacial Environment* (eds J. M. Gray and J. J. Lowe). Pergamon.

COOPE, G. R. AND BROPHY, J. A. (1972). Late Glacial environmental changes indicated by a coleopteran succession from North Wales. *Boreas*, **1**, 97–142.

MANGERUD, J., ANDERSEN, S. T., BERGLUND, B. E. AND DONNER, J. J. (1974). Quaternary stratigraphy of Norden, a proposal for terminology and classification. *Boreas*, **3**, 109–28.

PEARSON, R. G. (1962). The coleoptera from a late-glacial deposit at St. Bees, West Cumberland. *J. Anim. Ecol.*, **31**, 129–50.

PENNINGTON, W. AND BONNY, A. P. (1970). Absolute pollen diagram from the British Late-glacial. *Nature*, **226**, 871–3.

PENNY, L. F., COOPE, G. R. AND CATT, J. A. (1969). Age and insect fauna of the Dimlington silts, East Yorkshire. *Nature*, **224**, 65–7.

Flandrian

BUCKLAND, P. AND KENWARD, H. K. (1973). Thorne Moor: a palaeo-ecological study of a Bronze Age site. *Nature*, **241**, 405–6.

DENTON, G. H. AND KARLÉN, W. (1973). Holocene climatic variations – their pattern and possible cause. *Quat. Res.*, **3**, 155–205.

KELLY, M. R. AND OSBORNE, P. J. (1964). Two faunas and floras from the alluvium of Shustoke, Warwickshire. *Proc. Linn. Soc. Lond.*, **176**, 37–65.

OSBORNE, P. J. (1965). The effect of forest clearance on the distribution of the British insect fauna. *Proc. XII Int. Congr. Ent.*, 1964, 456–7.

OSBORNE, P. J. (1969). An insect fauna of Late Bronze Age date from Wilsford, Wiltshire. *J. Anim. Ecol.*, **38**, 555–66.

OSBORNE, P. J. (1972). Insect faunas of Late Devensian and Flandrian age from Church Stretton, Shropshire. *Phil. Trans. R. Soc.* B, **263**, 327–67.

OSBORNE, P. J. (1974). An insect assemblage of early Flandrian age from Lea Marston, Warwickshire, and its bearing on the contemporary climate and ecology. *Quat. Res.*, **4**, 471–86.

OSBORNE, P. J. (1976). Evidence from the insects of climatic variation during the Flandrian period: a preliminary note. *World Archaeology*, **8**, 150–8.

North America

ASHWORTH, A. C. (1977). A late Wisconsinan coleopterous assemblage from southern Ontario, and its environmental significance. *Can. J. Earth Sci.*, **14**, 1625–34.

ASHWORTH, A. C. AND BROPHY, J. A. (1972). Late Quaternary fossil beetle assemblage from the Missouri Coteau, North Dakota. *Bull. Geol. Soc. Amer.*, **83**, 2981–8.

ASHWORTH, A. C., CLAYTON, L. AND BICKLEY, W. B. (1972). The Mosbeck Site: a paleoenvironmental interpretation of the Late Quaternary history of Lake Agassiz based on fossil insect and mollusk remains. *Quat. Res.*, **2**, 176–88.

MATTHEWS, J. V. (1968). A paleoenvironmental analysis of three late Pleistocene coleopterous assemblages from Fairbanks, Alaska. *Quaest. Entomol.*, **4**, 202–24.

Relationships between British and Scandinavian insect faunas

COOPE, G. R. (1969). The contribution that the Coleoptera of Glacial Britain could have made to the subsequent colonization of Scandinavia. *Opus. Ent.*, **34**, 95–108.

LINDROTH, C. H. (1949). *Die fennoskandischen Carabidae. Eine tiergeographische Studie Vol. 3. Göteborgs Vetensk. Samh. Handl.*, **6B**, 4.

Quaternary vertebrates

General

CLARK, J. G. D. (1954). *Excavations at Star Carr*. Cambridge University Press.

DEGERBØL, M. (1964). Some remarks on late and post-glacial vertebrate fauna and its ecological relations in Northern Europe. *J. Anim. Ecol.*, **33** (Suppl.), 71–85.

DEGERBØL, M. AND IVERSEN, J. (1945). The bison in Denmark. *Danm. geol. Unders.* Ser. II, **73**, 62 pp.

DEGERBØL, M. AND KROG, H. (1951). Den europaeiske Sumpskildpadde (*Emys orbicularis* L.) i Danmark. *Danm. geol. Unders.* Ser. II, **78**, 130 pp.

KURTÉN, B. (1968). *Pleistocene Mammals of Europe*. Weidenfeld and Nicolson.

KURTÉN, B. (1972). *The Ice Age*. Hart-Davis.

MARTIN, P. S. AND WRIGHT, H. E. (1967). *Pleistocene Extinctions. The Search for a Cause*. Yale University Press.

STOCK, C. (1965). Rancho La Brea: a record of Pleistocene life in California. *Los Angeles Co. Mus. Sci. Ser.* 20, Publication **11**, 1–81.

Palaeoecology of Quaternary vertebrates in the British Isles

GIBBARD, P. L. AND STUART, A. J. (1975). Flora and vertebrate fauna of the Barrington Beds. *Geol. Mag.*, **112**, 493–501.

GOULD, S. J. (1974). The origin and function of 'bizarre' structures: antler size and skull size in the 'Irish Elk' *Megaloceros giganteus*. *Evolution*, **28**, 191–220.

LOCK, J. M. (1972). The effects of hippopotamus grazing on grasslands. *J. Ecol.*, **60**, 445–67.

MITCHELL, G. F. AND PARKES, H. M. (1949). The giant deer in Ireland. *Proc. R. Ir. Acad.*, **52B**, 291–314.

STUART, A. J. (1974). Pleistocene history of the British vertebrate fauna. *Biol. Rev.*, **49**, 225–66.

STUART, A. J. (1975). The vertebrate fauna of the type Cromerian. *Boreas*, **4**, 63–76.

STUART, A. J. (1976). The history of the mammal fauna during the Ipswichian/Last interglacial in England. *Phil. Trans. R. Soc.* B, **276**, 221–50.

STUART, A. J. (1977). The vertebrates of the Last Cold Stage in Britain and Ireland. *Phil. Trans. R. Soc.* B, **280**, 295–312.

STUART, A. J. (1979). Pleistocene occurrences of the European pond tortoise (*Emys orbicularis* L.) in Britain. *Boreas*, **8**, 359–71.

TURNER, C. (1975). Der Einfluss grosser Mammalier auf die interglaziale Vegetation. *Quartärpaläontologie Berlin*, **1**, 13–19.

WEST, R. G. (1977). *Pleistocene Geology and Biology* (2nd edition). Longman. (Chapter 13).

Population palaeoecology of vertebrates

DEEVEY, E. S. (1947). Life tables for natural populations of animals. *Quart. Rev. Biol.*, **22**, 283–314.

HARPER, J. L. (1977). *Population Biology of Plants*. Academic Press.

KURTÉN, B. (1953). On the variation and population dynamics of fossil and recent populations. *Acta Zool. Fennica*, **76**, 1–122.

KURTÉN, B. (1954). Population dynamics – a new method in paleontology. *J. Paleont.*, **28**, 286–92.

KURTÉN, B. (1957). A case of Darwinian selection in bears. *Evolution*, **11**, 412–16.

KURTÉN, B. (1958). Life and death of the Pleistocene cave bear, a study in palaeoecology. *Acta Zool. Fennica*, **95**, 1–59.

KURTÉN, B. (1964). Population structure in paleoecology. In *Approaches to Paleoecology* (eds J. Imbrie and N. Newell). J. Wiley and Sons.

KURTÉN, B. (1976). *The cave bear story*. Columbia University Press.

REYMENT, R. A. (1971). *Introduction to Quantitative Paleoecology*. Elsevier. (Chapter 5).

8

Principles and methods of pollen analysis

Introduction

Pollen analysis is the principal technique used to reconstruct Quaternary environments. This is so for several reasons.

1. Pollen is usually the most abundant fossil preserved in Quaternary sediments. Consequently it can be counted, and the resulting pollen spectrum should be a statistical representation of all the pollen grains in the sediment of that age at that place. Pollen assemblages can then be compared from different points in space and time.
2. Pollen grains are resistant to decay in non-oxidizing situations. Because they are abundantly produced by the plants, they tend to be deposited universally as the 'pollen rain', and thus they have the chance of being preserved in many sedimentary situations.
3. The taxonomy of pollen grains is relatively well known, and the major types are readily identifiable under the light microscope. Of course, refinements are constantly being made, especially with the use of the scanning electron microscope.
4. Because pollen grains are small (5–100 μm) and abundant, only small amounts of sediment are needed for an adequate sample. This contrasts with fossils such as beetles, molluscs, vertebrates, and plant macrofossils, where relatively large amounts of sediment are needed to provide an adequate sample of the fossil population.
5. Pollen grains originate from plants which originally grew together as the vegetation of an area. Therefore pollen can be used to reconstruct the vegetation, including both local vegetation, such as aquatic and wetland communities, and more distant, regional vegetation growing around the site of deposition. Vegetation is a feature of major importance in a landscape, because it largely controls the kinds of animals which live there. Because vegetation is responsive to environmental factors, these factors can be deduced from a reconstruction of the past vegetation. This can be done on a large scale, such as vegetational and hence climatic zones on a continent, right down to a small, local scale, where a mosaic of the different communities composing the vegetation may be detectable, and the influence of local environment factors on the past vegetation can be inferred, such as soil type, water level, or micro-climate.

Because pollen analysis is such an important technique, a large amount of work has been done on its various aspects. These include the study of pollen taxonomy and structure, pollen production, dispersal, and deposition, the representation of different plant species and vegetation types by contemporary pollen spectra, the preservation of fossil pollen, the interpretation of pollen diagrams in terms of flora, vegetation, and environment, and the comparison of pollen diagrams from different areas. Therefore five chapters will be devoted to a more detailed discussion of pollen analysis than has been possible with other groups of fossils.

Historical development of pollen analysis

The development of Quaternary pollen analysis parallels that of descriptive ecology. Qualitative descriptions of vegetation and pollen floras were made during the last century. At the beginning of this century, quantification in ecology developed, and vegetation began to be described in a quantitative way. Similarly, in his classic paper of 1916, (reprinted in English in 1967), the Swedish geologist Lennart von Post put forward the method of quantitative pollen analysis. He had the idea of presenting pollen spectra as percentages of the sum of the pollen grains counted, and of presenting the results as stratigraphic diagrams, with pollen spectra plotted against their stratigraphic position through the sediment. Von Post showed similarities in pollen diagrams from a small area, and differences between different areas. He was thus able to add a time dimension to the study of vegetation (see Fries, 1967). Although pollen analysis has been developed and refined through the years since 1916, the basic method remains the same.

Basic principles of pollen analysis

1. Pollen or spores are produced in great abundance by plants.
2. A very small fraction of these fulfil their natural reproductive function, and the majority fall to the ground.
3. Pollen and spores rapidly decay, unless the processes of biological decomposition are inhibited by lack of oxygen. This occurs in places such as bogs, lakes, fens, and the ocean floor, where pollen is preserved.
4. Before reaching the ground, pollen is well mixed by atmospheric turbulence, which results in a more or less uniform pollen rain over an area.
5. The proportion of each pollen type depends on the number of parent plants, and hence the pollen rain is a function of the composition of the vegetation. Therefore a sample of the pollen rain will be an index of the vegetation at that point in space and time.
6. Pollen is identifiable to various taxonomic levels.
7. If a sample of the pollen rain is examined from a peat or mud of known age, the pollen spectrum is an index of the vegetation surrounding that place at a point of time in the past.
8. If pollen spectra are obtained from several levels through the sediment, they provide a picture of the vegetation and its development at that place through the length of time represented by the sediments.
9. If two or more series of pollen spectra are obtained from several sites, it is possible to compare changes in vegetation through time at different places.

We have spelled out the rationale of pollen analysis, because all the stages have been subjected to detailed study, and it is sometimes difficult to keep an over-all view in mind when concentrating on one aspect.

Techniques of pollen analysis

After the site has been chosen, and a core or other samples obtained and the lithology described (see Chapter 3), the pollen sample must be treated in order to concentrate the pollen and to remove as much of the sediment matrix as possible.

Preparation of pollen samples

The matrix of the sediment is removed by physical and chemical processes, which, as far as possible, do not affect the pollen grains and spores. The standard method is described in Faegri and Iversen (1975). It is outlined below.

i) Boil sediment with 10% NaOH or KOH to remove soluble humic acids and to break the sediment down. Subsequent repeated washes in distilled water are essential.
ii) Sieve to remove coarse debris.
iii) Treat with cold 10% HCl to remove carbonates.
iv) Treat with either hot 60% HF for a short time (up to one hour) or cold 60% HF for a long time (up to 24 hours) to remove silicates (silt, clay, diatoms, etc.)
v) Treat with hot acetolysis mixture (9 parts acetic anhydride: 1 part concentrated H_2SO_4), to hydrolyse cellulose.
vi) After neutralization, stain the residue with safranin or basic fuchsin.
vii) Mount the residue in a suitable medium with a refractive index less than 1.55, which is that of pollen grains. An excellent medium is silicone oil (refractive index 1.4), because the slides last well, provided that the pollen has been thoroughly dehydrated, the grains can be turned readily for identification, and they hardly alter in size over time (Andersen, 1960). Glycerol and glycerine jelly are also in common use. The first is short lived, and the slides are easily damaged. The second is easy to use, but the grains are fixed and difficult to turn over, and there is a tendency for grains to swell after a time and lose their features for some unknown reason, making their identification difficult, and the use of size statistics impossible.

Other preparation techniques involving oxidation and the use of flotation liquids may be used for some sediments (see Faegri and Iversen, 1975), but in general we have found them to be less reliable than the standard procedure outlined above.

Morphology of pollen and spores

Before being able to identify pollen, the analyst should understand their basic function and structure.

1. *Function of pollen grains*

A pollen grain contains a male gamete of the plant, and its function is to transfer it to the female gamete *via* the stigma and style of an angiosperm, or the nucellus of a gymnosperm.

Pollen is formed in the anthers. Sporogenous tissue gives rise to pollen mother cells, each of which divides by meiosis into a tetrad of pollen grains. Most pollen is liberated as single grains, but in some families, such as Ericaceae and Orchidaceae, the grains remain in tetrads.

Pollen ranges in size between about 5–100 μm, the commonest size being about 30 μm. The winged grains of Pinaceae are some of the largest, apart from the 350 μm grains of Annonaceae. The important component of pollen to the palaeo-ecologist is the sporopollenin found in the pollen wall. Sporopollenin is a very inert substance, which accounts for the common preservation of pollen grains and the relative safety with which they can be treated chemically during preparation procedures. Spores of pteridophytes and some bryophytes, algae, and fungi also contain sporopollenin, and these are preserved along with pollen.

a) *General structure.* Pollen grains are produced in fours (tetrads). Gymnosperm pollen tends to be simply sphaerical, or invested with a pair of air sacs (Fig. 8.1) as an adaptation to wind dispersal. The wall structure of gymnosperm pollen tends to be poorly developed compared with angiosperm pollen.

Faegri and Iversen (1975) describe the structure of angiosperm pollen in detail and present a very useful key to pollen types found in north-west Europe. The key is not illustrated, but it is usefully complemented by the book of photographs by Erdtman *et al.* (1961). Moore and Webb (1978) present a similar key with illustrations for the British pollen flora. The first part of a well illustrated north-west European pollen flora arranged by morphological types of grains by Beug (1961) and the first part of a north-west European pollen flora edited by Punt (1976) are also useful aids to identification. Other relevant publications are given at the end of the chapter.

Due to the original formation of pollen grains in tetrads, the arrangement of their apertures follows a certain pattern (Fig. 8.2), related to the position within the tetrad. Because each grain touches three

Top view

Air sac

Body of pollen grain

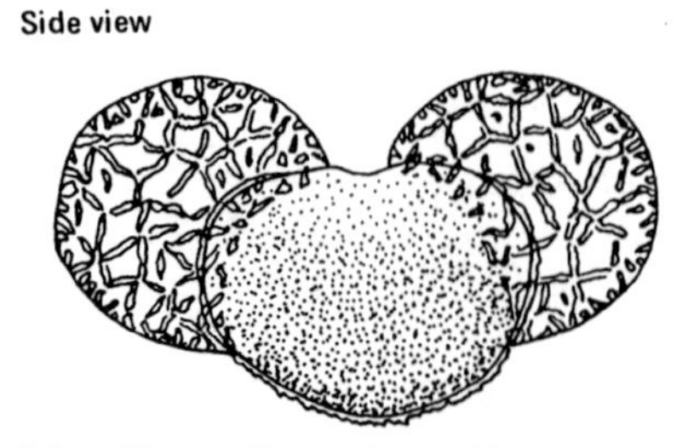

Fig. 8.1 *Pinus* pollen grain, typifying gymnosperm pollen with air sacs.

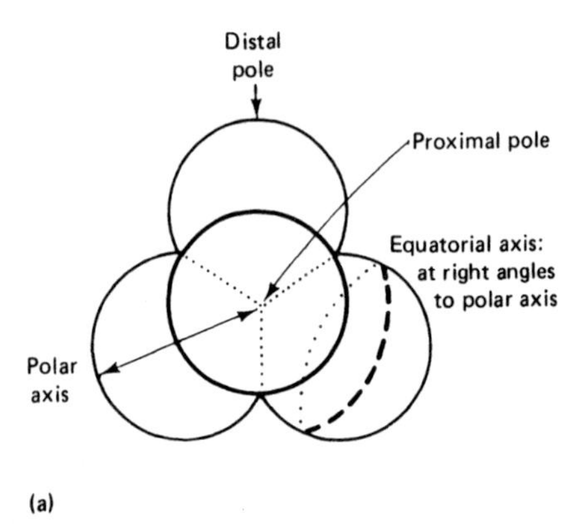

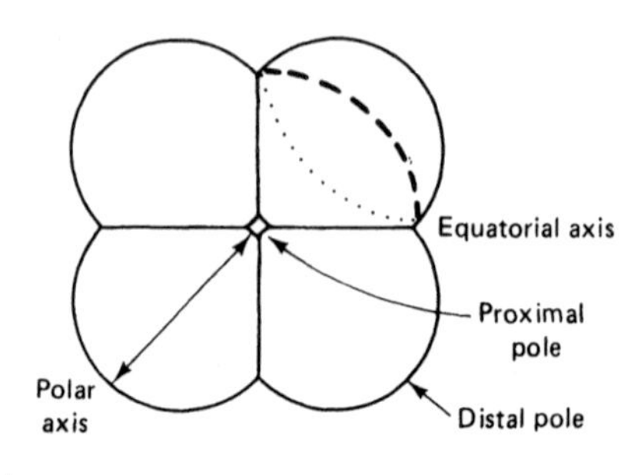

Fig. 8.2 Tetrad arrangement of pollen in the anther, showing the symmetry of the grains. (**a**) Dicotyledon arrangement and (**b**) Monocotyledon arrangement.

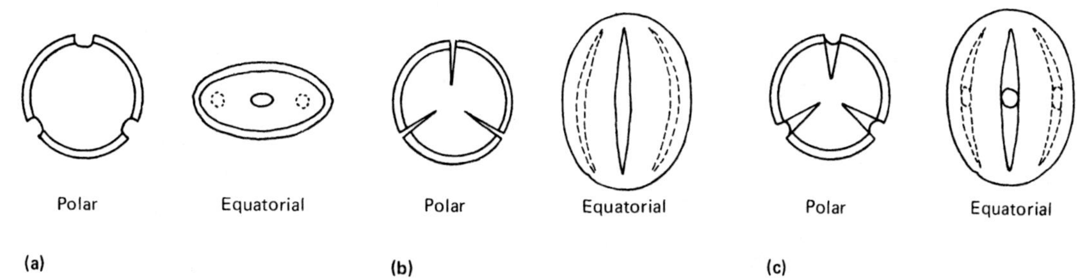

Fig. 8.3 Common three-apertured arrangements in pollen grains. (**a**) Triporate, e.g. *Corylus avellana*. Three pores arranged on equatorial axis. (**b**) Tricolpate, e.g. *Quercus petraea*. Three furrows (colpi) arranged longitudinally on equatorial axis. (**c**) Tricolporate, e.g. *Fagus sylvatica*. Three furrows each with a pore on the equatorial axis.

others, the apertures tend to be three or multiples of three. The apertures can be pores (***pori***) if both diameters are equal, or furrows (***colpi***) if one diameter exceeds the other. Pollen grains may have pores (***porate***), furrows (***colpate***), or a combination of both (***colporate***).

Figure 8.3 illustrates common three-apertured arrangements. Many other arrangements can occur, and the apertures may also vary in shape. Iversen and Troels-Smith (1950) described the variation, and invented a tidy shorthand descriptive code in which the shapes and sizes of pollen grains and their apertures are denoted. Walker (1974a) discusses the evolution of apertures in pollen of primitive angiosperms, and pollen and angiosperm phylogeny are reviewed by Walker and Doyle (1975).

b) *Structure of pollen walls*. The wall of angiosperm pollen has three basic layers. (i) The living cell membrane, which is lost after death. (ii) The middle wall, or ***intine***. This is composed primarily of cellulose, together with pectins, callose, proteins (which are responsible for allergic reactions in man, and for compatability systems of fertilization in the plant), polysaccharides, and antigens (Heslop-Harrison, 1968, 1976). It is also not usually preserved. (iii) The outer wall, or ***exine***. This is composed mainly of sporopollenin, whose chemistry has only recently been understood because of its great chemical inertness (Shaw, 1971; Brooks, 1971). It is one of the most resistant organic substances known, and it is not affected by hot hydrofluoric acid or concentrated alkali. It consists of complex polymers with a basic formula $[C_{90}H_{142}O_{36}]^n$. It is formed by the oxidative polymerization of carotenoids and carotenoid esters. It has a specific gravity of between 1.4 and 1.5 (Flenley, 1971).

The exine has a complex structure, which has an even more complex nomenclature! Different nomenclatural systems are compared in Fig. 8.4 (see Wittman and Walker, 1965). We will use the terminology of Faegri and Iversen in the following short account.

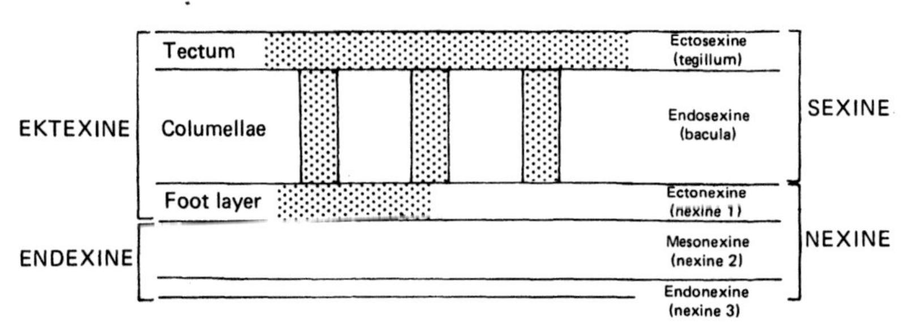

Fig. 8.4 The structure of an angiosperm pollen exine. The nomenclature of the layers used by Iversen and Troels-Smith (1950) and Faegri and Iversen (1975) is shown on the left. The more complex nomenclature of Erdtman (1952) is shown on the right.

The development of the exine has been described by Heslop-Harrison (1968), Godwin (1968), and Faegri and Iversen (1975) amongst others. The function of the various parts of the pollen grain has been discussed by Heslop-Harrison (1976).

Basically, the structure of the ektexine consists of three layers, the outer tectum, supported by columellae which rest on the foot layer. Some grains lack a tectum (***intectate***), some have a partial tectum (***semi-tectate***), and others are fully ***tectate***. Some intectate grains are clearly reduced tectate grains, because they still have columellae (for example, the aquatic plants *Potamogeton* and *Ruppia*). Others lack columellae, either secondarily (***etectate***), or as a primary primitive condition (***atectate***). Walker (1974b, 1976) deduces an evolutionary sequence from atectate, to granular with incipient columellae, to tectate-imperforate, to tectate-perforate, to semi-tectate, to intectate, and to etectate structure.

c) *Sculpture of pollen walls*. The elaboration of the exine superimposed on its basic structure is called the sculpture. It is either on the surface of the tectum (*tectate*) or on the endexine (*intectate*) (Fig. 8.5). The basic types of sculpture are psilate,

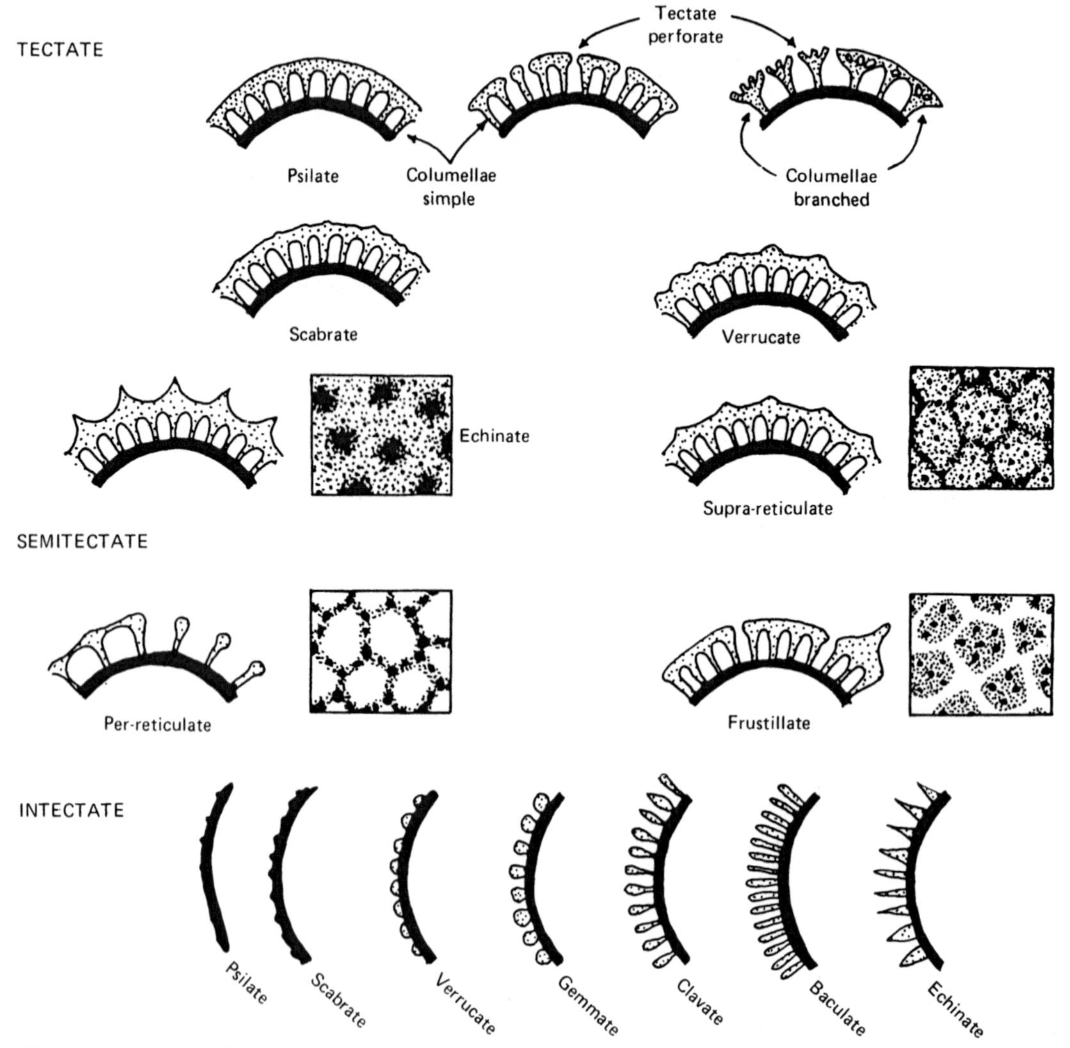

Fig. 8.5 Types of pollen grain structure and sculpture. Endexine black; ektexine dotted. (After Faegri and Iversen, 1975.)

scabrate, clavate, verrucate, baculate, echinate, striate, reticulate, and rugulate (see Faegri and Iversen, 1975, p. 38). The sculpturing of pollen grains is often related to their mode of dispersal. Smooth (psilate or scabrate) grains tend to be wind-dispersed, whilst rougher grains (e.g. echinate, clavate) tend to be insect-dispersed.

3. *Other features of pollen grains important in identification*

a) *Shape.* An important feature of pollen grains for their identification is shape. Whilst alive, pollen grains vary greatly in shape according to their water content, and their apertures enable the exine to accommodate the changes in volume. In the fossil condition, however, the shape is relatively constant. In order to obtain fresh reference pollen in a similar 'fossil' condition for comparison, it must be treated in a way similar to the fossil pollen sample, to remove the living parts and any waxes, etc. on the outer surface. It should be noted that shape can vary with chemical treatment and with the mounting medium used (see Praglowski, 1970). Christopher and Waters (1974) have used Fourier analysis (an expression of shape by means of sine and cosine wave functions) to describe the shape of pine pollen. They demonstrated that pollen of different species of pine had different shapes, that the shape could be varied by preparation procedure, and that these differences in preparation did not affect the shape of pollen of all the species in the same way.

b) *Size.* Grain size can be crucial in the distinction between some otherwise closely similar pollen grains. *Isoetes* microspores may be taken as an example. In Fig. 8.6 the size frequencies of microspores of *I. lacustris* and *I. echinospora* are plotted. Although morphologically identical, *I. lacustris* is bigger (mean length 33.9 ± 1.9 μm) than *I. echinospora* (mean length 24.4 ± 2.0 μm). Care must be taken, however, when employing size statistics for fine distinctions for several reasons.

1. Pollen can vary in size within a species. (i) This may be due to clinal variation over a geographical area. For example, *Pinus echinata* pollen is larger in the northern part of its range in eastern North America, and smaller in the south (Cain and Cain, 1948). (ii) It may be due to differentiation within the species into different geographical races. For example, *Picea abies* pollen from Scandinavia and Russia, from southern Europe, and from the Alps

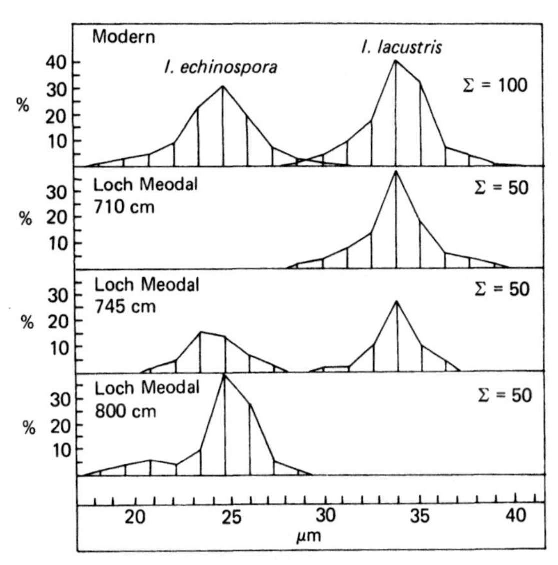

Fig. 8.6 Size frequencies of spores of *Isoetes echinospora* and *I. lacustris*, compared with fossil *Isoetes* spores from Devensian late-glacial sediments from Loch Meodal, Isle of Skye. (After Birks, 1973.)

have different sizes and shapes which may reflect their different migrational histories during the post-glacial period (Birks, 1978). (iii) Size differences may be caused by the physiological state of the plant, for example whether sufficient nutrients or water are available. Such effects have been observed in *Corylus avellana*, *Lythrum salicaria*, and in cereals.

2. Pollen size may vary within a single plant (van der Pluym and Hideaux, 1977).

3. Preparation procedures may affect pollen size (Reitsma, 1969). Treatment with potassium hydroxide produces little change, whereas the grains swell by 20% after acetolysis, and by 40% after potassium hydroxide plus acetolysis. However, treatment by potassium hydroxide, hydrofluoric acid, and acetolysis leads to only a 10% increase in size (see Andersen, 1960; Faegri and Deuse, 1960).

4. The mounting medium also affects size. Pollen grains swell over a period of time in glycerol or glycerine jelly by 20–30%, but little variation has been observed in silicone oil mounts (Andersen, 1960, 1978).

Bearing all these possibilities in mind, it is important when making fine distinctions using size to use identical preparation procedures and mounting media for fossil and modern material, and to make slides of comparable thickness. Whitehead (1964), Gordon and Prentice (1977), and Andersen

(1978) discuss fully the problems involved in the use of size statistics.

c) *Numbers of apertures.* Otherwise similar periporate pollen grains may be distinguished by the number of their pores. This exercise has been useful in the Plantaginaceae (Faegri and Iversen, 1975), *Juglans nigra* and *J. cinerea* (Whitehead, 1963), Caryophyllaceae, Chenopodiaceae and Amaranthaceae (McAndrews and Swanson, 1967), and *Ulmus* (Stockmarr, 1970).

4. *Scanning electron microscopy*

The scanning electron microscope (SEM) has added a new dimension to the study of pollen surfaces, and some most striking and beautiful patterns have been revealed. Specific distinctions can be made between different types of cereal pollen (Andersen and Bertelsen, 1972) and *Quercus* pollen (Smit, 1973). However, these distinctions are difficult, if not impossible, to see and interpret under the light microscope. It may become possible in the future to isolate individual fossil grains for SEM examination.

Identification of fossil pollen

The identification of fossil pollen is obviously of crucial importance, because all further palaeoecological interpretations will hang upon the identifications. Pollen and spores can be grouped into two main categories; determinable and indeterminable:

1. *Indeterminable pollen*

The essential morphological features of a grain may be obscured and prevent its assignment into a known pollen type. This may be due to various causes.

i) Concealment. Other debris on the slide may conceal the grain and prevent the critical morphological features being observed.
ii) Unsuitable orientation. The grain may be in an unsuitable position to show its diagnostic features, and it may not be possible to turn it over.
iii) Deterioration. The exine may have undergone some change, such as corrosion, degradation, breakage, or crumpling (see Chapter 9).
iv) Unknown. Although the features of the grain may be well preserved and visible, the analyst may not be able to identify it. It may be an aberrant grain, or it may not be represented in the reference collection. There is a finite possibility that a grain from anywhere in the world can be transported to any particular place.

The recording of indeterminable pollen is important, because it is a measure of the accuracy of the total pollen count. If a large proportion of the pollen is indeterminable, the numbers of the determined grains are likely to be less reliable in palaeoecological terms.

2. *Determinable pollen*

Identification can be approached in two ways.

i) Fossils can be classified directly into morphological units without reference to any pre-existing modern taxonomy. This is often the only way to classify pre-Quaternary pollen and spores (see Chapter 2).
ii) The fossil grains can be compared with and assigned to modern taxa using modern reference material (see Chapter 2). If the fossil grain is similar to one modern taxonomic category and dissimilar to all the others, the name of the modern taxon is applied to the fossil grain. In doing this the assumption is made that morphological evolution since the time the fossil was deposited has not been significant. We shall discuss this later.

The Quaternary pollen analyst attempts to match his pollen grains with those of modern taxa of the lowest possible rank whose range of variation uniquely includes the fossil grains. It is sometimes possible to assign the grain to a species, but more often, only generic rank can be determined, or, in some cases, only family rank. Successful identifications depend upon several factors.

i) The range of the reference material used. Obviously, all likely taxa on phytogeographical criteria should be considered.
ii) Variability within species should be represented in the reference material, by having several collections of each from different localities, which cover the range of its natural variation.
iii) The morphology of the grain. Some grains are very distinctive, whereas others have to be studied very carefully to find differential characters.
iv) The preservation of the fossil grains. Some grains can be identified even though they are badly deteriorated, but others lose their definitive characteristics readily. Thus a count from badly preserved material will tend to be biased towards the more robust and distinctive pollen types.
v) The experience and knowledge of the analyst.

vi) The selection and scoring of morphological characters.

In practice, the pollen analyst identifies grains using their over-all appearance, rather as we recognize different people. Only when there is difficulty in distinguishing between closely similar grains, are individual characters and perhaps size examined. An analyst can remember up to about 2000 pollen types, and keys such as those of Faegri and Iversen (1975) and Moore and Webb (1978) for north-west European types are useful leads to the appropriate reference material for unknown grains. In very large floras, where keys would become cumbersome, computer-based methods may be useful. Morphological information on each modern reference pollen type can be stored and the analyst can feed in morphological descriptions of his unknown grains. The computer will draw up a 'short-list' of modern grains with similar morphology, and the analyst can now compare his fossil grains with reference material for the species on the short-list. Such a data storage and retrieval system has been developed in Australia by Walker *et al.* (1968). The system is simple, flexible, and fast, and it is particularly useful in coping with the problems of pollen identification in areas of very diverse floras such as Australia, south-east Asia, and equatorial Africa.

Computers can also be used to perform numerical methods for identifying, in statistical terms, morphologically similar pollen types. The most detailed study of this nature is that by Hansen and Cushing (1973) on *Pinus* pollen in lake sediments in the Chuska Mountains of New Mexico. Two species of pine grow there today, but at least three morphologically distinguishable pollen types occur in the sediments. They recorded eight qualitative and three quantitative characters for a total of 150–220 grains per species in eighteen reference pollen collections of five pine species which, on phytogeographical and ecological criteria, could have contributed pollen to the sediments. Depending upon the state of preservation, as many as possible of these eleven attributes were recorded on 2107 fossil grains. Each fossil grain was then compared with the reference material for the five modern species (Fig. 8.7). By knowing the 'norm' of each character in each species, the probability that a particular fossil grain was drawn from that species can be calculated. If the probability exceeds a preselected value, the fossil grain is identified as being derived from the species concerned.

The measure of similarity used was Goodall's (1966) deviant index (d_i) for a single attribute. It is simply the proportion of the modern reference material that is further from the 'norm' than the

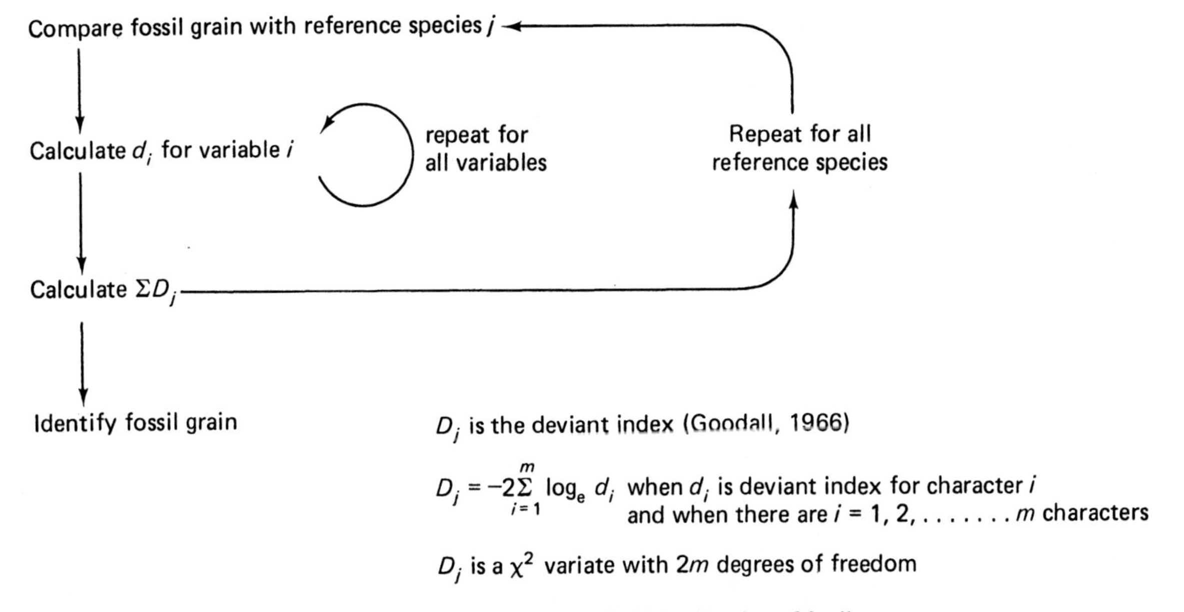

Fig. 8.7 Summary of the processes involved in the numerical identification of fossil pine pollen by the method of Hansen and Cushing (1973).

individual fossil grain in question. The deviant indices for each character can be summed to form D_j (see Fig. 8.7). D_j can be used as a measure of whether a particular fossil grain could or could not be drawn from a particular reference species.

Of the 2107 fossil grains, 44.6% were assigned to three species, *Pinus ponderosa* and *P. edulis* which both grow in the Chuska Mountains today, and *P. flexilis*, a more northern species. Of the rest, 31.7% were unknown and 21.7% were indeterminable.

This method has several advantages over conventional identification procedures. The results obtained depend solely on the available reference material, all the characters scored are used, and all the reference material examined is considered in the identifications. The method incorporates the inherent variability of the reference material, avoids the inevitable bias in scoring characters, and takes into account the problems of preservation. Because the method is mathematical, the pollen analyst has to be explicit about the bases for identification. It also has an 'in-built' weighting. If a fossil grain has an attribute confined to one species, it can be assigned to that species with a high probability. However, if diagnostic characters are not visible, the grain will be indeterminable. Four hypotheses were proposed to explain the high proportion (31.7%) of the fossils that were unknown. Species other than the five examined may have contributed pollen to the fossil pollen rain; the reference material may not adequately represent the intraspecific variation; morphological changes in the fossil pollen may have occurred due to diagenesis; and the original morphology of the pollen may have changed with time due to evolution (see p. 165).

The disadvantages of the method are that it is very time consuming, that the validity of the results relies on the choice of modern species to be considered, that the reference material may be inadequate, and that the choice of attributes may be inappropriate. Hansen and Cushing considered that the method was justified in this case because *Pinus* was the dominant pollen type in both the Holocene and Pleistocene sediments from the Chuska Mountains.

Other numerical approaches have been used to assist in identifying morphologically similar fossil pollen grains. Birks and Peglar (1980) used linear discriminant analysis (see Chapter 2) to discriminate between pollen of *Picea glauca* and *P. mariana* using six morphological variables. Each fossil *Picea* grain can then be positioned on the discriminant function, and assigned to one of the two species. Gordon and Prentice (1977) have used size-frequency data for modern pollen of different Scandinavian *Betula* species as a basis for identifying fossil grains. The sizes of the modern species overlap considerably, but by using a mathematically complex procedure called maximum likelihood estimation, the observed size histograms of the fossil *Betula* grains can be used to derive an estimate of the frequencies of the different species of *Betula* contributing pollen to the fossil assemblage.

Evolution in vascular plants during the Quaternary

It is often assumed that no significant evolution of vascular plants has occurred during the Quaternary. However, when Tertiary and early Quaternary fossils are compared with modern material, convincing morphological differences may be found. Mastrogiuseppe *et al.* (1970) used canonical variates analysis (a form of discriminant analysis) to demonstrate that Miocene *Ginkgo* wood from the western United States differed from modern wood. Similarly, Hills *et al.* (1974) used a variety of univariate and multivariate methods, including canonical variates analysis, to show that nuts of Miocene *Juglans* from Arctic Canada were very different from modern nuts of *J. nigra*, and similar but by no means identical to nuts of *J. cinerea*. Therefore the Tertiary *Juglans* nuts should be assigned to a distinct fossil species, *J. eocinerea*.

It is difficult to use pollen for such studies because of the problems of preservation and size statistics. Macrofossils are easier to work with. For example, Jentys-Szaferowa (1958) investigated *Carpinus* and *Menyanthes* macrofossils in different interglacials back to the Pliocene. She found no anatomical differences in *Carpinus*, but the *Menyanthes* seeds fell into two types, those with low wide epidermal cells, and those with tall narrow epidermal cells. The Pliocene seeds had very small square epidermal cells, sufficiently different for her to name them *Menyanthes carpatica*.

Kokawa (1959) also published measurements on *Menyanthes* seeds in Japan from the Pliocene through the interglacials to the present day. The first two components of a principal components analysis of these data (H. J. B. Birks, unpublished data) are shown in Fig. 8.8, plotted against time. The first component accounts for 66% of the variance and shows a striking morphological

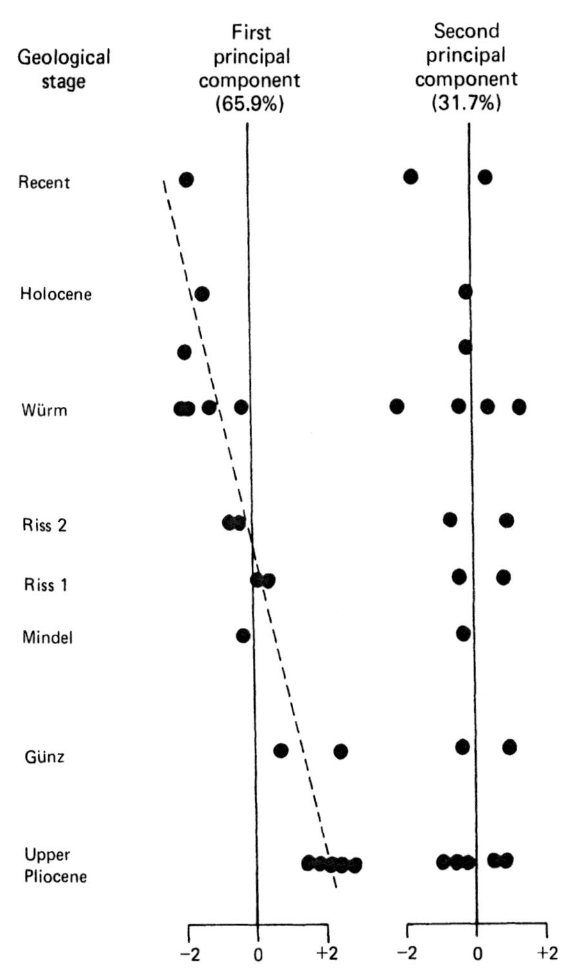

Fig. 8.8 The first two principal components plotted against age, of a principal components analysis using five variables, of *Menyanthes* fossil seed data from Japan of Kokawa (1959). The first component shows a chronocline. See text for explanation.

chronocline from seeds with positive values from the upper Pliocene to those with negative values at the present day. The second component, accounting for a further 32% of the variance, shows no clear pattern with time, suggesting a random variation with time, quite possibly due to changes brought about by different environments of deposition and conditions of preservation. Similar morphological variations with time can be demonstrated using fossil *Pinus* cones and *Dulichium* fruits.

The fact that variations can be demonstrated in some species through the Quaternary raises important doubts about the implicit assignment of fossils to modern taxa. The dangers of being misled decrease with decreasing age of the flora. However, investigators of early Quaternary floras should look out for subtle morphological differences and bear in mind the possibility that morphological evolution may have occurred before assigning modern names to their fossils.

Pollen counting

A few words on pollen counting are appropriate here, because numeracy is a basic feature of pollen analysis, being one of the main reasons why pollen analysis is such an important palaeoecological method.

Counting should be carried out along regular traverses of the microscope slide, most conveniently at a magnification of ×300 or ×400. Higher magnification (×1000) using an oil-immersion objective is necessary for the identification of some grains.

The number of grains counted depends on the problem being investigated. However, enough should be counted to achieve a random sample of the pollen grains present, otherwise the count will not be reproducible. Therefore traverses should be positioned evenly over the whole slide and not concentrated near the edge or the middle. This is because smaller pollen grains tend to travel towards the edges of the coverslip more readily than larger ones (Brookes and Thomas, 1967). Enough grains should be counted to maintain constant percentages of the pollen sum. In Fig. 8.9 the number of grains indicated by the arrow should be reached at least. This is usually in the region of 300–500. It depends largely on the number of taxa in the sample.

No pollen count is absolutely reproducible from another pollen slide of the same sample. However, the probability of achieving an answer within 95% of the original count can be calculated (see Mosimann, 1965). The size of the 95% confidence limit depends on the number of grains counted in the pollen sum (Table 8.1). Mosimann (1965) discusses the basic theory of such confidence limits, and Maher (1972) demonstrates their importance in determining which fluctuations in a pollen diagram are due to statistical variations, and which are significant statistically.

Pollen counting is a time-consuming process, and recently efforts have been made to automate it. Mirkin and Bagdasaryan (1972) used optical analys-

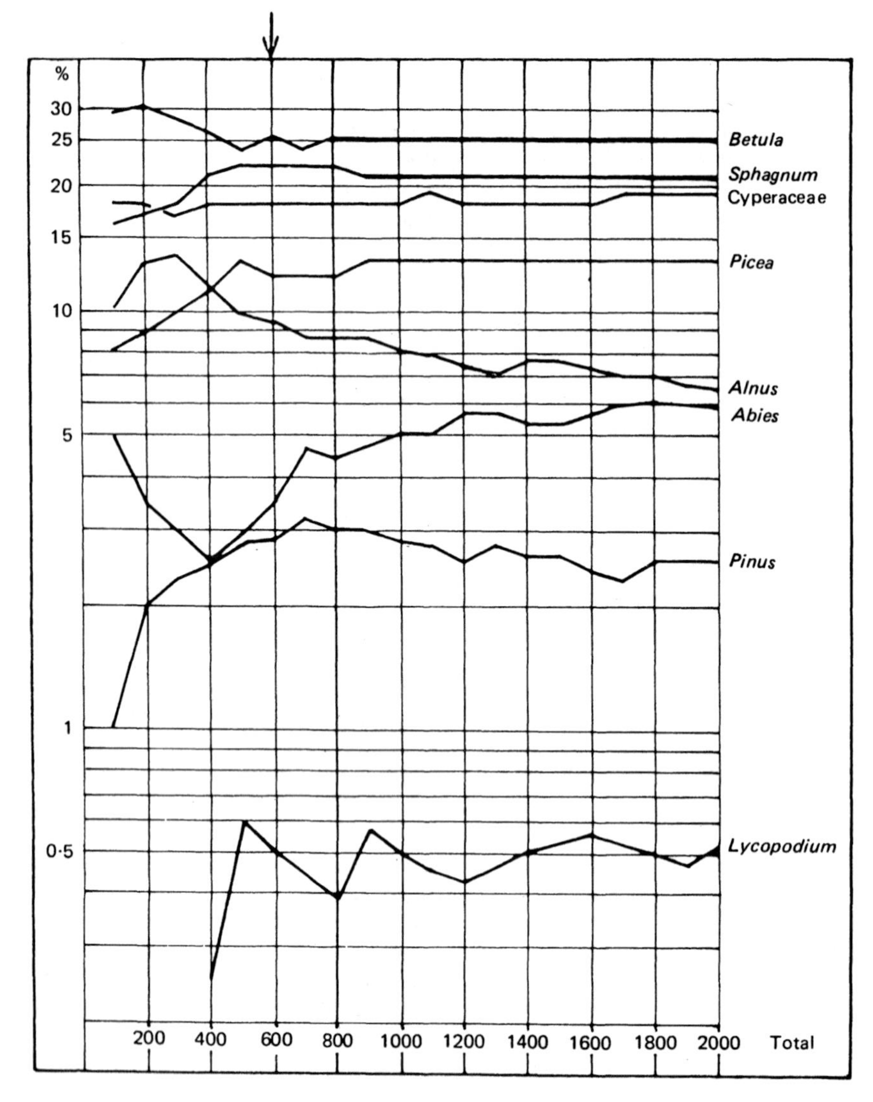

Fig. 8.9 Curves showing how the percentages of various kinds of pollen and spores counted in one sample on one slide vary as the count of the total number of pollen and spores is increased. The arrow indicates the onset of relatively constant percentages.

ing systems to describe reference material as Fourier holograms. Fossil grains were described in the same way, and the holograms were compared with the reference holograms. They were able to distinguish 20 different pollen types in this manner, thus opening up exciting possibilities for the future.

Presentation of pollen analytical data

Having obtained a stratigraphical sequence of pollen counts, the next stage is to present the results in the form of a graph or other figure, prior to interpretation of the data. Pollen analytical data are complex and are most effectively presented in the form of a pollen diagram. This is a series of graphs of the values for different pollen and spore taxa plotted against their stratigraphic depth or, sometimes, against their age.

Having identified and counted all the pollen and spores in the different levels of the sequence, the investigator then has to decide what to plot on his pollen diagram. Mostly, pollen analysts consider the

Table 8.1 95% confidence intervals for percentages of increasing total pollen counts, assuming a binomial distribution. (Data from Faegri and Iversen, 1975.)

	Number of pollen grains counted				
Percentage	50	100	200	500	1000
1 ±	4.6	2.7	1.7	1.2	0.6
3 ±	5.8	3.8	2.5	1.6	1.1
5 ±	6.9	4.6	3.2	2.0	1.4
10 ±	8.7	6.0	4.3	2.7	1.9
20 ±	11.1	7.9	5.6	3.6	2.5
30 ±	12.5	9.0	6.4	4.1	2.9
50 ±	13.6	9.6	7.0	4.4	3.1
70 ±	12.5	9.0	6.4	4.1	2.9
80 ±	11.1	7.9	5.6	3.6	2.5
90 ±	8.7	6.0	4.3	2.7	1.9
95 ±	6.9	4.6	3.2	2.0	1.4
97 ±	5.8	3.8	2.5	1.6	1.1
99 ±	4.6	2.7	1.7	1.2	0.6

relative proportions of the different pollen types, which are generally expressed as percentages of a pollen sum. In addition it is now possible to estimate 'absolute' numbers of pollen grains, in which case the numbers of different taxa are independent of each other. We shall discuss the estimation of absolute pollen frequencies (APF) in Chapter 10.

The pollen sum

The choice of which taxa should be included in the pollen sum should be based on the principle that all the members of the population under study should be included. At first pollen analysts were basically interested in forest history, and so their pollen sums consisted of tree pollen. Any other pollen types were calculated as percentages of the tree pollen sum (often called the sum of arboreal pollen, or ΣAP). However, the modern approach to vegetational history includes a study of the herb and shrub components, especially in situations with few trees, such as northern latitudes, high altitudes, prairies and savannas, and land artificially cleared of forest by man. Hence all the pollen and spore types which have originated from the upland vegetation should be included in the pollen sum. Usually pollen and spores of obligate aquatic plants are excluded from the pollen sum because they are locally produced from a different environment from that in which the investigator is primarily interested. Special pollen sums can be used based on selected taxa only, if particular aspects of the past environment are being distinguished (e.g. Wright and Patten, 1963; H. H. Birks, 1972).

Faegri and Iversen (1975, p. 194) discuss the calculation of taxa outside the pollen sum. 'It should be the general rule in pollen analysis that the occur-

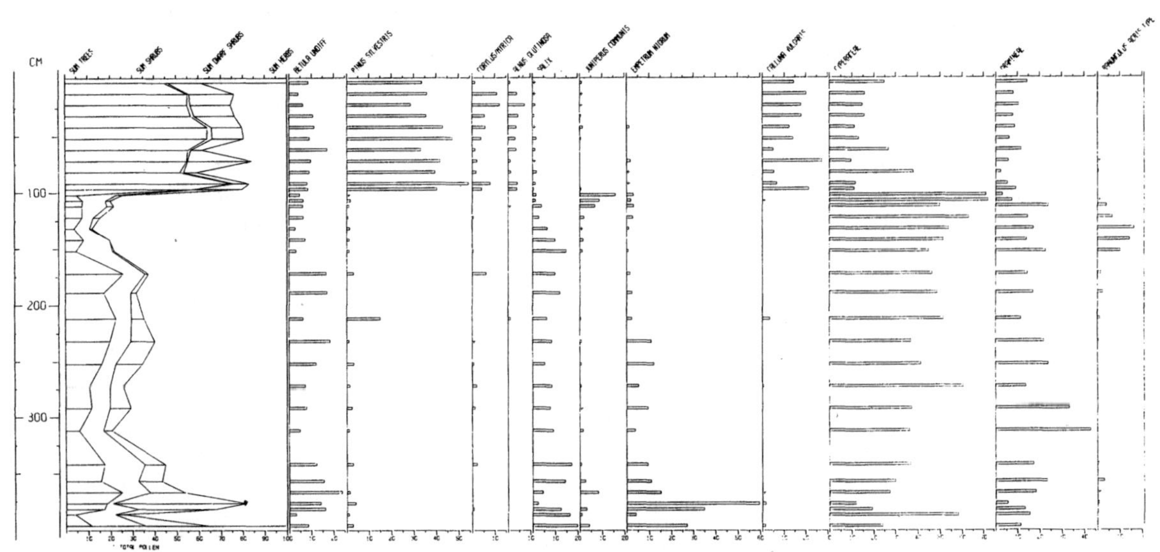

Fig. 8.10 A computer-drawn pollen diagram from Morrone Birkwoods, Scotland. Drawn by the University of Cambridge Calcomp plotter and IBM 370/165 computer using the program POLLDATA, written by H. J. B. Birks and B. Huntley.

rence of any pollen category should be expressed in percentages of a universe of which it forms a part'. This is also statistically valid (see Mosimann, 1965). For example, if obligate aquatics are excluded from the basic sum, their percentages should be calculated as percentages of the basic sum (ΣP) plus their own sum, i.e. ΣP +ΣAquatics. In this way, one avoids getting greater than 100% of any taxon. Faegri (1966) discusses features of the pollen sum and the constraints imposed by percentage calculations.

After the percentages have been calculated, the diagram can be drawn. There are very many different styles of pollen diagram. It should be remembered that the data should be visible in the clearest possible way to the reader. The symposium volume of the British Ecological Society *Quaternary Plant Ecology* edited by Birks and West (1973) contains a wide selection of different styles of pollen diagram. Some of the clearest are from the Minnesota school, e.g. Bradbury and Waddington (1973).

Nowadays, computers can be used to speed up the calculations of pollen data and the drawing of pollen diagrams. The calculations take about ten seconds, and it is easy to recalculate the data using different pollen sums. An example of a computer-drawn pollen diagram is shown in Fig. 8.10. Such diagrams are very useful for preliminary arrangement and interpretation of the data as the order of the curves can be easily changed to improve the diagram. Several different arrangements and pollen sums can be tried in a matter of minutes; a process which would take several weeks by hand. However, computer-drawn diagrams are not sufficiently good for publication, and diagrams for publication must still be drafted by hand.

Zonation of pollen diagrams

Pollen diagrams are complex, consisting of many levels and many taxa. Some form of simplification is often needed. This can be most readily achieved by drawing horizontal lines, separating sections or zones of differing pollen composition. Such a subdivision aids in the description of the diagram, the discussion and interpretation of its contents, and its comparison and correlation with other diagrams.

1. *Description of pollen zones*

The most useful subdivision is the ***pollen zone***. This has been defined as 'a body of sediment with a consistent and homogeneous fossil pollen and spore content that is distinguished from adjacent sediment bodies by differences in the kind and frequencies of its contained fossil pollen and spores' (Gordon and Birks, 1972).

A pollen zone falls into the category of an ***assemblage zone*** (see Chapter 2) which is a biostratigraphic unit. Cushing (1967) first introduced this strict concept of the pollen zone into Quaternary palaeoecology. Previously, there had been much (mostly sub-conscious) confusion about the criteria used in pollen zonation. Pollen zones have, in the past, been defined on various criteria.

a) The observed pollen stratigraphy.
b) Units of inferred past vegetation.
c) Units of inferred past climate.
d) Units of sediment lithology.
e) Units of inferred time.

Only (a) should be used in strict pollen zonation, as it is free of interpretive overtones (Chapter 10 in Birks (1973) discusses this topic fully).

If strict pollen assemblage zones are to be used, they should be properly defined. The definition should include, where appropriate:

a) Type locality and section.
b) Description of the fossils in the zone.
c) Description of the contacts with other zones.
d) The thickness of the zone and its age, if known.
f) The name of the zone, e.g. *Betula-Juniperus* regional pollen assemblage zone.
g) Other occurrences and general notes.

A pollen diagram should be divided as carefully as possible into pollen assemblage zones. As such, the zones will be unique to that site, and they can be called site zones or local zones. If they can be matched at other sites, regional pollen assemblage zones can be defined, and fully described as outlined above.

If the regional pollen assemblage zones are adequately radiocarbon-dated, they can be mapped in space and time. This was first done by Cushing (1967) for pollen zones in the Wisconsinan late-glacial of Minnesota (Fig. 8.11), and then by McAndrews (1966) for post-glacial (Holocene) pollen zones in north-west Minnesota, by Birks (1973) for the Devensian late-glacial of the Isle of Skye, Scotland, and by Birks and Berglund (1979) for southern Sweden.

If the regional pollen assemblage zones can be interpreted in terms of past vegetation, correlation charts of the vegetation changes through space and

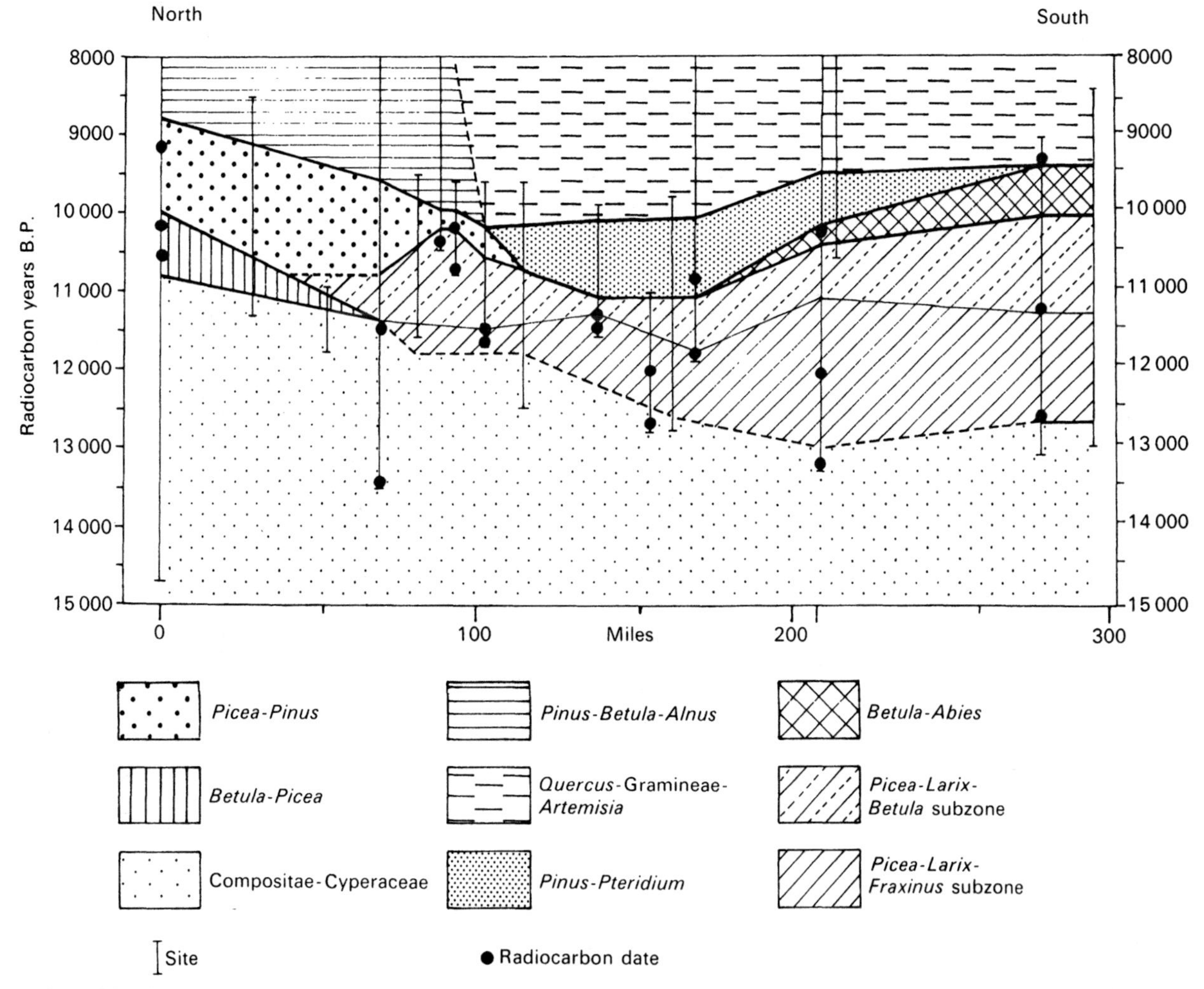

Fig. 8.11 Correlation chart of pollen stratigraphy along a north–south transect in eastern Minnesota between 8000 and about 15 000 radiocarbon years B.P. (●) indicates radiocarbon dates used to determine the ages of the zone boundaries. (– – –) indicates that the position of the zone boundary is uncertain. (After Cushing, 1967.)

time can be constructed, as in Fig. 8.12 (see also McAndrews, 1966; Birks, 1973). In this way, it is easier to see the past patterns of vegetation, and the pattern of changes within an area over time. In due course, the climatic implications of the inferred vegetation may also be mapped, to provide some insight into the past climatic patterns, and the patterns of climatic change. Hence it is very important that pollen zones should be properly described in an unbiased way.

2. *Numerical definition of pollen zones*

However strict he may be with himself, a pollen analyst, being human, will tend to divide his pollen diagram into zones, not only by considering the changes in the diagram, but also by drawing on his previous experience, and looking for important changes which have been recognized in other pollen diagrams. A good example of this sub-conscious bias is the drawing of the zone boundary at the decline of *Ulmus* pollen in the mid-Flandrian of the British

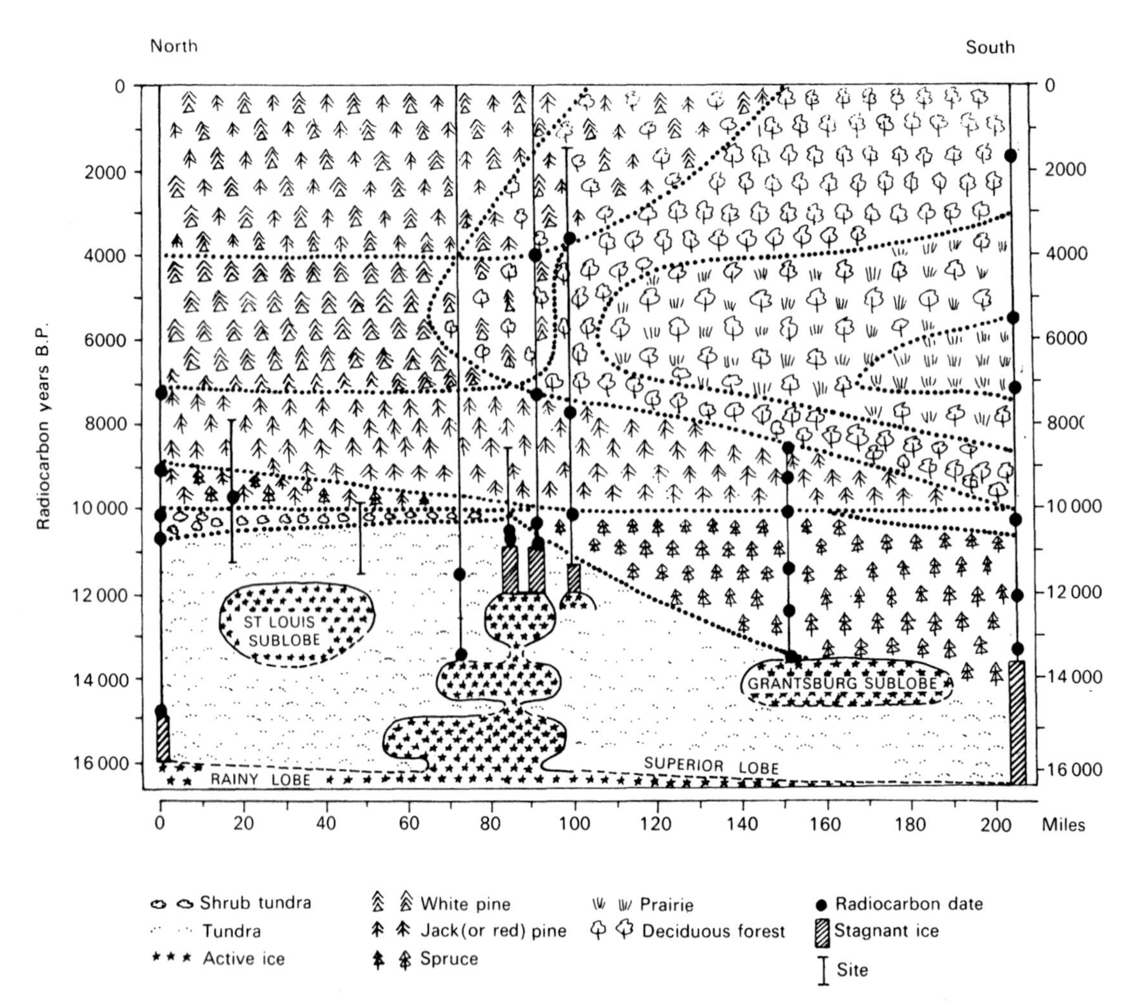

Fig. 8.12 A reconstruction of vegetation patterns in time and space in part of the north–south transect in eastern Minnesota in Fig. 8.10 between the present-day and about 16 000 B.P. The extents in space and time of the Wisconsinan ice advances in the area after 16 000 B.P. are also shown.

Isles (Birks, 1974). Where it has been dated, this boundary consistently falls at about 5000 B.P., and it is generally interpreted as corresponding to the first major influence of man in disturbing the natural forest ecosystem. When pollen diagrams are zoned using numerical criteria, the elm-decline horizon is frequently not distinguished by the purely numerical methods, because the total change in the diagram is so slight. Gordon and Birks (1972) discuss a similar phenomenon in the Wisconsinan late-glacial of Minnesota, U.S.A. In contrast, the computer may distinguish horizons where there are large changes, but these may be ignored by the pollen analyst, who, in the light of his previous experience, interprets the changes as purely local events; for example, a large rise in Gramineae pollen at the top of the profile from Wolf Creek, Minnesota, was distinguished as a zone by the computer (Gordon and Birks, 1972) but it was ignored in the final zonation because it could be related to the local overgrowth of the site by wild rice and other grasses (Birks, 1976).

If allowances are made for such situations, the zonation of pollen diagrams by numerical methods implemented by computers is highly desirable, for several reasons.

i) The methods are quick to implement, and several methods can be used on the same diagram to find the numerically most consistent zonation.
ii) The zones can only be distinguished on the observed fossil content. They are therefore strict pollen assemblage zones.
iii) Each method uses consistent criteria, and there is no bias on the part of the investigator as to which are the most important pollen types.
iv) Because the computer programs require the investigator to be explicit in the criteria to be used, the investigator has to define the criteria, which is valuable to all who may be interested in his results and in his methods.
v) The results of the analysis are repeatable by different investigators provided with the same data.
vi) The investigator may benefit because he is forced to look at the main numerical changes in his diagram, and to interpret them, as well as looking for correlative features with other diagrams, such as a statistically insignificant elm-decline.

In general, computer methods give results which agree broadly with the zonation done by eye. Any discrepancies are interesting because they demand an explanation. For example, Gordon and Birks (1972) showed that the lower part of the pollen diagram from Kirchner Marsh, Minnesota, U.S.A. is divided quite differently by numerical procedures from the published zonation by Wright *et al.* (1963).

3. *Methods of numerical zonation of pollen diagrams*

Because zonation is a form of classification and the data are quantitative and multivariate, numerical methods of classification can be applied to pollen data. Recently, there have been several methods proposed.

i) Dale and Walker (1970) used an information statistic as a dissimilarity coefficient (DC) and an agglomerative clustering technique to group the samples from Scaleby Moss, Cumbria, England. However, they ignored the stratigraphic order, and hence found that some stratigraphically separated samples were grouped together in clusters. Although their method has interesting ecological implications, the stratigraphic order of the samples must be maintained to provide a useful zonation.
ii) Gordon and Birks (1972) developed a constrained single-link clustering method (CONSLINK) where the stratigraphic order was imposed, and only adjacent levels could be combined. In this method, measures of dissimilarity between all pairs of samples are initially calculated. The dissimilarity measure used by Gordon and Birks (1972) was

$$DC_{ij} = \sum_{k=1}^{m} | pk_i - pk_j |$$

where DC_{ij} is the dissimilarity coefficient between samples i and j, and pk_i is the proportion of pollen type k in sample i when there are $k = 1, 2, \ldots, m$ pollen types. The computer then finds the two samples which are stratigraphically adjacent with the lowest dissimilarity. These two samples are grouped together, and the procedure is repeated until all the samples are grouped together. The clusters obtained of samples of similar pollen composition can be viewed as pollen assemblage zones.
iii) Two main divisive methods were developed by Gordon and Birks (1972). The methods divide a pollen diagram in such a way that the total numerical information it contains is maximally reduced at each division. Again the stratigraphic order is maintained. SPLITINF divides the data in terms of the total information content, and SPLITLSQ divides the data on the sum of least squares deviations.
iv) Ordination rather than classification methods were also used by Gordon and Birks (1974). The method of non-metric multi-dimensional scaling (MDSCAL) ordinates samples and finds the lowest 2-dimensional representation with the minimal stress function. If the levels are joined in stratigraphic order on the 2-dimensional plot, zone boundaries can be placed between groups of samples of similar pollen composition that have been positioned together on the plot. This method gave similar results to the classification methods.
v) Principal components analysis (PCA) is another ordination method which has successfully been used in pollen diagram zonation (Birks, 1974; Adam, 1974; Pennington and Sackin, 1975). The results of PCA and MDSCAL can either be plotted as 2-dimensional scatter diagrams, or the component scores can be plotted stratigraphically. In this way, many pollen curves can be represented as two or three curves which summarize the major patterns of variation in the diagram. Figure 8.13 from Birks (1974) shows the results of the zonation of the diagram from Scaleby Moss, Cumbria, England (Walker, 1966) by Walker's Cumbrian zones, the 'standard British zones' (Godwin, 1940), by the classification methods of CONSLINK, SPLITINF, and SPLITLSQ, and by the ordination methods of PCA and MDSCAL. In general, the numerical methods conform

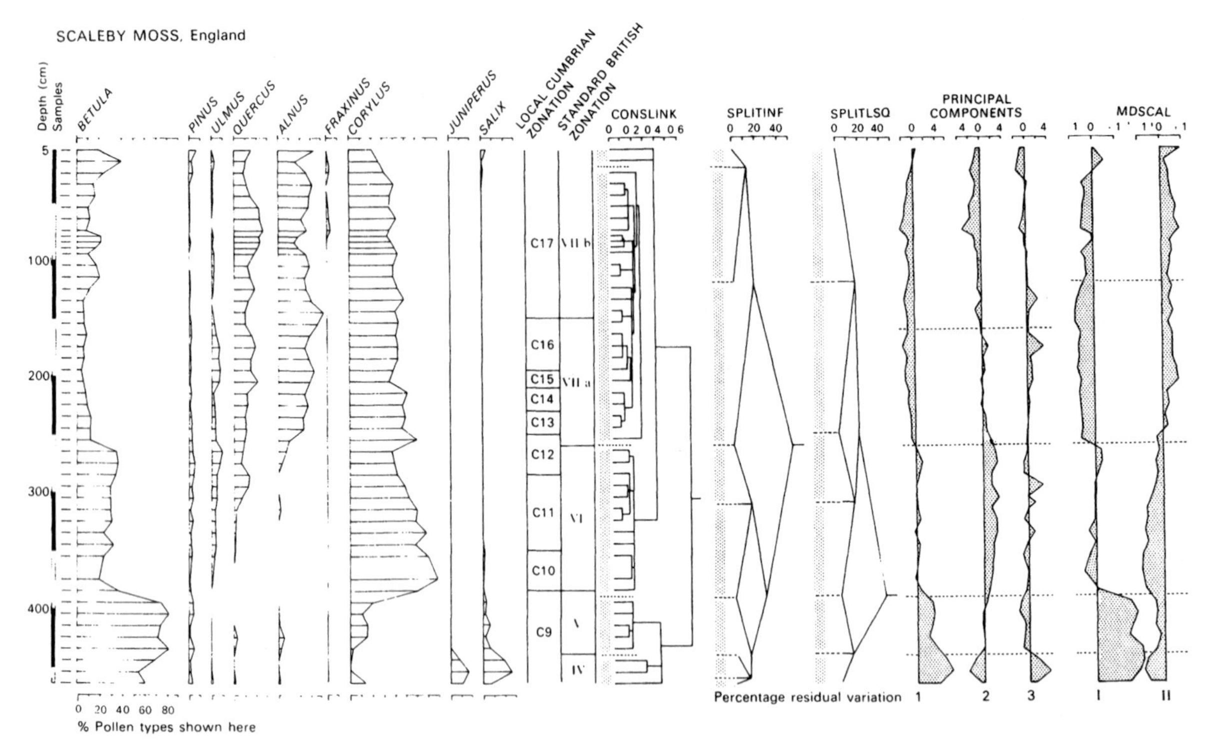

Fig. 8.13 The results of different methods of zoning the pollen diagram from Scaleby Moss, Cumbria, England. The pollen types shown are those used in the numerical analyses. Values are percentages of the sum of the total pollen types shown, and the curves are drawn to a constant scale. The Cumbrian and Standard British pollen zones are shown, together with dendrograms for agglomerative and divisive analyses, plots of component scores for the first three principal components, and sample coordinates derived from non-metric multi-dimensional scaling. (After Birks, 1974.)

with the standard British zonation. However, the 'elm decline' at 150 cm is only distinguished by PCA, showing that the detection of minor changes is dependent upon the degreee of weighting of the minor pollen types by the different numerical methods. This does not, of course, deny that the 'elm decline' is an ecologically significant and important boundary, only that it is of low numerical importance. Walker's Cumbrian zonation differs considerably from all the other schemes, but this is not surprising because his criteria of definition were not only homogeneity of pollen composition, but included features of both pollen stratigraphical stability and stratigraphical change of a restricted and selected group of taxa.

Zonation using other fossils

These numerical methods can also be used to zone stratigraphic sequences characterized by other types of fossils. Gordon and Birks (1972) used the plant macrofossils from Kirchner Marsh, Minnesota, U.S.A., and compared the zonation of macrofossils with that of the pollen diagram from the same core. Although some zone boundaries coincided, several zones were delimited in the macrofossil diagram which were not present in the pollen diagram. This demonstrates the ecological difference between pollen, which is mostly derived from the upland, and plant macrofossils, which are mostly locally derived (see Chapter 5). Some environmental factors affect both, whereas other factors have a more localized effect on one or the other. Other comparisons of pollen and macrofossil zonations have been made by Birks (1976) at Wolf Creek, Minnesota, U.S.A., and by Birks and Mathewes (1978) at Abernethy Forest, Scotland.

Comparison of pollen diagrams

Zonation of a pollen diagram is not an end in itself. The zones which are delimited are site or local zones. The pollen analyst is usually interested in

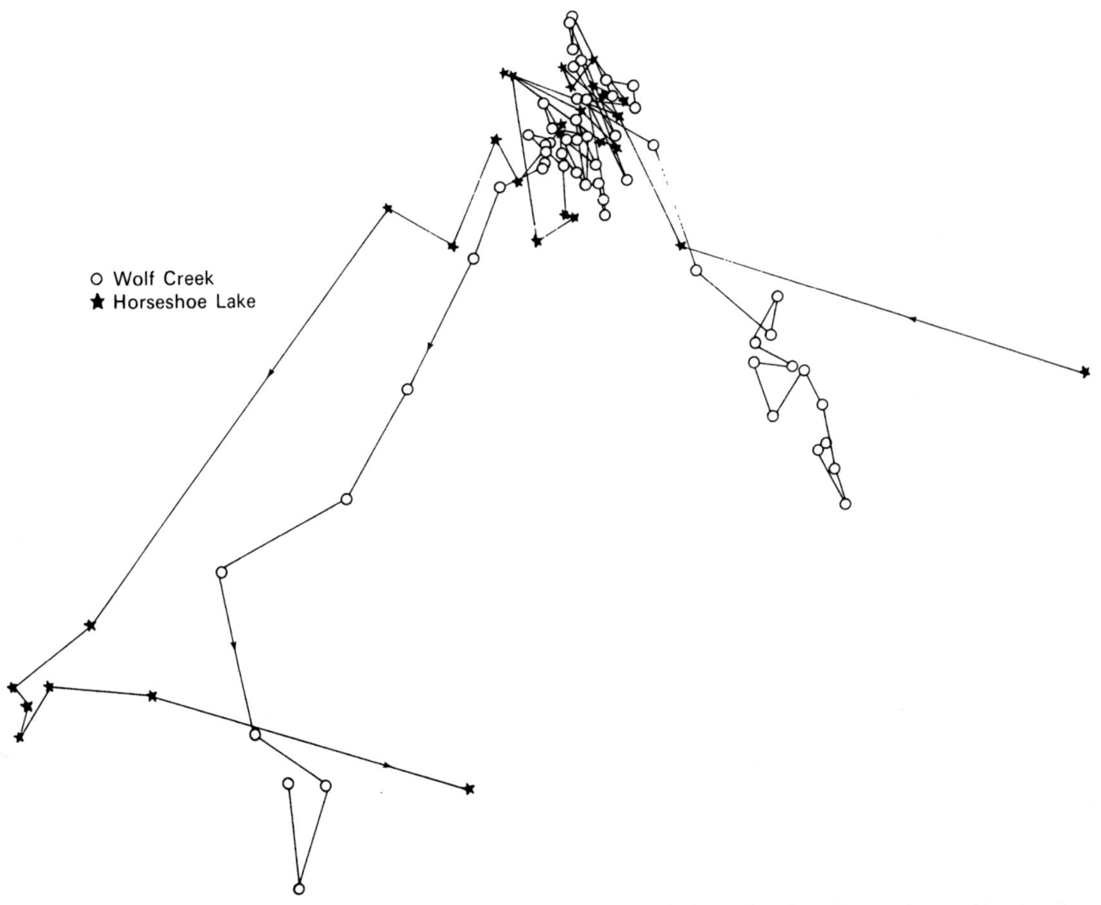

Fig. 8.14 The configuration obtained by applying non-metric multi-dimensional scaling to the combined pollen stratigraphic data from Wolf Creek (O) and Horseshoe Lake (★). Neighbouring levels in each core are joined by a line. (After Gordon and Birks, 1974.)

comparing his sequence with others. If the zones can be matched, regional pollen assemblage zones can be defined, and mapped temporally and geographically (see Cushing, 1967, and p. 169).

Usually, pollen zones have been compared visually, and either matched or correlated using the concept of 'regional parallelism' (von Post, 1946; Godwin, 1940). Gordon and Birks (1974) devised several numerical methods of comparing pollen diagrams, which involved no bias from the investigator, which were repeatable, and which were quick to implement by computer.

1. Pollen zones were numerically defined from two sites and were compared as such, using the method of zone-by-zone comparison with information radii and concordance as measures of zone similarity.

2. a) Individual pollen samples were ordinated from two sites at the same time using the method of non-metric multi-dimensional scaling. When the samples on the 2-dimensional plot (Fig. 8.14) were joined in stratigraphic order, there was a considerable overlap between the sites, showing the similarity between the various groups of samples. The local zones containing these samples can now be combined into regional pollen assemblage zones.

b) Birks and Berglund (1979) compared two diagrams from south-east Sweden (Blekinge) using PCA ordination techniques. These are computationally easier than MDSCAL. Regional pollen assemblage zones could be delimited. However, a similar comparison between two sites from south-east and south-central Sweden (Scania) showed little resemblance suggesting that there had been marked vegetational differentiation between the two areas in the past. A similar result was obtained with SLOTSEQ (see below). Previously, the pollen zones from these two areas had been correlated and fitted into the standard Swedish zonation sequence. The numerical methods demonstrated that many assumptions and interpretive overtones had been made during the primary data simplification using the standard Swedish zonation scheme.

3. With the stratigraphic constraint maintained, the samples from two sites can be slotted together according to their degree of similarity, in the best possible mathematical fit, using the method of sequence slotting and the program called SLOTSEQ. Gordon and Birks (1974) illustrated the use of this method on the sites of the Itasca transect of McAndrews (1966) (see Chapter 11). They showed the correspondence of the early zones and the subsequent divergence as the vegetational belts differentiated around each site, from prairie in the west to coniferous–deciduous forest in the east.

All these methods give broadly similar results with the same data sets, and the results make good sense when inspected visually. When radiocarbon dates are available, the comparisons between sites are generally confirmed. The first major use of these methods has been by Birks and Berglund (1979) in southern Sweden, where they have highlighted several problems and limitations in the definition and use of the standard Swedish zonation system (Mangerud *et al.*, 1974).

References

Pollen analysis – general principles

ERDTMAN, G. (1943). *An Introduction to Pollen Analysis.* Chronica Botanica. Waltham.

ERDTMAN, G. (1969). *Handbook of Palynology.* Munksgaard.

FAEGRI, K. AND IVERSEN, J. (1975). *Textbook of Pollen Analysis* (3rd edition). Blackwell.

FRIES, M. (1967). Lennart von Post's pollen diagram series of 1916. *Rev. Palaeobot. Palynol.*, **4**, 9–13.

GODWIN, H. (1934). Pollen analysis: an outline of the problems and potentialities of the method. I. Technique and interpretation. II. General applications of pollen analysis. *New Phytol.*, **33**, 278–305, 325–58.

TSCHUDY, R. H. AND SCOTT, R. A. (1969). *Aspects of Palynology.* Wiley – Interscience.

VON POST, L. (1967). Forest tree pollen in South Swedish peat bog deposits. *Pollen Spores*, **9**, 375–401.

WEST, R. G. (1971). *Studying the Past by Pollen Analysis.* Oxford Biology Reader 10.

Pollen morphology and development

ANDERSEN, S. T. (1960). Silicone oil as a mounting medium for pollen grains. *Danm. geol. Unders.*, Ser. IV, **4**(1), 24 pp.

ANDERSEN, S. T. (1978). On the size of *Corylus avellana* L. pollen mounted in silicone oil. *Grana*, **17**, 5–13.

ANDERSEN, S. T. AND BERTELSEN, F. (1972). Scanning electron microscope studies of pollen of cereals and other grasses. *Grana*, **12**, 79–86.

BIRKS, H. J. B. (1978). Geographic variation of *Picea abies* (L.) Karsten pollen in Europe. *Grana*, **17**, 149–60.

BROOKS, J. (1971). Some chemical and geochemical studies on sporopollenin. In *Sporopollenin* (eds J. Brooks *et al.*). Academic Press.

CAIN, S. A. AND CAIN, L. G. (1948). Size-frequency characteristics of *Pinus echinata* pollen. *Bot. Gaz.*, **110**, 325–30.

CHRISTENSEN, B. B. (1946). Measurement as a means of identifying fossil pollen. *Danm. geol. Unders.*, Ser. IV, **3**(2), 22 pp.

CHRISTOPHER, R. A. AND WATERS, J. A. (1974). Fourier series as a quantitative indicator of miospore shape. *J. Paleontol.*, **48**, 697–709.

CUSHING, E. J. (1961). Size increase in pollen grains mounted in thin slides. *Pollen Spores*, **3**, 265–74.

ERDTMAN, G. (1952). *Pollen Morphology and Plant Taxonomy.* Almqvist and Wiksell.

FAEGRI, K. AND DEUSE, P. (1960). Size variations in pollen grains with different treatments. *Pollen Spores*, **2**, 293–8.

FLENLEY, J. R. (1971). Measurements of the specific gravity of the pollen exine. *Pollen Spores*, **13**, 179–86.

GODWIN, H. (1968). The origin of the exine. *New Phytol.*, **67**, 667–76.

HESLOP-HARRISON, J. (1968). Pollen wall development. *Science*, **161**, 230–7.

HESLOP-HARRISON, J. (1976). The adaptive significance of the exine. In *The Evolutionary Significance of the Exine* (eds I. K. Ferguson and J. Muller). Academic Press.

IVERSEN, J. AND TROELS-SMITH, J. (1950). Pollenmorfologiska definitioner og typer. *Danm. geol. Unders.*, Ser. IV, **3**(8), 52 pp.

MCANDREWS, J. H. AND SWANSON, A. R. (1967). The pore number of periporate pollen with special references to *Chenopodium*. *Rev. Palaeobot. Palynol.*, **3**, 105–17.

PRAGLOWSKI, J. (1970). The effects of pre-treatment and the embedding media on the shape of pollen grains. *Rev. Palaeobot, Palynol.*, **10**, 203–8.

REITSMA, TJ. (1969). Size modification of recent pollen grains under different treatments. *Rev. Palaeobot, Palynol.*, **9**, 175–202.

SHAW, J. (1971). The chemistry of sporopollenin. In *Sporopollenin* (eds J. Brooks *et al.*), Academic Press.

SMIT, A. (1973). A scanning electron microscopical study of the pollen morphology in the genus *Quercus*. *Acta bot. Neerl.*, **22**, 655–65.

STOCKMARR, J. (1970). Species identification of *Ulmus* pollen. *Danm. geol. Unders.*, Ser. IV, **4**(11), 19 pp.

VAN DER PLUYM, A. AND HIDEAUX, M. (1977). Numerical analysis study of pollen grain populations of *Eryngium maritimum* L. (Umbelliferae). *Rev. Palaeobot. Palynol.*, **24**, 119–39.

WALKER, J. W. (1974a). Aperture evolution in the pollen of primitive angiosperms. *Amer. J. Bot.*, **61**, 1112–36.

WALKER, J. W. (1974b). Evolution of exine structure in the pollen of primitive angiosperms. *Amer. J. Bot.*, **61**, 891–902.

WALKER, J. W. (1976). Evolutionary significance of the exine in the pollen of primitive angiosperms. In *The Evolutionary Significance of the Exine* (eds I. K. Ferguson and J. Muller). Academic Press.

WALKER, J. W. AND DOYLE, J. A. (1975). The bases of angiosperm phylogeny: palynology. *Annals Miss. Bot. Garden*, **62**, 664–723.

WHITEHEAD, D. R. (1963). Pollen morphology in the Juglandaceae I. Pollen size and pore number variation. *J. Arnold Arbor.*, **44**, 101–10.

WHITEHEAD, D. R. (1964). Fossil pine pollen and full-glacial vegetation in southeastern North Carolina. *Ecology*, **45**, 767–77.

WITTMAN, G. AND WALKER, D. (1965). Towards simplification in sporoderm description. *Pollen Spores*, **7**, 443–56.

Pollen identification manuals

BEUG, H.-J. (1961). *Leitfaden der Pollenbestimmung*. Fischer.

ERDTMAN, G., BERGLUND, B. E. AND PRAGLOWSKI, J. (1961). *An Introduction to a Scandinavian Pollen Flora*. Almqvist and Wiksell.

FAEGRI, K. AND IVERSEN, J. (1975). *Textbook of pollen analysis* (3rd edition). Blackwell.

HYDE, H. A. AND ADAMS, K. F. (1958). *An Atlas of Airborne Pollen Grains*. Macmillan.

MCANDREWS, J. H., BERTI, A. A. AND NORRIS, G. (1973). Key to the Quaternary pollen and spores of the Great Lakes Region. *Life Sci. Misc. Publ. R. Ont. Mus.*

MOORE, P. D. AND WEBB, J. A. (1978). *An Illustrated Guide to Pollen Analysis*. Hodder and Stoughton.

PILCHER, J. R. (1968). Some applications of scanning electron microscopy to the study of modern and fossil pollen. *Ulster J. Archaeol.*, **31**, 87–91.

PUNT, W. (ed.) (1976). *The North-West European Pollen Flora*. I. Elsevier.

Pollen identification procedures

BIRKS, H. J. B. AND PEGLAR, S. M. (1980). Identification of *Picea* pollen of late Quaternary age in eastern North America: a numerical approach. In press.

GOODALL, D. W. (1966). Deviant index: a new tool for numerical taxonomy. *Nature*, **210**, 216.

GORDON, A. D. AND PRENTICE, I. C. (1977). Numerical methods in Quaternary palaeoecology. IV. Separating mixtures of morphologically similar pollen taxa. *Rev. Palaeobot. Palynol.*, **23**, 359–72.

HANSEN, B. S. AND CUSHING, E. J. (1973). Identification of pine pollen of late Quaternary age from the Chuska Mountains, New Mexico. *Bull. Geol. Soc. Amer.*, **84**, 1181–200.

WALKER, D., MILNE, P., GUPPY, J. AND WILLIAMS, J. (1968). The computer assisted storage and retrieval of pollen morphological data. *Pollen Spores*, **10**, 251–62.

Morphometric variation in Quaternary fossils

HILLS, L. V., KLOVAN, J. E. AND SWEET, A. R. (1974). *Juglans eocinerea* n.sp., Beaufort Formation (Tertiary), southwestern Banks Island, Arctic Canada. *Can. J. Bot.*, **52**, 65–90.

JENTYS-SZAFEROWA, J. (1958). Importance of Quaternary materials for research on the historical evolution of plants. *Veröff. Geobot. Inst. Rübel*, **34**, 67–73.

KOKAWA, S. (1959). Morphometry of *Menyanthes* seed remains in Japan. *J. Inst. Polyt. Osaka City Univ.* D, **10**, 44–63.

MASTROGIUSEPPE, J. D., CRIDLAND, A. A. AND BOGYO, T. P. (1970). Multivariate comparison of fossil and recent *Ginkgo* wood. *Lethaia*, **3**, 271–7.

Pollen counting and presentation of pollen analytical data

BIRKS, H. H. (1972). Studies in the vegetational history of Scotland. II. Two pollen diagrams from the Galloway Hills, Kirkcudbrightshire. *J. Ecol.*, **60**, 183–217.

BIRKS, H. J. B. AND WEST, R. G. (1973). *Quaternary Plant Ecology*. Blackwell.

BRADBURY, J. P. AND WADDINGTON, J. C. B. (1973). The impact of European settlement on Shagawa Lake, Northeastern Minnesota, U.S.A. In *Quaternary Plant Ecology* (eds H. J. B. Birks and R. G. West). Blackwell.

BROOKES, D. AND THOMAS, K. W. (1967). The distribution of pollen grains on microscope slides I. The non-randomness of the distribution. *Pollen Spores*, **9**, 621–9.

FAEGRI, K. (1966). Some problems of representativity in pollen analysis. *Palaeobotanist*, **15**, 135–40.

MAHER, L. J. (1972). Nomograms for computing 0.95 confidence limits of pollen data. *Rev. Palaeobot. Palynol.*, **13**, 85–93.

MIRKIN, G. R. AND BAGDASARYAN, L. L. (1972). The feasibility of identifying paleontological objects with the aid of optical analysing systems. *Paleontol. J.*, **6**, 103–8.

MOSIMANN, J. E. (1965). Statistical methods for the pollen analyst: multinomial and negative multinomial techniques. In *Handbook of Paleontological Techniques* (eds B. Kummel and D. M. Raup). W. H. Freeman.

WRIGHT, H. E. AND PATTEN, H. L. (1963). The pollen sum. *Pollen Spores*, **5**, 445–50.

Pollen zonation

ADAM, D. P. (1974). Palynological applications of principal component and cluster analyses. *J. Res. U.S. Geol. Survey*, **2**, 727–41.

BIRKS, H. H. AND MATHEWES, R. W. (1978). Studies in the vegetational history of Scotland. V. Late Devensian and early Flandrian pollen and macrofossil stratigraphy at Abernethy Forest, Inverness-shire. *New Phytol.*, **80**, 455–84.

BIRKS, H. J. B. (1973). *Past and Present Vegetation of the Isle of Skye – a palaeoecological study*. Cambridge University Press (Chapter 10).

BIRKS, H. J. B. (1974). Numerical zonations of Flandrian pollen data. *New Phytol.*, **73**, 351–8.

BIRKS, H. J. B. (1976). Late-Wisconsinan vegetational history at Wolf Creek, central Minnesota. *Ecol. Monogr.*, **46**, 395–429.

BIRKS, H. J. B. AND BERGLUND, B. E. (1979). Holocene pollen stratigraphy of Southern Sweden: a reappraisal using numerical methods. *Boreas*, **8**, 257–79.

CUSHING, E. J. (1967). Late-Wisconsin pollen stratigraphy and the glacial sequence in Minnesota. In *Quaternary Palaeoecology* (eds E. J. Cushing and H. E. Wright). Yale University Press.

DALE, M. B. AND WALKER, D. (1970). Information analysis of pollen diagrams. *Pollen Spores*, **12**, 21–37.

GODWIN, H. (1940). Pollen analysis and forest history of England and Wales. *New Phytol.*, **39**, 370–400.

GORDON, A. D. AND BIRKS, H. J. B. (1972). Numerical methods in Quaternary palaeoecology. I. Zonation of pollen diagrams. *New Phytol.*, **71**, 961–79.

GORDON, A. D. AND BIRKS, H. J. B. (1974). Numerical methods in Quaternary palaeoecology. II. Comparison of pollen diagrams. *New Phytol.*, **73**, 221–49.

MCANDREWS, J. H. (1966). Postglacial history of prairie, savanna, and forest in northwestern Minnesota. *Mem. Torrey Bot. Club.*, **22**, 72 pp.

MANGERUD, J., ANDERSEN, S. T., BERGLUND, B. E. AND DONNER, J. J. (1974). Quaternary stratigraphy of Norden, a proposal for terminology and classification. *Boreas*, **3**, 109–28.

PENNINGTON, W. AND SACKIN, M. J. (1975). An application of principal components analysis to the zonation of two Late-Devensian profiles. *New Phytol.*, **75**, 419–53.

VON POST, L. (1946). The prospect for pollen analysis in the study of the Earth's climatic history. *New Phytol.*, **45**, 193–217.

WALKER, D. (1966). The late Quaternary history of the Cumberland Lowland. *Phil. Trans. R. Soc.* B, **251**, 1–210.

WRIGHT, H. E., WINTER, T. C. AND PATTEN, H. L. (1963). Two pollen diagrams from southeastern Minnesota: problems in the regional late-glacial and post-glacial vegetational history. *Bull. Geol. Soc. Amer.*, **74**, 1371–96.

9

Pollen production, dispersal, deposition, and preservation

In the previous chapter, we described the techniques involved in the production and zonation of a pollen diagram. Now we must consider how to interpret the pollen diagram in terms of palaeoecology. In order to do this it is important to have a factual background of how pollen is produced, liberated from the parent plant, dispersed in the atmosphere, deposited and preserved in the sediment, and how it is likely to be redeposited. We shall discuss these processes in this chapter.

Pollen production and liberation

Pollen production

It is well known that different plants produce different amounts of pollen. In general, wind-pollinated species produce very much more than insect-pollinated species, whereas cleistogamous species, such as *Viola*, in which the flowers do not open, have a very low pollen production indeed. Entomophilous species often have highly ornamented pollen grains which tend to stick together, whereas anemophilous species tend to have smooth, light, and round or winged pollen grains. However, some entomophilous species are high pollen producers, for example *Calluna vulgaris* and *Tilia cordata*. These are both good sources of honey.

It is difficult to estimate the actual pollen production of a species, but this was attempted by Pohl (1937). He counted the number of pollen grains in an anther, and multiplied this by the number of anthers in a flower, the number of flowers in an inflorescence, the number of inflorescences on a herbaceous plant or on a ten-year old branch of a tree, and the number of branch systems on the tree. Some of his results are shown in Table 9.1. The figures probably contain large inaccuracies. However, other studies of estimating pollen production using pollen traps and surface samples (mainly moss cushions) in forests confirm the order of magnitude of Pohl's calculations. The major

Table 9.1 Pollen production figures for selected European plants. (From Pohl, 1937.)

Species	Pollen stamen^{-1}	Pollen flower^{-1}	Pollen inflorescence^{-1}	Pollen 10 yr. branch^{-1}
Rumex acetosa	30 125	180 750	392 950 500	–
Secale cereale	19 103	57 310	4 240 940	–
Fraxinus excelsior	12 525	25 050	1 605 705	–
Betula verrucosa	10 072	20 145	5 452 500	118 502 500
Arrhenatherum elatius	6187	18 562	3 730 962	–
Quercus robur	5146	41 168	554 368	110 984 474
Calluna vulgaris	2216	17 714	–	–
Vallisneria spiralis	36	72	144	–
Pinus sylvestris	–	157 661	5 773 445	346 412 700
Fagus sylvatica	–	12 214	173 976	28 010 130
Tilia cordata	–	43 500	200 100	89 044 500
Populus canadensis	–	–	5 813 333	100 373 008
Alnus glutinosa	–	–	4 445 000	302 266 000
Polygonum bistorta	710	5678	2 861 712	–
Picea excelsa	–	589 500	–	106 699 500
Aesculus hippocastanum	25 828	180 000	764 800	18 543 571

forest trees fall in order of decreasing pollen production: *Pinus sylvestris*, *Alnus glutinosa*, *Corylus avellana*, *Betula verrucosa*, *Quercus robur*, *Picea abies*, *Populus canadensis*, *Tilia cordata*, *Fagus sylvatica*, and *Aesculus hippocastanum*. It has been estimated that *Pinus sylvestris* produces 10–80 kg of pollen per hectare annually.

Some trees in particular do not seem to produce the same amount of pollen each year. Andersen (1974) demonstrated a biennial pattern in Denmark, in which *Tilia cordata* seemed to alternate with *Quercus*, *Betula*, and *Fagus* in good flowering years. There may be some inherent flowering rhythm, and the climate of the preceding year may also influence the formation of flower primordia. Grosse-Brauckmann (1978), working at several sites in Germany, has shown considerable year-to-year variation in pollen production that appears to be related primarily to climate.

Pollen liberation

Most studies on pollen liberation have been connected with studies on aerobiology and hay fever.

1. *Seasonal variation*

There is a well marked seasonal variation in pollen liberation. Hyde and Williams (1944) recorded daily variation in pollen content of the atmosphere over a year at Cardiff, Great Britain in 1942. Figure 9.1 shows that in March, the main pollen types were *Corylus*, *Taxus*, and *Alnus*. In April *Ulmus* and *Betula* predominated, followed by *Quercus* in May. In June and July, grass pollen became abundant, and in August and September grass pollen was largely replaced by pollen of other herbs. Stix and Grosse-Brauckmann (1970) demonstrated a similar seasonal pattern at Darmstadt in Germany.

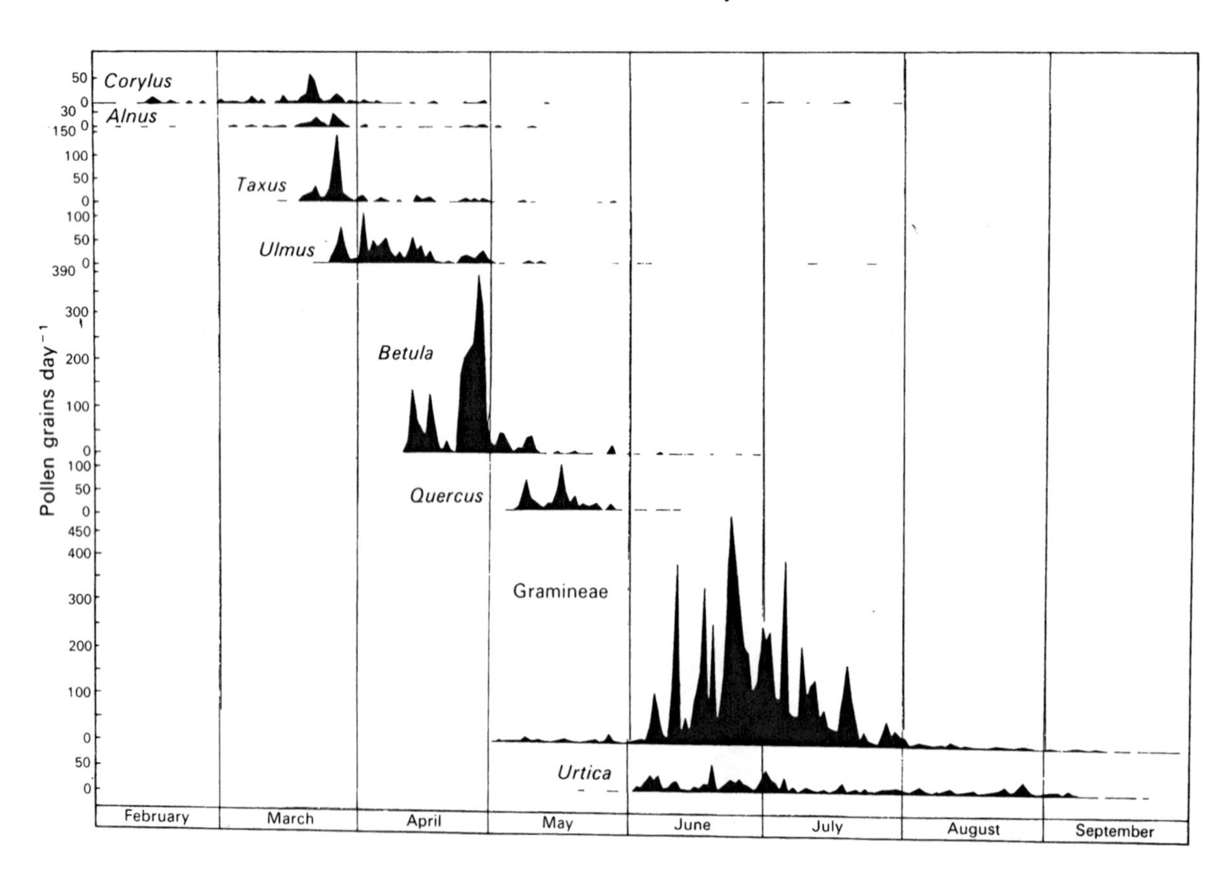

Fig. 9.1 Graphs showing the day-to-day variation in deposition of the principal types of pollen at Llandough, near Cardiff, Wales, in 1942. The graphs are set out in order of the first effective appearance of each type. (After Hyde and Williams, 1944.)

2. *Diurnal variation*

Studies on pollen liberation of two grasses *Holcus lanatus* and *Festuca rubra* through the day have shown that early morning and evening are often times of anthesis, related to air temperature, light intensity, and low relative humidity (RH) (Liem and Groot, 1973). Pollen liberation into the atmosphere is related to RH and atmospheric turbulence, suggesting that anthesis is an active process, whereas liberation is purely mechanical. Ogden *et al.* (1969), working on *Ambrosia* (ragweed) and the grasses *Phleum* and *Zea*, and Holmes and Bassett (1963) working on *Ambrosia* showed similar effects. Davies and Smith (1973) have shown that grass-pollen concentration in the atmosphere in London is, as elsewhere, positively correlated with air temperature and hours of bright sunshine.

Pollen dispersal and deposition

Pollen dispersal into the atmosphere

Apart from those grains transported by insects or water, or expelled forcibly by some active process, pollen grains are dispersed in the atmosphere following the laws governing all small particles (Tauber (1965) elaborates the theory and presents the equations). The dispersion of pollen depends principally on:

i) turbulence of the atmosphere;
ii) wind speed and direction;
iii) terminal falling velocity, which depends on the weight and shape of the pollen (Dyakowska, 1936; Dyakowska and Zurzycki, 1959);
iv) height and strength of the pollen source.

The *theoretical* curve of pollen deposition from an elevated point source under average meteorological conditions is shown in Fig. 9.2. There is no deposition near the source due to the wind carrying pollen laterally away from the source. Maximum deposition occurs at about 600 m from the source, and it falls off inversely proportional to the distance until about 2000 m, when a low constant deposition is maintained.

Raynor *et al.* (1970) confirmed this theoretical pattern experimentally using ragweed pollen. However, the natural situation is obviously very much more complex. Turner (1964) used a relatively simple situation by analysing surface samples across a raised bog at different distances from a line of pine trees. Her results are shown in Fig. 9.3. There are high 'local' values only near the source, and an exponential fall off to low 'regional' values over the relatively short distance of 400 m.

Direct application of theoretical models to natural situations is very difficult for three main reasons.

i) Natural sources of pollen are not point or line sources, but are areas of vegetation of varying height and pollen production. This explains why the theoretical area of zero deposition is rarely, if ever, found, because pollen is blown in from nearby sources.
ii) Meteorological conditions vary so much over a period of time that extrapolation from a surface sample to a source is impossible.

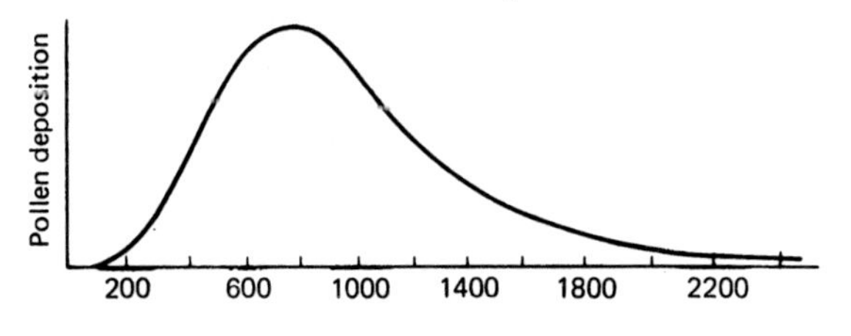

Fig. 9.2 Theoretical curve of pollen deposition from an elevated point source under average meteorological conditions.

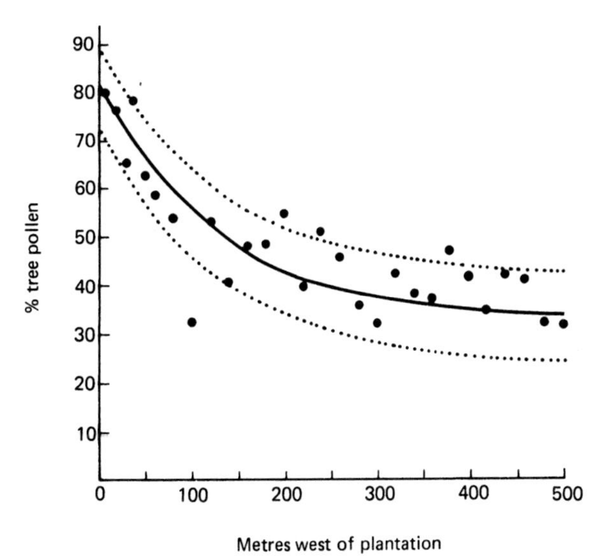

Fig. 9.3 *Pinus* pollen frequencies in surface samples in a transect west of the Cameron's Moss plantation, expressed as percentages of total tree pollen, together with the estimated curve $y(x) = a + \beta e^{\gamma x}$ and the 80% confidence limits on either side. (After Turner, 1964.)

iii) The density and structure of the vegetation causes air turbulence, and pollen is also filtered by the vegetation, often sticking to it, only to be washed or blown off later.

Therefore pollen deposition cannot be related directly to pollen production because of all the complex and poorly understood processes occurring in between.

Long-distance pollen dispersal in the atmosphere

Pollen can be transported great distances in the atmosphere. Samples of air over the oceans many hundreds of miles from land contain pollen grains. Samples 450 km from the Norwegian coast deposited in the summer of 1949 contained the following numbers of pollen

Pinus 5.5 grains cm^{-2}
Betula 4.7 grains cm^{-2}
Quercus 0.8 grains cm^{-2}
Gramineae 5.5 grains cm^{-2}

This assemblage was recorded some 1700 km from the nearest oak tree. Many studies of pollen deposition in arctic regions have recorded pollen of temperate trees, long distances from the nearest trees (see Nichols *et al.*, 1978).

The mechanism of long-distance transport in the atmosphere was discussed by Christie and Ritchie (1969) and Tyldesley (1973). From Tyldesley's work on tree pollen arriving in the Shetlands from mainland Britain and Scandinavia, it appears that the transport of tree pollen can be related to the trajectories of air masses with their contained particles. The occurrence of long distance transport must be borne in mind, particularly when interpreting pollen assemblages from environments with a low local pollen production (e.g. Maher, 1964).

Pollen deposition within a temperate deciduous forest

Tauber (1965) proposed a theoretical model of pollen dispersal within a forest and from a forest into a basin of deposition. The model was based on recent work on the dispersal of aerosols, radioactive debris, and other pollutants. Tauber proposed that there were three major dispersal components (Fig. 9.4). The first component, the trunk space (C_t on Fig. 9.4) contains pollen that is deposited within the forest and that is moved by wind through the trunk space. Trees and shrubs such as *Corylus avellana*, *Alnus glutinosa*, *Populus*, *Ulmus*, and *Salix* emit their pollen before the forest is in leaf when wind speeds in the trunk space are relatively high. Pollen of these plants may thus be dispersed some distances in the trunk-space component. In contrast pollen of trees such as *Fagus sylvatica* and *Tilia*, which flower when the forest is in leaf and horizontal wind velocities are low, are poorly dispersed within the trunk space.

The second component that Tauber recognized was the canopy component (C_c). This contains pollen blown above the canopy by convective turbulence. Theoretical calculations suggest that 50% of

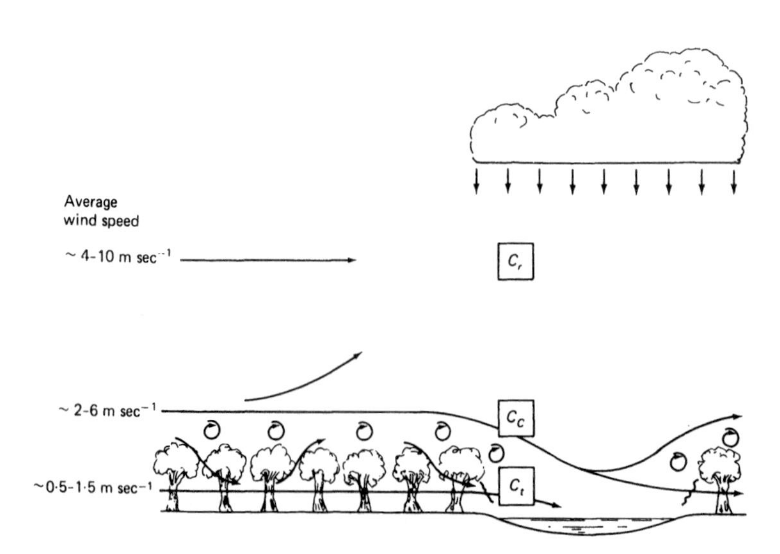

Fig. 9.4 Tentative model of pollen transfer in a forested area. Cr, rain-out component. Cc, canopy component. Ct, trunk space component. (After Tauber, 1965.)

the pollen in this component comes from a distance of 250 m, 75% from 5 km, and only 5% originates from 5–10 km. As one moves away from the forest edge, the role of long-distance dispersal increases. For example, at a distance of 500 m from the forest edge, it is predicted that 50% of the pollen comes from over 7 km, whereas at a distance of 5 km from the forest edge 50% of the pollen may originate from over 30 km away.

The third component in Tauber's model (Fig. 9.4) is the rain-out component (C_r). This contains pollen blown considerable distances which is brought to earth by impact with rain drops. The three components of the model all have different origins and hence may have different pollen compositions. In theory, Tauber's model predicts that the contribution of each of these components will vary with the size of the basin of deposition. For a small bog, 100–200 m in diameter, the expected percentages of the different components would be 80% C_t, 10% C_c, and 10% C_r. For a large bog (1 km or more in diameter), the expected percentages are 10% C_t, 70% C_c, and 20% C_r. It should be emphasized that these predicted values assume that all the pollen input into a site is by aerial transport. In the case of lakes this is not so (see below), as pollen is transported in large amounts by run-off and by steams.

Experimental pollen trapping in a small lake (200 × 100 m) within a forest in Denmark by Tauber (1967, 1977) has largely confirmed his model, with a C_r of 5%, C_c of 35%, and C_t of 60%. Work by Currier and Kapp (1974) in Michigan, U.S.A. has similarly shown that the trunk-space component to be smaller than is predicted by Tauber's model and that the canopy component is correspondingly higher than is predicted. Tauber (1977) also showed that up to 50% of the pollen entering his small lake was water-borne rather than aerial in its transport.

The reasons why the observed trunk-space component is less important than was predicted are not fully understood. One hypothesis is that a considerable amount of pollen produced by trees falls directly below the parent tree, either as individual grains or in detached catkins and anthers (Andersen, 1974) and is not moved laterally within the trunk space. Handel (1976) using neutron-activation techniques has shown that in the case of two forest herbs, the pollen dispersal is very local (1–2 m) within the forest. This is because the wind speeds within a forest are less than 10% of the wind speeds outside the forest. A second hypothesis is that considerable amounts of pollen produced within the forest get trapped on leaves and twigs, so-called filtration of Tauber (1965), and this pollen is eventually washed off by rain or by leaf fall and deposited vertically below. Tauber has shown that nearly 60% of the total pollen was deposited between August and November due to resuspension of the filtration component. This process may be so important that Krzywinski (1977) has proposed that pollen deposition in a forest comprises three components: (i) deposition during flowering, (ii) redeposition of pollen by rain, and (iii) deposition of pollen with litter in the autumn. The overriding importance of local, near vertical pollen deposition within forests has been shown experimentally by Raynor *et al.* (1974, 1975) using stained ragweed pollen.

Pollen deposition within a tropical forest

Kershaw and Hyland (1975) studied the pollen deposition in a small lake within a tropical forest in Queensland, Australia. They demonstrated several differences with a temperate forest, largely due to the differing biology of the two systems.

i) There is a very low pollen production, equivalent to that in arctic tundra sites.
ii) The pollen rain is dominated by species with a high canopy, suggesting that the canopy component is the most important.
iii) Seasonal variation is pronounced, as in a temperate forest, both in the taxonomic composition of the pollen rain, and in its amount. Highest deposition occurs at the end of the dry season.
iv) There is little evidence for filtration, although it undoubtedly occurs, because flowering is continuous throughout the year.

Pollen dispersal in water

1. *Freshwater*

Although it has long been realized that pollen can be transported in suspension by streams and rivers, it is only recently that attempts have been made to construct a pollen budget for a lake.

Peck (1973) studied a catchment in Yorkshire, England containing an upper, rather exposed lake, and a lower sheltered lake surrounded by forest. She placed pollen traps at various depths in the lakes, on their surfaces, and in the inflow and outflow streams, and took monthly samples over 14 months. On the upper lake, the yearly aerial pollen deposition was 4542 grains cm^{-2}, but in the lake it was

169 736 grains cm^{-2} $year^{-1}$. In the lower lake, aerial deposition was 3705 grains cm^{-2} $year^{-1}$, and in the lake it was 182 231 grains cm^{-2} $year^{-1}$. Therefore, aerial deposition is a small proportion of the total, and the rest must have come *via* the streams and from slope run-off. Additional apparent deposition can also come from pollen resuspended from the upper sediments due to water turbulence. In order to evaluate this process, Peck analysed cores from the lake sediments and calculated the pollen influx over the last 80 years. In the lower, sheltered lake, where turbulence was minimal, the annual influx to the sediment was 124 443 grains cm^{-2} $year^{-1}$, which is somewhat less than the influx to the traps. However, in the upper lake in open country, the annual influx to the sediment was 49 071 grains cm^{-2} $year^{-1}$, which was considerably less than the 169 736 grains cm^{-2} $year^{-1}$ in the traps. With greater turbulence, much more pollen has been resuspended. In a comparable study, Davis (1968) demonstrated that two to four times as much pollen was deposited in sediment traps than was deposited annually at the sediment surface in Frains Lake, Michigan, U.S.A. which is a small lake with no inflow or outflow streams. She demonstrated that the greatest resuspension of sediment occurred at the autumnal overturn with a lesser amount in the spring (Fig. 9.5).

Peck's lakes did, however, have inflow and outflow streams, and she estimated the amount of pollen carried by them. When the sediment influx values were compared with the influx values of aerial deposition, she found that the aerial deposition in the upper lake was 9% of the total influx,

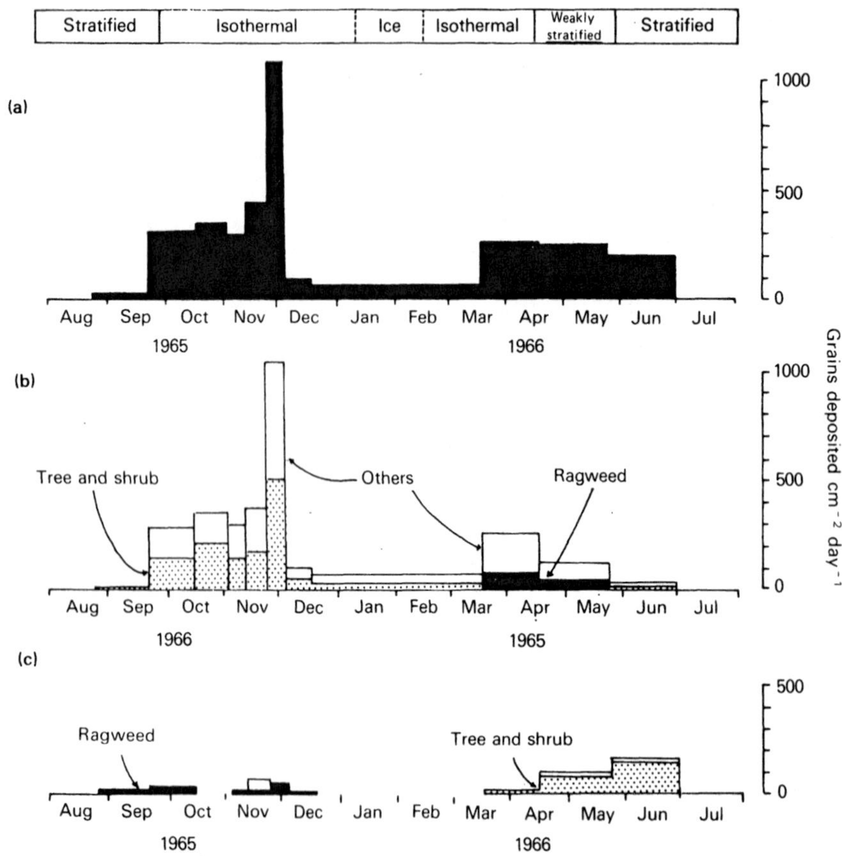

Fig. 9.5 (**a**) Pollen deposition rates (grains deposited cm^{-2} day^{-1}) in sediment traps in Frains Lake, from August 1965 through July 1966. (**b**) Redeposited pollen in sediment traps. Stippled bars, tree and shrub pollen; black bars, ragweed pollen (the out-of-season category within the redeposited pollen component). (**c**) Input of pollen from the air, Stippled bars, tree and shrub pollen; black bars, ragweed pollen. (After Davis (1968). © 1968 by the American Association for the Advancement of Science.)

and in the lower lake it was 3%. Therefore between 91 and 97% of the pollen influx into the lakes was by water transport.

The pollen spectra in traps in the streams were different from those in the air traps. Fern spores were 20% of the total compared with 1% in the air traps, and *Calluna* pollen reached 8–13%, compared with 2% in the air traps. Pollen or spores of these taxa are poorly dispersed in air, but large amounts are distributed by water. In addition, the streams carried a more or less constant amount of pollen throughout the year, of a consistent taxonomic composition, whereas air traps showed marked seasonal variations, depending upon what was in flower. The main sources of stream-borne pollen are:

i) direct fall of pollen from plants growing along the banks;
ii) bank erosion. This becomes important during floods;
iii) surface run-off. This appears to be the major source. During floods, pollen influx to streams increased 142–311-fold, from about 150 grains l^{-1} to 22–48 000 grains l^{-1}.

Bonny (1976) obtained essentially similar results from a study of the pollen budget of Blelham Tarn, English Lake District, with a water-borne component of about 90% and an aerial component of 10%. Some differential input of pollen and spores was demonstrated with significantly larger proportions of *Pinus*, *Betula*, and *Urtica* pollen in the aerial component, and of *Corylus*, *Alnus*, ferns, and *Calluna* in the water-borne component.

2. *The sea*

Marine palynology has been reviewed by Stanley (1969). In the oceans, pollen is preserved in the fine-grained sediments deposited offshore. It is chiefly water transported, although there is a small aerial component (Muller, 1959). Ocean currents and turbulence are of great importance in distributing the pollen. There is a tendency for winged conifer pollen to be over-represented in the spectra because it tends to float for a long time before being deposited, and it also tends to be transported large distances in the air (Traverse and Ginsburg, 1967).

Heusser and Balsam (1977) examined the pollen spectra in 61 core tops from the north-east Pacific off the west coast of North America. They found the highest concentration of pollen opposite river mouths and in fine-grained sediments, indicating that rivers were a major source of pollen input to the sea. A principal components analysis of the samples divided them into three main groups related to the vegetation of the adjacent land; an *Alnus/Picea* group off the Alaskan coast, a *Tsuga heterophylla* group off the coast of Oregon, Washington, and British Columbia, and a *Quercus-Sequoia-Pinus* group off the California coast. High amounts of *Pinus* pollen a long distance off-shore demonstrated its long distance dispersal.

Pollen sedimentation

When a pollen grain lands on a surface, it is then subject to the processes of deposition, before it finally becomes incorporated and preserved in the sediment.

Pollen sedimentation in bogs

It is often assumed that pollen landing on the surface of a bog is simply incorporated into the peat. However, some studies have shown that there is both vertical and horizontal movements of pollen within a bog surface. Rowley and Rowley (1956) marked *Pinus* pollen and a small sphaeroidal pollen grain (*Dodonaea*) and deposited them on the surface of a bog. Within a year, the *Dodonaea* pollen had percolated downwards in the interstices of the living *Sphagnum*, whereas the larger winged *Pinus* had been effectively trapped at the surface. Compaction of the *Sphagnum* at lower levels restricted the downward movement. Data on the growth of *Sphagnum* (Clymo, 1973) indicate that the compact layer is formed in 5–10 years. Therefore, the downward mixing of pollen is not generally important on the time scale usually involved in pollen analytical studies.

On a large raised bog in Finland, Salmi (1962) showed that there is differential horizontal deposition in bog pools (Fig. 9.6). *Pinus* pollen percentages were higher near the edge of the pool due to its long flotation time and hence being blown towards the edge. The ability of pine pollen to float for long periods of time has been investigated experimentally by Hopkins (1950). As bog pools may persist for long periods of time (see Chapter 4), this differential deposition of pine pollen within bog pools may lead to distortion of the pollen record from raised bogs.

Pollen sedimentation in lakes

Pollen can be redistributed both horizontally and vertically in lake sediments. We have already

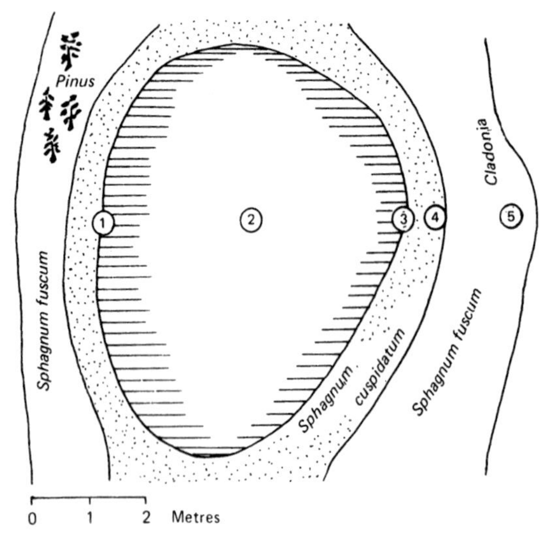

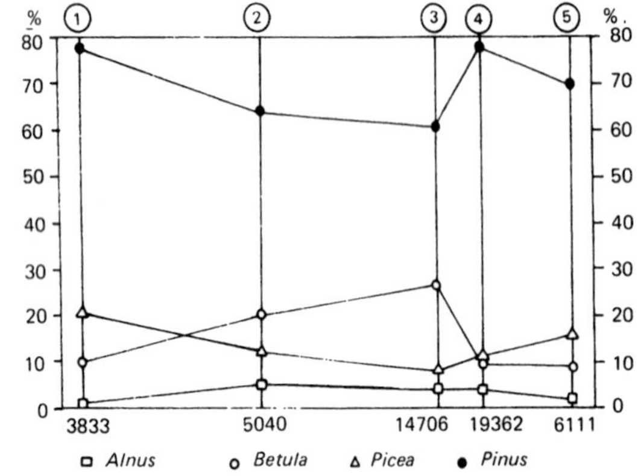

Fig. 9.6 Map of a bog pool in Finland, and a transect of pollen samples across it showing the concentration of *Pinus* pollen at the edges of the pool. Numbers below the pollen diagram are pollen grains in 100 mg of peat. (After Salmi, 1962.)

mentioned the studies by Peck, Davis, and Bonny on the resuspension of material from the recent sediment (p. 182). Davis *et al.* (1971) carried out a comprehensive study of pollen sedimentation in Frains Lake, a small shallow lake in Michigan, U.S.A.

They showed that medium to large sphaeroidal pollen grains which are typified by *Quercus* (oak), but include the majority of deciduous trees, were evenly distributed over the lake (Fig. 9.7a). When oak was taken as constant, the ratios of winged *Pinus* pollen, *Ambrosia*, and other herbs (excluding Cyperaceae) showed higher values in shallow water (Fig. 9.7 b–e). All these types were regarded as non-local in origin, and therefore they would have fallen evenly onto the lake surface. Locally produced pollen such as that of *Salix* and aquatic plants were deposited in higher amounts near the source, in the shallow water.

Similar patterns were found in other lakes (Fig. 9.8). However, in shallow, unstratified Sayles Lake, little evidence of this pattern was found (Fig. 9.8). In Blind Lake with areas of greater than 10 m water depth, the ratio of ragweed: oak pollen rose again, perhaps suggesting the differential deposition of the smallest particles in the deepest water. Davis *et al.* (1971) concluded that internal limnological features determined the dispersal and sedimentation of pollen within a lake. To investigate what these limnological features might be, Davis and Brubaker (1973) set out sediment traps at different water depths in Frains Lake (Fig. 9.9).

During the flowering period of oak, oak pollen was distributed more or less evenly down the water column. During the flowering period of ragweed, ragweed pollen was concentrated in the epilimnion where it tended to be swept near the shore as a result of wind action. It was then differentially deposited in littoral sediments, covered by epilimnetic water only. Table 9.2 shows the size, densities, and settling velocities of *Quercus* and *Ambrosia* pollen. Such a seemingly small difference can apparently determine whether the pollen can sink through the thermocline into the hypolimnion.

During overturn in spring and autumn, Davis (1968) showed that about four times the amount of the net annual sediment increment was redeposited in Frains Lake. It is mixed and deposited evenly over the lake, with no preferential particle sorting. This leads to a net accumulation of sediment in deeper water, because most of the redeposited material originates in the littoral zone. The effects this has on oak and ragweed pollen are shown in Fig. 9.10. The net result is that the final annual influx of oak pollen is greatest in deeper water, whereas the net annual influx of ragweed pollen tends to be the same in shallow and deep water. The movement of littoral sediment rich in ragweed pollen to the lake centre tends to equalize the absolute influx of ragweed pollen throughout the basin.

These differences within a lake are the expression of differences between different pollen types, and of natural processes occurring within the lake. Each lake is liable to be somewhat different, but the overall process of sediment redistribution appears to

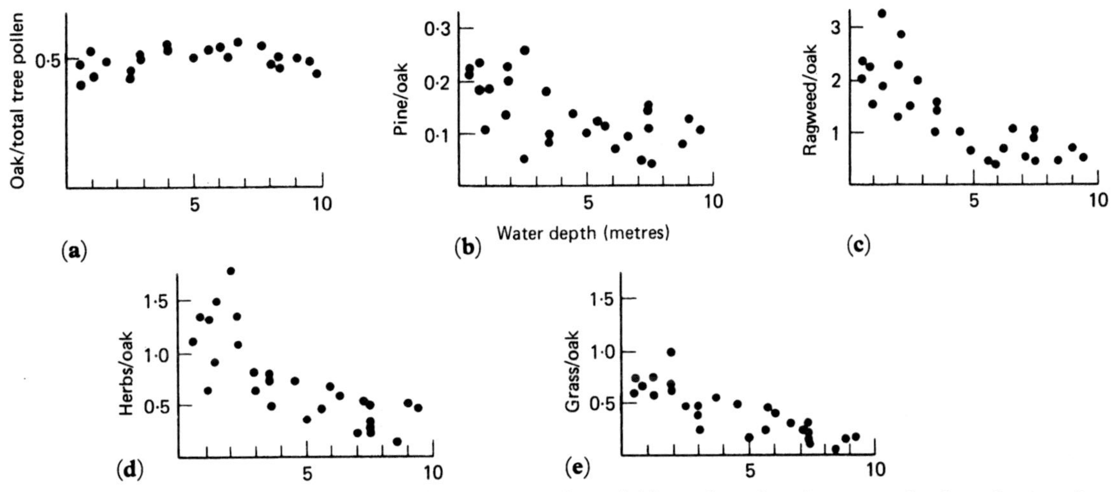

Fig. 9.7 Pollen ratios in surface samples from Frains Lake, Michigan plotted against water depth at the sampling stations. (After Davis *et al.*, 1971.)

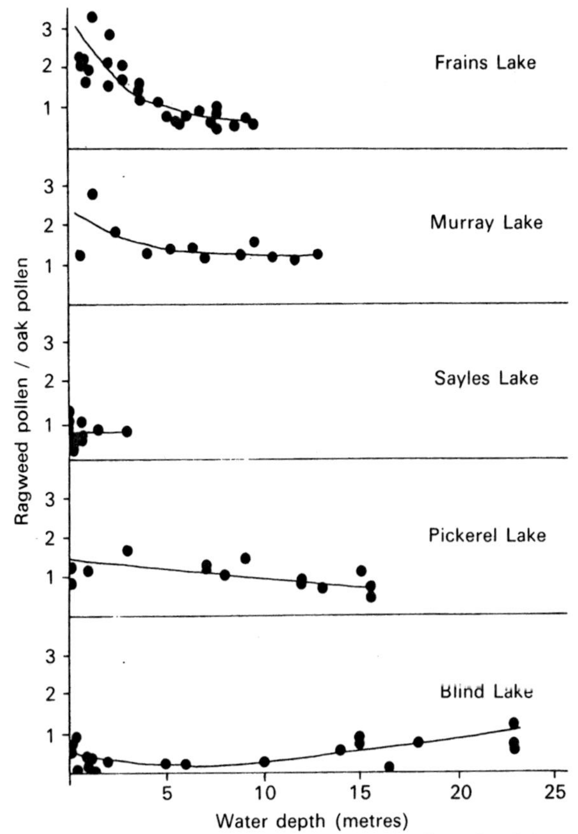

Fig. 9.8 Pollen ratios in surface samples from five lakes in Michigan, plotted against water depth at the sampling stations. The lines have been fitted to the points by eye. (After Davis *et al.*, 1971.)

be common to all lakes so far studied (e.g. Davis, 1973; R. B. Davis *et al.*, 1969). Therefore any pollen analyst should be aware of these potential differences in pollen assemblages from different parts of the same lake. The effects of different sedimentation patterns in lakes are discussed in more detail in Chapter 10.

The relevance of this modern, experimental work to palaeoecology is well shown in the interpretation of four pollen diagrams from the same lake in western Norway (Sønstegaard and Mangerud, 1977). The pollen diagram from the deep, central part of the lake differed from the other diagrams in having low herb pollen and higher *Pinus* pollen than elsewhere in the lake. The low herb values in the centre may result from the absence of any locally growing fringing vegetation at the deep site, whereas the high *Pinus* values in the centre may be a result of over-representation of the regional pollen component in the centre compared with cores nearer the shore.

In addition to horizontal movement of pollen in lakes, pollen in lake sediments is usually mixed vertically by the activities of burrowing animals, such as midge larvae, worms, etc. R. B. Davis (1967, 1974) has demonstrated experimentally that such animals can redistribute pollen over a vertical distance of up to 15 cm, although most disturbance is between 3–4 cm. He showed that about 36% of the pollen at the surface was more than 30 years old, and that about 5% was more than 90 years old. As might be expected, smaller grains (<40 μm)

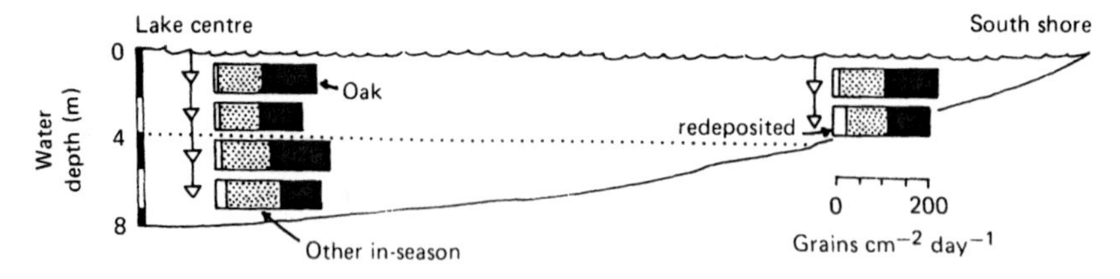

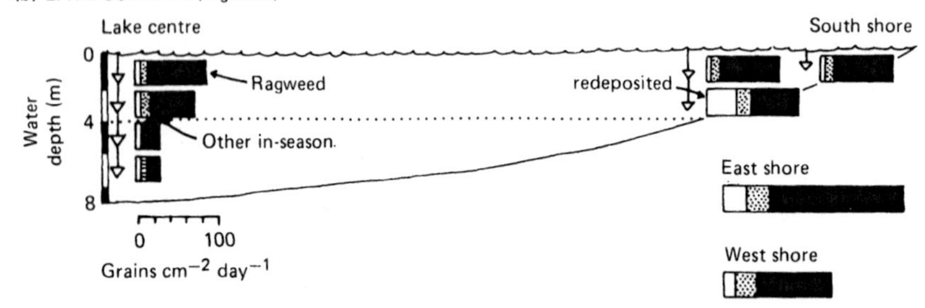

Fig. 9.9 Numbers of pollen grains deposited in sediment traps across Frains Lake. The diagram depicts half of a cross section of the lake. The dotted line indicates the top of the hypolimnion. (**a**) *Quercus* (oak) in late spring. (**b**) *Ambrosia* (ragweed) in late summer. (After Davis and Brubaker, 1973.)

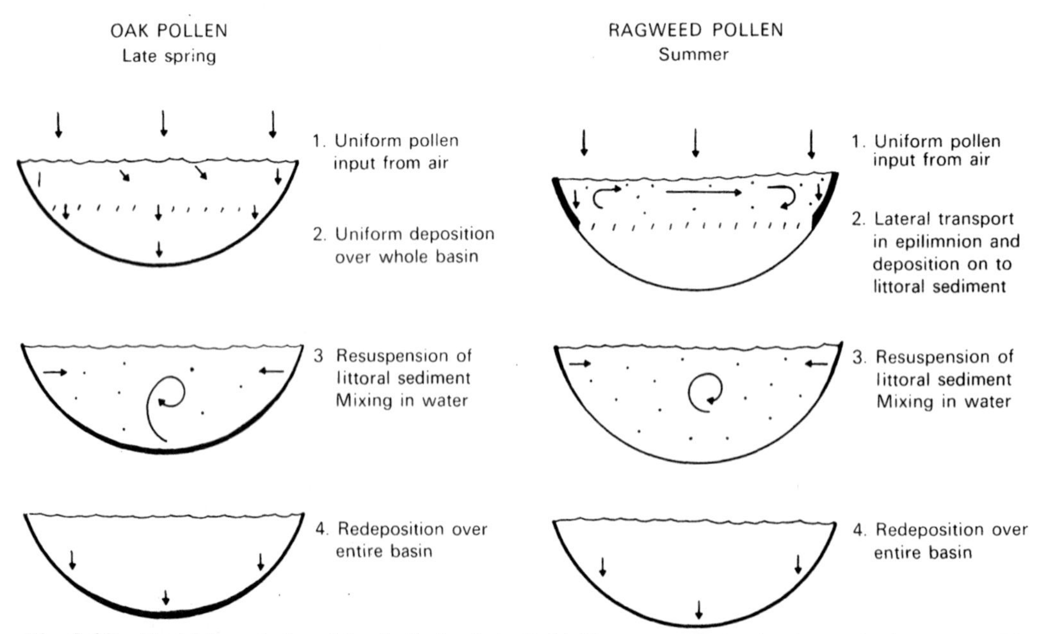

Fig. 9.10 Model for oak deposition in Frains Lake (left). Net accumulation is greatest in deep water. Model for ragweed pollen deposition in Frains Lake (right). The dashed line indicates the position of the thermocline in summer when ragweed pollen enters the lake. Net accumulation is more or less uniform at all water depths. (After Davis and Brubaker, 1973.)

Table 9.2 Size, density, and settling velocity of *Quercus* and *Ambrosia* pollen. (From Davis and Brubaker, 1973.)

Genus	Size (μm)	Density	Settling velocity (cm hr^{-1})
Quercus	28	1.16	24.6
Ambrosia	18	1.20	12.7

moved the most. In general, the worms were more effective at moving pollen upwards than downwards. When Davis constructed a mathematical model of the mixing process, and applied it to a pollen diagram from a long core, he showed the degree of smoothing of the curves which may be expected. Animal activity in the sediments varies from lake to lake. In meromictic lakes with a permanently anaerobic bottom layer the sediment remains undisturbed and laminations may be preserved (Chapter 6).

Pollen preservation and deterioration

Physical, chemical, and biological processes affect pollen grains from the moment they are liberated from the plant. Although the pollen exine is very resistant to decay, due to its sporopollenin content, it can be destroyed in various ways before being examined by the pollen analyst. Cushing (1967) distinguished five main categories of pollen preservation.

i) Well preserved.
ii) Corroded. Exines have distinctive etching, generally affecting the ektexine only.
iii) Degraded. The exine has undergone structural changes, such that individual elements become fused and blurred. This process affects the whole exine, which becomes opaque, and has been described as looking like a 'warmed-up wax ball'.
iv) Broken. The exine is split, rupturing the pollen grain, or dividing it into two or more pieces.
v) Crumpled. The grains are folded, wrinkled, or collapsed.

Badly preserved or deteriorated pollen may still be identifiable, but in some cases it is too badly damaged to be recognizable. Various studies have indicated that different types of deterioration may predominate in different sediment types. Figure 9.11 is part of the pollen diagram from White Lily Lake, Minnesota, U.S.A. prepared by Cushing (1967) which shows the proportion of the preservation classes in the determinable and indeterminable

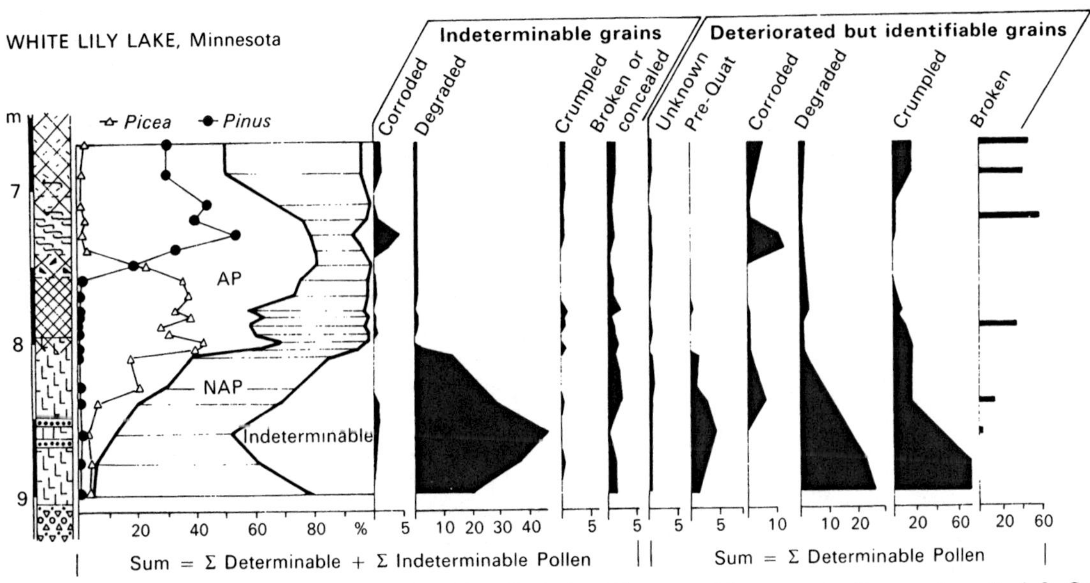

Fig. 9.11 Part of the pollen diagram from White Lily Lake, Minnesota. The summary diagram is on the left. Curves of preservation classes for all the indeterminable pollen, based on the same sum as the summary diagram, and curves of preservation classes for all identifiable pollen, based on a sum of all pollen grains and spores except indeterminable grains are shown. Note the variation in the percentage scales. (After Cushing, 1967.)

categories. Degraded grains are mainly found in the silts, crumpled grains predominate in the silts and the lake muds, whereas corroded grains mainly occur in moss peats. Cushing went on to demonstrate differential deterioration of different pollen taxa in the different sediment types.

Experimental work on pollen preservation

It appears that corrosion is partly due to biological oxidation of the ektexine by bacteria, fungi, and yeasts etc, and partly due to direct chemical oxidation in aerobic sediments. Havinga (1964, 1967, 1971) investigated the susceptibility of different pollen types to corrosion in various sediments. When placed in order of susceptibility, there was a distinct relationship with sporopollenin content (Table 9.3) (see also Sangster and Dale, 1961, 1964).

Spores such as *Lycopodium* have a high sporopollenin content and resist corrosion, whereas pollen such as *Acer* and *Ulmus* have lower sporopollenin contents and are susceptible to corrosion. The precise organisms responsible for corrosion are unknown, but they must possess effective sporopolleninases. They are characteristic of aerobic environments, particularly soils. It is of interest that abundant, well-preserved pollen was recovered from archaeological sites in Illinois, U.S.A. (King *et al.*, 1975) associated with copper artifacts. The copper was acting effectively as a fungicide.

Table 9.3 Order of susceptibility to corrosion in relation to sporopollenin content.

		Sporopollenin content (%)
Low	*Lycopodium clavatum*	23.4
	Polypodium vulgare	–
	Pinus sylvestris	19.6
	Tilia cordata	14.9
	Alnus glutinosa	8.8
	Alopecurus pratensis	–
	Corylus avellana	8.5
	Betula verrucosa	8.2
	Calluna vulgaris	–
	Carpinus betulus	8.2
	Ulmus carpinifolia	7.5
	Populus sp.	5.1
	Quercus robur	5.9
	Taxus baccata	6.3
	Fagus sylvatica	–
↓	*Fraxinus excelsior*	–
High	*Acer pseudoplatanus*	7.4
↓	*Salix* sp.	–

No experimental work has been done on degradation or crumpling of pollen grains. Relationships with sediments indicate that they are probably mechanical effects caused either by stream transportation, or by diagenetic processes occurring within silts and clays, such as pressure and physical abrasion or auto-oxidation. Similarly, the mechanism of pollen breakage is unclear. It may be partly a function of the preparation procedure, when sediments which are difficult to break down are stirred very vigorously.

Soil pollen analysis

The problems of pollen preservation combined with vertical mixing are most acute in the pollen analysis of mineral soils, particularly those with a pH greater than 5.5. Such soils often contain high percentages of pollen and spore types with a large content of sporopollenin, such as *Tilia* and *Polypodium*, which are differentially preserved (e.g. Dimbleby and Evans, 1974). Vertical mixing is also a problem in mildly acid soils where earthworms are still active (Walch *et al.*, 1970). Soil pollen analysis is reviewed by Dimbleby (1957, 1961) and Godwin (1958). The latter emphasizes the difficulties of interpretation of soil pollen profiles (see also Havinga, 1974). Another problem of soil pollen analysis was encountered by Bottema (1975) in the Mediterranean. He discovered that abnormally high (90%) Compositae pollen values were due to pollen brought in by burrowing digger bees. Analysis of modern nests and bees yielded 85% Compositae pollen, whereas modern surface samples from nearby only contained 20% Compositae pollen.

Only in very acid soils with little biological activity are the problems reduced sufficiently to obtain reliable profiles. In suitable conditions, mor humus accumulates that is not broken down by biological activity. In forests, mor humus may accumulate to depths of about 1 m. In Draved Forest, Denmark, Iversen (1958) was the first to recognize the potential of mor humus for providing very local profiles from inside the forest, so that the history of very small areas could be reconstructed. This work was followed up by Stockmarr (1975) and several other workers.

Pollen redeposition

Having discussed the liberation, transportation, deposition, and diagenesis of pollen (see Fig. 1.5) the last process to affect a fossil pollen assemblage is redeposition. Redeposition is the process by which fossils that originated at one time may be deposited and preserved with fossils originating from a different time.

Redeposited fossils are easily recognized if they are of a different morphological nature. For example, Phillips (1974) described a microfossil assemblage consisting entirely of Jurassic gymnosperms and pteridophytes from the Tertiary leaf beds on Mull. Similarly, Jurassic pollen and spores were found in Devensian late-glacial deposits from the Isle of Skye (Birks, 1973) and Cretaceous pollen and spores were found in Wisconsinan late-glacial silts in Minnesota, U.S.A. (Cushing, 1964; Birks, 1976).

Redeposition results from the erosion of polleniferous deposits and the subsequent deposition of the pollen with the contemporary pollen. Such erosion frequently occurs in glacial environments, and hence redeposited pollen is often found in silts and clays. A major source of such pollen is boulder clays, which often contain pre-Quaternary pollen and spores, and also Quaternary pollen eroded from interglacial deposits.

It is often impossible to distinguish redeposited pollen when it is of similar morphology to the contemporary pollen. However, in certain situations it has been possible, as first shown by Iversen (1936). At Nørre Lyngby in N. Jutland, Denmark, a cliff exposure occurs of Weichselian late-glacial sands and clays with thin layers of moss peat contained within the sands and clays. The moss peat contained a pollen assemblage typical of the Danish late-glacial, containing pollen of *Salix*, *Betula*, *Pinus*, and abundant herbs. However, the sands contained an abundance of pollen of thermophilous trees, such as *Alnus*, *Ulmus*, *Tilia*, *Carpinus*, *Picea*, *Corylus*, *Pinus* Haploxylon-type, *Tsuga*, *Carya*, *Rhus*, and *Pterocarya*. When Iversen analysed unweathered boulder clay from the cliff, he found that it contained a similar assemblage of thermophilous tree pollen. He divided this assemblage into

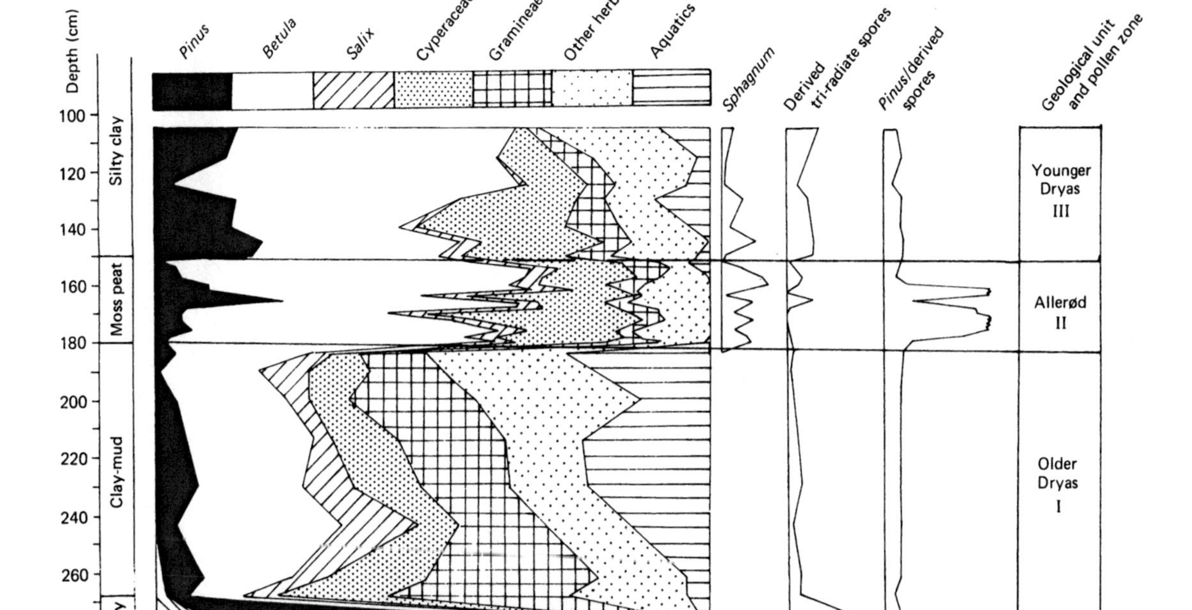

Fig. 9.12 Pollen diagram from Devensian late-glacial sediments at Aby Grange, Lincolnshire, England, showing all curves as percentages of total pollen in relation to the stratigraphy. (After Suggate and West, 1959.)

group A, which contained species of undoubtedly secondary origin, such as *Tsuga*, *Rhus*, and *Carya*, and group B, which contained pollen of possibly secondary origin, such as *Betula* and *Pinus*. He showed that the ratio of the types within group A remained constant in the boulder clay, clays, sands, and silts. He assumed therefore that the ratio of A:B was also constant. Therefore, by knowing the amount of group A taxa in the late-glacial silts and sands, he could subtract the appropriate proportion of the group B taxa from the pollen assemblage, and thus correct the late-glacial assemblage for the redeposited pollen. The final result was a typical late-glacial assemblage lacking thermophilous pollen, and dominated by herb pollen. This classic piece of work was only possible because the exotic types were in sufficiently large quantity to be evaluated, and could be independently assessed from the boulder clay.

A similar approach was used by Suggate and West (1959) at Aby Grange, Lincolnshire, England. Pollen analyses of the typically Devensian late-glacial clay -moss peat- clay stratigraphy yielded an irregular pollen diagram containing pollen of *Pinus*, *Picea*, *Abies*, and *Corylus*, together with large amounts of Carboniferous triradiate spores (Fig. 9.12). Pollen analyses of unweathered boulder clay nearby yielded pollen of *Pinus*, *Picea*, *Abies*, and Carboniferous spores. Suggate and West used the ratio of *Pinus*: Carboniferous spores in the till to subtract the supposed secondary pollen in the late-glacial samples. The resulting diagram (Fig. 9.13) is much more typical of the Devensian late-glacial with *Pinus* now only in Zone II. A bulk sample of the moss peat was radiocarbon dated to 12 870 ± 180 B.P. However, the sample contained tiny fragments of coal. When individually picked moss fragments were dated, the result was 11 205 ± 120 B.P. which is the expected age of an Allerød depos

There are numerous reports in the literature of

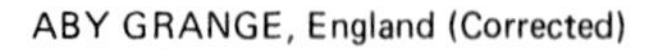

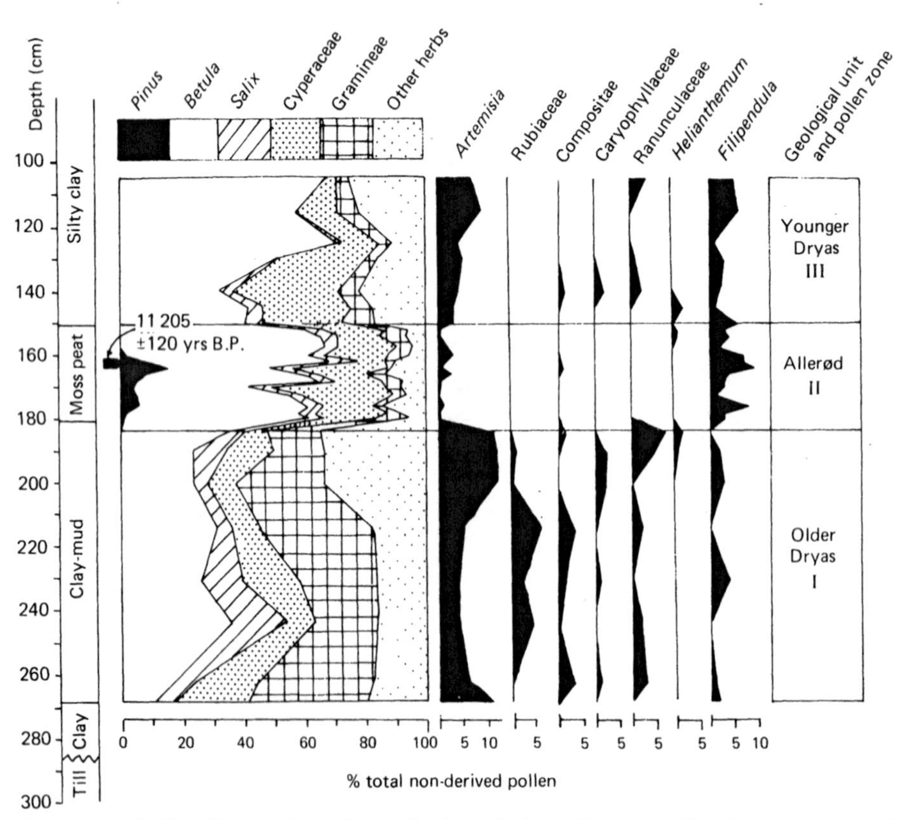

Fig. 9.13 Pollen diagram from Devensian late-glacial sediments at Aby Grange, corrected by subtraction of pollen spectra from adjacent unweathered boulder clay. Percentages are of total non-derived pollen. (After Suggate and West, 1959.)

people 'correcting' for redeposited pollen simply by subtracting pollen types which appear to be ecologically incompatible. An example is the unexpected but quite consistent occurrence of thermophilous deciduous trees in the Wisconsinan late-glacial of the eastern and central United States. These types have been subtracted without the demonstration that they have been secondarily derived from nearby deposits. The environment of the Wisconsinan late-glacial appears to have been different from any known environment today, and it may be totally wrong to subtract what appear to be ecologically incompatible pollen types (see Cushing (1964), who discusses in detail the problem of redeposited pollen in the Wisconsinan late-glacial of Minnesota, U.S.A.).

Other attempts at recognizing secondary pollen have utilized the different staining properties of pollen of different ages. Similarly, the fluorescent properties of pollen exines change over time, and this has also been used to identify secondary pollen (Phillips, 1972).

Redeposition of pollen may occur over a relatively small time scale, for example, when recently deposited pollen is washed into a sediment from the surrounding soil. Cushing (1964) calls this *penecontemporaneous* pollen, as distinct from secondary pollen. Penecontemporaneous pollen can generally be recognized by its corroded nature, and its association with fungal remains inwashed from the soil.

Birks (1970) demonstrated the occurrence of penecontemporaneous pollen in Loch Fada, Isle of Skye, Scotland. He analysed contiguous 1 cm samples of a thin lens of moss in fine-grained silts of Devensian late-glacial age. The mosses were all terrestrial species of *Pohlia*, *Bryum*, and *Drepanocladus*. Figure 9.14 shows that the majority of the pollen in the moss layer was corroded, whether it was determinable or indeterminable. There were also abundant fungal remains and other remains of terrestrial organisms, such as rhizopods and monocotyledon fragments. Certain pollen types were restricted to the moss layer, such as *Saxifraga stellaris*, *Caltha*-type, *Epilobium*-type, and *Stellaria*-type. Birks interpreted the moss layer as representing a moss-dominated spring community (Philonoto-Saxifragetum stellaris) which had been washed into

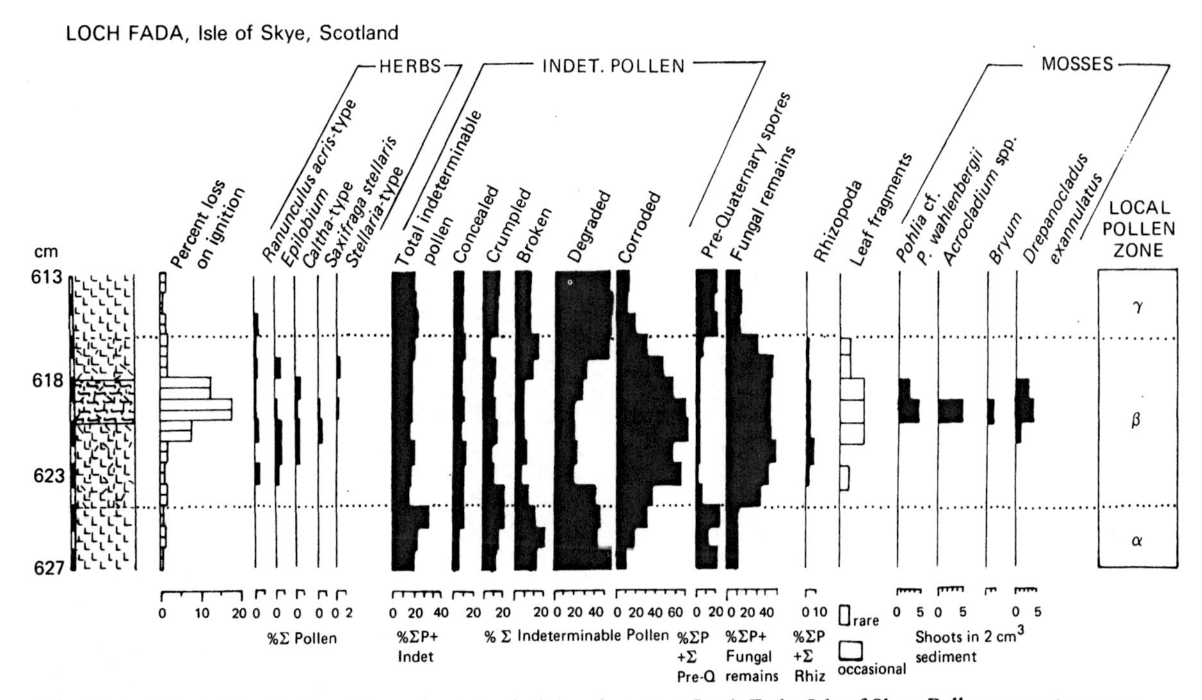

Fig. 9.14 Selected fossils from Devensian late-glacial sediments at Loch Fada, Isle of Skye. Pollen percentages are based on total dry land pollen (ΣP) as indicated. The stratigraphy is illustrated using the symbols of Troels-Smith (Chapter 3), and the inwashed moss layer is at 618–620 cm. P, pollen; indet, indeterminable; undiff, undifferentiated; Pre-Q, Pre-Quaternary microfossils; rhiz, rhizopoda. (After Birks, 1970.)

the lake during a flood. It contained its own local pollen flora, which had been subject to corrosion in the relatively oxidative environment of the spring. Soil fungi and rhizopods were also inwashed with this moss layer. This local event gives valuable information about the local environment, which would not have been available if the spring community had not been buried more or less intact.

References

Pollen production

ANDERSEN, S. T. (1974). Wind conditions and pollen deposition in a mixed deciduous forest. II. Seasonal and annual pollen deposition 1967–1972. *Grana*, **14**, 64–77.

POHL, F. (1937). Die Pollenerzeugung der Windbluter. *Botanisch. Centralblatt*, **56A**, 365–470.

GROSSE-BRAUCKMANN, G. (1978). Absolute jährliche Polleniederschlagsmengen an verschiedenen Beobachtungsorten in der Bundesrepublik Deutschland. *Flora*, **167**, 209–47.

Pollen liberation

DAVIES, R. R. AND SMITH, L. P. (1973). Weather and the grass pollen content of the air. *Clinical Allergy*, **3**, 95–108.

HOLMES, R. M. AND BASSETT, I. J. (1963). Effect of meteorological events on ragweed pollen count. *Int. J. Biometeor.*, **7**, 27–34.

HYDE, H. A. AND WILLIAMS, D. A. (1944). Studies in atmospheric pollen. I. A daily census of pollens at Cardiff, 1942. *New Phytol.*, **43**, 49–61.

LIEM, A. S. N. AND GROOT, J. J. (1973). Anthesis and pollen dispersal of *Holcus lanatus* L. and *Festuca rubra* L. in relation to climate factors. *Rev. Palaeobot. Palynol.*, **15**, 3–16.

OGDEN, E. C., HAYES, J. V. AND RAYNOR, G. S. (1969). Diurnal patterns of pollen emission in *Ambrosia*, *Phleum*, *Zea*, and *Ricinus*. *Amer. J. Bot.*, **56**, 16–21.

STIX, E. AND GROSSE-BRAUCKMANN, G. (1970). Der Pollen – und Sporengehalt der Luft und seine tages – und jahreszeitlichen Schwankungen unter mitteleuropäischen Verhaltnissen. *Flora*, **159**, 1–37.

Pollen dispersal and deposition

ANDERSEN, S. T. (1974). Wind conditions and pollen deposition in a mixed deciduous forest. I. Wind conditions and pollen dispersal. *Grana*, **14**, 57–63.

ANDERSEN, S. T. (1974). Wind conditions and pollen deposition in a mixed deciduous forest. II. Seasonal and annual pollen deposition 1967–1972. *Grana*, **14**, 64–77.

BONNY, A. P. (1976). Recruitment of pollen to the seston and sediment of some English Lake District lakes. *J. Ecol.*, **64**, 859–87.

CHRISTIE, A. D. AND RITCHIE, J. C. (1969). On the use of isentropic trajectories in the study of pollen transports. *Natur. can.*, **96**, 531–49.

CURRIER, P. J. AND KAPP, R. O. (1974). Local and regional pollen rain components at Davis Lake, Montcalm County, Michigan. *Mich. Academician*, **7**, 211–25.

DAVIS, M. B. (1968). Pollen grains in lake sediments: redeposition caused by seasonal water circulation. *Science*, **162**, 796–9.

DYAKOWSKA, J. (1936). Researches on the rapidity of the falling down of pollen of some trees. *Bull. Acad. Pol. Sci. Letters* B, (1936), 155–68.

DYAKOWSKA, J. AND ZURZYCKI, J. (1959). Gravimetric studies on pollen. *Bull. Acad. Pol. Sci. Letters*, **7**, 11–16.

HANDEL, S. N. (1976). Restricted pollen flow of two woodland herbs determined by neutron-activation analysis. *Nature*, **260**, 422–3.

JANSSEN, C. R. (1973). Local and regional pollen deposition. In *Quaternary Plant Ecology* (eds H. J. B. Birks and R. G. West). Blackwell.

KERSHAW, A. P. AND HYLAND, B. P. M. (1975). Pollen transfer and periodicity in a rain-forest situation. *Rev. Palaeobot, Palynol.*, **19**, 129–38.

KRZYWINSKI, K. (1977). Different pollen deposition mechanism in forest: a simple model. *Grana*, **16**, 199–202.

MAHER, L. J. (1964). *Ephedra* pollen in sediments of the Great Lakes region. *Ecology*, **45**, 391–5.

NICHOLS, H., KELLY, P. M. AND ANDREWS, J. T. (1978). Holocene palaeo-wind evidence from palynology in Baffin Island. *Nature*, **273**, 140–2.

PECK, R. M. (1973). Pollen budget studies in a small Yorkshire catchment. In *Quaternary Plant Ecology* (eds H. J. B. Birks and R. G. West). Blackwell.

RAYNOR, G. S., OGDEN, E. C. AND HAYES, J. V. (1970). Dispersion and deposition of ragweed pollen from experimental sources. *J. Appl. Meteor.*, **9**, 885–95.

RAYNOR, G. S., HAYES, J. V. AND OGDEN, E. C. (1974). Particulate dispersion into and within a forest. *Boundary-Layer Meteorology*, **7**, 429–56.

RAYNOR, G. S., HAYES, J. V. AND OGDEN, E. C. (1975). Particulate dispersion from sources within a forest. *Boundary-Layer Meteorology*, **9**, 257–77.

TAUBER, H. (1965). Differential pollen deposition and the interpretation of pollen diagrams. *Danm. geol. Unders.* Ser. II, **89**, 69 pp.

TAUBER, H. (1967). Investigations of the mode of pollen transfer in forested areas. *Rev. Palaeobot. Palynol.*, **3**, 277–86.

TAUBER, H. (1977). Investigations of aerial pollen transport in a forested area. *Dansk Bot. Arkiv.*, **32**, 121 pp.

TURNER, J. (1964). Surface sample analyses from Ayrshire, Scotland. *Pollen Spores*, **6**, 583–92.

TYLDESLEY, J. B. (1973). Long-range transmission of tree pollen to Shetland. I, II, III. *New Phytol.*, **72**, 175–81, 183–90, 691–7.

Marine palynology

HEUSSER, L. AND BALSAM, W. L. (1977). Pollen distribution in the northeast Pacific Ocean. *Quat. Res.*, **7**, 45–62.

MULLER, J. (1959). Palynology of recent Orinoco delta and shelf sediments. *Micropaleontology*, **5**, 1–32.

STANLEY, E. A. (1969). Marine palynology. *Oceanogr. Mar. Biol. Ann. Rev.*, **7**, 277–92.

TRAVERSE, A. AND GINSBURG, R. N. (1967). Pollen and associated microfossils in the marine surface sediments of the Great Bahama Bank. *Rev. Palaeobot. Palynol.*, **3**, 243–54.

Pollen sedimentation

CLYMO, R. S. (1973). The growth of *Sphagnum*: some effects of environment. *J. Ecol.*, **61**, 849–69.

DAVIS, M. B. (1967). Pollen deposition in lakes as measured by sediment traps. *Bull. Geol. Soc. Amer.*, **78**, 849–58.

DAVIS, M. B. (1968). Pollen grains in lake sediments: redeposition caused by seasonal water circulation. *Science*, **162**, 796–9.

DAVIS, M. B. (1973). Redeposition of pollen grains in lake sediment. *Limnol. Ocean.*, **18**, 44–52.

DAVIS, M. B. AND BRUBAKER, L. B. (1973). Differential sedimentation of pollen grains in lakes. *Limnol. Ocean.*, **18**, 635–46.

DAVIS, M. B., BRUBAKER, L. B. AND BEISWENGER, J. M. (1971). Pollen grains in lake sediments: pollen percentages in surface sediments from Southern Michigan. *Quat. Res.*, **1**, 450–67.

DAVIS, R. B. (1967). Pollen studies of near-surface sediments in Maine lakes. In *Quaternary Paleoecology* (eds E. J. Cushing and H. E. Wright). Yale University Press.

DAVIS, R. B. (1974). Stratigraphical effects of tubificids in profundal lake sediments. *Limnol. Ocean.*, **19**, 466–88.

DAVIS, R. B., BREWSTER, L. A. AND SUTHERLAND, J. (1969). Variation in pollen spectra within lakes. *Pollen Spores*, **11**, 557–71.

HOPKINS, J. S. (1950). Differential flotation and deposition of coniferous and deciduous tree pollen. *Ecology*, **31**, 633–41.

ROWLEY, J. R. AND ROWLEY, J. (1956). Vertical migration of spherical and aspherical pollen in a *Sphagnum* bog. *Proc. Minn. Acad. Sci.*, **24**, 29–30.

SALMI, M. (1962). Investigations on the distribution of pollens in an extensive raised bog. *Comp. Rend. Soc. geol. Finland*, **34**, 159–93.

SØNSTEGAARD, E. AND MANGERUD, J. (1977). Stratigraphy and dating of Holocene gully sediments in Os, Western Norway. *Norsk Geol. Tidskr.*, **57**, 313–46.

Pollen preservation and deterioration

CUSHING, E. J. (1967). Evidence for differential pollen preservation in late Quaternary sediments in Minnesota. *Rev. Palaeobot. Palynol.*, **4**, 87–101.

HAVINGA, A. J. (1964). Investigation into the differential corrosion susceptibility of pollen and spores. *Pollen Spores*, **6**, 621–35.

HAVINGA, A. J. (1967). Palynology and pollen preservation. *Rev. Palaeobot. Palynol.*, **2**, 81–98.

HAVINGA, A. J. (1971). An experimental investigation into the decay of pollen and spores in various soil types. In *Sporopollenin* (eds J. Brooks *et al.*). Academic Press.

KING, J. E., KLIPPEL, W. E. AND DUFFIELD, R. (1975). Pollen preservation and archaeology in eastern North America. *Amer. Antiquity*, **40**, 180–90.

SANGSTER, A. G. AND DALE, H. M. (1961). A preliminary study of differential pollen grain preservation. *Can. J. Bot.*, **39**, 35–43.

SANGSTER, A. G. AND DALE, H. M. (1964). Pollen grain preservation of underrepresented species in fossil spectra. *Can. J. Bot.*, **42**, 427–49.

Soil pollen analysis

BOTTEMA, S. (1975). The interpretation of pollen spectra from prehistoric settlements (with special attention to Liguliflorae). *Palaeohistoria*, **17**, 17–35.

DIMBLEBY, G. W. (1957). Pollen analysis of terrestrial soils. *New Phytol.*, **56**, 12–28.

DIMBLEBY, G. W. (1961). Soil pollen analysis. *J. Soil Sci.*, **12**, 1–11.

DIMBLEBY, G. W. AND EVANS, J. G. (1974). Pollen and land-snail analysis of calcareous soils. *J. Arch. Sci.*, **1**, 117–33.

GODWIN, H. (1958). Pollen analysis in mineral soil. An interpretation of a podzol pollen-analysis by G. W. Dimbleby. *Flora*, **146**, 321–7.

HAVINGA, A. J. (1974). Problems in the interpretation of pollen diagrams from mineral soils. *Geol. Mijn.*, **53**, 449–53.

IVERSEN, J. (1958). Pollenanalytischer Nachweis des Reliktencharakters eines jüstischen Linden-Mischwaldes. *Veröff. Geobot. Inst. Rubel.*, **33**, 137–44.

STOCKMARR, J. (1975). Retrogressive forest development, as reflected in a mor pollen diagram from Mantingerbos, Drenthe, The Netherlands. *Palaeohistoria*, **17**, 38–51.

WALCH, K. M., ROWLEY, J. R. AND NORTON, N. J. (1970). Displacement of pollen grains by earthworms. *Pollen Spores*, **12**, 39–44.

Pollen redeposition

BIRKS, H. J. B. (1970). Inwashed pollen spectra at Loch Fada, Isle of Skye. *New Phytol.*, **69**, 807–20.

BIRKS, H. J. B. (1973). *Past and Present Vegetation of the Isle of Skye – a Palaeocological Study*. Cambridge University Press (Chapter 9).

BIRKS, H. J. B. (1976). Late-Wisconsinan vegetational history at Wolf Creek, central Minnesota. *Ecol. Monogr.*, **46**, 395–429.

CUSHING, E. J. (1964). Redeposited pollen in Late-Wisconsin pollen spectra from East-Central Minnesota. *Am. J. Sci.*, **262**, 1075–88.

DAVIS, M. B. (1961). The problem of rebedded pollen in late-glacial sediments at Taunton, Massachusetts. *Am. J. Sci.*, **259**, 211–22.

IVERSEN, J. (1936). Secondary pollen as a source of error. *Danm. geol. Unders.* Ser. IV(2), **15**, 1–24 (English translation by E. J. Cushing).

PHILLIPS, L. (1972). An application of fluorescence microscopy to the problem of derived pollen in British Pleistocene deposits. *New Phytol.*, **71**, 755–62.

PHILLIPS, L. (1974). Reworked Mesozoic spores in Tertiary leaf-beds on Mull, Scotland. *Rev. Palaeobot. Palynol.*, **17**, 221–32.

SUGGATE, R. P. AND WEST, R. G. (1959). On the extent of the last glaciation in eastern England. *Proc. R. Soc.* B, **150**, 263–83.

10

The reconstruction of past floras and past plant populations

Stages in the interpretation of pollen analytical data

Once a pollen diagram has been efficiently divided into several homogeneous pollen zones, it is ready to be interpreted. In general, any interpretation should follow the sequence: past flora, past vegetation, past environment. Basically there are six questions which can be asked of a pollen diagram.

1. What taxa were present?
2. What were the relative abundances of the taxa present in the past?
3. What plant communities were present?
4. What space did each community occupy?
5. What time did each community occupy?
6. What were the other factors operating in the ecosystem in that time and space?

It is only when we reach question 6 that we are attempting to reconstruct past ecosystems, for which the answers to the previous question are all important. In order to answer several of these questions, more information is necessary than is available in a single pollen diagram. Most of the questions have initiated separate lines of research, which we will now discuss in this chapter, and in the next two.

What taxa were present?

This seems to be a simple question. However, the fossil record has limitations, which must be considered. A complete species list for the past flora can never be obtained, for several reasons.

1. Some plants produce pollen that is not preserved. A good example of a common group of plants in this category is the Juncaceae (e.g. *Juncus*, *Luzula*). Other examples are principally aquatic plants, such as *Najas flexilis*.
2. Many plants produce pollen which is not specifically distinct from pollen of other members of the same genus, or even the same family. Many tree pollen taxa cannot be readily distinguished within the genera, and Gramineae and Cyperaceae are good examples of families with remarkably homogeneous pollen.
3. Some plants produce such small quantities of pollen, that the chance of it becoming incorporated into organic sediments is very slight (see Chapter 9).

If a detailed floristic list is required, it is obviously important to identify each pollen grain to as low a taxonomic level as possible. As we have discussed in Chapter 8, this depends on various factors, such as the state of preservation, and the variability of the reference material. An additional study of any macrofossils present in the sequence will add valuable floristic information, because macrofossils (seeds etc.) can often be identified to lower taxonomic levels than pollen (see Chapter 5).

The contribution of pollen transported from long distances also has to be considered when making a floristic analysis. There is a small probability of finding a pollen grain from anywhere in the world. Some taxa are notorious for their long distance transport. *Pinus* pollen travels long distances, and it can be numerically abundant in treeless situations, such as the Devensian late-glacial, where local pollen production is very low. Pollen of *Ephedra* spp. and *Ambrosia* have frequently been found in British sites (e.g. Birks (1973) on the Isle of Skye). Although *Ephedra* pollen could originate from southern Europe, *Ambrosia* pollen probably crossed the North Atlantic. There are many examples of exotic pollen found in arctic regions (e.g. Fredskild (1973) in Greenland) and in isolated oceanic islands (e.g. Churchill (1973) on Signy Island, S. Orkneys (lat. 60°40′S); Hafsten (1960) on Tristan da Cunha). We have already mentioned the problem of pollen of thermophilous trees in the Wisconsinan late-glacial of North America (Chapter 9, p. 191). Macrofossils can again be of use in confirming the local presence of plants. However, macrofossils of many of the so-called 'long distance taxa' are rarely found, even when they are undoubtedly local (e.g. *Ambrosia* macrofossils have rarely been recorded although its pollen is plentiful in sites in North America).

A floristic approach is obviously important in the subsequent stages of reconstructing the past vegetation and environment. It is also valuable in throwing light on the phytogeographical affinities of past floras (see Birks, 1973), in investigating changes in floristic diversity with time, and in studying the history of particular phytogeographical elements.

What were the relative abundances of the taxa in the fossil flora?

By attempting to answer this question, we are attempting to reconstruct plant populations of the past. We have already discussed attempts to reconstruct fossil animal populations (Chapters 2 and 7). A somewhat different approach is required to reconstruct plant populations, because plant fossils do not each represent one individual. It is theoretically possible that all the fossils of one taxon in a sample may have originated from an individual plant, or that they may have each originated from one plant. The truth lies somewhere in between, and the problem is, where?

The pollen productivity of different species and its transport have already been discussed in Chapter 9. There is some very complex function which relates the number of pollen grains in a sample to the number of parent plants in the vegetation surrounding the site of deposition. It contains the independent variables which we considered in Chapter 9 which affect pollen production all the way through to pollen preservation.

Pollen analysts have long been aware that some species are more abundantly represented by pollen than others. Iversen (1947) in Faegri and Iversen (1975), first attempted to quantify this relationship by dividing European trees into three groups:

1. Over-represented compared with their abundance in the vegetation: *Betula*, *Alnus*, *Corylus*, and *Pinus*. To obtain a more accurate picture of the past vegetation, he suggested that the number of pollen counted of these taxa should be divided by four before calculating percentages.

2. Equally represented in pollen and vegetation; *Picea*, *Quercus*, *Fraxinus*, *Fagus*.

3. Under-represented in the pollen spectrum; *Tilia*, *Hedera*. Iversen suggested that grains of these taxa should be multiplied by four before calculating percentages.

This approach is discussed by Faegri and Iversen (1975, p. 156). Andersen (1970) calculated correction factors for spectra from measurements of pollen production and vegetation *within* forests, which were similar to Iversen's, but he suggested that the pollen productivity of *Quercus* was higher than previously supposed, and that of *Fraxinus* was less (see Table 10.1).

Table 10.1 Correction factors for pollen spectra of North European trees within forests. (From Andersen, 1970.)

Quercus, *Betula*, *Alnus*, *Pinus*	1:4
Carpinus	1:3
Ulmus, *Picea*	1:2
Fagus, *Abies*	1:1
Tilia, *Fraxinus*, *Acer*	1 × 2
Corylus as an understory shrub	1:1
Corylus as a canopy tree	1:4

The R-value model

M. B. Davis (1963) introduced the concept of the R-value, by attempting to quantify more exactly Iversen's correction factors.

$$\text{For species a, } R_a = \frac{\text{Pollen percentage a}}{\text{Vegetation percentage a}} = \frac{P_a}{V_a}$$

If the taxon is over-represented in pollen spectra, the R-value will be high. The modern R-value, R_m, of a taxon can be estimated by measuring its pollen percentage in a modern surface sample, P_m, and its percentage in the vegetation around the sampling site, V_m. If it is assumed that the R-value has not changed through time, the fossil vegetation, $\hat{V}_f$, can be estimated from the fossil pollen percentage, P_f, and R_m:

$$\hat{V}_f = \frac{P_f}{R_m}$$

Davis demonstrated, using a hypothetical model, that R-values vary because of the constraint imposed by percentage calculations. However, if the R-value of a particular taxon is taken as 1, the others are always in the same ratio (see Table 10.2). These relative R-values are termed R_{rel} values. Absolute R-values can only be calculated from

Table 10.2 Hypothetical comparisons of modern vegetational percentages and modern pollen percentages in sediments. (From Table 1, Davis, 1963.)

	Species	Vegetation (%)	Pollen (%)	R	Ra:Rb:Rc
Site 1	a	40	80	2	10
	b	10	10	1	5
	c	50	10	0.2	1
Site 2	a	20	50	2.5	10
	b	30	37.5	1.25	5
	c	50	12.5	0.25	1
Site 3	a	68	85	1.25	10
	b	22	13.75	0.625	5
	c	10	1.25	0.125	1

pollen influx data. However, such data from contemporary situations are rare, and we mostly have to deal with percentage data.

The modern R-values are determined by measuring both pollen percentages in surface samples and vegetation percentages. Surface pollen samples can be taken from moss cushions, surface soil, or surface mud in lakes. In all these cases percentage data are obtained. Pollen influx can be measured by collecting pollen in a trap or on a sticky slide of a known area over a known period of time. Any quantitative measure of vegetation can be estimated, such as frequency, density, cover, basal area, and crown area. Mapping from aerial photographs is also a useful technique.

Davis demonstrated how R-values are used to calculate fossil vegetation percentages by extending her hypothetical model (see Table 10.3) to fossil pollen data.

Measurement of pollen production and representation factors

The most detailed and comprehensive attempt to estimate production and representation factors is the work of Andersen (1970) on pollen deposition within two Danish forests. Andersen (1974) extended this study by collecting absolute pollen influx data in pollen traps.

Andersen proposed that pollen deposited on the forest floor was largely derived from trees within 20–30 m of the sampling site, and that pollen deposition, p, was proportional to the tree crown area, a, depending upon a constant production factor, P, thus:

$$p = P \times a \qquad \text{and} \qquad P = \frac{p}{a}$$

If some of the pollen, p_0, in p was derived from beyond the vegetation plot:

$$p = P \times a + p_0$$

If the species is not present in the vegetation plot, then $p = p_0$. From the above equation, P, the production factor is

$$P = \frac{p - p_0}{a}$$

Andersen measured a by estimating the crown area of each species in 20 m and 30 m radius plots. From the centre of each plot he counted the pollen spectrum from a moss cushion, and determined p. In order to eliminate percentage constraints, Andersen related his pollen values to the exotic pollen rain. This was composed of taxa from outside the whole forest (e.g weeds, crops) and he demonstrated statistically that it was proportionally constant throughout the forest.

Hence, relative pollen deposition, p_r, relative to the exotic pollen, could be calculated from

$$p_r = \frac{\text{number of tree grains cm}^{-2}\text{ year}^{-1}}{\text{number of exotic grains cm}^{-2}\text{ year}^{-1}} = \frac{\text{tree pollen counted}}{\text{exotic pollen counted}}$$

Table 10.3 Hypothetical example of interpretation of fossil pollen percentages by correction for differences in representation. (From Table 2, Davis, 1963.)

Species	Fossil pollen percentage	Ra:Rb:Rc	Corrected pollen number (P/R)	Corrected pollen percentage = past vegetational %
a	64	10	6.4	32
b	28	5	5.6	28
c	8	1	8.0	40
	100			100

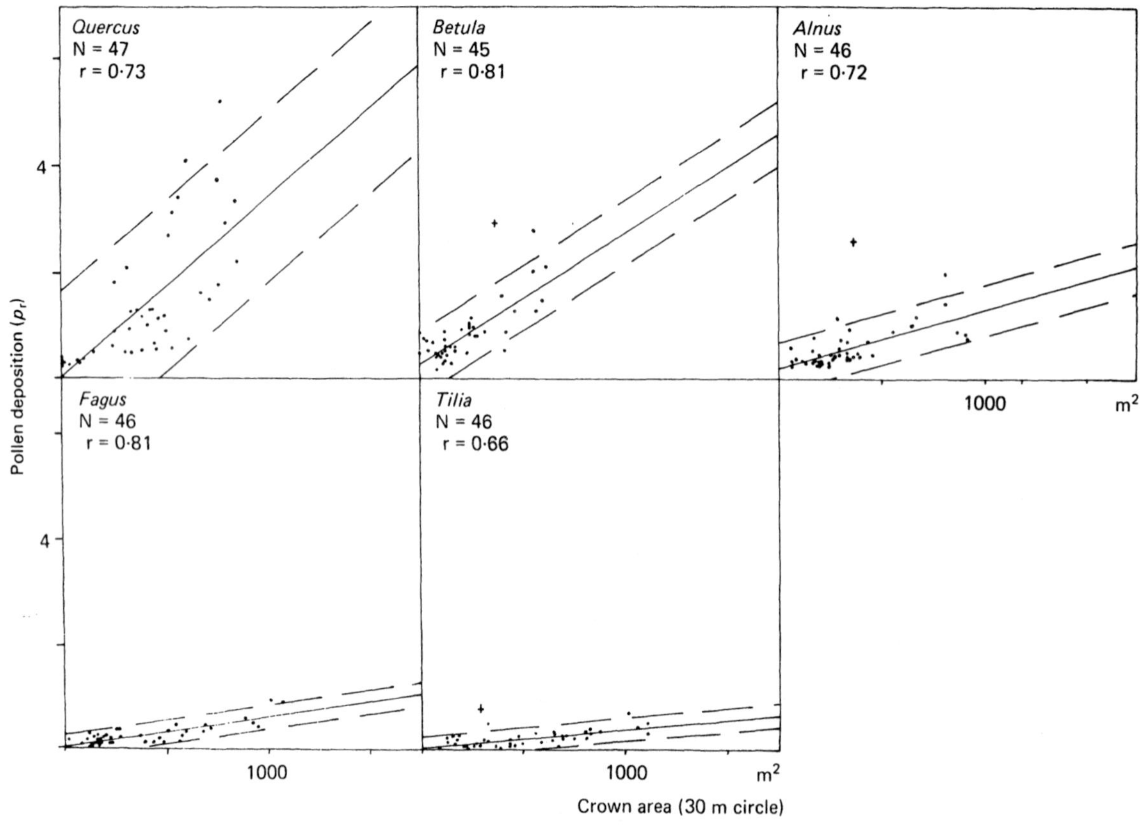

Fig. 10.1 The relationship between semi-absolute pollen deposition and crown areas in the 30 m radius sample plots in Draved Forest 386, Denmark. Calculated regression lines with two standard error limits. P_r, relative pollen deposition. N, number of samples; r, correlation coefficient; +, sample with 'over representation'. (After Andersen, 1970.)

Therefore, the equation now reads:

$$p_r = P \times a + p_{r(0)}$$

where $p_{r(0)}$ is the relative pollen value from beyond the vegetation plot.

This equation is a linear regression equation. When the data were plotted as in Fig. 10.1, linear regression lines could be fitted. In all cases the correlation coefficient was high, suggesting that there was a significant relationship between crown area and pollen deposition. Where the regression line meets the vertical axis above the origin, the intercept indicates the value of $p_{r(0)}$. Now that $p_{r(0)}$, p_r, and a are known, P the production factor, can be calculated for each species in each forest. The figures for various forest stands can be compared if they are normalized relative to the P-values of a reference species. Andersen chose *Fagus sylvatica*, because it is likely to flower equally well on various soils, and there is no problem of species identification.

The relative P-values (P_{rel}-values) with that of *Fagus* taken as 1 are shown for two stands in Draved Forest and one in Longelse Forest, in Table 10.4.

Using the same data, Andersen (1970) then calculated R-values, the ratio of tree pollen percentage to vegetation area percentage, and converted them to R_{rel}-values relative to *Fagus*. The results are compared with the P_{rel}-values in Table 10.4. The values are similar, and show that pollen production seems to be the main controlling factor, rather than dispersal. Further work with pollen traps (Andersen, 1974) rather than moss cushions confirmed that most of the tree pollen fell more or less vertically downwards, either in the air, or washed down by rain. Apart from a fascinating study on

Table 10.4 Relative pollen production (P_{rel}) and pollen representation (R_{rel}) factors from Draved and Longelse Forests, Denmark, based on analysis of moss cushions (Andersen, 1970) and of pollen traps (Andersen, 1974).

	P_{rel}-values			R_{rel}-values			P_{rel} traps
	Draved			Draved			Draved
	1	2	Longelse	1	2	Longelse	(Andersen, 1974)
Quercus	5.8	6.0	5.5	4.6	3.8	3.3	3.9
Betula	4.4	4.2	–	4.8	4.6	–	2.4
Alnus	1.9	3.9	–	2.3	3.6	–	5.8
Carpinus	–	–	2.4	–	–	2.5	–
Ulmus	–	–	2.0	–	–	1.7	–
Fagus	1.0	1.0	1.0	1.0	1.0	1.0	1.0
Tilia	0.6	–	–	0.6	–	–	0.9
Fraxinus	–	0.4	0.4	–	0.4	0.5	0.4

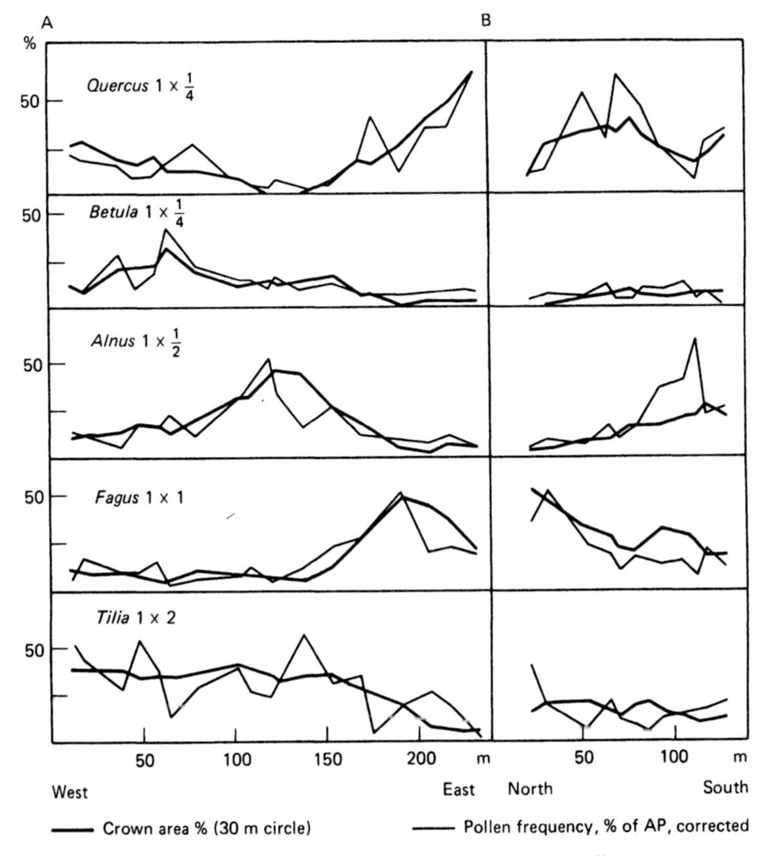

Fig. 10.2 Plot of vegetation percentage compared with pollen percentage along two transects, A and B, in Draved Forest. Corrected pollen frequencies are as percentages of corrected tree pollen totals, and crown area frequencies are as percentages of the total crown areas in the 30 m radius sample plots. (After Andersen, 1970.)

seasonal deposition of pollen and differences in flowering from year to year (see Chapter 9), Andersen calculated P_{rel} figures from the pollen trap data. These are shown as the last column in Table 10.4. They are generally similar to the moss cushion data, and Andersen attributes the relatively minor differences to differences in flowering between the different years and to inherent errors in the methods.

Having shown satisfactorily that P_{rel} and R_{rel} were relatively consistent within Danish forests, Andersen then used them to calculate the correction factors shown in Table 10.1. A plot of corrected pollen percentage against vegetation area percentage (Fig. 10.2) along transects in Draved forest shows a good relationship, and if the two values are graphed (Fig. 10.3) the points mostly fall near the 45° line of 1 : 1 correspondence.

Similar work was done by Dabrowski (1975) on the natural forest of Białowieza in north-east Poland. He set out pollen traps in 1960, 1962, and 1963, and compared the pollen deposition values with the forest composition. He found a considerable variation within a species from year to year, but could not relate this to climatic factors as Andersen (1974) was able to do. The pollen deposition values also varied from one forest community to another within Białowieza, trees in bogs and poor fens producing less pollen than those of the same species on richer soils.

Relative R-values can be calculated from Dabrowski's data, but because *Fagus* is absent from Białowieza, they have been related to *Ulmus* in Table 10.5 and compared with R_{rel}-values calculated in the same way for Draved forest.

In all cases except for *Fraxinus*, the R_{rel}-values are lower in Białowieza. This may be because

Table 10.5 Comparisons of R_{rel} values from Białowieza Forest, Poland, and Draved Forest, Denmark. (Data from Dabrowski, 1975, and Andersen, 1970.)

Genus	Białowieza	Draved
Quercus	0.8	2.7
Betula	1.1	2.8
Alnus	0.4	2.1
Carpinus	0.5	1.5
Ulmus	1.0	1.0
Tilia	0.06	0.4
Fraxinus	0.4	0.2

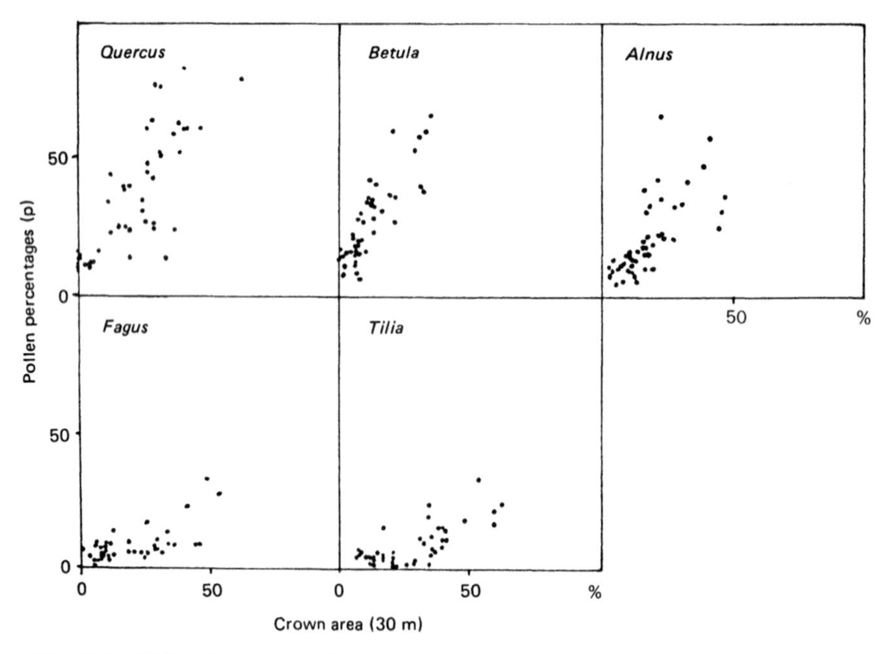

Fig. 10.3 Pollen frequencies plotted as percentages of the tree pollen totals, and crown area frequencies plotted as percentages of the total crown areas in 30 m radius sample plots in Draved forest 386. p, pollen frequency. (After Andersen, 1970.)

flowering is less in the natural forest with a closed canopy than in the somewhat disturbed, more open Danish forests.

The use of correction factors and pollen representation factors

1. *Studies within forests*

a) Iversen (1964) first used correction factors to obtain a more realistic picture of the past vegetation of Draved Forest, Denmark, from pollen spectra in mor humus. Andersen (1973) used his more sophisticated correction factors to correct a pollen diagram from a very small hollow within Eldrup Forest, Denmark. Figure 10.4 shows the effect of the correction. *Tilia*, rather than *Quercus* was the major component of the forest after its initial development *via* a phase of birch and pine. After the onset of human interference in stage III, *Tilia* was gradually replaced by *Fagus*, and *Quercus* remained as a subordinate member of the forest community.

b) Andersen (1975) also used his correction factors on a pollen diagram from the Eemian (last) interglacial, from deposits in a 5 m diameter hollow in boulder clay. Andersen assumed that this would have resembled a small, modern, woodland hollow in the Eemian, and that the correction factors, which are valid within a 30 m radius could therefore be used. The percentage diagram in Fig. 10.5 shows a succession from a *Betula-Pinus-Ulmus* phase to a *Quercus-Fraxinus*-dominated phase. This passes into a *Quercus-Corylus* phase with *Tilia*, which gives way finally to a *Carpinus-Tilia* phase. When the data are transformed, it can be seen that *Fraxinus* was dominant in phase 3. It was replaced by *Corylus*, and then *Tilia* became dominant in phase 4b. It is unexpected, in the light of previous interpretations

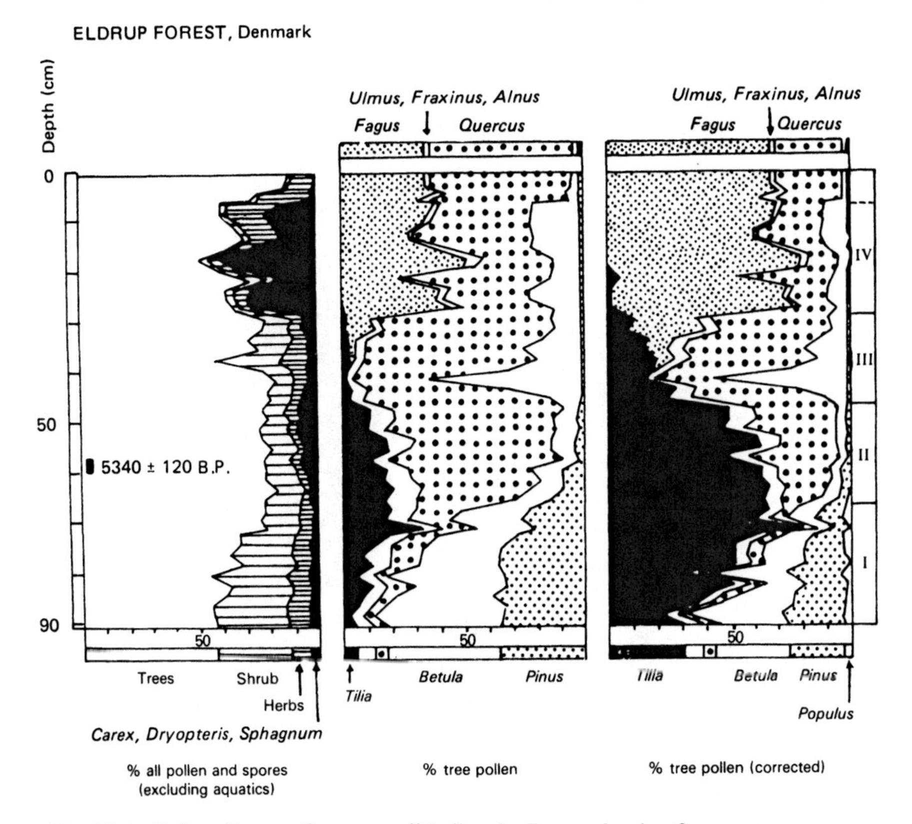

Fig. 10.4 Pollen diagram from a small hollow in *Fagus sylvatica-Quercus petraea* forest of Eldrup Forest, Denmark. The tree pollen numbers were corrected before the percentage calculations using the correction factors in Table 10.1. Correction factor for *Populus* = 1. (After Andersen, 1973.)

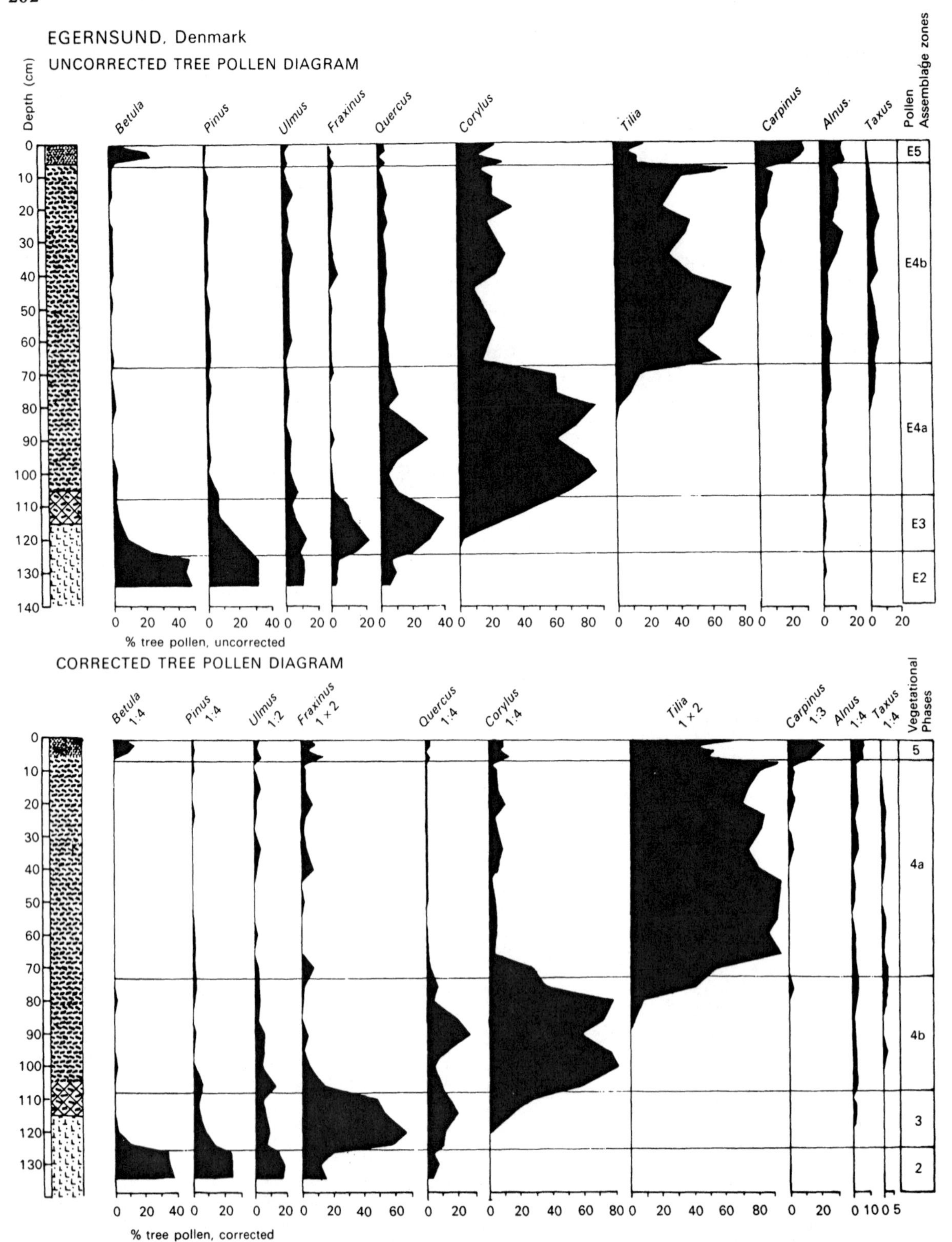

Fig. 10.5 Pollen diagram from Eemian (last) interglacial deposits in a small hollow at Egernsund, Denmark. (**a**) Tree pollen diagram based on the tree pollen total, with tree pollen frequencies unchanged. (**b**) Tree pollen diagram based on the corrected tree pollen total, according to the correction factors in Table 10.1. (After Andersen, 1975.)

of the Eemian vegetation that *Tilia* was the dominant tree in south Denmark in the last interglacial on mull soils. The later phases reflect gradual soil acidification, with the increase of *Carpinus*, *Betula*, and *Picea*.

c) In Britain, Baker *et al.* (1978) published a pollen diagram from a very small hollow in Epping Forest, just north of London. Today the forest is dominated by *Fagus*, with some *Quercus* and *Betula* on acid soils. In Fig. 10.6 the diagram shows relatively high percentages of *Tilia*, which decline at 1350 B.P. and are replaced by *Betula* and *Fagus* pollen. If the data are recalculated using Andersen's correction factors, it appears that *Tilia* formed 80–90% of the original forest, which after the influence of man, was replaced largely by *Fagus* with *Betula* and *Quercus*, and also some *Carpinus*.

2. *Studies from bogs and lakes larger than 30 m radius*

So far we have discussed studies from very small basins or humus profiles within forests, for which the correction factors calculated by Andersen for inside a forest are valid. However, most pollen diagrams are from the deposits of medium or large size lakes and bogs. How can we interpret pollen frequencies from these in terms of past population size?

a) After proposing her hypothetical R-value model, Davis (1963) went on to test it. She compared pollen percentages from the surface mud of Browninigton Pond, Vermont, U.S.A. with tree percentages, which she determined by basal area, for a sample of the vegetation around the pond. She then calculated the modern R-values, which are shown in Table 10.6. They range from 240 for *Quercus* and 200 for *Pinus*, to 0.0083 for *Larix*. She then assumed that the R-values had remained the same in the past, and corrected the fossil pollen percentages for the zone in the early Holocene, about 8000 years ago, which was dominated by pine pollen. She reached the spectacular conclusion that the pollen spectra dominated by pine pollen in fact came from a vegetation containing only 0.3% pine, and 42.8% larch. Because the local vegetation, dominated by larch, produced such low amounts of pollen, the long-distance component, mostly consisting of pine pollen, was responsible for the overall pollen spectrum, and pine forest did not, as originally thought, surround the lake. Because this was such an unexpected conclusion, Davis (1967a) later tested it independently by counting absolute pollen frequencies (APF) from the pine pollen zone (see p. 208).

The major difficulty in using R-values for larger basins is that in the light of what is known about pollen dispersal, filtration, etc (Chapter 9) R-values can be expected to be different for the same species in different areas (Faegri, 1966). This will depend upon:

i) The forest composition and structure.

ii) The transport efficiency of the different pollen types (see Tauber, 1965). For example, pine pollen will be blown long distances, representing the vegetation over a very large area, whereas larch pollen is only locally dispersed, and represents a relatively small area of vegetation.

iii) Basin size. A small basin, as we have seen, collects pollen largely from the surrounding vegetation, predominantly from the trunk space. Larger basins collect more of the canopy and regional rain-out components.

iv) The regional vegetation type, which will influence the composition of the pollen originating beyond the immediate vicinity of the sample.

b) The variation in R-values for the same species occurring in different vegetation types in Minnesota, U.S.A. was demonstrated by Janssen (1967). His Clearwater site was situated in a region of predominantly xeric oakwoods, whereas the Lake Sylvia site was in an area of mesic mixed deciduous forest. In neither area do conifers occur. In contrast, his site at Lake Itasca is in mixed coniferous–deciduous forest, and the site at Myrtle Lake is in an area of low-lying predominantly bog vegetation in northern Minnesota. He calculated the vegetation percentages over large, 7 mile square areas, and for Myrtle Lake, over 206 square miles (6 townships). He selected pollen samples which would primarily represent the regional pollen rain, and which would not be unduly influenced by locally growing trees. He then calculated R-values for each area. When these are standardized relative to *Ulmus*, the results are as shown in Table 10.7.

The differences between the four sites are clear. In general, high R-values tend to be found where the species is rare in the vegetation. For example, at Myrtle Lake, the R-value of *Fraxinus* is 2.1, whereas *Fraxinus* is extremely rare in the vegetation. This is due to long distance transport of the pollen from regions beyond that used for recording the vegetation percentages. Conversely, a species tends to have its lowest R-values where it is most abundant in the vegetation.

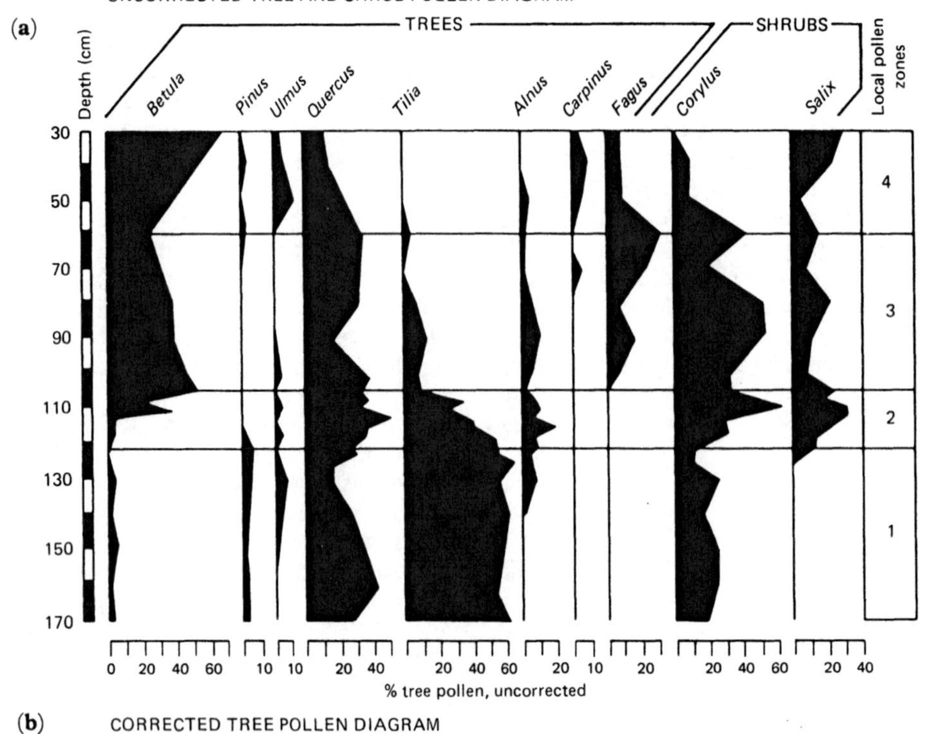

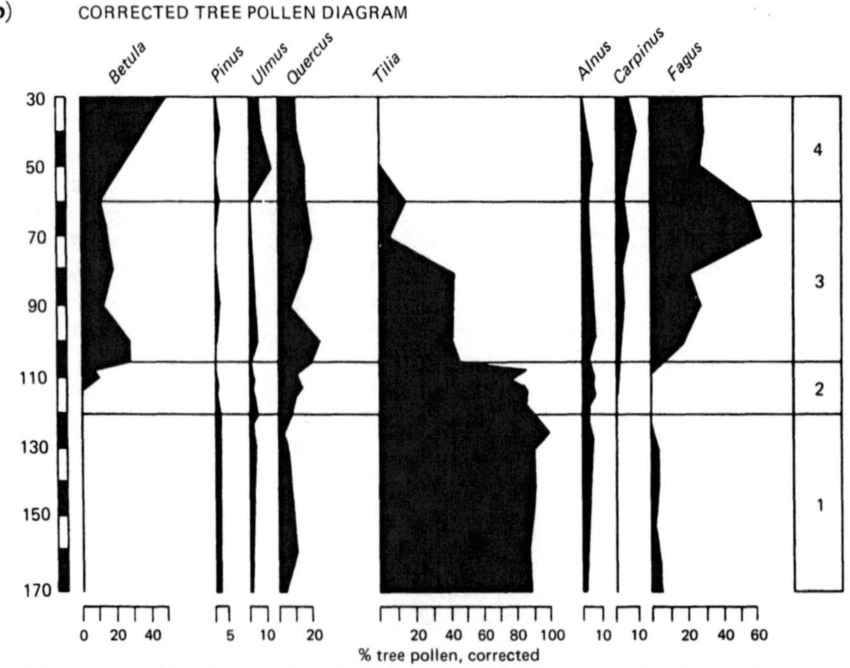

Fig. 10.6 Pollen diagram from late Flandrian deposits in a small hollow at Lodge Hill, Epping Forest, London. (**a**) Tree and shrub pollen diagram based on the tree pollen total, with tree pollen frequencies unchanged. (**b**) Tree pollen diagram based on the corrected pollen total, according to the correction factors in Table 10.1. (After Baker *et al.*, 1978.)

Table 10.6 Tree and shrub percentages of basal areas in the forest surrounding Brownington Pond (column 1), tree and shrub pollen percentages in the surface sediment of Brownington Pond (column 2), R-values based on these data (column 3), fossil pollen, numbers counted (column 4), relative pollen number (column 5), and calculated forest percentage in the past (column 6). (From Davis, 1963.)

Genus	1 Percentages of basal area (total 409.53 ft^2) in sample of modern forest	2 Percentages of tree and shrub pollen (total 1253 grains) in sample of modern sediment	3 R	4 Relative numbers of pollen counted total 1462 grains) in sample of sediment at 8.5 m depth (pine zone)	5 Corrected relative number $\left(\frac{\text{pollen number}}{R}\right)$	6 Corrected percentage = forest percentage corresponding to pollen frequencies in column 4
Oak (*Quercus*)	0.01 (est.)	2.4	240	194	0.8	0.1
Pine (*Pinus*)	0.04	8.2	200	829	4.1	0.3
Alder (*Alnus*)	0.2	4.2	21	12	0.6	0.04
Birch (*Betula*)	8.8	48.9	5.6	202	36	2.6
Beech (*Fagus*)	4.5	6.7	1.5	1	0.7	0.05
Hemlock (*Tsuga*)	5.5	6.8	1.2	17	1.4	0.1
Elm (*Ulmus*)	2.4	2.7	1.1	33	30	2.1
Spruce (*Picea*)	11.3	6.5	0.58	23	40	2.8
Ash (*Fraxinus*)	2.9	1.4	0.48	34	71	5.1
Hop-hornbeam (*Ostrya*)	2.1	0.9	0.43	18	42	3.0
Arbor vitae (*Thuja*)	15.9	4.5	0.28	18	64	4.6
Sugar and striped maple (*Acer*)	17.6	3.3	0.19	29	153	10.9
Red maple (*Acer*)	2.8	0.5	0.18	1	5.6	0.4
Balsam fir (*Abies*)	11.8	1.6	0.14	35	250	17.8
Poplar and aspen (*Populus*)	7.7	0.3	0.039	4	100	7.1
Basswood (*Tilia*)	2.8	0.1 (est.)	0.036	0	0	0.0
Larch (*Larix*)	2.4	0.02 (est.)	0.0083	5	600	42.8
Others (Rosaceae, *Ilex*, *Salix* etc.)	1.3	1.5	1.2	7	5.8	0.4
					Total 1405.0	

Table 10.7 R_{rel} values for four different areas in Minnesota. (Data from Table VI in Janssen, 1967.)

Genus	Clearwater	Sylvia	Itasca	Myrtle Lake
Pinus	–	–	0.7	2.3
Betula	14.9	2.1	1.2	0.5
Quercus	2.9	1.1	1.2	1.4
Ulmus	1.0	1.0	1.0	1.0
Fraxinus	1.6	0.4	0.8	2.1
Tilia	1.5	0.2	0.4	0.08
Populus	0.07	0.2	0.02	0.01

Janssen's study highlights the limitations in the use of R-values in larger basins.

i) The areal extent of the vegetation contributing the pollen is unknown, and varies in different vegetation types. The presence of pollen from beyond the area may make a substantial contribution to the pollen spectrum. This is the reason why Davis' (1963) results were wrong (see Livingstone, 1968).

ii) The R-value of a particular species depends not only on its abundance within the community, but also on the structure and composition of the vegetation in which it is growing. This was also demonstrated by Janssen by his Myrtle Lake site, where much of the vegetation was treeless bog, allowing non-local trees with good pollen production and dispersal, such as pine, to be well represented, even though they were relatively rare in the vegetation. He also demonstrated the differences between different forest types. Similarly, differences were found between Białowieza, Poland, and Draved, Denmark (p. 200).

iii) R-values for a species vary from one environment to another because of the effects of factors controlling flowering, such as climate, soil, and competition from other plants.

These limitations impose severe restrictions on the use of R-values in reconstructing past populations from fossil pollen spectra from lakes and bogs. It is highly unlikely that R-values have stayed constant over time in view of the changing vegetational environments which are undoubtedly recorded in pollen diagrams. In spite of these limitations, Livingstone (1968) calculated and applied R-values to diagrams from the maritime states of south-east Canada. He further used absolute pollen concentration estimations based on an assumption of constant sedimentation rate, to convert vegetational percentages into the volume of wood per unit area. This approach carries the use of R-values to its logical conclusion. However, the inaccuracies are so large, as Livingstone points out, that the results are only accurate to an order of magnitude. Recent work by R. W. Parsons and I. C. Prentice, as yet unpublished, has involved the detailed mathematical derivation of pollen representation factors using pollen percentages but in which the role of long-distance pollen is considered. Parsons has also developed important methods for deriving error estimates for the use of R-values. When this work is published and the methods applied to a range of data sets, the real potential of pollen correction factors in reconstructing past populations from fossil pollen data from lakes and bogs can be assessed.

It appears at present that R-values can only be used in an accurate way to determine vegetational percentages from humus within forests, or from basins of less than 30 m radius. An alternative method of reconstructing past populations is needed for larger lakes or bogs. This has been provided in the form of absolute pollen frequency (APF) measurements.

Absolute pollen frequency

The determination of absolute pollen frequency (APF) avoids the constraints imposed by percentage calculations (Faegri and Iversen, 1975). The values for each pollen type are independent of each other. As Maher (1972) and Colinvaux (1978) have pointed out ***absolute pollen frequency*** is a misleading phrase. However, many pollen analysts have referred to ***absolute pollen diagrams*** as opposed to percentage pollen diagrams. It is more informative to refer explicitly to pollen concentration and pollen influx diagrams. These and other necessary terms are defined as follows by Davis (1969b).

Pollen concentration is the number of grains per unit volume or mass of wet or dry sediment. This is often referred to as ***absolute pollen frequency*** or APF. The units are grains cm^{-3} or grains gm^{-1}.

Pollen influx, ***pollen accumulation rate***, or ***pollen deposition rate*** is the net number of grains accumulated per unit area of sediment surface per unit time. The units are grains cm^{-2} $year^{-1}$.

Sediment matrix accumulation rate is the net thickness of sediment accumulated per unit time, after compaction and diagenesis. The units are cm $year^{-1}$.

Deposition time is the amount of time per unit thickness of sediment. The units are years cm^{-1}.

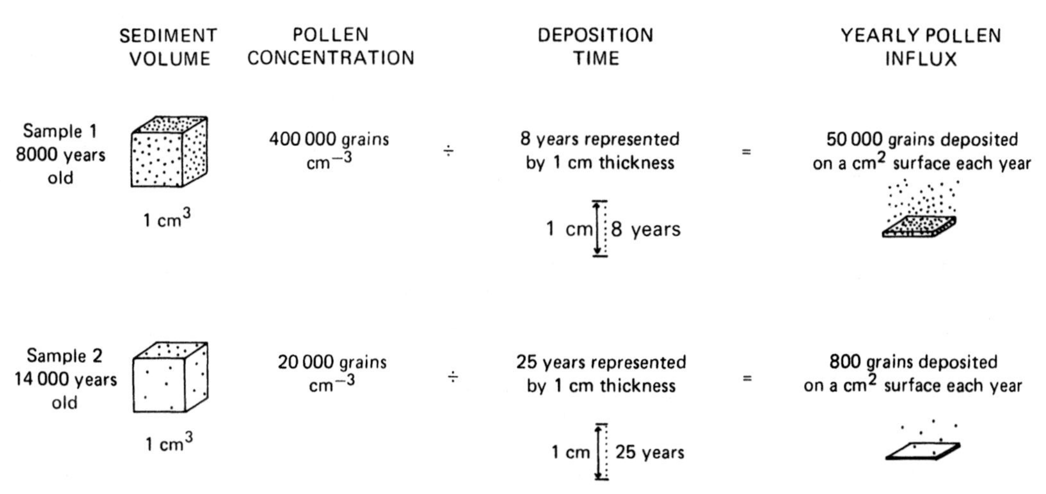

Fig. 10.7 Two examples of the calculation of pollen influx from measurements of pollen concentration and deposition time. (After Davis, 1969a.)

Pollen influx = Pollen concentration × Sediment matrix accumulation rate
= Pollen concentration/Deposition time.

From these definitions, we can see that two parameters have to be known before pollen influx can be calculated, namely pollen concentration and the sediment matrix accumulation rate or deposition time. This is illustrated in Fig. 10.7.

The determination of pollen concentration

The methods for the determination of pollen concentration fall into two main groups, the direct methods which estimate the actual amount of pollen in a sample, and the indirect methods which rely on the addition of a known concentration of exotic pollen.

1. *Direct methods*

a) Davis (1965, 1966) described a method of determining pollen concentration based on the principles of the methods of Erdtman (1943) and Muller (1959). She took a known volume of wet sediment, suspended it in a known volume of liquid, prepared it by the necessary chemical treatment, withdrew aliquots of known volume from the final suspension, and mounted them on a slide. The error involved in the withdrawal of a small volume was lessened by the addition of a large amount of a volatile liquid, such as tertiary butyl alcohol. The aliquot was dropped carefully on to silicone oil on a slide and the TBA was allowed to evaporate, leaving only the pollen residue. The whole slide had then to be counted to determine all the pollen within the aliquot. The pollen concentration in the sediment could then be calculated. Davis measured the statistical variance of her method, and calculated a confidence interval for the counts.

b) Jørgensen (1967) described a similar direct method, which relied on the weight of a sample, rather than its volume. A similar method was described by Borowik-Dabrowska and Dabrowski (1972) in which the density of the sediment (mass per unit volume) was also estimated, thus allowing the calculation of the amount of pollen in a volume of sediment from a sample of known mass.

2. *Indirect method*

Benninghoff (1962) first briefly described this method, which was subsequently tested and described fully by Matthews (1969) and Bonny (1972). The concentration of a suitable exotic pollen suspension is determined using a haemocytometer. Known volumes of this suspension are then added to known volumes of sediment, and the sample prepared in the usual way. The numbers of both native and exotic pollen are counted. By simple proportion:

$$\frac{\text{Fossil pollen concentration}}{\text{Exotic pollen concentration}} = \frac{\text{Fossil pollen counted}}{\text{Exotic pollen counted}}$$

Hence the fossil pollen concentration can be calculated as

$$\text{Fossil pollen concentration} = \frac{\text{Fossil pollen counted}}{\text{Exotic pollen counted}} \times \text{Exotic pollen concentration added}$$

Various species have been used as exotic pollen, such as *Nyssa sylvatica*, *Ailanthus glandulosa*, and *Eucalyptus globulus*. To overcome the difficulty of adding a known volume of hopefully homogeneous exotic pollen suspension, Stockmarr (1971) used tablets based on readily soluble calcium carbonate, which contained a known number of *Lycopodium clavatum* spores. Because these are previously acetolysed, they can be easily distinguished microscopically from any native *L. clavatum* spores in the sediment. Stockmarr estimated the errors involved in this procedure. Errors in the exotic pollen method have also been estimated theoretically and experimentally by Maher (1972), Bonny (1972), and Birks (1976).

The determination of pollen influx

This involves the determination of the sediment matrix accumulation rate or sediment deposition time. This is most usually achieved by a series of radiocarbon dates at different depths. Occasionally, annually laminated sediments are available (see Chapter 6), in which case the various problems involved with radiocarbon dating are avoided. These problems include the following:

i) There are errors in radiocarbon dates from lake sediments, either because old ^{14}C-deficient carbon has been derived from calcareous bedrock, such as limestone, or that ^{14}C-deficient material is washed into the lake from the catchment. An approximate correction can be made if the surface muds are dated. Their radiocarbon age (e.g. 700 years) may then be subtracted from all the other samples, assuming the contamination processes have been constant (Davis, 1967a). This is not necessarily a valid procedure (Brubaker, 1975) as it presupposes a constant supply of ^{14}C-deficient carbon during the history of the lake.
ii) It is often difficult to obtain enough carbon in a sufficiently narrow stratigraphic thickness, especially from sediment cores. It is generally desirable that the dated sediment should occupy no more than 10 cm depth.
iii) It is now clear that radiocarbon years are not directly equivalent to calendar years (Suess, 1970) especially before 2000 B.P., due to fluctuations in the cosmic-ray influx which creates the ^{14}C in the atmosphere. (See Renfrew (1973) for a good account of this phenomenon.)
iv) There are inherent errors in the sampling, sample preparation, and counting of all radiocarbon samples. The latter two are usually given as one standard deviation of the age (68% confidence) after a radiocarbon date, e.g. 4570 $\pm$ 80 radiocarbon years B.P.

Laminated sediments have been used to estimate pollen influx, for example by Craig (1972) and Swain (1973). However, most studies rely on radiocarbon dates to establish the sediment accumulation rate.

Pollen influx analysis from Rogers Lake, Connecticut, U.S.A.

Davis and Deevey (1964) and Davis (1967a, 1969b) presented the first pollen influx data from Rogers Lake as an exploration of the potentialities of the method of absolute pollen analysis, and as a reinvestigation of the nature of the New England *Pinus* pollen zone (see p. 203). They estimated the pollen influx by dividing the pollen concentration in 1 ml sediment samples by the numbers of years each sample represented (deposition time) which they calculated from the sediment matrix accumulation rate derived from a series of 24 radiocarbon samples down the core.

The percentage pollen diagram is compared with the pollen influx diagram in Fig. 10.8. The pollen curves are plotted against the estimated age of each sample, and the diagrams are divided into the traditional New England pollen zones.

Above 8000 B.P., the sediment accumulation rate is virtually constant, and the diagrams differ little therefore. But below this level, they differ significantly. Before 12 000 B.P., the percentage diagram shows a dominance of herb pollen. However, tree pollen reaches 40% and *Pinus* is the most abundant tree pollen type. Subsequently, during zones A2 and A3, the percentages of *Picea*, *Pinus*, and *Quercus* increase. At about 10 000 B.P., *Quercus* pollen percentages decline to a minimum during zone A4, whereas values of *Picea*, *Pinus*, and *Alnus* increase. At about 9000 B.P., *Pinus* pollen percentages increase markedly, and it becomes the most abundant pollen type, giving its name to zone B. The *Pinus* pollen zone lasted until about 7000 B.P.,

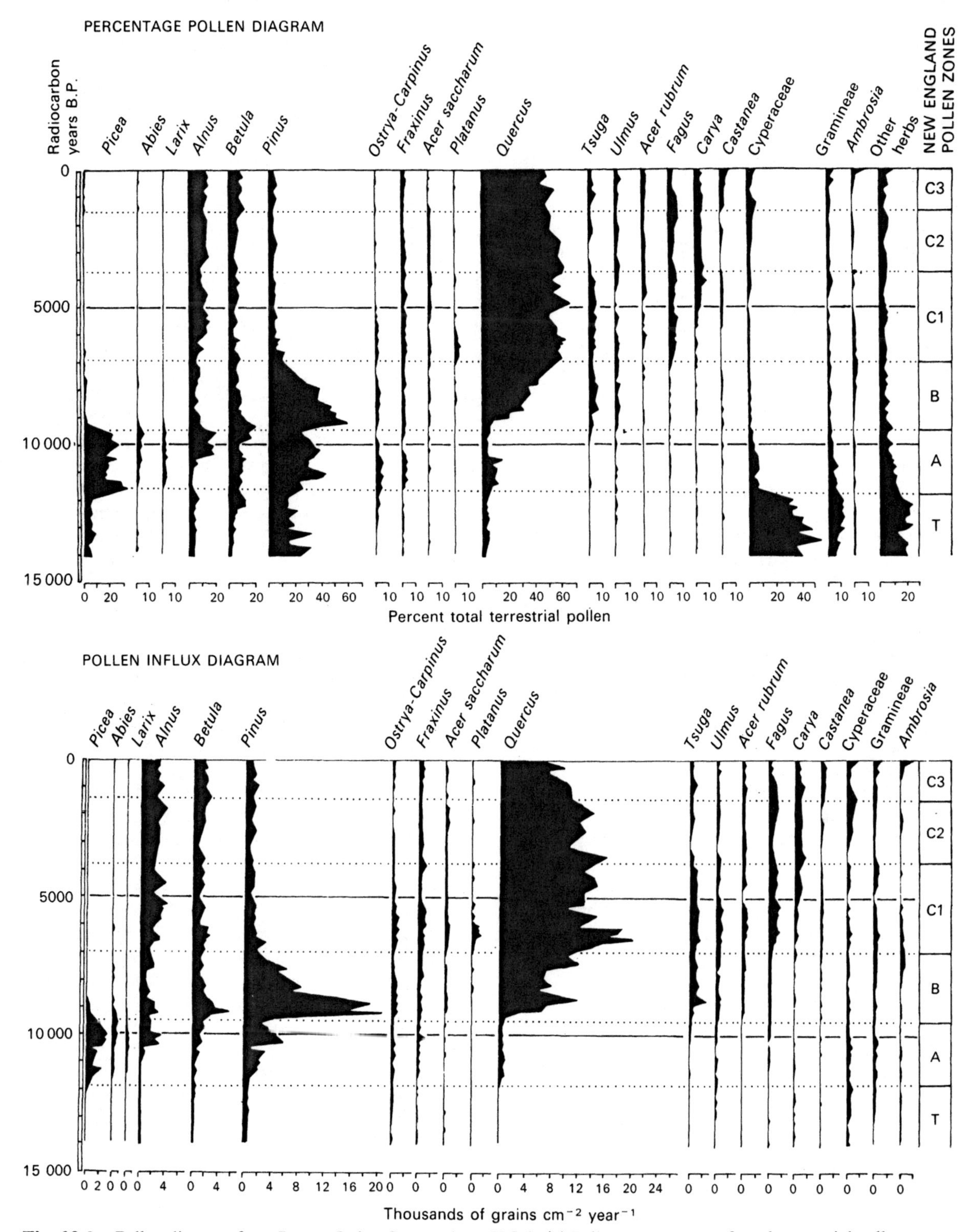

Fig. 10.8 Pollen diagram from Rogers Lake, Connecticut, U.S.A. (**a**) Pollen percentages of total terrestrial pollen. (**b**) Pollen influx diagram in thousands of grains cm^{-2} $year^{-1}$. (After Davis, 1967a.)

when *Pinus* pollen percentages decline to below 10%, and *Quercus* becomes the over-all dominant pollen type.

Comparison with the pollen influx diagram shows that these percentage changes are not a true reflection of past vegetational changes (see Davis, 1969a). A very small number of pollen grains of any kind were deposited before 12 000 B.P., and in reality, tree pollen grains were very rare, and were probably deposited in the lake from far distant sources. At 12 000 B.P., the tree pollen influx increased, but the herb pollen influx remained more or less constant. This suggests that trees had moved nearer the site at this time, but that large areas of treeless vegetation persisted. From this time, the influx of conifer pollen gradually increases up to about 9000 B.P. indicating the development of open spruce woodland with some pine, and later with increased amounts of alder and birch. The minimum in the *Quercus* pollen percentages is an artifact of the increased influx of conifer pollen. The influx of *Quercus* and other deciduous tree pollen remains uniformly low.

After 9000 B.P., the influx of all tree pollen types increases, with that of *Pinus* reaching some 18 times higher than its subsequent post-glacial influx values. This led Davis (1969a,b) to conclude that pine had indeed been an important member of the forest during the *Pinus* pollen zone, but not so dominant as had previously been assumed, and that her previous conclusions that the high *Pinus* pollen percentages were an artifact caused by a low pollen production by the rest of the vegetation was incorrect. She also demonstrated that the high *Pinus* percentages corresponded to an importance of pine in the vegetation by comparison with surface samples in the boreal forest regions of Canada.

Above this level, both diagrams are similar, and interpretation of the percentage diagram is not modified by the new influx data.

The use of pollen influx diagrams in palaeoecology

Since the original pollen influx diagram from Rogers Lake was published, numerous pollen influx diagrams have been prepared. Their aims have been various; some have been produced to supplement percentage pollen data, and others have been directed to special palaeoecological problems. We shall now discuss some of these investigations to demonstrate how pollen influx data have thrown a new light on some problems, and we shall keep in mind how, if at all, absolute pollen analyses can lead to an estimate of past plant populations. Most investigators only discuss past populations indirectly in terms of past vegetation: only a few have been concerned with the reconstruction of the numbers or of the areas occupied by different species.

1. *The interpretation of percentage data, especially from late-glacial sediments*

The pollen influx diagram from Rogers Lake demonstrates the importance of influx data in the interpretation of pollen percentage data. All influx diagrams can be used for this purpose, thus yielding a more realistic reconstruction of past vegetation and how it has changed. This property has been particularly useful in late-glacial sediments, for interpreting changes from periods of very low pollen influx in glacial or tree-less conditions to conditions of shrub or tree vegetation with much higher pollen influx. The first British pollen influx diagram was from the Devensian late-glacial sediments of Blelham Bog, in the English Lake District (Pennington and Bonny, 1970). It is partly reproduced in Fig. 10.9. In the zone of very low influx at the base of the sequence (zone Ba, 440–450 cm), pollen percentages give a false impression of abundance of all the pollen types, particularly *Betula*, *Juniperus*, *Salix*, *Rumex*, and Gramineae. Their influx is very small indeed, and there was probably very little vegetation on the landscape. During the overlying interstadial, pollen influx increased considerably, with the development of juniper scrub and then open birch woodland during zones Bb, Bc, and Bd. The sudden return of cold conditions is reflected in zones Be and Bf by a marked decrease in pollen influx affecting all taxa except Filicales (fern spores, probably washed in as a result of soil erosion) and of *Artemisia* and Caryophyllaceae, which are characteristic of open, treeless vegetation. The opening of the Flandrian (post-glacial) in zones Bg and Bh repeats the previous rise in influx resulting from the redevelopment of juniper scrub and birch forest.

Craig (1978) showed that there was a similar pattern of pollen influx and record of vegetation development in late-glacial sediments from two lakes in south-east Ireland, although birch forest did not develop during the Allerød interstadial period, and the pollen influx did not fall as much as might have been expected in the subsequent Younger Dryas cold stadial. The two lakes had very different influx rates, and Craig concluded that pollen concentration may be more useful in making

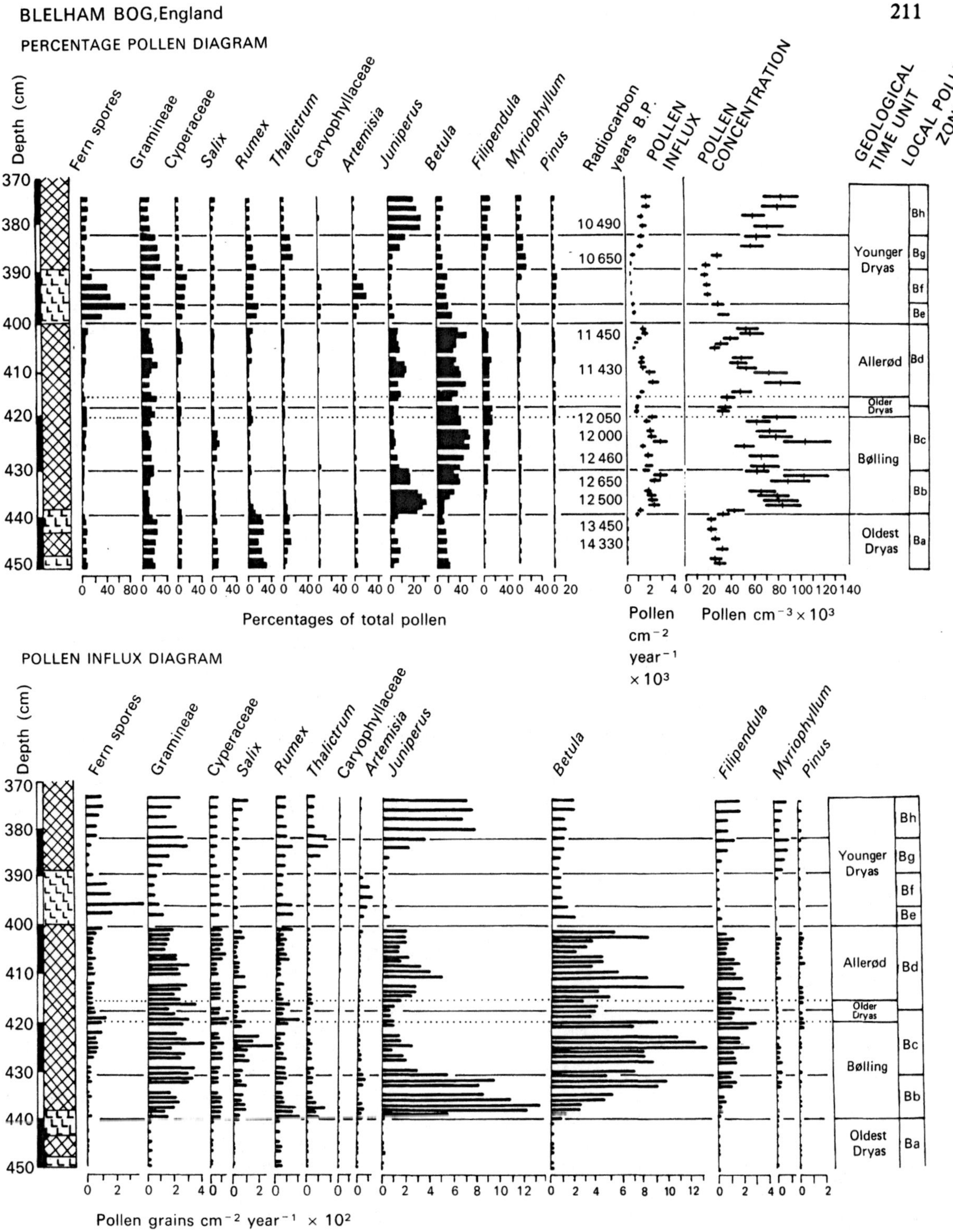

BLELHAM BOG, England

PERCENTAGE POLLEN DIAGRAM

Depth (cm)

Fern spores, Gramineae, Cyperaceae, Salix, Rumex, Thalictrum, Caryophyllaceae, Artemisia, Juniperus, Betula, Filipendula, Myriophyllum, Pinus, Radiocarbon years B.P., POLLEN INFLUX, POLLEN CONCENTRATION, GEOLOGICAL TIME UNIT, LOCAL POLLEN ZONE

370, 380, 390, 400, 410, 420, 430, 440, 450

10 490
10 650
11 450
11 430
12 050
12 000
12 460
12 650
12 500
13 450
14 330

Younger Dryas — Bh, Bg, Bf, Be
Allerød — Bd
Older Dryas
Bølling — Bc, Bb
Oldest Dryas — Ba

0 40 80 0 40 0 40 0 40 0 40 0 40 0 40 0 40 0 20 0 40 0 40 0 40 0 20

Percentages of total pollen

0 2 4

Pollen cm^{-2} year^{-1} $\times 10^3$

0 20 40 60 80 100 120 140

Pollen cm$^{-3} \times 10^3$

POLLEN INFLUX DIAGRAM

Depth (cm)

Fern spores, Gramineae, Cyperaceae, Salix, Rumex, Thalictrum, Caryophyllaceae, Artemisia, Juniperus, Betula, Filipendula, Myriophyllum, Pinus

370, 380, 390, 400, 410, 420, 430, 440, 450

Younger Dryas — Bh, Bg, Bf, Be
Allerød — Bd
Older Dryas
Bølling — Bc, Bb
Oldest Dryas — Ba

0 2 0 2 4 0 0 2 0 0 0 0 0 2 4 6 8 10 12 0 2 4 6 8 10 12 0 2 4 0 0 2

Pollen grains cm^{-2} year^{-1} $\times 10^2$

Fig. 10.9 Pollen diagram from Blelham Bog, English Lake District. (**a**) Selected taxa as percentages of total pollen, radiocarbon dates, and total pollen influx and pollen concentration, with approximate 90% confidence intervals. (**b**) Pollen influx diagram showing annual deposition rates of selected taxa. (After Penington and Bonny, 1970.)

comparisons of past vegetational history because of inter-lake variability (see p. 222).

In North America, there has been considerable argument as to whether there was a climatic deterioration in the Wisconsinan late-glacial period equivalent to the Younger Dryas of north-west Europe (e.g. Davis, 1967b). The very detailed pollen influx diagram from the Wisconsinan late-glacial sediments at Wolf Creek, Minnesota (Birks, 1976) fails to indicate any revertence to cooler conditions with lowered pollen influx after the initial treeless phases. A study of macrofossils at the site supports this evidence, and no other site in North America has provided any influx data to contradict the conclusions drawn from Wolf Creek. Indeed, they were strikingly supported by R. B. Davis *et al.* (1975) with a pollen percentage and influx diagram from Moulton Pond, Maine, U.S.A. No climatically attributable changes occurred during the Wisconsinan late-glacial in pollen composition or influx, although a glacial advance occurred within that time period which terminated only 50 km away. Their diagram also illustrates the importance of influx estimates in vegetation interpretation from percentage diagrams. A marked increase in *Picea* pollen percentages could not be interpreted as a spruce forest period, because the total influx was so low that only tundra could have been growing round the site. Although there is good evidence of spruce forest further south, *Picea* failed to colonize northern New England before deciduous trees and pines reached it at about 10 000 B.P.

Today, there are many pollen influx diagrams from all over the world, and we cannot discuss them all here. We have, however, selected a few to demonstrate the uses of pollen influx in solving particular palaeoecological problems.

2. *Vegetation on different soils*

Brubaker (1975) constructed three pollen diagrams from areas of distinct soil types in northern Michigan, U.S.A. which support distinct vegetation at the present day. The well-drained sandy soils of the Yellow Dog outwash plain are occupied by *Pinus banksiana* (jack pine) forest, *Pinus strobus* (white pine) is most common in the mixed deciduous–coniferous woods on the moister soils of the Michigamme outwash, and dense deciduous–coniferous forest occupies the heavy soils of the Michigamme till.

In Fig. 10.10, the percentage pollen diagrams show that jack pine expanded at about 9000 B.P. in the area, and always remained most abundant on the Yellow Dog outwash. This conclusion is supported by the consistently high influx of jack pine pollen at Yellow Dog Pond. White pine, together with deciduous trees, expanded at about 8000 B.P. and replaced jack pine everywhere except on the outwash plain. There is little differentiation between the percentage curves for white pine pollen, but the influx data demonstrate that white pine was uncommon on the Yellow Dog outwash and that it became more abundant on the Michigamme outwash at Lost Lake, than on the till at Camp 11 Lake. Deciduous trees increased primarily on the moisture-retaining soils of the till during the dry mid-post-glacial time, but since 3000 B.P. they have colonized the till and Michigamme outwash equally. This demonstrates the effect of the same climatic change on communities on soils of different moisture-retaining capacity (see Fig. 12.1).

3. *Fire history*

Craig (1972) constructed a pollen influx diagram from the annually laminated sediments of Lake of the Clouds, northern Minnesota, U.S.A., and showed that there was little change in the influx values throughout the Holocene (post-glacial). Fire is an important ecological factor in the northern coniferous forests today, and Swain (1973) took advantage of the annual laminations to investigate its importance in the past. In Fig. 10.11 the historical occurrence of fires is indicated on the curve of charcoal influx, and previous fires are indicated with a star. It would appear that fires occurred about every 60–70 years during the last 1000 years. There is more charcoal between A.D. 1000 and 1400 than in the last 500 years. This may be a reflection of the climatic deterioration known as the Little Ice Age, between about A.D. 1550 and 1850. The varves tend to be thicker at peaks of charcoal influx, probably due to soil erosion and greater sediment movement induced by wind turbulence after the shelter of the trees had been removed. The conifer:'sprouter' pollen ratio decreased after fires, as the quickly regenerating birch and aspen recolonized the fire site before being suppressed by the slower conifer regeneration.

A similar study using laminated sediments was made by Cwynar (1978) in the Algonquin National Park, Ontario, Canada. He also demonstrated increased soil erosion following a fire by showing an increase in the influx of aluminium and vanadium in the thicker varves which contained high amounts

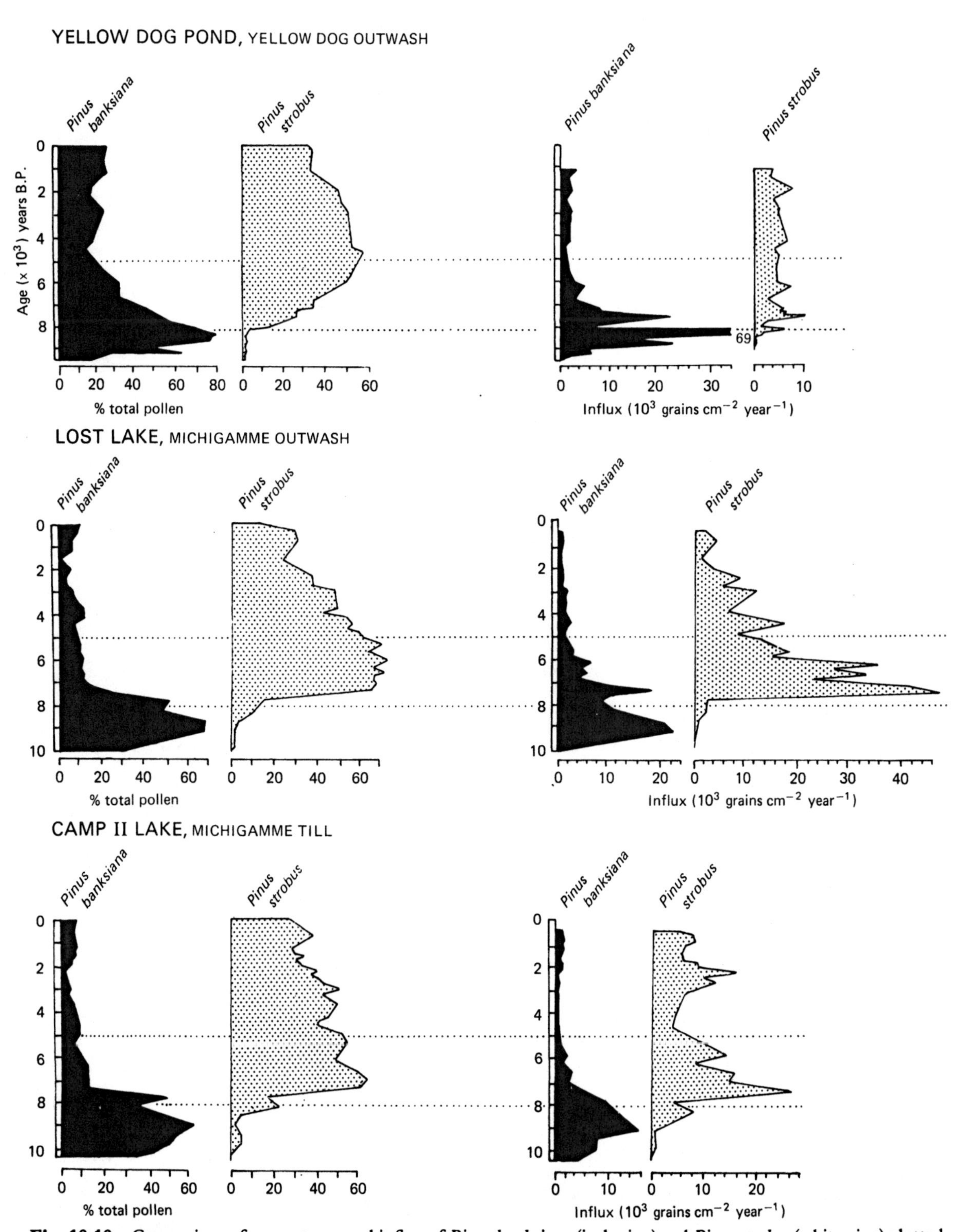

Fig. 10.10 Comparison of percentages and influx of *Pinus banksiana* (jack pine) and *Pinus strobus* (white pine) plotted against radiocarbon age at three sites on different soils in northern Michigan. Percentages of total pollen of terrestrial plants. Influx values are 10^3 grains cm^{-2} $year^{-1}$. Horizontal lines at 5000 and 8000 B.P. delimit the 'prairie period'. (After Brubaker, 1975.)

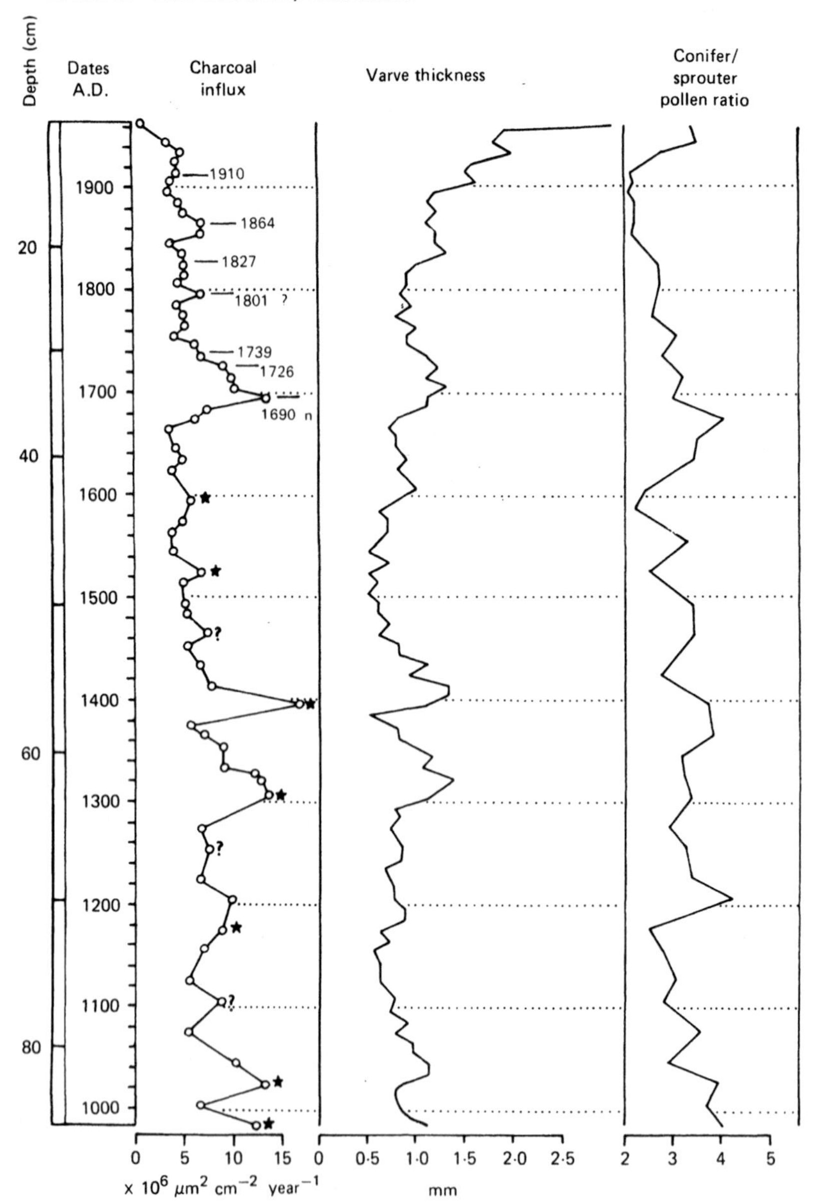

Fig. 10.11 Charcoal and pollen influx diagram for the last 1000 years at Lake of the Clouds, Minnesota. The date at the top of the diagram is 1970. Peaks in charcoal influx are marked with ★. (After Swain, 1973.)

of charcoal, and also pollen redeposited from the soil. Swain (1978) has studied the pollen stratigraphy and charcoal stratigraphy in a laminated profile from Wisconsin, and he suggests a climatic history for the last 2000 years on the basis of fire history and vegetation dynamics.

4. *Vegetation history in the arctic*

Fredskild (1969) presented a pollen influx diagram from the arctic desert of Peary Land, in North Greenland. He measured the modern pollen influx using a trap, and caught an average of only 2.5 grains cm^{-2} $year^{-1}$. The fossil influx showed greater values during the mid-post-glacial 'climatic optimum' when plants were able to flower better, but the vegetation was essentially unchanged. This feature was not detectable on the percentage pollen diagram.

Fredskild (1973) also worked extensively in south-west Greenland. He used pollen influx data to demonstrate the relative importance of long-distance pollen in an area of very low local pollen productivity. The changing composition of the exotic pollen indicated changes of wind direction during the melting of the North American Laurentide ice sheet. The exotic pollen was most prominent near the coast where the local pollen production was 177–327 grains cm^{-2} $year^{-1}$. In the more sheltered interior the more luxuriant vegetation produced 1710–2650 grains cm^{-2} $year^{-1}$. His work demonstrated the sensitivity of the magnitude of pollen influx in environments marginal to plant growth, and where trees are rare or absent. The vegetation reacts to climatic changes, and the reaction is reflected in the pollen influx.

5. *Forest clearance by man*

As in the Devensian late-glacial, forest clearance produces a contrast between a forested and relatively treeless landscape. However, forest clearance progressed over many centuries in Europe, beginning in the Neolithic period, some 5000 years ago. This date coincides with a widespread decline of *Ulmus* pollen which has frequently been attributed to man's activity. Pennington (1975) used pollen influx data to establish that the percentage fall in elm pollen was a *true* decrease in elm pollen influx to between a fifth and a tenth of its previous values in the English Lake District. In the uplands, near Neolithic axe factories, influx measurements suggest that forests of pine and birch were cleared and that, although elm pollen declined, other deciduous trees remained unaffected on the lower slopes. By contrast, on the coastal plain of Cumbria, a decrease in the pollen influx of *Quercus*, together with an increase in pollen of cereals and weeds indicated that oakwoods had been cleared for agricultural purposes. Here the elm decline could well have resulted from this activity. In the valleys, oak pollen remained abundant at the time of the elm decline, indicating that the valley oakwoods were not cleared then. However, an increase in sediment accumulation rate in the lakes, coupled with evidence from chemical analyses, showed that there was increased run-off and soil erosion at this time, perhaps due to soil disturbance on the upper slopes, or perhaps to a climatic deterioration. The latter could be a possible explanation for the widespread decline in elm, although disease could be another factor involved, and the activities of early man cannot be excluded in some areas.

The process of European settlement and forest clearance in North America also increased run-off and sediment deposition rates in lakes. The event is marked by an increase in *Ambrosia* (ragweed) pollen, and because it is historically dated, it has been used as a datum for studies in changes in lake sedimentation associated with forest clearance (see p. 103).

6. *Movement of tree lines and migration*

Hyvärinen (1975, 1976) followed the migration of *Pinus* and *Betula* in northernmost Fennoscandia. A transect of four sites spanned the present limits of *Pinus* and *Betula*. From surface samples, he determined that local presence of pine was shown by influx values of 500 grains cm^{-2} $year^{-1}$ or more in the three southernmost lakes. The northernmost lake proved to be a pollen-rich site (*sensu* Pennington, 1973). In Fig. 10.12 it can be seen that birch had colonized the whole area by about 9000 B.P. and that pine began to expand at about 8500 B.P. in the north, but later in the south, where glacial retreat had been most recent. At all sites except Domsvatnet, birch influx declines as pine values rise, suggesting that pine had replaced birch. At Domsvatnet birch is not replaced and Hyvärinen suggests that pine was not locally present, all the pine pollen being derived from regions to the south. At Bruvatnet, pine was probably scattered in the birch forest, as it is today at Suovalampi. Further south, pine formed dense stands. Pine started to decline at about 5000 B.P., and birch showed a parallel decline at Domsvatnet. By about 3000 B.P., birch had retreated from Domsvatnet, leaving a treeless landscape, and pine had retreated from Bruvatnet, leaving birch forest; pine became scattered at Suovalampi, although it remained dominant at Akuvaara. These conclusions could only be reached using pollen influx data, as the percentage diagrams resembled each other, and the actual amounts of the trees on the landscape could not be envisaged.

Pollen influx was used to demonstrate the altitudinal movement of the treeline in Colorado, U.S.A. by Maher (1972). In his usual careful and thoughtful way, he constructed the first influx diagram from the western U.S.A., from Redrock

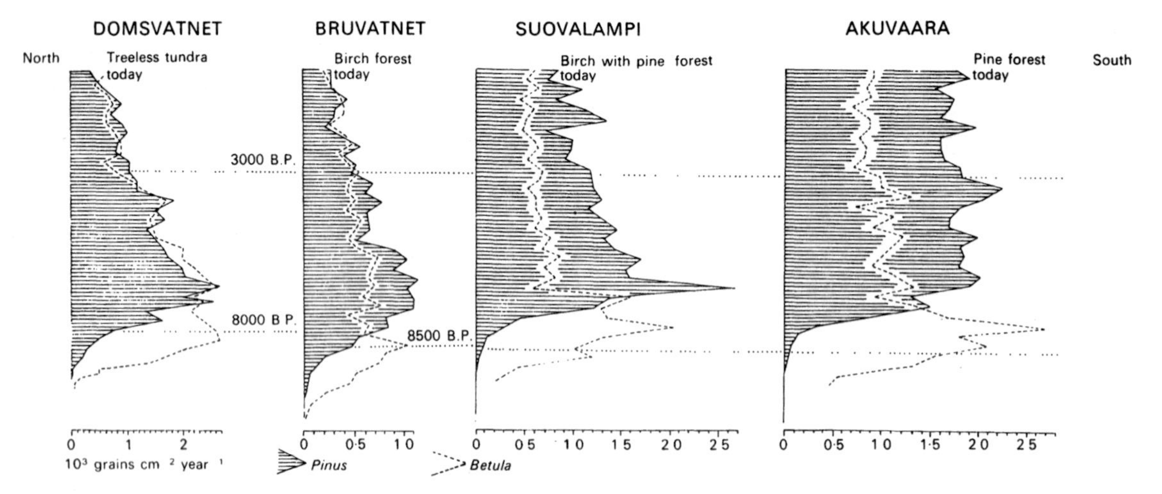

Fig. 10.12 Pollen influx of *Betula* and *Pinus* in four lakes in a north–south transect in northern Finland, which crosses the present northern limits of birch and pine. (After Hyvärinen, 1975, 1976.)

Lake. *Pinus* pollen influx increased gradually through the post-glacial, suggesting a gradual upward movement of the vegetation zones. This suggestion was supported by the use of modern surface samples.

Watts (1973) attempted to reconstruct the mechanism and rate of colonization of tree species using influx data. He used Craig's (1972) Lake of the Clouds data, and also other data from both post-glacial and interglacial sites in the U.S.A. and Europe. He concluded that the migration of any species into an already stable community, with no change in external factors, such as climatic change or the influence of man, took a considerable time, from 500–1000 years. This time covers several generations of the invading species, suggesting that its population built up from chance-establishment of a few individuals in natural openings, until the total available niche was occupied, and intraspecific competition became the factor controlling any further increase in abundance.

Davis, Brubaker, and Webb (1973) estimated the level of pollen influx of tree species in Michigan U.S.A. which indicated absence, nearby presence, and local presence. These data could be used to reconstruct past forest communities. Davis (1976) applied the data to influx diagrams from North America, and was able to draw tentative migration maps for the different trees (Fig. 10.13). She demonstrated that they moved independently across the continent from different glacial refugia. In some cases, e.g. *Pinus strobus*, westward migration was temporarily halted at the Great Lakes region because of the eastwards expansion of prairie between 8000 and 5000 B.P.; see also Wright (1968) and Jacobson (1975). In other cases, e.g. *Pinus banksiana/P. resinosa*, the occupied area moved rapidly as a band, expanding northwards and retreating from the south, until the present area was occupied before interference from the prairie expansion. Some species, such as *Carya* migrated from the south-west, and others such as *Pinus strobus*, *Tsuga canadensis*, and *Castanea* migrated from the east.

7. *Constructing a time scale*

a) Because interglacials are beyond the range of radiocarbon dating, the only reliable time scales have been constructed from annually laminated sediments. However, Dabrowski (1971) ingeniously constructed a relative time-scale for an Eemian (last interglacial) site in Poland. He used modern pollen influx measurements from Białowieza Forest (see p. 200) to estimate the average number of pollen grains falling on 1 cm^2 in one year. He then measured the depth of Eemian sediment of section 1 cm^2 which contained this number of pollen grains, and therefore how much sediment was deposited in one year. Because of the inaccuracy of the estimate, he preferred to call the time units palynochrones. His Eemian diagram contained 18 000 palynochrones, which is longer than Müller's (1974) estimate of the length of the Eemian of 11 000 years, based on counts of annually laminated sediments.

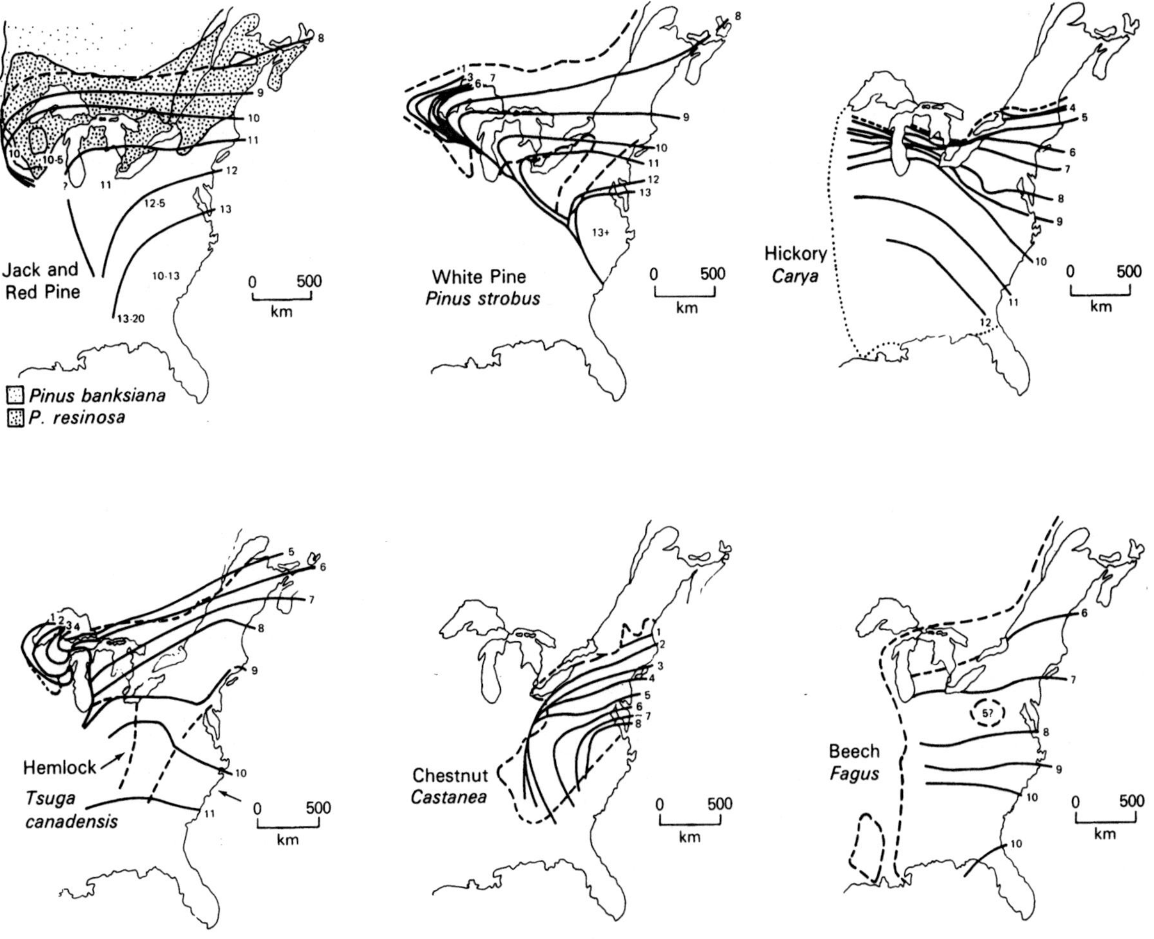

Fig. 10.13 Migration of some trees in North America during the Holocene or post-glacial. Contours show the advancing front at 1000-year intervals. Dashed line surrounds the modern range of the species, except for *Pinus banksiana* and *P. resinosa*, whose present ranges are indicated by stippling. (After Davis, 1976.)

Mehringer *et al.* (1977) used a similar technique to determine the number of years represented by volcanic ash falls in Montana, U.S.A. Because the pollen influx above and below the ash layers was similar, they proposed that 4000 pollen grains cm^{-2} were deposited in one year, assuming that the pollen productivity of the surrounding vegetation was unaffected, which seemed to be the case. They ingeniously showed (Fig. 10.14) that 4.6 cm of ash fell during autumn and winter, when pine was not flowering, 1 cm during the next year, and 1.7 cm in the following year. Although the ash fall reduced aquatic productivity, and seemed to stimulate the flowering of steppe elements such as *Artemisia* and Chenopodiaceae, these effects were short-lived, as shown by a quick return to previous conditions in the upper mud (see also Mack *et al.*, 1978).

Hicks (1974) also measured modern influx in the northern coniferous forest in Finland, and used the results to construct a time scale for a pollen concentration diagram from a peat deposit in the area. She emphasized that the modern influx estimates must be made in the same area and from the same vegetation type for the method to be at all plausible.

8. *Plant macrofossil influx*

Plant macrofossil influx was estimated by H. H. Birks (in Birks, 1976) from sediments of Wolf Creek, Minnesota, U.S.A. Previously, macrofossils had usually been represented as concentration values (see Chapter 5). The technique was further developed by Birks and Mathewes (1978) from Devensian late-glacial and early post-glacial sediments at Abernethy Forest, Scotland, when pollen and macrofossil influxes were compared directly (see Chapter 5).

9. *Other pollen influx studies*

a) Hicks (1975) presented the first pollen influx diagram from a peat deposit at Kangerjoki, Finland. Pollen influx increased as trees colonized the landscape early in the post-glacial, but it fell again when trees occupied the bog surface, and effectively filtered the pollen from the surrounding landscape. Following the decline of trees on the bog surface, the pollen influx rose again.

b) The first pollen influx diagram from the tropics was produced by Kendall (1969) from Lake Victoria, Central Africa. Although he carried out a large and detailed study of this difficult area, the pollen influx measurements added little to his over-all conclusions.

c) Similarly, although Fredskild's (1975) concentration diagram from Flådet Lake, Langeland, Denmark is interesting from the vegetational history point of view, the concentration and estimated influx data add little to the interpretation of the results, apart from demonstrating that the depression of *Pinus* and *Betula* pollen percentages in the early post-glacial was an artifact of the increasing influx from other deciduous trees as they arrived.

d) Pennington (1975) compared pollen influx diagrams from Devensian late-glacial sediments at Blelham Bog, English Lake District, and Cam Loch, north-west Scotland. Although the vegetation differed throughout the late-glacial at these two contrasting sites, she demonstrated that the vegetational changes recorded occurred at the same times, and that they were parallel expressions of the same climatic changes. The changes in pollen influx at both sites confirmed these conclusions.

e) Pennington and Sackin (1975) subjected pollen percentage and concentration data from Blea Tarn, English Lake District, and Cam Loch, north-west Scotland, to principal components analysis (PCA) in order to compare the numerical divisions of the data. Similar stratigraphic divisions of the two sites were indicated by both data sets, the boundaries being somewhat clearer with the concentration data. A PCA of chemical data also led to a similar division, suggesting that the observed changes were connected with the reaction of vegetation and soils to climatic changes. They concluded that the only additional use of concentration data was the clarification of situations with a changing pollen influx, where percentages could give an erroneous impression of the true situation, for example in a treeless environment.

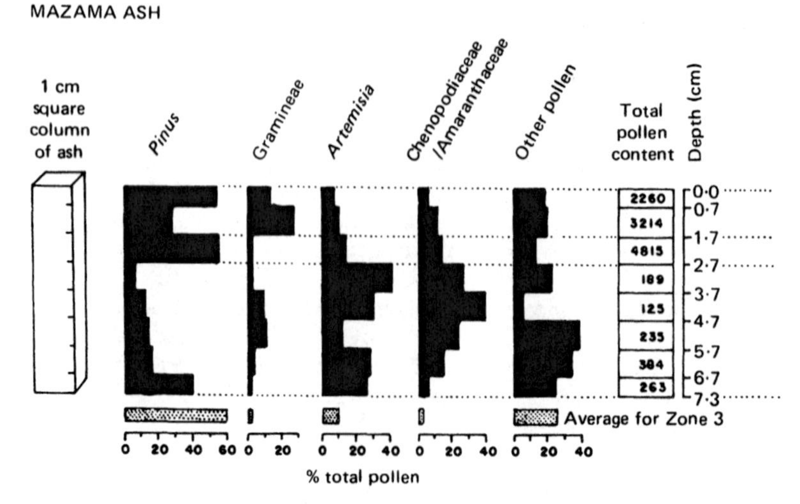

Fig. 10.14 Pollen percentages and total pollen content of eight samples of Mazama ash compared with the average percentages for the organic lake muds of the rest of pollen zone 3. (After Mehringer *et al.*, 1977.)

The calibration of pollen influx

Pollen influx diagrams are expensive and time consuming to prepare, and so we may ask if they are of value in reconstructing past vegetation and past populations of species. Clearly they are of considerable value when there are great changes in pollen production by the local vegetation, such as the transition from tundra to forested communities in the late-glacial, the clearance of forest by man, and the climatically controlled retreat of forest at its northern limit (Hyvärinen, 1975, 1976).

However, in many lakes, the sedimentation rate and pollen influx remain fairly constant during the post-glacial, as in zone C at Rogers Lake (p. 210) and Lake Vakojärvi in Finland (Donner, 1972). In such cases, percentage data convey almost as much information about changes in the numbers of the taxa on the landscape as do influx data.

In their discussion of the first pollen influx diagram from Rogers Lake, Davis and Deevey (1964) remarked, 'The possibility that (pollen) accumulation rates provide reliable indices of the absolute abundances of plants on the landscape is worth pursuing; it would allow a completely objective interpretation of results such as those from Rogers Lake'. If pollen influx data could be interpreted in terms of the abundance of individual taxa in the vegetation, we would then have an approximate population curve for different taxa through time. This would provide new and exciting information for understanding vegetation dynamics in the past, over periods of time much longer than is possible by direct observation. Watts (1973) approached this possibility through influx and percentage data, but as yet there are no direct conversion factors, similar to R-values, relating influx to numbers of plants. There are many variables operating between pollen production and its final sedimentation (Chapter 9) so perhaps a precise calibration of pollen influx will never be possible. However, Davis *et al.* (1973) attempted to calibrate pollen influx by comparing recent pollen influx values directly with the numbers of trees recorded in the forests of Michigan, U.S.A.

Davis *et al.* (1973) considered two basic questions:

i) What is the variation in pollen influx to be expected due to within-lake processes, and to differences between lakes in the same vegetational region?
ii) Is total pollen input related to the distribution of vegetation and to the abundance of its individual components?

Davis and her associates (see Chapter 9) have demonstrated the complexity of sedimentation processes within a lake, and so it was obviously important to assess the variability introduced into influx measurements from these causes. Any remaining variation could be attributed to variation in the vegetation composition.

1. *Variation of modern and fossil influx within a lake*

Davis *et al.* (1973) first investigated the variation in influx within Frains Lake and Blind Lake, Michigan. From influx measurements from cores taken from varying depths of water, they concluded that influx to deeper water was within three times the influx to shallower water, provided that very deep and very shallow areas were excluded. Similarly, a comparison of the fossil influx in two cores from Rogers Lake showed a variation of no more than three times, especially when errors in the radiocarbon time-scale were considered. This inherent variability is due to the sedimentary processes within a lake, and the tendency for fine sediment to accumulate in deeper water, where the rate may be ten times the rate in the rest of the lake.

Several investigators have described variations in pollen influx in fossil cores or modern samples due to changes in sedimentation rates and processes rather than to vegetational changes. They are generally interpretable in terms of changes in sedimentation patterns as the lake basin becomes filled with sediment, or changes, such as deforestation, occurring in the watershed.

i) Waddington (1969) demonstrated at Rutz Lake, on the forest/prairie boundary in Minnesota, U.S.A., that pollen influx decreased during the time of climatic dryness which resulted in expansion eastwards of the prairie between 8000 and 5000 B.P. However, as the climate became cooler and moister again and forest re-occupied the landscape, pollen influx did not increase as expected, but it increased suddenly somewhat later (Fig. 10.15). Waddington suggested that lake level was low during the prairie period, zone Rz III, and hence there was a relatively large amount of marginal sediment transferred to the lake centre by erosion. During zone Rz IVA when forest became re-established, the lake became deeper, and sediment was retained beneath the thermocline at the margins. Therefore the pollen reaching the sediment in the centre most closely resembled that falling on to the lake surface from the air. As sedimentation continued at the edges, the sediment surface was raised to the level of

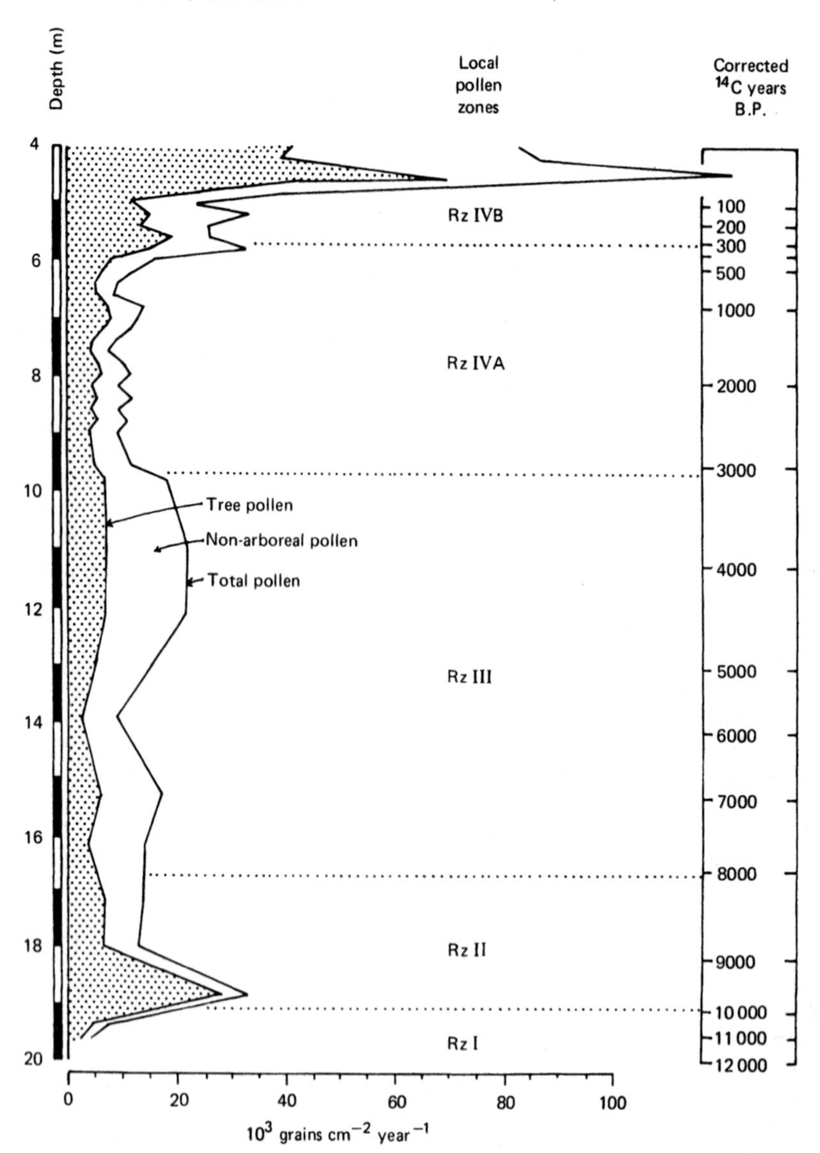

Fig. 10.15 Total pollen influx in Rutz Lake, Minnesota. (After Waddington, 1969.)

the epilimnion, which resulted in the rapid redeposition of material to the centre during zone Rz IVB, by the mechanism proposed by Davis (1968).
ii) A similar phenomenon was described from the deepest part of Frains Lake, Michigan, U.S.A. by Davis *et al.* (1973). Sedimentation appears to have been uniform throughout the lake up to mid-postglacial times. Just before European settlement, sediment started to accumulate 2–3 times faster in the centre, and this rate increased to about ten times following settlement. Pollen percentages did not differ from marginal sediments, suggesting that there had been a mass movement of total sediment to the centre of the lake (Davis, 1973).
iii) Maher (1977) demonstrated that similar processes occur even in the biggest lakes. He analysed seven cores from the western arm of Lake Superior, and found the highest pollen concentration in the

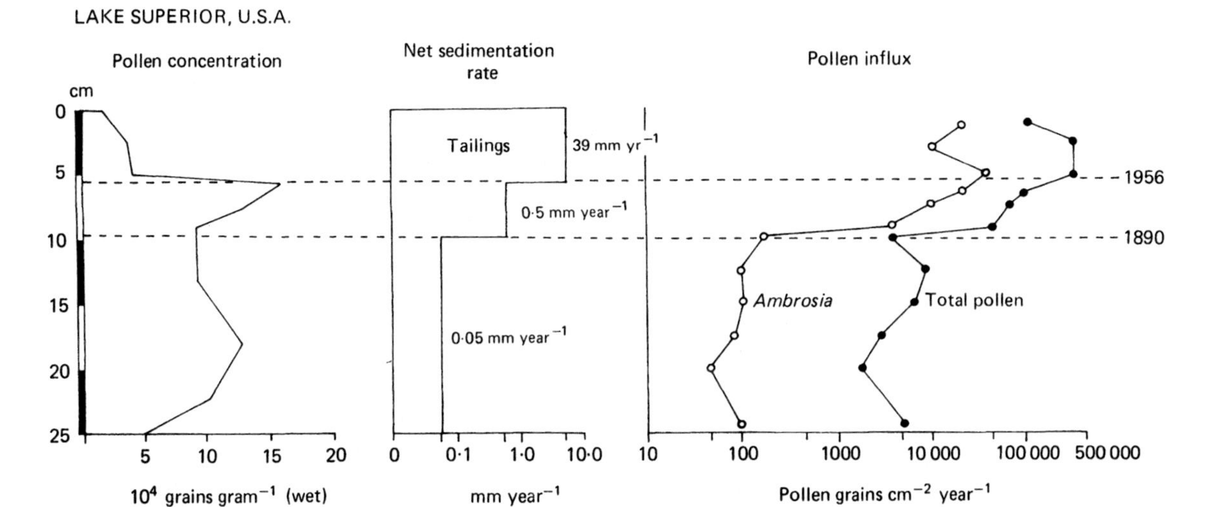

Fig. 10.16 Pollen concentration, net sedimentation rate, and pollen influx in the upper sediments of the western arm of Lake Superior, showing the influence of European settlement in 1890, and the addition of non-polleniferous taconite tailings since 1956. (After Maher, 1977.)

deepest water. This was due to selective movement of the finest particles. In all the cores (Fig. 10.16) European settlement of the area in 1890 was marked by increases in pollen concentration and sediment accumulation rate, as a result of soil disturbance leading to inwashing of soils and their contained pollen. A great increase in sediment accumulation rate accompanied by a fall in pollen concentration followed the input of non-polleniferous taconite tailings from the iron ore processing plant on the north shore. However, the net pollen influx remained similar to the post-settlement level.

iv) McAndrews and Power (1973) systematically investigated pollen concentration in surface sediments of another of the Great Lakes, Lake Ontario. They demonstrated the movement of finer particles, including pollen, to deeper water. Coarser sediments with little or no pollen were deposited in shallower areas. However, the percentages of the pollen types were not related to water depth, showing that all sediments contained a reasonable representation of the regional pollen rain. Differences within the lake were attributable to the inflow of rivers draining large areas of differing vegetation, such as coniferous/deciduous forests to the north, and deciduous forests and agricultural land to the south. McAndrews and Power concluded that the majority of the pollen in Lake Ontario originated from river transport, rather than direct aerial deposition.

v) Birks (1976) described a period of abnormally high pollen influx at the beginning of the *Picea* zone in the Wisconsinan late-glacial sediments at Wolf Creek, Minnesota. Because the percentage composition and pollen concentration did not change from previous values, he concluded that there had been mass transport of sediment from the lake margins. This conclusion was supported by finds of a few pollen grains of taxa characteristic of the treeless environment of the previous zone, suggesting reworking and redeposition of older sediments.

vi) R. B. Davis *et al.* (1975) compared post-glacial pollen influx from Moulton Pond, Maine, and Rogers Lake, Connecticut, and demonstrated an over-all similarity (Fig. 10.17). However, they proposed that because the pollen percentages were so consistent during the mid- and late-post-glacial, the changes in influx could be ascribed to limnological changes, which were probably similar in the two closely similar lakes.

vii) Lehman (1975) modelled patterns of sediment deposition for different shapes of lake basin, and showed (Fig. 10.18) how the different shapes accumulated sediment at different rates depending upon water depth and total basin depth. At Mirror Lake, New Hampshire, U.S.A., Likens and Davis (1975) found that sediment accumulation rate declined after 4500 B.P. even though other evidence suggested stable conditions in the watershed. Lehman fitted Mirror Lake into his models, and concluded that the sediment influx to the lake had indeed remained constant; it had only appeared to decline at the coring point because the gradual filling of the basin had resulted in a more even deposition of sediment over the basin.

2. *Variation in pollen influx between lakes*

a) *Variation in modern influx between lakes in the same vegetation type.* Davis *et al.* (1973) selected 29 lakes from the five major forest regions of Michigan, from the deciduous forests in the south to the deciduous/coniferous forests in the north. They converted pollen concentration measurements from short cores into influx data, using the time-scale provided by the *Ambrosia* pollen rise, which in that region corresponds to European settlement at about 1830. The results are shown in Table 10.8.

Within each group of lakes, the variation in influx is due to differences unrelated to variation in vegetation. In general, the between-lake variability is three or four times, with a few lakes differing by up to six times. This variation is of the same order of magnitude as that within a single lake (p. 219). Within these limits, lakes surrounded by the same vegetation type record similar influx values.

Davis (1967c) recorded a similar level of variation in pollen influx collected in underwater traps in three lakes of widely differing morphometry in New England. Within one lake, the traps caught comparable numbers of pollen grains in one year. However, there could be considerable variability from year to year, perhaps depending on the amount of sediment stirred up at overturn, and other factors in the watershed. Between the lakes, the highest pollen influx was recorded in the smallest lake, and the lowest in the largest, most complex lake. This direction of variation may be related to the greater length of shore per unit area of surface. Davis demonstrated that the proportions of the pollen types in all the samples were similar, supporting the assumption that a pollen spectrum from a core is representative of the pollen spectrum for the whole region.

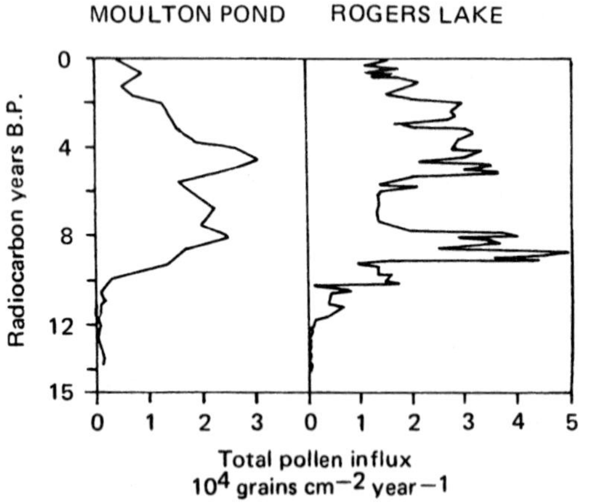

Fig. 10.17 A comparison of total pollen influx in postglacial sediments at Moulton Pond, Maine, and Rogers Lake, Connecticut. (After R. B. Davis *et al.*, 1975.)

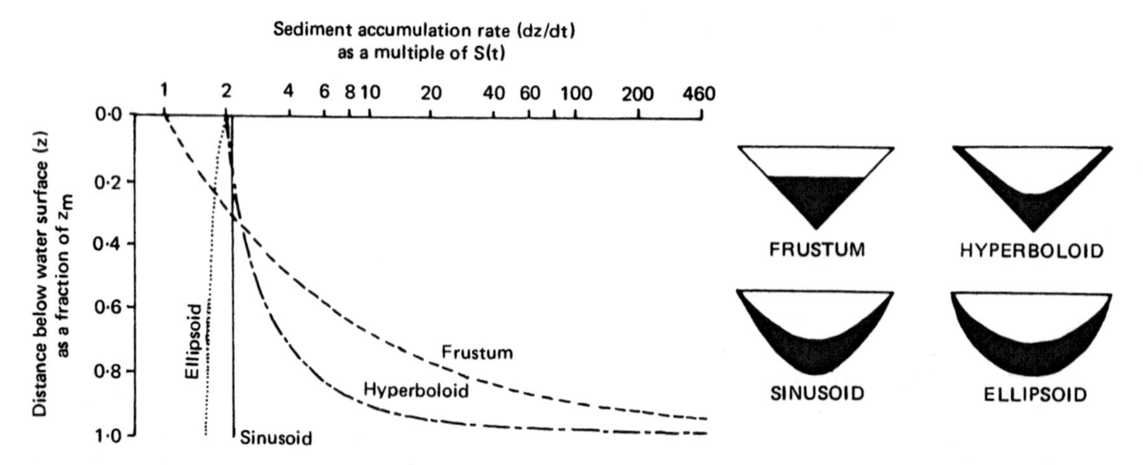

Fig. 10.18 Predictions for rates of sediment accumulation for four basin shapes shown on right, as a function of the distance of the sediment surface below the water surface (Z) as a fraction of total basin depth (Z_m) when the influx rates of sediment per unit lake surface ($S_{(t)}$) are constant with respect to time. Note the large variations in sediment accumulation rates in frustrum and hyperboloid basins even though sediment influx is constant. (After Lehman, 1975.)

Table 10.8 Total pollen influx at 29 lakes in 5 different vegetation regions of Michigan. (From Table 3, Davis *et al.*, 1973.)

Vegetation	Total pollen influx ($\times 10^3$) (grains cm^{-2} year^{-1})	$\bar{x}$ ($\times 10^3$)	S.D. ($\times 10^3$)
Northern hardwoods (Upper Peninsula)	49.2	34.2	27.4
	14.2		
	75.3		
	19.7		
	12.4		
Northern hardwoods (Lower Peninsula)	28.4	40.5	23.0
	25.1		
	24.2		
	46.2		
	78.3		
Pine region	44.3	46.8	28.7
	34.8		
	49.2		
	17.5		
	101.0		
	33.8		
Oak–hickory region	46.9	33.7	17.1
	22.5		
	27.1		
	68.8		
	35.2		
	31.1		
	22.8		
	15.0		
Beech–maple and ash–elm regions	20.6	18.3	2.2
	15.6		
	20.6		
	17.3		
	17.4		

b) *Variations in fossil influx between lakes.* Pennington (1973) discusses differences in pollen influx to different lakes in the English Lake District and Scotland. Four lakes are compared in Fig. 10.19 for their shape and size, and for their total pollen influx and influx of elm pollen just before the elm decline at 5500 B.P. It is clear that small lakes collect more pollen than larger ones. However, there is no exact relationship with size because other features interact, such as the degree of through-flow of water. The fact that Windermere and Loch a'Chroisg are essentially river lakes with a large through-flow of water probably explains the relatively low pollen influx to the sediments, because much of the pollen is washed through. In Windermere the influx of *Ulmus* pollen is much lower than in Loch Clair in Scotland. However, pollen percentages indicate that *Ulmus* formed a significant proportion of the forests around Windermere, whereas it was unlikely to have been locally present at all at Loch Clair. The *Ulmus* pollen deposition rate is thus a function of lake size and morphometry, and not a function of the size of the local population of *Ulmus* trees.

Pennington's study emphasized that, if comparisons between lakes are to be made on the basis of pollen influx, lakes of comparable size and morphometry should be selected, in order to minimize variations not due to the vegetation. However, Craig (1978) discovered that of two similar lakes in south-east Ireland, one accumulated pollen much faster than the other. This was probably a sedimentation feature, because the pollen concentration and percentages were both similar at the two sites.

3. *Variation in influx between lakes in different vegetation types*

a) *Calibration of total pollen influx.* Having assessed the variability in pollen influx which is due to factors other than the composition of the vegetation around the lake, Davis *et al.* (1973) were in a position to estimate any variation which *could* be attributed to the vegetation. Table 10.8 shows that, although there are general trends in influx, lakes in the pine region of Michigan having the highest values and lakes in the beech–maple and ash–elm regions having the lowest, the variation is so great that no statistically significant differences can be found. Total pollen influx can, however, distinguish between different vegetation formations, as shown in Table 10.9.

b) *Calibration of pollen influx of individual species.* Although not distinguishable by total pollen influx, different vegetation communities can be distinguished within one vegetation formation by examining the composition of the pollen spectra (see Chapter 11). Davis *et al.* (1973) attempted to reconstruct population sizes using the recent pollen influx values of individual taxa from their 29 lakes in Michigan, and comparing them with modern tree counts from plots (to give basal area) in an area of 576 square miles (14 mile radius) around each lake. Their results are shown in Fig. 10.20. It can be seen that there is a general tendency for the highest

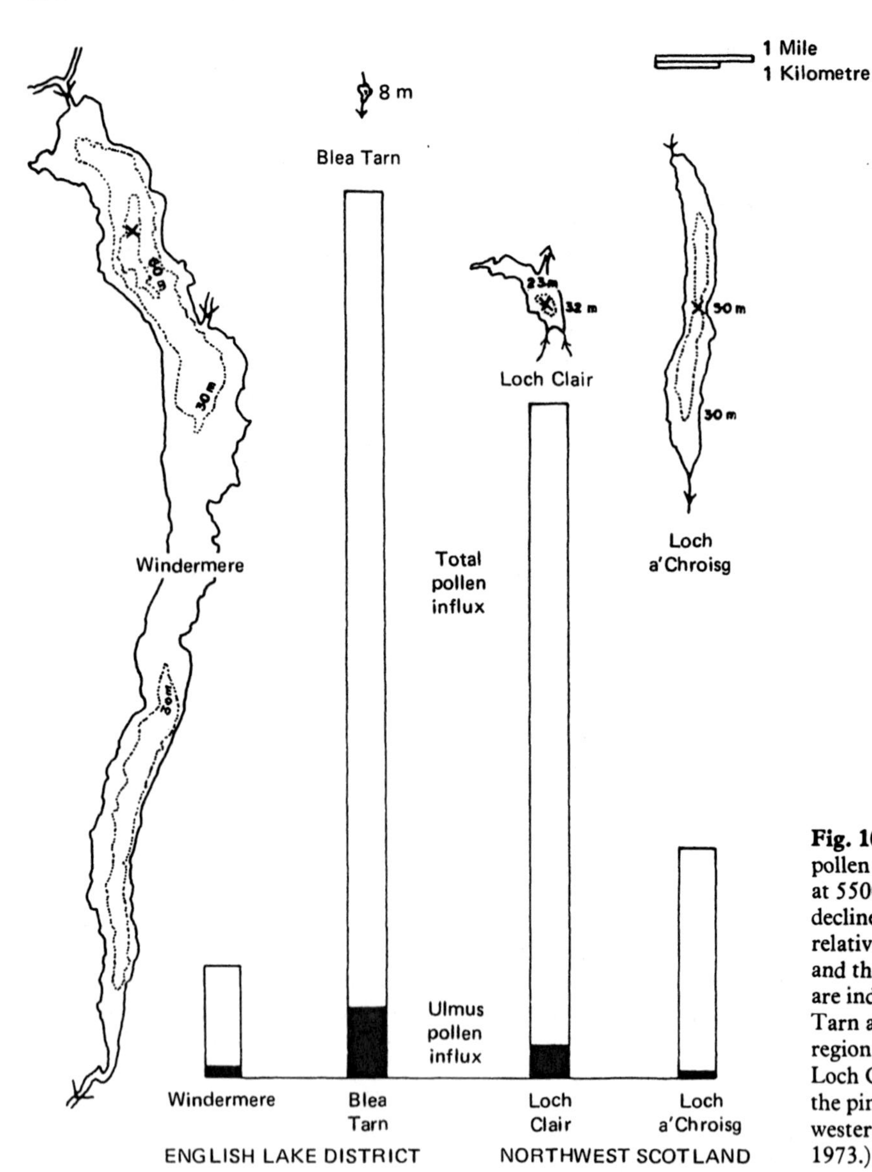

Fig. 10.19 Comparison of total pollen influx and *Ulmus* pollen influx at 5500 B.P. (just before the *Ulmus* decline) in four contrasting lakes. The relative sizes of the lakes are shown, and their depths and morphometry are indicated. Windermere and Blea Tarn are in the deciduous oak forest region of the English Lake District. Loch Clair and Loch a'Chroisg are in the pine forest with birch region of western Scotland. (After Pennington, 1973.)

pollen influx to be found in the regions of greatest basal area, with *Carya*, *Fagus*, and *Quercus* highest in the south, and *Betula*, *Tsuga*, *Acer*, and *Pinus* highest in the north. However, when the data were examined statistically, no significant correlations were found. This was disappointing, because a significant relationship would have then indicated a direct conversion factor between influx and population size. There may be several reasons for the lack of correlation; the high variance of pollen influx within and between lakes may mask the relationship; the influx values are averaged over 1830–1970, whereas the forests were surveyed in 1966, and there may have been changes in the forests since 1830, such as an increase in oak in the oak–pine areas due to fires; the high variance in the forest data may mask the relationship; trees in the open produce more pollen per crown than in closed forest, and hence the relationship will be altered in open areas, such as the agricultural land in the

Table 10.9 Summary of modern pollen influx values (grains cm^{-2} $year^{-1}$) for different vegetation types.

Vegetation	Source 1 Mean	Range	2 Mean	Range	3 Mean	Range
Tundra (sedge-moss)	34	5–65	–	–	–	–
Tundra (dwarf-shrub)	335	53–763	1500	1120–2000	1440	100–500
Forest-tundra	1680	275–6425	3710	1720–8870	6620	1000–15 000
Boreal forest	6800	2688–11 570	17 450	6520–30 500	13 700	2000–40 000
Mixed pine forest	–	–	–	–	46 800	17 500–101 000
Deciduous forest	–	–	8760	1980–50 160	37 300	14 200–78 300
Prairie	–	–	–	–	9500	1000–15 000

Sources: 1, Ritchie and Lichti-Federovich (1967). 2, R. B. Davis *et al.* (1975) Table 1. 3, M. B. Davis *et al.* (1973) Table 4.

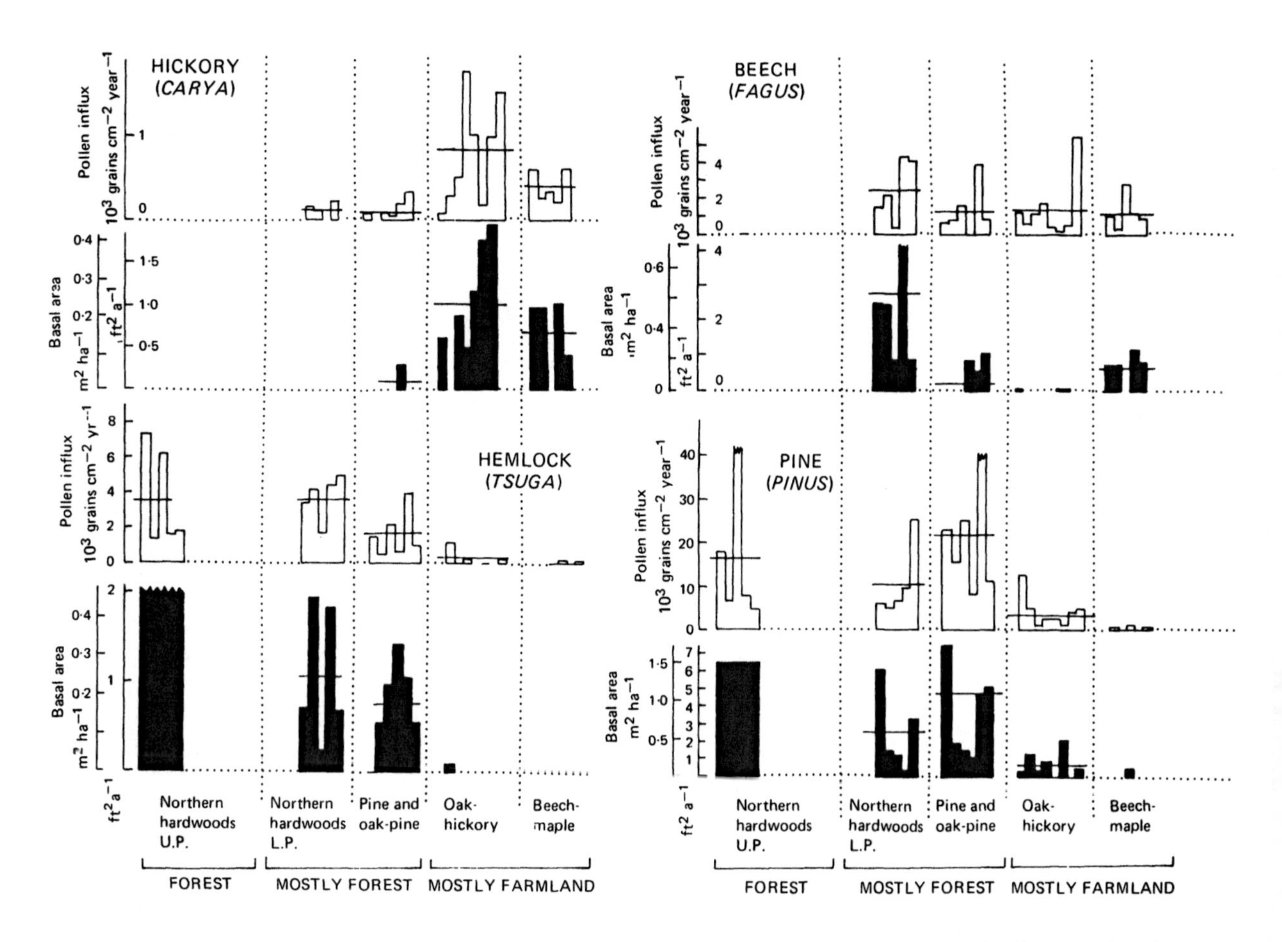

Fig. 10.20 Modern pollen influx for *Carya*, *Tsuga*, *Fagus*, and *Pinus* at 29 lakes in Michigan (U.P., Upper Peninsula; L.P., Lower Peninsula) compared with the basal area of the species per unit area of the landscape in the vicinity of each lake. Mean values for each group of lakes are indicated by a horizontal line. (After Davis *et al.*, 1973.)

south. So although some relationship is evident from visual inspection of Fig. 10.20, it cannot be quantified, and further work is required before population sizes can be reconstructed with any accuracy.

The data of Davis *et al.* (1973) also give information on pollen dispersal, particularly for *Tsuga*, *Fagus*, *Carya*, and *Pinus*, which all have range limits across Michigan. Their pollen is dispersed beyond their limits causing substantial pollen influx where trees are absent. *Carya* pollen shows a clear gradient north of its limit in Fig. 10.20. These influx numbers could be calibrated to estimate the distance of a site from the edge of a species boundary during the past.

Reconstruction of fossil populations at Rogers Lake

Within the limits of their data, Davis *et al.* (1973) attempted to reconstruct past populations of different tree species at Rogers Lake by comparing the fossil pollen influx values with the modern values obtained from Michigan. They had varying degrees of success, depending on the preciseness of the calibration for each taxon. ***Pinus*** and ***Betula*** were problematical, because several species could have contributed to each pollen type. Examples are shown in Fig. 10.21.

After its pollen influx maximum, ***Picea*** values fall to about 300 grains cm^{-2} $year^{-1}$ after 8000 B.P. Its population can be approximately estimated from the Michigan data, where *Picea* influx is about 600 grains cm^{-2} $year^{-1}$ where the tree has a frequency of about 1.2 m^2 ha^{-1} in the Upper Peninsula of Michigan, and about 200 grains cm^{-2} $year^{-1}$ in lakes where *Picea* occupies about 0.05 m^2 ha^{-1}. The influx curves for *Fagus* and *Carya* illustrate the increase in pollen influx before the tree becomes locally present, when pollen values pass the 'rational limit'. Between 12 400 and 8000 B.P. *Fagus* influx is 50–100, which resembles values in the northern hardwoods of the Upper Peninsula of Michigan, 50 km west of the present range of *Fagus*. Its influx rose gradually to 200 grains cm^{-2} $year^{-1}$ at about 8000 B.P., still lower than values indicating its local presence. Influx rose gradually at 6000 B.P. reaching a maximum of 1500 grains cm^{-2} $year^{-1}$ at

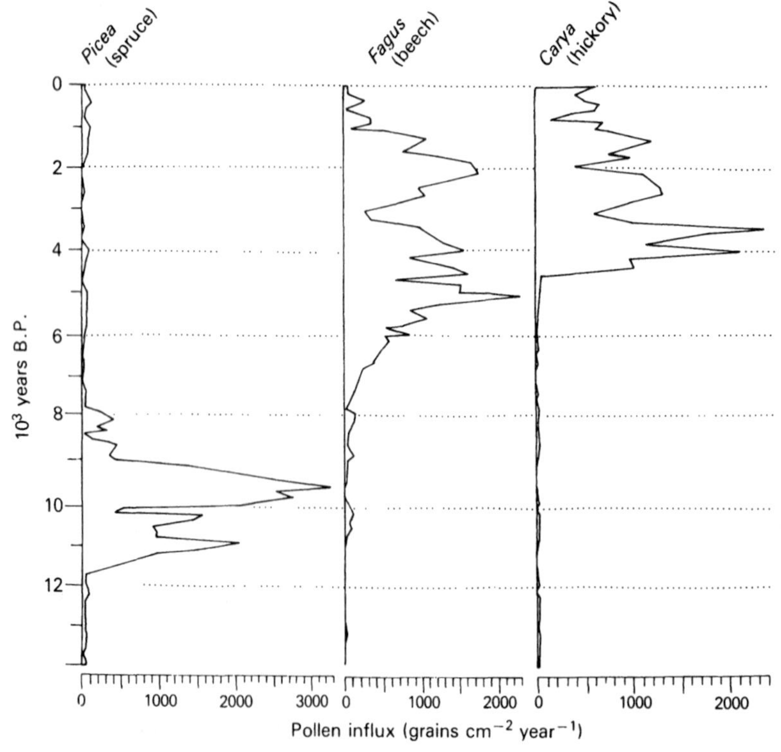

Fig. 10.21 Curves of pollen influx for *Picea*, *Fagus*, and *Carya* at Rogers Lake, Connecticut. For discussion, see text.

5000 B.P. This value corresponds to the abundance of *Fagus* in the northern hardwoods region of lower Michigan. *Carya* immigrates more rapidly. Before 6000 B.P. its influx is very low, about 50 grains cm^{-2} $year^{-1}$, similar to values in northern Michigan, 250 km north of its modern limit. By 5500 B.P. its influx had risen to 200–300 grains cm^{-2} $year^{-1}$, slightly higher than in central Michigan, just north of its present limit. By 4000 B.P., its influx had risen steeply to 1000–2000 grains cm^{-2} $year^{-1}$. This is higher than in the oak–hickory region of Michigan today, where the average is 800 grains cm^{-2} $year^{-1}$. The rational limit for *Carya* was passed at about 4000 B.P. indicating its local arrival at this time. The time of local arrival at Rogers Lake can be estimated for various taxa.

Picea . . . . . 11 500 B.P.
Pinus . . . . . . 8500 B.P.
Tsuga . . . . . . 8500 B.P.
Quercus . . . . 8500 B.P.
Fagus . . . . . 5000 B.P.
Carya . . . . . 4000 B.P.
Castanea . . . 2000 B.P.

Davis (1976) extended this approach to other sites in North America, and was able to reconstruct migration maps for the various trees (Fig. 10.13), as we have discussed earlier in this chapter.

An evaluation of absolute pollen frequency

There are several disadvantages and difficulties inherent in the study of pollen concentration and pollen influx.

1. The method is expensive if many radiocarbon dates are required.
2. It is more time consuming than percentage pollen analysis. However, the extra time involved is relatively small, particularly using the methods of addition of exotic pollen.
3. The fossil time scale is difficult to calibrate in terms of calendar years, partially because of inherent inaccuracies in the radiocarbon measurements, and partly because of the, as yet, not fully understood variations in cosmic ray activity and hence variation in the production of ^{14}C in the atmosphere, particularly before 2000 B.P.
4. Pollen concentration and influx can vary widely between adjacent samples in a core, and it can be difficult to detect significant trends in the curves, especially when no standard errors are given for the measurements.
5. Pollen influx in lakes varies with factors not associated with the surrounding vegetation. These include sedimentary processes within a lake, and differences between lakes due to lake size and morphometry. Some lakes are efficient pollen collectors (pollen-rich lakes) and others are less efficient (pollen-poor lakes). This variation has to be assessed before comparisons can usefully be made between sites.
6. Because of the inherent limnological variation, it is difficult to calibrate pollen influx in terms of vegetation and populations of individual species (cf. Walker and Wilson, 1978).

To balance these disadvantages, the study of absolute pollen frequency has been successful in illuminating several situations where pollen percentages were inadequate.

1. Pollen influx is independent for each species, unlike percentages. This has been invaluable in interpreting percentage diagrams from treeless situations, where low local pollen production leads to over-representation of distant taxa in the percentage diagram, which gives a false impression of the past vegetation. Differences in forest vegetation on different soils can also be detected where pollen percentages give little indication of any variation. Similarly, changes in the luxuriance and flowering of arctic vegetation can be shown, whilst the percentage pollen composition remains constant.
2. The study of pollen influx has led to a greater understanding of limnological sedimentary processes.
3. Because of the independence of the measurements, pollen influx values should be proportional to the number of plants on the landscape. Studies on modern influx have encountered difficulties in calibrating pollen influx, but a general relationship has been found to exist, and fossil diagrams can tentatively be interpreted in terms of past plant populations.
4. The relationship of influx to population allows the discernment of local, nearby, and distant presence of a taxon in relation to the site. Therefore the migration of species across a country or a continent can be traced, and changes in treelines can be assessed.
5. Pollen influx throws light on past vegetation dynamics, showing how a population can build up from a few pioneer plants to a maximum abundance. Such long-term studies of forest dynamics are otherwise impossible.
6. Pollen influx can be used to construct a time

scale if this is otherwise impossible (e.g. for interglacials and volcanic ash horizons).

In spite of the inaccuracies and complications involved, the study of pollen concentration and influx has added a new dimension to the study of vegetation history. Important new information can be gained from carefully designed investigations.

References

Long distance pollen and the reconstruction of past floras

BIRKS, H. J. B. (1973). *Past and Present Vegetation of the Isle of Skye – a Palaeoecological Study*. Cambridge University Press (Chapter 12).

CHURCHILL, D. M. (1973). The ecological significance of tropical mangroves in the early Tertiary floras of southern Australia. *Spec. Publ. Geol. Soc. Australia*, **4**, 79–86.

FREDSKILD, B. (1973). Studies in the vegetational history of Greenland. *Meddr. om Grønland*, **198**(4), 245 pp.

HAFSTEN, U. (1960). Pleistocene development of vegetation and climate in Tristan da Cunha and Gough Island. *Årbok Univ. Bergen, Mat-Naturv.*, Serie Nr **20**, 48 pp.

The measurement of correction factors, pollen production factors, and pollen representation factors (R-values)

ANDERSEN, S. T. (1967). Tree-pollen rain in a mixed deciduous forest in South Jutland (Denmark). *Rev. Palaeobot. Palynol.*, **3**, 267–75.

ANDERSEN, S. T. (1970). The relative pollen productivity and representation of North European trees, and correction for tree pollen spectra. *Danm. geol. Unders.*, Ser. II, **96**, 99 pp.

ANDERSEN, S. T. (1974). Wind conditions and pollen deposition in a mixed deciduous forest. II. Seasonal and annual pollen deposition 1967–1972. *Grana*, **14**, 64–77.

DABROWSKI, M. J. (1975). Tree pollen rain and the vegetation of the Białowieza National Park. *Biul. Geolog. Polon.*, **19**, 157–72.

DAVIS, M. B. (1963). On the theory of pollen analysis. *Am. J. Sci.*, **261**, 897–912.

FAEGRI, K. AND IVERSEN, J. (1975). *Textbook of Pollen Analysis* (3rd edition). Blackwell.

The use of correction factors and representation factors in Quaternary palaeoecology

ANDERSEN, S. T. (1973). The differential pollen productivity of trees and its significance for the interpretation of a pollen diagram from a forested region. In *Quaternary Plant Ecology* (eds H. J. B. Birks and R. G. West). Blackwell.

ANDERSEN, S. T. (1975). The Eemian freshwater deposit at Egernsund, South Jylland, and the Eemian landscape development in Denmark. *Danm. geol. Unders. Årbog*, 1974, 49–70.

ANDERSEN, S. T. (1978). Local and regional vegetational development in eastern Denmark in the Holocene. *Danm. geol. Unders. Årbog*, 1976, 5–27.

BAKER, C. A., MOXEY, P. A. AND OXFORD, P. M. (1978). Woodland continuity and change in Epping Forest. *Field Studies*, **4**, 645–69.

DAVIS, M. B. (1967a). Pollen accumulation rates at Rogers Lake, Connecticut, during late- and post-glacial time. *Rev. Palaeobot. Palynol.*, **2**, 219–30.

FAEGRI, K. (1966). Some problems of representativity in pollen analysis. *Palaeobotanist*, **15**, 135–40.

IVERSEN, J. (1964). Retrogressive vegetational succession in the post-glacial. *J. Ecol.*, **52** (Suppl.), 59–70.

JANSSEN, C. R. (1967). A comparison between the recent regional pollen rain and the sub-recent vegetation in four major vegetation types in Minnesota (U.S.A.). *Rev. Palaebot. Palynol.*, **2**, 331–42.

LIVINGSTONE, D. A. (1968). Some Interstadial and Postglacial pollen diagrams from Eastern Canada. *Ecol. Monogr.*, **38**, 87–125.

LIVINGSTONE, D. A. AND ESTES, A. H. (1967). A carbon-dated pollen diagram from the Cape Breton Plateau, Nova Scotia. *Can. J. Bot.*, **45**, 339–59.

TAUBER, H. (1965). Differential pollen dispersion and the interpretation of pollen diagrams. *Danm. geol. Unders.*, Ser. II, **89**, 69 pp.

Absolute pollen frequency – general considerations

BIRKS, H. J. B. (1976). Late-Wisconsinan vegetational history at Wolf Creek, Central Minnesota. *Ecol. Monogr.*, **46**, 395–429.

COLINVAUX, P. (1978). On the use of the word 'absolute' in pollen statistics. *Quat. Res.*, **9**, 132–3.

DAVIS, M. B. (1969a). Palynology and environmental history during the Quaternary Period. *Amer. Sci.*, **57**, 317–32.

LIVINGSTONE, D. A. (1969). Communities of the past. In *Essays in Plant Geography and Ecology* (ed. K. M. H. Greenidge). Nova Scotia Museum.

MAHER, L. J. (1972). Absolute pollen diagram of Redrock Lake, Boulder County, Colorado. *Quat. Res.*, **2**, 531–53.

Methods of determining pollen concentration and pollen influx

BENNINGHOFF, W. S. (1962). Calculation of pollen and spore density in sediments by addition of exotic pollen in known quantities. *Pollen Spores*, **4**, 332–3.

BONNY, A. P. (1972). A method for determining absolute pollen frequencies in lake sediments. *New Phytol.*, **71**, 391–403.

BOROWIK-DABROWSKA, M. AND DABROWSKI, M. J. (1972). Absolute pollen concentration. *Bull. l'Acad. Sci. Polon.*, **20**, 129–32.

BRUBAKER, L. B. (1975). Post-glacial forest pattern associated with till and outwash in Northcentral Upper Michigan. *Quat. Res.*, **5**, 499–527.

CRAIG, A. J. (1972). Pollen influx to laminated lake sediments: a pollen diagram from northeastern Minnesota. *Ecology*, **53**, 46–57.

DAVIS, M. B. (1965). A method for determination of absolute pollen frequency. In *Handbook of Paleontological Techniques* (eds B. Kummel and D. M. Raup). W. H. Freeman.

DAVIS, M. B. (1966). Determination of absolute pollen frequency. *Ecology*, **47**, 310–11.

DAVIS, M. B. (1967a). Pollen accumulation rates at Rogers Lake, Connecticut, during late- and postglacial time. *Rev. Palaeobot. Palynol.*, **2**, 219–30.

ERDTMAN, G. (1943). An introduction to pollen analysis. *Chronica Botanica*. Waltham, Mass.

JØRGENSEN, S. (1967). A method of absolute pollen counting. *New Phytol.*, **66**, 489–93.

MATTHEWS, J. (1969). The assessment of a method for the determination of absolute pollen frequencies. *New Phytol.*, **68**, 161–6.

MULLER, J. (1959). Palynology of recent Orinoco delta and shelf sediments. *Micropaleontology*, **5**, 1–32.

RENFREW, C. (1973). *Before Civilisation*. Penguin Books.

STOCKMARR, J. (1971). Tablets with spores used in absolute pollen analysis. *Pollen Spores*, **13**, 615–21.

SUESS, H. E. (1970). Bristlecone pine calibration of the radiocarbon time scale 5200 B.C. to the present. In *Radiocarbon Variations and Absolute Chronology* (ed. I. U. Olsson). Almqvist and Wiksell.

SWAIN, A. M. (1973). A history of fire and vegetation in northeastern Minnesota as recorded in lake sediments. *Quat. Res.*, **3**, 383–96.

Pollen influx at Rogers Lake, Connecticut

DAVIS, M. B. (1967a). Pollen accumulation rates at Rogers Lake, Connecticut during late- and postglacial time. *Rev. Palaeobot. Palynol.*, **2**, 219–30.

DAVIS, M. B. (1969b). Climatic changes in southern Connecticut recorded by pollen deposition at Rogers Lake. *Ecology*, **50**, 409–22.

DAVIS, M. B. AND DEEVEY, E. S. (1964). Pollen accumulation rates: estimates from late-glacial sediment of Rogers Lake. *Science*, **145**, 1293–5.

The use of absolute pollen frequency in palaeoecological studies

BIRKS, H. H. AND MATHEWES, R. W. (1978). Studies in the vegetational history of Scotland. V. Late-Devensian and early Flandrian pollen and macrofossil stratigraphy at Abernethy Forest, Inverness-shire. *New Phytol.*, **80**, 455–84.

BIRKS, H. J. B. (1976). Late-Wisconsinan vegetational history at Wolf Creek, Central Minnesota. *Ecol. Monogr.*, **46**, 395–429.

BRUBAKER, L. B. (1975). Post-glacial forest patterns associated with till and outwash in Northcentral Upper Michigan. *Quat. Res.*, **5**, 499–527.

CRAIG, A. J. (1972). Pollen influx to laminated sediments: a pollen diagram from northeastern Minnesota. *Ecology*, **53**, 46–57.

CRAIG, A. J. (1978). Pollen percentage and influx analyses in south-east Ireland: a contribution to the ecological history of the late-glacial period. *J. Ecol.*, **66**, 297–324.

CWYNAR, L. C. (1978). Recent history of fire and vegetation from laminated sediment of Greenleaf Lake, Algonquin Park, Ontario. *Can. J. Bot.*, **56**, 10–21.

DABROWSKI, M. J. (1971). Palynochronological materials – Eemian interglacial. *Bull. l'Acad. Sci. Polon.*, **19**, 29–36.

DAVIS, M. B. (1967b). Late-glacial climate in northern United States: a comparison of New England and the Great Lakes Region. In *Quaternary Palaeoecology* (eds E. J. Cushing and H. E. Wright). Yale University Press.

DAVIS, M. B. (1976). Pleistocene biogeography of temperate deciduous forests. *Geoscience and Man*, **13**, 13–26.

DAVIS, M. B., BRUBAKER, L. B. AND WEBB, T. (1973). Calibration of pollen influx. In *Quaternary Plant Ecology* (eds H. J. B. Birks and R. G. West). Blackwell.

DAVIS, R. B., BRADSTREET, T. E., STUCKENRATH, R. AND BORNS, H. W. (1975). Vegetation and associated environments during the past 14 000 years near Moulton Pond, Maine. *Quat. Res.*, **5**, 435–65.

FREDSKILD, B. (1969). A postglacial standard pollen diagram from Peary Land, North Greenland. *Pollen Spores*, **11**, 573–85.

FREDSKILD, B. (1973). Studies in the vegetational history of Greenland. *Meddr. om Grønland*, **198**(4), 245 pp.

FREDSKILD, B. (1975). A late-glacial and early post-glacial pollen-concentration diagram from Langeland, Denmark. *Geol. Fören. Förhandl. Stock.*, **97**, 151–61.

HICKS, S. (1974). A method of using modern pollen rain values to provide a time-scale for pollen diagrams from peat deposits. *Memor. Soc. Fauna Flora Fennica*, **49**, 21–33.

HICKS, S. (1975). Variations in pollen frequency in a bog at Kangerjoki, N. E. Finland during the Flandrian. *Commen. Biol.*, **80**, 28 pp.

HYVÄRINEN, H. (1975). Absolute and relative pollen diagrams from northernmost Fennoscandia. *Fennia*, **142**, 23 pp.

HYVÄRINEN, H. (1976). Flandrian pollen deposition rates and tree-line history in northern Fennoscandia. *Boreas*, **5**, 163–75.

JACOBSON, G. L. (1975). *A Palynological Study of the History and Ecology of White Pine in Minnesota*. Ph. D. thesis, University of Minnesota.

KENDALL, R. L. (1969). An ecological history of the Lake Victoria basin. *Ecol. Monogr.*, **39**, 121–76.

MACK, R. N., RUTTER, N. W. AND VALASTRO, S. (1978). Late Quaternary pollen record from the Sanpoil River Valley, Washington. *Can. J. Bot.*, **56**, 1642–50.

MAHER, L. J. (1972). Absolute pollen diagram of Redrock Lake, Boulder County, Colorado. *Quat. Res.*, **2**, 531–53.

MEHRINGER, P. J., BLINMAN, E. AND PETERSEN, K. L. (1977). Pollen influx and volcanic ash. *Science*, **198**, 257–61.

MÜLLER, H. (1974). Pollenanalytische Untersuchungen und Jahresschichtenzählungen an der eem – zeitlichen Kieselgur von Bispingen/Luhe. *Geol. Jb.*, **A21**, 149–69.

PENNINGTON, W. (1975). The effect of Neolithic man on the environment of north-west England: the use of absolute pollen diagrams. In *The Effect of Man on the Landscape: the Highland zone* (eds J. G. Evans, S. Limbrey and H. Cleere). Council for British Archaeology.

PENNINGTON, W. AND BONNY, A. P. (1970). Absolute pollen diagram from the British Late-glacial. *Nature*, **226**, 871–3.

PENNINGTON, W. AND SACKIN, M. J. (1975). An application of principal components analysis to the zonation of two Late-Devensian profiles. *New Phytol.*, **75**, 419–53.

SWAIN, A. M. (1973). A history of fire and vegetation in northeastern Minnesota as recorded in lake sediments. *Quat. Res.*, **3**, 383–96.

SWAIN, A. M. (1978). Environmental changes during the past 2000 years in North-Central Wisconsin: analysis of pollen, charcoal, and seeds from varved lake sediments. *Quat. Res.*, **10**, 55–68.

WATTS, W. A. (1973). Rates of change and stability in vegetation in the perspective of long periods of time. In *Quaternary Plant Ecology* (eds H. J. B. Birks and R. G. West). Blackwell.

WRIGHT, H. E. (1968). The roles of pine and spruce in the forest history of Minnesota and adjacent areas. *Ecology*, **49**, 937–55.

Past and present pollen deposition rates and sedimentary processes, and the calibration of pollen influx

CRAIG, A. J. (1978). Pollen percentage and influx analyses in south-east Ireland: a contribution to the ecological history of the late-glacial period. *J. Ecol.*, **66**, 297–324.

DAVIS, M. B. (1967c). Pollen deposition in lakes as measured by sediment traps. *Bull. Geol. Soc. Amer.*, **78**, 849–58.

DAVIS, M. B. (1968). Pollen grains in lake sediments: redeposition caused by seasonal water circulation. *Science*, **162**, 796–9.

DAVIS, M. B. (1973). Redeposition of pollen grains in lake sediment. *Limnol. Ocean.*, **18**, 44–52.

DAVIS, M. B., BRUBAKER, L. B. AND WEBB, T. (1973). Calibration of pollen influx. In *Quaternary Plant Ecology* (eds H. J. B. Birks and R. G. West). Blackwell.

DAVIS, R. B. AND WEBB, T. (1975). The contemporary distribution of pollen in Eastern North America: a comparison with the vegetation. *Quat. Res.*, **5**, 395–434.

DONNER, J. J. (1972). Pollen frequencies in the Flandrian sediments of Lake Vakojärvi, South Finland. *Commen. Biol.*, **53**, 19 pp.

LEHMAN, J. T. (1975). Reconstructing the rate of accumulation of lake sediment: the effect of sediment focusing. *Quat. Res.*, **5**, 541–50.

LIKENS, G. E. AND DAVIS, M. B. (1975). Post-glacial history of Mirror Lake and its watershed in New Hampshire, U.S.A.: an initial report. *Verh. Int. Verein. Limnol.*, **19**, 982–93.

MAHER, L. J. (1977). Palynological studies in the Western arm of Lake Superior. *Quat. Res.*, **7**, 14–44.

MCANDREWS, J. H. AND POWER, D. M. (1973). Palynology of the Great Lakes: the surface sediments of Lake Ontario. *Can. J. Earth Sci.*, **10**, 777–92.

PENNINGTON, W. (1973). Absolute pollen frequencies in the sediments of lakes of different morphometry. In *Quaternary Plant Ecology* (eds H. J. B. Birks and R. G. West). Blackwell.

RITCHIE J. C. AND LICHTI-FEDEROVICH, S. (1967). Pollen dispersal phenomena in Arctic-Subarctic Canada. *Rev. Palaeobot. Palynol.*, **3**, 255–66.

WADDINGTON, J. C. B. (1969). A stratigraphic record of the pollen influx to a lake in the Big Woods of Minnesota. *Geol. Soc. Amer. Spec. Paper*, **123**, 263–82.

Interpretation of A.P.F. data in terms of plant population sizes and population dynamics

DAVIS, M. B. (1976). Pleistocene biogeography of temperate deciduous forests. *Geoscience and Man*, **13**, 13–26.

WALKER, D. AND WILSON, S. R. (1978). A statistical alternative to the zoning of pollen diagrams. *J. Biogeogr.*, **5**, 1–21.

WATTS, W. A. (1973). Rates of change and stability in vegetation in the perspective of long periods of time. In *Quaternary Plant Ecology* (eds H. J. B. Birks and R. G. West). Blackwell.

11

The reconstruction of past plant communities

Introduction

In this chapter we shall discuss various approaches which have been developed to answer our third question in the interpretation of pollen diagrams: what plant communities were present?

A plant community can be defined as 'an assemblage of plants living together in the same place during the same time in relatively stable proportions'. Pollen analysis can be used primarily for reconstructing communities in terms of their species composition. However, it is unusual to be able to reconstruct individual communities because fossil pollen spectra are usually integrated from an area containing several different communities (see Chapter 1).

Past plant communities can be reconstructed in three main ways.

1. Direct evidence may be obtained when a community has been buried *in situ*. Such situations are rare. Examples include buried forests, either covered by a catastrophe such as a landslide, or by overgrowth by other plant communities (e.g. Munaut in Godwin, 1968; Birks, 1975), waterlogged communities, where peat formation has preserved the succession of plant communities (e.g. Birks and Mathewes, 1978; Casparie, 1972; Rybníček and Rybníčková, 1968), and burial of communities by volcanic ash or lava.
2. Indirect evidence of past communities can be obtained by establishing the numerical relationships in time between fossil taxa, and defining ***recurrent groups*** (see Chapter 2). If the fossils are associated in time, it is possible that they were also associated in space as communities.
3. Indirect evidence can also be obtained by finding modern analogues for past assemblages, either in the ecological behaviour of individual species or groups of species, or by comparison with pollen spectra produced by present vegetation units.

It is clear that past communities can not generally be reconstructed directly from pollen spectra. An exceptional case was the inwashing of a complete plant community, a moss-dominated spring community, into the sediments of Loch Fada, Isle of Skye (Birks, 1970). The inwashed layer contained a pollen spectrum resembling that from similar modern communities (Chapter 9, p. 191). Because of this major limitation, reconstruction of past communities from pollen analyses must necessarily be indirect, using one or more of the following broad methods.

i) The statistical approach, leading to the delimitation of ***recurrent groups***.
ii) The application backwards in time of known ecological and sociological preferences of taxa. Those with a well-defined narrow ecological tolerance can be used as ***indicator species***.
iii) The comparison of fossil pollen spectra with modern spectra from known vegetation types.

Approaches (ii) and (iii) depend upon finding modern analogues so that the present can be extrapolated back into the past, i.e. methodological uniformitarianism (see Chapter 1). We shall now discuss these three approaches in some detail.

The statistical approach

The model underlying this approach supposes that an assemblage of fossils consistently occurring together, or numerically correlated in a series of samples within and between stratigraphical sequences, represents a past life assemblage, or community. Such groups of fossil taxa are called ***interspecific associations*** or ***recurrent groups***. A mathematical method (see Chapter 2) is employed to calculate some measure of interspecific association for presence or absence data, or of interspecific correlation for quantitative data. From these measures, groups of associated or correlated taxa can be delimited.

This approach is used widely in pre-Quaternary palaeoecological studies particularly in the reconstruction of fossil animal communities, e.g. Reyment (1963). The applicability of various mathematical methods is reviewed by Buzas (1969). We have already discussed the delimitation of recurrent groups of macrofossils from Bláto mire, Czechoslovakia (Rybníček and Rybníčková, 1968) in Chapter 5. Plant macrofossils and animal fossils are frequently fossilized at or near the place they lived, and therefore the composition of recurrent groups is likely to resemble the original communities fairly closely, and meaningful palaeoecological conclusions can be drawn. In the case of pollen analysis, a pollen spectrum is a homogenized mixture of pollen from a range of communities near to and distant from the sampling site (Janssen, 1970). Therefore the delimitation of recurrent groups will only tend to distinguish assemblages of taxa in time, and there will be little spatial association. However, in pre-Quaternary pollen studies, the investigator has little option but to try every method at his disposal to unravel the palaeoecology of the past, and varying degrees of success have been attained.

Uses in pre-Quaternary pollen analysis

Oltz (1971) used the sum-of-squares agglomeration clustering procedure of Orloci (1967) using presence/absence data on the occurrence of pollen types in five cores from the Cretaceous and early Tertiary of Montana, U.S.A. Dendrograms from each site revealed clusters of similar composition in each, and he concluded that these groups may have ecological significance. Clapham (1970a,b, 1972) used different mathematical techniques to analyse pollen and spore data from several Upper Permian sites in Oklahoma. He calculated correlation bonds at 99% and 95% significance levels and was able to separate an upland and a lowland community from the Flowerpot formation. In the overlying Blaine and Dog Creek formations, the lowland community remained constant but the upland community changed gradually in the abundance of taxa and the addition of new ones. Harris and Norris (1972) tested the recurrent group approach by applying it to data from Quaternary sediments in New Zealand, in which the parent plants which produced the pollen and spores were known. They obtained several groups which had ecological and climatological meaning, and so they concluded that recurrent groups in pre-Quaternary sediments, where the parent plants are largely unknown, may be useful in community reconstruction.

Uses in Quaternary pollen analysis

The delimitation of recurrent groups is less frequently necessary in Quaternary studies, because the method of analogy with modern situations is

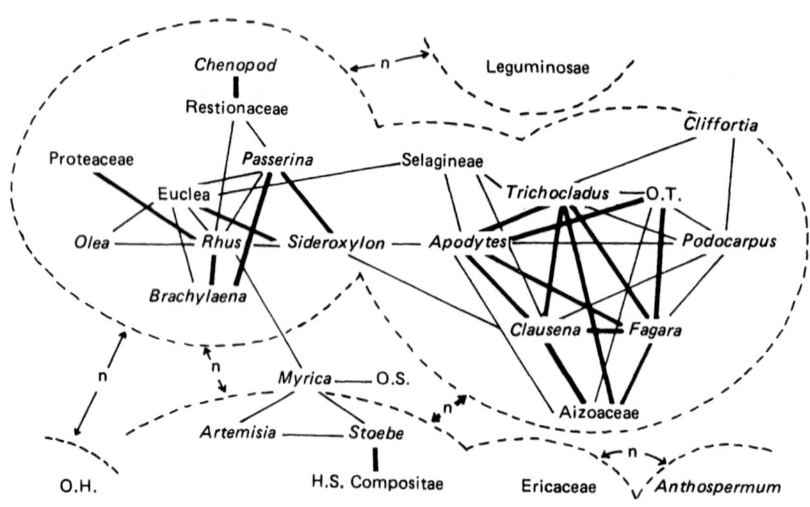

Fig. 11.1 Correlation net for pollen types at Groenvlei, S. Africa, based on product/moment correlation of the most abundant non-local pollen taxa. The remaining taxa are grouped as 'other trees' (O.T.), 'other shrubs' (O.S.), and 'other herbs' (O.H.). Leguminosae = Mimosaceae + Fabaceae; H.S. Compositae = high-spine Asteraceae. **—** strong positive correlation. — positive correlation. ⟨—n—⟩ negative correlation. ⟨**—n—**⟩ strong negative correlation. (After Martin, 1968.)

available. However, P. S. Martin and Mosimann (1965) attempted to establish correlations between pollen types within a 42 m core from south-eastern Arizona, which spanned at least the last glaciation. Mosimann (1964) overcame the problem of establishing correlations between pollen percentages, which are interdependent, by the use of the multivariate β distribution. In any sample the sum of the values of all the pollen types equalled one, which represents the percentage constraint. Thus, if one variable increases, another must decrease to compensate, and it will appear to be negatively correlated. The magnitude of this constraint is estimated by a constraint coefficient. This is compared with the observed correlation coefficient. If there is a significant difference, the two variables can be said to be positively or negatively correlated. As a result, groups of taxa which vary together can be delimited. Martin and Mosimann showed that *Pinus* was positively correlated only with *Picea* + *Abies* + *Pseudotsuga* pollen. *Artemisia*, Compositae, and Cheno-Am pollen formed another group. The pattern of correlation was interpreted as shifts between woodland and yellow-pine parkland through the Wisconsinan glaciation.

A. R. H. Martin (1968) delimited recurrent groups in a species-rich pollen diagram from South Africa, placed on an ecotone between forest and Cape heath, semi-arid vegetation. Figure 11.1 shows the resulting correlation net. The two largest groups represent *Podocarpus* forest and *Rhus* scrub. An *Artemisia*–*Stoebe*–high spine Compositae group representing sand-dune communities is strongly negatively correlated with the first two groups. The three groups characterized by Leguminosae, Ericaceae, and *Anthospermum* are all negatively correlated with all other groups. These taxa nowadays combine to form the interesting Cape heath community. Martin's analysis shows how communities have evolved over the last 8000 years, and that modern communities in South Africa may be of quite recent origin (see also Janssen, 1970).

The studies mentioned above have all used cluster analysis or correlation techniques. The ordination method of principal components analysis (PCA) has also been used to find which taxa co-vary. Reyment (1963) used this method to simplify ostracod assemblages from boreholes in Nigeria. H. J. B. Birks (unpubl.) applied it to pollen analytical data from Devensian late-glacial sediments at Hatchmere, Cheshire, England. Figure 11.2 is a plot of the species loadings on the first two axes. Three main groups can be separated which can be interpreted ecologically. The Cyperaceae–Caryophyllaceae group are mainly plants of open disturbed soils, and they are characteristic of the Younger Dryas stadial sediments, when cold conditions eliminated the woodland of the underlying Allerød interstadial period. The group characterized by *Betula*, *Populus*, and *Filipendula* are woodland plants and tall herbs characteristic of moist fertile soils. They occur mainly in the Allerød period and again at the beginning of the Flandrian, as a response to improved climatic conditions. The group characterized by Gramineae, *Rumex*, and *Potentilla*-type includes common herbs of grasslands, and they are mainly found in the Older Dryas sediments. A few taxa such as *Pinus*, *Salix*, and *Equisetum* occur throughout the profile. The net result of the analysis was mainly stratigraphic groups, suggesting that different communities grew around Hatchmere at different times.

Disadvantages of the approach

The major disadvantage of this approach is that the unit of study, the fossil pollen assemblage, is not closely related in space and time to the death assemblage, and thus to the life assemblage. Pollen is derived from an undefinable source area, and the pollen assemblage which enters a lake or bog may undergo many changes before being finally preserved (see Fig. 1.5). Recurrent groups of fossil pollen taxa are therefore only associated in time. More locally derived fossils, including plant macrofossils and many animal fossils will tend to form fossil assemblages more closely linked to the death and life assemblages from which they were derived (see Chapter 5). Despite this limitation, the approach has been of some use in pollen analysis, and in pre-Quaternary situations where the ecology of the plants is unknown, it may be the only method available with which to attempt community reconstruction.

The indicator-species approach

This second approach to community reconstruction involves the extension backwards in time of known sociological and ecological preferences of individual taxa. Communities can best be reconstructed by grouping together taxa with similar modern ecological preferences, especially if there is some indication of their past abundance.

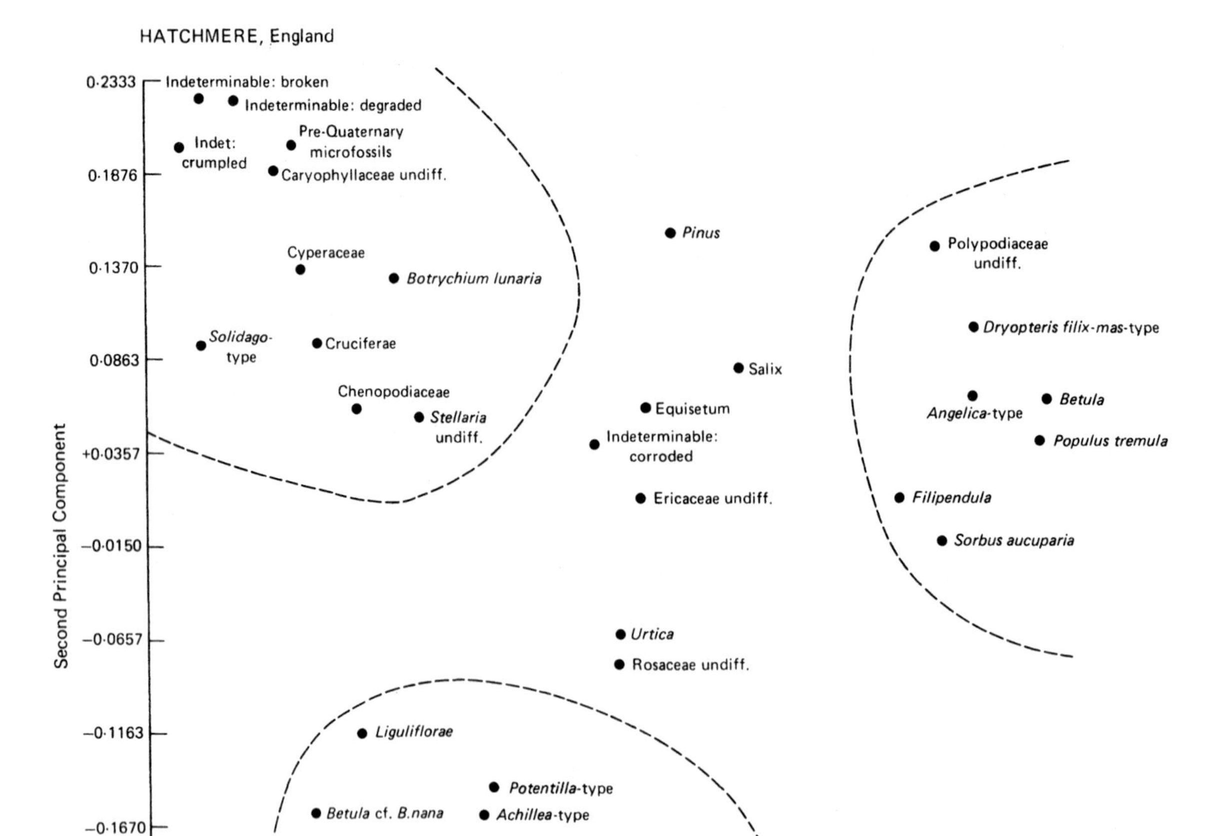

Fig. 11.2 Results of principal components analysis of pollen types from the Devensian late-glacial sediments at Hatchmere, Cheshire. The distribution of the pollen types on the first two principal components is shown.

Iversen (1964) evaluated the usefulness of different types of plant as indicators of climate, soil, and other factors. He concluded that trees were good climatic indicators, but that many had the disadvantage of a slow migration rate across continents after the initial climatic warming at the beginning of any interglacial, and therefore their arrival in an area was not an indication of climatic change at that time, but rather of the climate that was already existing there. Iversen concluded that aquatic plants were the most sensitive climatic indicators because they multiply rapidly, and are readily dispersed by waterfowl. Aquatic plants do not depend on soil maturation, and are not restricted by forest growth. They react particularly to thermal climate, as it affects the temperature of stagnant water.

Iversen (1954) used indicator species to reconstruct past climates, without reconstructing past communities apart from broad vegetation types such as forest, grassland, aquatics, etc. Janssen (1967, 1972) used pollen assemblages to reconstruct past communities in some detail, using the indicator-species approach. The pollen types from Stevens Pond, a small hollow in Minnesota, U.S.A. were arranged by Janssen (1967) into modern sociological groups, that is groups of taxa with similar

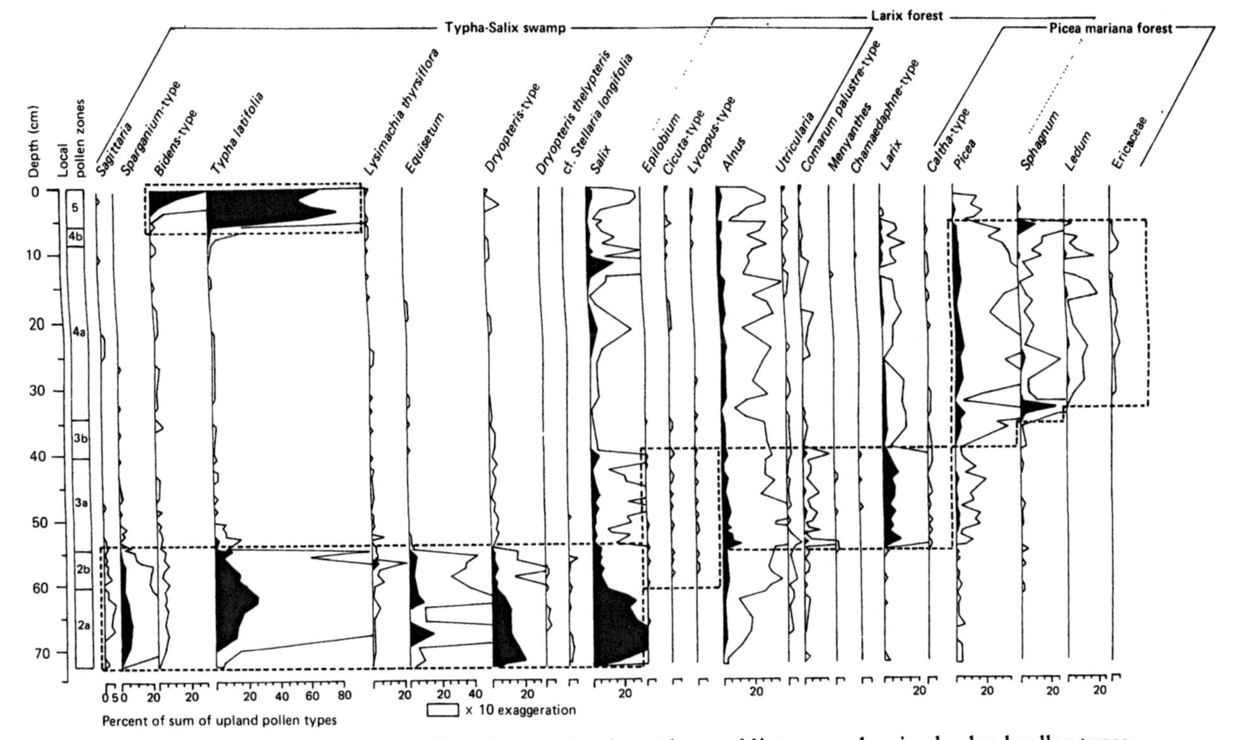

Fig. 11.3 Percentage pollen diagram from Stevens Pond, north-west Minnesota, showing lowland pollen types arranged into ecological groups. Those thought to have formed local communities are blocked, and the succession can be traced. (After Janssen, 1967.)

ecological and sociological preferences, which he had described already from a detailed study of the vegetation of the area. The main groups of pollen types are shown in Fig. 11.3. Because the pond was only 80 m in diameter, most of the pollen was local or from nearby vegetation (extra-local) derived largely through the trunk space (Chapter 9). However, little information was gained about the local forest composition, probably because the pollen of forest herbs does not get dispersed even as far as the middle of Stevens Pond, and also because the local forest was probably not very different from the regional forest. In contrast, changes in the lowland communities were quite distinct (Fig. 11.3). The pollen spectra could be compared with modern aquatic and swamp vegetation. The pond started to become overgrown by a *Typha–Salix* community during the dry prairie period, leading to a *Larix* swamp forest, and then to a *Picea mariana* swamp forest characteristic of peaty soils in the area today. Logging caused this community to be replaced once more by a *Typha* mat in about 1900, some eutrophic influence being represented by an abundance of *Bidens*-type pollen.

Janssen (1972) extended this approach to organic deposits in very small hollows in the Netherlands. In rather the same way that Rybníček and Rybníčková (1968) reconstructed communities from Bláto Mire, Czechoslovakia using macrofossils, Janssen used pollen types as evidence of the past occurrence of distinct plant communities which have been well described from Europe by phytosociologists.

Rybníčková and Rybníček (1971) used the indicator species approach to determine local pollen and spore types that were produced in a range of organic sediment types in Czechoslovakia. They subtracted them from the total pollen spectra in order to lead to a more accurate reconstruction of the upland vegetation. Each sediment type produced its own characteristic local spectrum, and this had to be taken into account when correcting a profile containing different sediment types.

Evaluation of the indicator-species approach

1. The key assumption in this approach is that the ecological tolerances of the taxa involved have not changed over time. It is difficult to test this assumption, but it must be borne in mind that species frequently consist of several ecotypes, each of which is adapted to slightly different conditions (see Iversen, 1964). Ecotypes cannot be distinguished morphologically from their fossils, and therefore any reconstructions must be made using information from the whole ecological range of the morphological species, bearing in mind that ecological studies usually only investigate one ecotype.

2. Conditions of competition have changed over time, as communities have evolved. A species' ecological tolerance in the past may be different from its tolerance today in the presence of competition. This reservation is particularly important when reconstructing communities from late-glacial assemblages, where conditions were open and soils were unleached in contrast to today (Birks, 1973a, p. 326).

3. Plants usually occur in several communities, and therefore it is desirable to reconstruct a community from an assemblage of several taxa with similar ecological behaviour. If *groups* of taxa occur together in the past in the same way as they do in the present, it is fairly safe to assume that their ecological behaviour has not changed significantly (Andersen, 1961; Iversen, 1954, 1964; Janssen, 1972; Birks, 1973a).

4. It is difficult to obtain an estimate of the past population size of any taxon (see Chapter 10). It is a

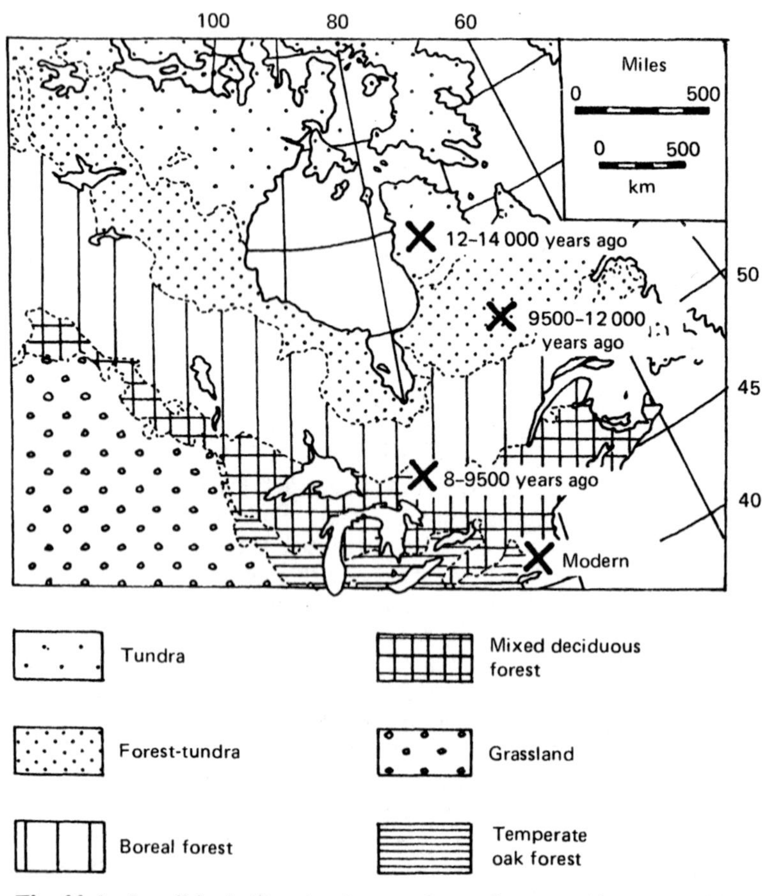

Fig. 11.4 Localities in Canada where surface pollen assemblages resemble fossil assemblages in southern New England. The sites are indicated by X. The age of the analogous fossil spectra is indicated for each locality. (After Davis, 1969.)

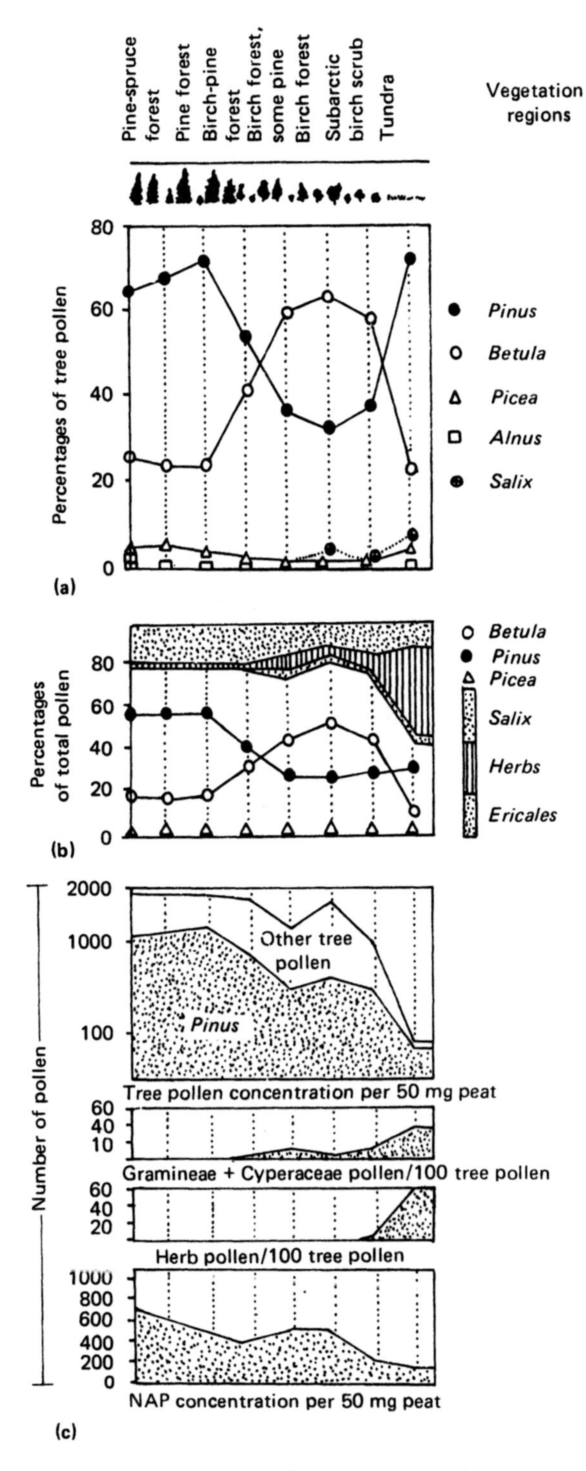

Fig. 11.5 Pollen spectra from surface samples along a transect through the vegetational regions of northern Finland. (**a**) The percentage frequencies of the principal tree pollen types expressed as percentages of total tree pollen. (**b**) The percentage frequencies of the principal pollen types expressed as percentages of total pollen. (**c**) The pollen concentration of various pollen types expressed as number of grains in 50 mg of air-dried peat. (Data of Aario (1940) redrawn from Birks, 1973b.)

huge task to calibrate the influx of each taxon in the way that Davis *et al.* (1973) attempted for the trees of Michigan (Chapter 10). However, this may be the only way in which past communities can be reconstructed which appear to have no modern analogue today. Davis *et al.* reconstructed the forest of zone C-1a at Rogers Lake, Connecticut by estimating the population size of each component from its pollen influx, which they had calibrated in terms of the modern influx of the taxon concerned. No other calibrations have so far been attempted.

The comparative approach

This third approach to community reconstruction involves the characterization of a range of modern vegetation types by means of contemporary pollen spectra (usually from surface samples), and then the comparison of these spectra with fossil pollen spectra. If the two spectra are similar, then it can be concluded that they were produced by similar vegetation. A modern analogue can thus be suggested for the fossil vegetation. This approach has been well reviewed by Wright (1967).

If similar pollen assemblages can be recognized in a stratigraphical and hence temporal sequence of fossil pollen spectra, changes in vegetation in the past over time can be interpreted in terms of present-day vegetational differences in space. Davis (1969) illustrated this (Fig. 11.4) by comparing fossil spectra from southern New England with modern spectra from different vegetation types in eastern North America.

If, however, no modern analogue for the fossil spectrum can be found, it may either be concluded that the past vegetation has no modern analogue, or that the modern information is inadequate, and analogues should be sought elsewhere. To exclude the latter possibility, a comprehensive characterization of modern pollen spectra produced by all types of modern vegetation is required, and nowadays, many surface sample studies have been carried out, particularly in North America. In Europe, man has interfered and altered the natural vegetation so extensively that it has been argued that fossil vegetation before man's influence can never be characterized by a modern analogue approach. However, there are a few relatively undisturbed forests still available for study. It would also appear that boreal forests and treeless alpine and tundra habitats have been altered relatively little by man and that modern pollen spectra from them do have relevance to past situations (e.g. Birks, 1973a,b; Hicks, 1977; Prentice, 1978).

Some examples of the comparative approach

1. *Modern pollen samples from arctic and boreal regions*

a) Pioneer work on the characterization of modern vegetation by pollen spectra was done in Finnish Lappland by Auer (1927), Firbas (1934), and Aario (1940). They all illustrated the high percentages of *Pinus* pollen in tundra well to the north of the nearest pine trees. Aario (1940) also counted the number of pollen grains in 50 mg of air dried peat. Figure 11.5 shows the comparison of his pollen percentage data with his pollen concentration data. It can be seen that the high percentages of *Pinus* pollen in tundra samples represent quite low concentrations of *Pinus* pollen compared with its concentration in pine forests, and that its high percentages are due to the over-all low pollen productivity of the tundra vegetation.

b) Hicks (1977) enlarged upon Aario's work in northern Finland. She collected moss samples of equal surface area and mass, and was thus able to calculate concentration figures. The three major forest types she studied are delimited in Fig. 11.6 according to the pollen composition of her samples, which are positioned on the grid. The three forest types are adequately separated on their pollen percentages, but samples from the tundra fall within the range of samples from the mixed pine/birch/spruce forest. However, the tundra samples can be readily distinguished from the mixed coniferous forest samples by the much higher proportion of non-arboreal pollen, and the occurrence of pollen of indicator species of tundra. The separation is confirmed by the lower pollen concentration in the tundra samples, especially of the tree pollen which had blown in from the south. Hicks tested the usefulness of her model by adding independently-collected samples from pollen traps. They all fell within the correct vegetation type, thus showing that the pollen rain of northern Finland specifically represents the different vegetational types.

c) Ritchie and Lichti-Federovich (1967) collected modern influx data from the forest transition in Canada. They exposed Petri-dish samplers within standard meteorological screens at 20 stations in the high-arctic rock desert, the mid-arctic sedge–moss tundra zone, the low-arctic dwarf-shrub tundra, the forest-tundra transition, and the northern coniferous forest belt. The relative percentages of the various taxa and the total pollen influx along their transect are shown in Fig. 11.7. The most noticeable

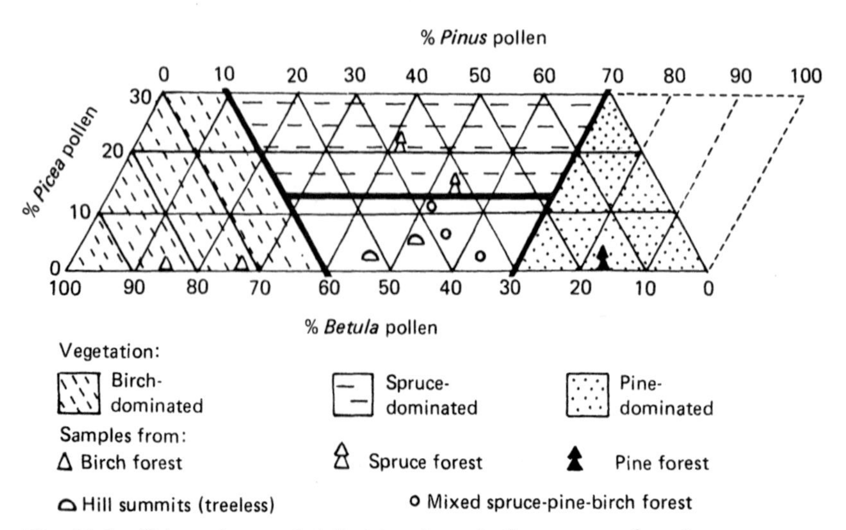

Fig. 11.6 Triangular graph delimiting the main forest types of northern Finland on the basis of their tree pollen composition. The vegetation types from which each surface sample was taken are indicated. (After Hicks, 1977.)

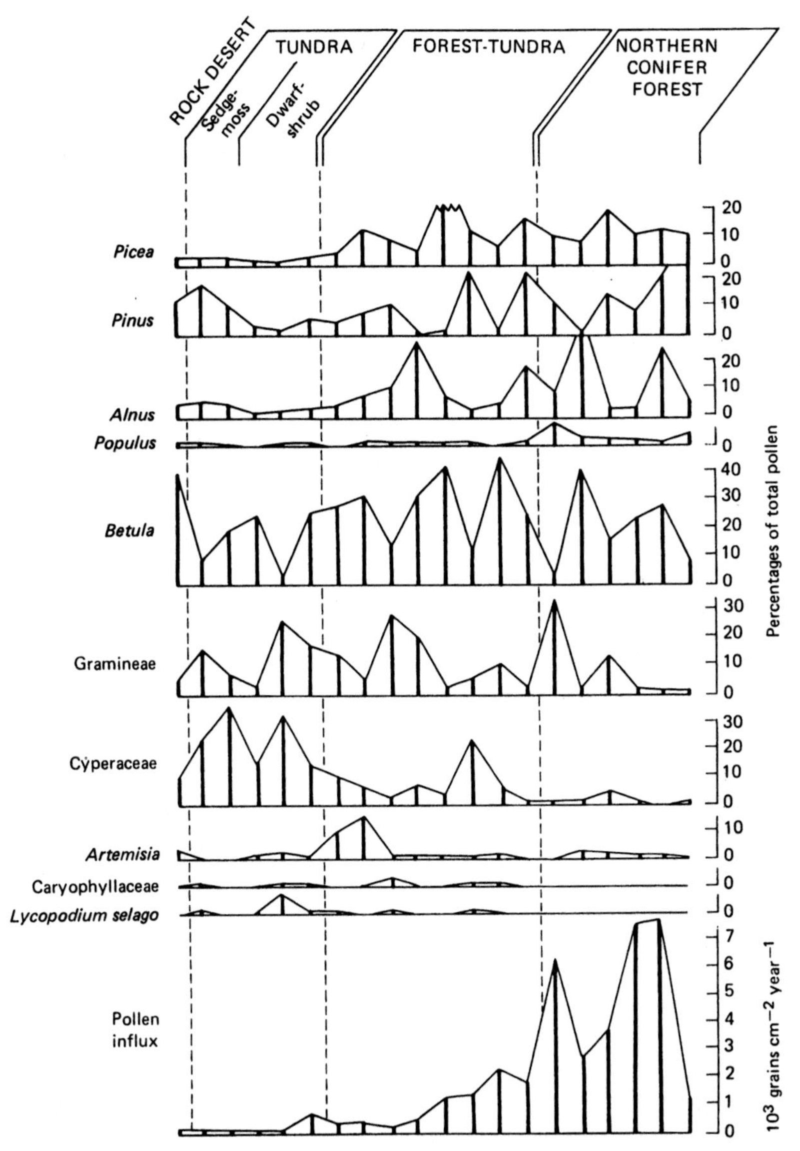

Fig. 11.7 Pollen spectra obtained from open Petri-dish samplers in various vegetational zones in the Arctic and Sub-arctic of Canada. Only principal pollen and spore types are shown. Results are expressed as percentages of total pollen and spores. The estimated annual influx in the vegetational zones is also shown. (Data of Ritchie and Lichti-Federovich (1967) redrawn from Birks, 1973b.)

feature is the low pollen influx in the tundra, ranging from 5 to 335 grains per year. These values are comparable to those obtained by Fredskild (1969, 1973) from Greenland. Similar percentage results were obtained by Lichti-Federovich and Ritchie (1968) from surface muds from arctic lakes to those from the Petri-dish samples, except for consistently lower values of Gramineae.

d) Because moss samples contain a high proportion of local pollen, which is not desirable in the reconstruction of regional vegetational history from lake deposits (Birks, 1977), Prentice (1978) collected surface mud samples from 67 lakes evenly scattered through Finland. He analysed pollen percentages and concentrations, and compared the results with the frequency of tree species in the surrounding

Pinus pollen %

Betula pollen %

Picea pollen %

Fig. 11.8 Isopollen maps for percentages of ***Pinus***, ***Betula***, and ***Picea*** pollen in modern pollen spectra from Finland compared with isofrequency maps of the trees. (After Prentice, 1978.)

forests. Figure 11.8 shows the excellent correspondence between the mapped pollen percentages (isopolls; Szafer, 1935) and the isofrequencies of the trees for *Pinus*, *Picea*, and *Betula*. Frequencies of other pollen taxa also agreed well with the occurrences of the parent plants. An ordination of the percentages of total pollen by the method of non-metric multidimensional scaling (MDSCAL) is shown in Fig. 11.9. The four clusters of samples correspond to the mixed and boreal forest, pine forest, birch forest, and tundra, placed in their correct north–south sequence. Prentice found, like previous authors, that a consideration of non-arboreal pollen was necessary to distinguish the tundra samples from mixed coniferous forest spectra, and better discrimination still is obtained using pollen concentration data. Because Finnish vegetation can be so well characterized, Prentice predicted that his data could be compared with fossil data (by numerical methods) to give a satisfactory reconstruction of fossil communities. They could also be used to provide more detailed R-values than those of Donner (1972) in order to reconstruct the populations of taxa in the past, and they could perhaps be applied to the fossil isopollen maps of Finland by Birks and Saarnisto (1975) to produce maps of past tree abundance in Finland.

e) Birks (1973a,b) took advantage of the fact that local pollen is well represented in samples from moss cushions in his attempt to characterize different mountain vegetation types in Scotland by modern pollen spectra. Although it has been considerably disturbed by man, Scottish mountain vegetation has certain floristic affinities with Devensian late-glacial pollen spectra and can provide some modern analogues. In addition, surface samples can provide much information on the pollen representation of individual taxa. Birks demonstrated by observation and by the use of numerical methods that characteristic pollen spectra were produced by the vegetation of the alpine zone of the summits, by sub-alpine vegetation, including grasslands, tall-herb communities, and willow scrub, by the sub-alpine *Juniperus communis* communities, by the *Betula pubescens* woods, and by the *Betula–Corylus avellana* woods (Table 11.1). However, it was difficult to distinguish communities within these broad physiognomic types because of the poor pollen representation of diagnostic taxa such as *Salix*. Birks (1973a) used the modern pollen spectra to interpret Devensian late-glacial spectra from the Isle of Skye, Scotland, in vegetational terms.

f) Birks (1977) undertook a similar survey of vegetation at and above the treeline in the St. Elias

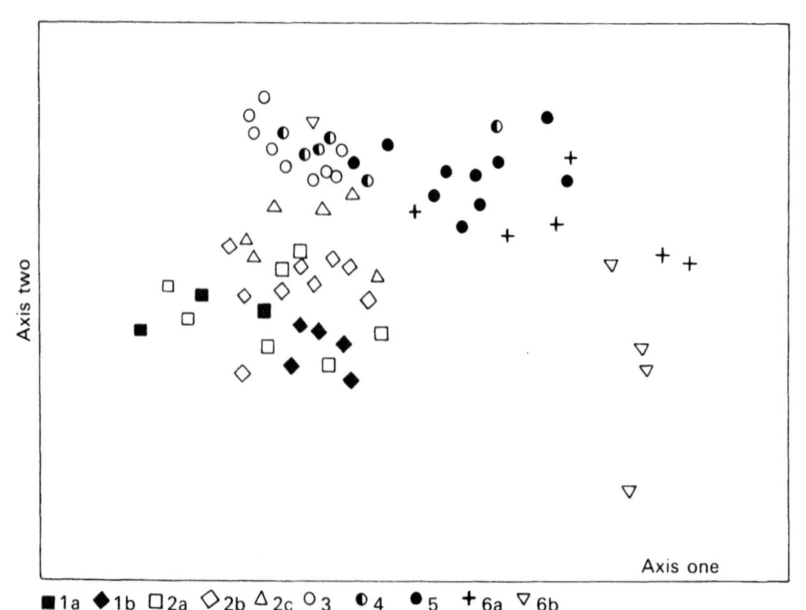

Fig. 11.9 Plot of modern pollen spectra (as percentages of total pollen) from Finland ordinated by MDSCAL (non-metric multidimensional scaling). 1a and 1b, mixed deciduous forest; 2a, southern boreal forest; 2b, middle boreal forest; 2c, northern boreal forest; 3, pine forest; 4, birch-pine forest; 5, birch forest; 6a, tundra near the timber line, with scattered birch trees; 6b, tundra distant from the timber line. (After Prentice, 1978.)

Table 11.1 Summary of modern pollen assemblages in Scotland. (From Birks, 1973a.)

Vegetation type	*Corylus*	*Betula*	*Juniperus*	*Salix*	Gramineae	Cyperaceae	*Rumex*	*Filipendula*	*Ranunculus acris* type	*Thalictrum*	*Lycopodium selago*
Betula/Corylus woods	38	40	0	0.5	8.5	3.4	0.3	0.4	0.5	0	0
Corylus woods with 'tall herbs'	31	26	0	0.7	7.8	3.1	0.5	3.1	1.1	0.1	0
Betula woods with 'tall herbs'	1.5	62	0.2	0.8	18.1	5.7	0.3	2.6	2.6	0.1	0
Betula woods	0.6	61	2	0.1	22	2.7	0.1	0.1	0.2	0	0
Juniper scrub	0	0.7	35	0.2	32	25	0.4	0	0.2	0.1	0.4
Subalpine grassland	0	0	0.1	0.1	37	15	3.2	0.6	3.3	1.6	0.1
'Tall herb' communities	0	0	0.1	0.4	37	11	4.8	3.9	3.1	0.9	0.2
Willow communities	0	0	0.4	1.6	48	9	4.5	0.4	4.9	1.2	2.0
Snow-beds	0	0	0	0.2	24	35	0.4	0	0.9	0	14.6
Summit-heaths	0	0	0.3	1.4	24	37	0.4	0	1.0	0.2	7.2
'Fell-field'	0	0	0.3	0.9	21	48	0.2	0	0.9	0.7	2.9

Expressed as mean percentages of total pollen with selective exclusions.

Mountains, Yukon Territory, Canada. He described four main vegetation types; *Picea glauca* forest, *Populus balsamifera* forest, *Betula glandulosa* shrub-tundra, and *Dryas integrifolia* tundra. The *Populus* forests and *Dryas* tundra produced distinctive pollen spectra, but the spectra from the *Picea* forests and *Betula glandulosa* shrub-tundra were, surprisingly, indistinguishable, even when the data were subjected to principal co-ordinates analysis. Instead, as Fig. 11.10 shows, the samples are separated on the basis of their sedimentary origin, from moss cushions, moss in sedge swamps, or lake muds, showing the influence of local vegetation in modifying the regional pollen spectrum. However, even when the lake mud samples only were analysed, the *Picea* forests and *Betula* scrub were still indistinguishable. Davis (1969) and Davis and Webb (1975) discuss the difficulties of separating pollen spectra from boreal forest and the forest-tundra using percentage data. Usually, the two communities are not far apart spatially and the large amounts of tree pollen blown into the shrub or open forest communities have a large percentage effect, because the pollen production of these communities is low. The data of Ritchie and Lichti-Federovich (1967) indicate that the two vegetation types can probably be separated in terms of pollen influx.

g) Ritchie (1974) drew similar conclusions from a survey of modern pollen spectra from moss cushions and surface lake sediments in the Mackenzie Delta region of northern Canada. Lake samples gave consistent spectra for forest and for tundra communities, but the forest-tundra samples varied between the two extremes. Samples for moss cushions varied as much within a vegetation type as between different vegetation types because local pollen deposition masked the regional pollen rain. However, the local pollen rain relates to the local vegetation, and these samples are valuable in interpreting fossil data from peats.

In spite of several modern pollen surveys of arctic vegetation, no satisfactory modern analogue has yet been found for the unique Wisconsinan pollen assemblages found in the Yukon Territory, North West Territories, and Alaska (Matthews, 1970, 1974; Colinvaux, 1964; Rampton, 1971; Ritchie, 1977) and in the Great Lakes Region (Birks, 1976).

Most of the modern pollen surveys have concentrated on large vegetation units, for example the rock desert, sedge-moss and heathy tundra, and forest tundra of Lichti-Federovich and Ritchie (1968). They were termed ***vegetation-landform units***. Even when individual plant associations were sampled by Birks (1973a,b), little distinction could be made below the vegetation-landform unit level because of the limitations of pollen production, preservation, and identification. Ritchie (1974) and Janssen (1973, 1979) have shown that samples from the centre of a lake contain a regional pollen spectrum, integrated from the mosaic of communities surrounding the lake, and therefore pollen diagrams from such situations are unlikely to be able to be interpreted in terms of plant associations.

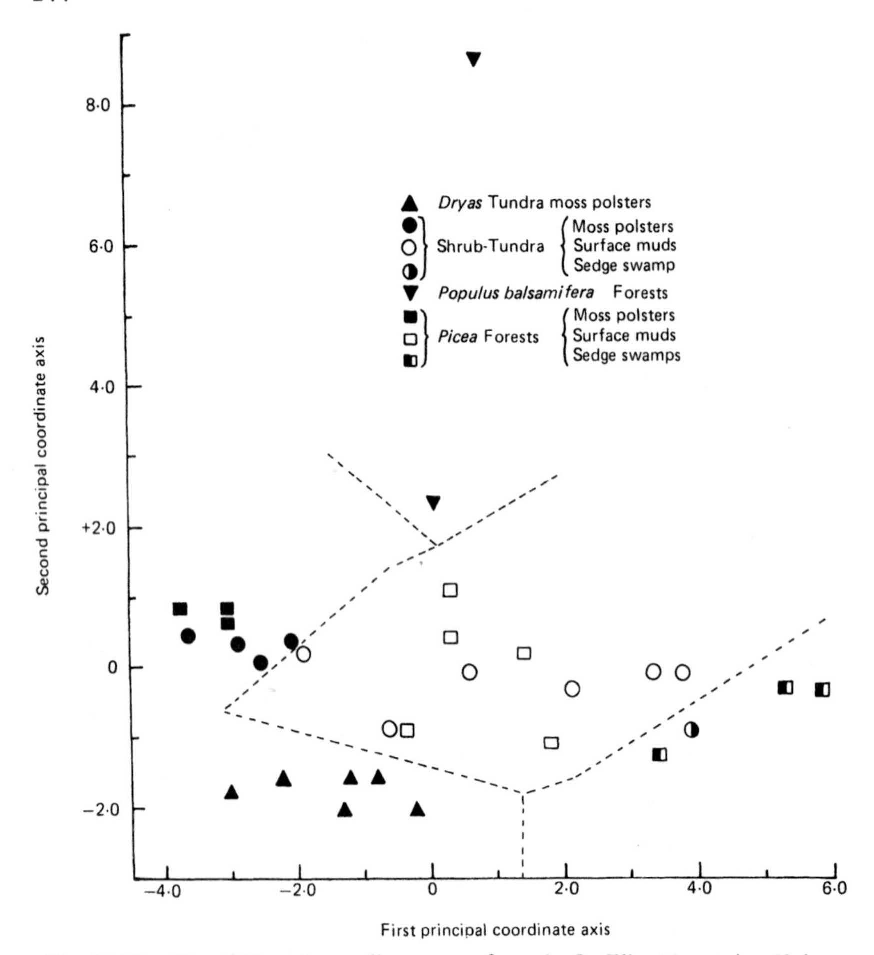

Fig. 11.10 Plot of 30 modern pollen spectra from the St. Elias Mountains, Yukon Territory, Canada, on the first and second principal coordinate axes. Open symbols from surface lake muds, halved symbols from sedge swamps, black symbols other than triangles from moss polsters. (After Birks, 1977.)

2. *Modern pollen samples from forested regions*

Modern pollen surveys from forested areas have also led to vegetation characterization at the formation or at the vegetation-landform unit level rather than at the association level. The characterization of forests has been largely developed in North America, where large tracts of relatively undisturbed vegetation still occur, and there is frequently a good record of the forests before extensive modification by European man occurred.

a) McAndrews (1966, 1967) studied the present and past vegetation differentiation along an east–west transect 66 miles long and 6 miles wide through the main vegetation types in north-west Minnesota near Lake Itasca. The relation of the vegetation to landform is shown in Fig. 11.11, from prairie in the west, through oak savanna on the Big Stone Moraine, to deciduous forest on fine-textured soils, to mixed pine–hardwood forest on coarse-textured soils in the east. McAndrews took short cores and obtained pollen spectra just before the *Ambrosia* pollen rise indicated the onset of European disturbance. He reconstructed the vegetation at this time from historical accounts (Fig. 11.11) and found that the different vegetation types had distinct pollen spectra (Table 11.2). He then used this information to reconstruct the past vegetation from four long profiles as indicated in Fig. 11.11. (Thompson Pond from the prairie, Terhell Pond from the deciduous forest, and Bog D and Martin Pond from the pine–hardwood forest). He construc-

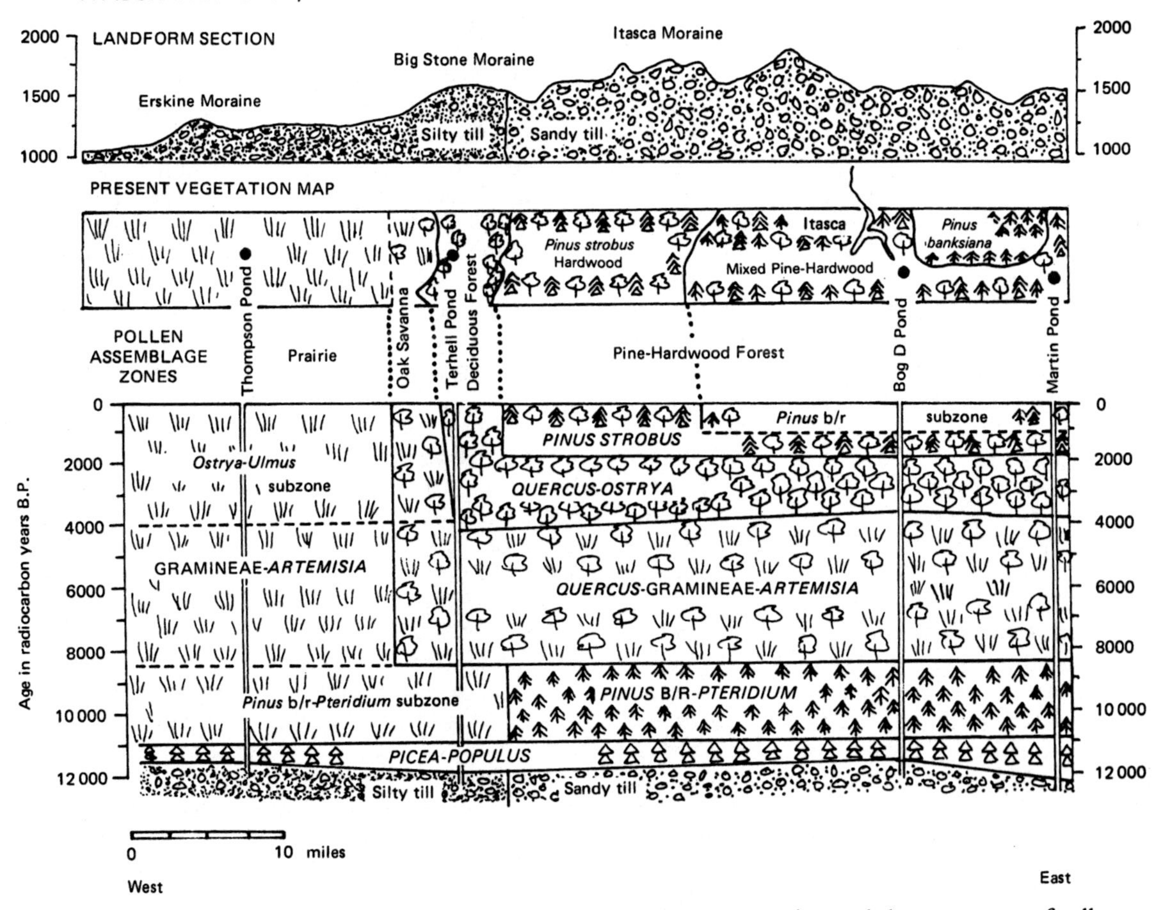

Fig. 11.11 The Itasca transect, north-western Minnesota; land-forms, vegetation, and chronosequence of pollen assemblage zones. The pollen zones are shaded the same as their modern vegetation equivalent where possible. The four long cores are indicated. *Pinus b/r*, *Pinus banksiana/P. resinosa*. (After McAndrews, 1966.)

Table 11.2 Summary pollen assemblages for vegetation types in N.W. Minnesota. (From McAndrews, 1966, 1967.)

Vegetation type	Gramineae	*Artemisia*	*Quercus*	*Ulmus*	*Pinus*	*Ostrya*
Pine–hardwood forest	<5	<5	<10	<5	>35	<5
Mixed deciduous forest	<10	<5	15–30	10	<30	>10
Oak savanna	10–50	5–20	>10	<5	<20	<5
Prairie	10–50	5–20	<10	<5	<20	<5

Expressed as percentages of total pollen excluding Cyperaceae and obligate aquatic taxa.

ted a time-space diagram of the regional pollen assemblage zones, using a chronology based on radiocarbon dates.

From 12 000–11 000 B.P. the whole transect was occupied by vegetation producing a *Picea–Populus* pollen assemblage, for which no modern analogue has been found. At about 11 000 B.P. prairie spread from the west, and the area round Thompson Pond has been prairie ever since. To the east, a *Pinus banksiana/P. resinosa-Pteridium* assemblage replaced the *Picea–Populus* assemblage, the coarser soils presumably being colonized by jack pine, *P.*

banksiana. At about 8500 B.P. prairie or oak savanna expanded throughout the transect, represented by the Gramineae–*Artemisia* and *Quercus*–Gramineae–*Artemisia* assemblage zones. A dry warm period is recorded throughout the midwest United States at this time.

At about 4000 B.P. savanna was replaced by deciduous forest on the fine textured soils, represented by the *Quercus–Ostrya* assemblage zone, as a response to a change to a cooler, wetter climate.

At about 2000 B.P. *Pinus strobus* immigrated into the deciduous forest and became dominant on coarser soils. It was followed by *Pinus banksiana* and *Pinus resinosa* at about 1000 B.P. in the east, and the mixed pine forests of today became established.

b) We have discussed McAndrews elegant study in some detail, but there have been other studies following a similar approach. In Iran, Wright *et al.* (1967) characterized the modern vegetation of Iran from the arid steppe of the Mesopotamian lowlands, through the *Amygdalus–Pistacia* savanna of the foothills, the oak woodlands of the steep slopes of the Zagros mountains with their higher precipitation, and the upper savanna to the *Artemisia*–Chenopodiaceae interior steppe of the arid eastern Iranian plateau. Each vegetation type produced a distinctive pollen assemblage, as summarized in Table 11.3. Wright *et al.* used these spectra to interpret a long pollen diagram from Lake Zeribar, within the Zagros mountains. Arid conditions with steppe vegetation in the full-glacial before 11 000 B.P. gradually developed into a savanna type of vegetation with pollen of *Pistacia*, *Plantago*, and some *Quercus*, but lower values of *Artemisia*, which appears to have no exact modern analogue. At about 5500 B.P. the climate became sufficiently moist for the development of oak woodland, resembling that of the present day. These changes were corroborated by independent evidence from plant and animal macrofossils and chemical analyses (see Chapter 6, p. 108).

c) Davis (1967) in an attempt to interpret the late- and post-glacial pollen assemblages found in southern New England constructed a north–south transect of surface samples from the tundra of Canada, through the boreal forest, the mixed deciduous/coniferous forest, to the deciduous forest of eastern North America. She found a consistent and recognizable assemblage in each vegetation formation and showed that differences between formations were greater than the variation within formations, except for the forest-tundra zone between the tundra and boreal forest. However, this zone could be distinguished by using its pollen influx characteristics. Davis was then able to reconstruct the fossil vegetation, with the limitation that the older pollen spectra had no exact modern analogues, because of subsequent migration of trees (see Chapter 10, p. 226). She was able to relate vegetation changes in time to present spatial vegetation distribution (Fig. 11.4).

d) R. B. Davis and Webb (1975) extended M. B. Davis' (1967) study in eastern North America, with a total of 478 surface samples from all the major vegetation formations. Like M. B. Davis (1967) they found that each formation could be characterized by a pollen assemblage. They mapped each pollen type, and showed, in general, a close correspondence of the pollen abundance (both percentage and concentration) of a taxon with the vegetation formations in which it occurred. A similar approach was used by Webb and McAndrews (1976) using surface samples from central North America, and similar conclusions were reached.

e) Webb (1974a) used the mapping approach in Lower Michigan, U.S.A. to compare pollen spectra and vegetation in more detail. The boundary between the mixed coniferous/deciduous forest and the deciduous forest formations crosses Michigan today, and there are several forest community types present in each formation. Surface samples from 64 lakes were compared with forest inventory data, and

Table 11.3 Summary of modern pollen assemblages in western Iran. (From Wright *et al.*, 1967.)

Vegetation type	*Quercus*	*Pistacia*	Gramineae	*Plantago*	Chenopodiaceae	*Artemisia*
Lowland steppe	0	0	18	26	18	4
Lower savanna	1	+	15	35	10	5
Zagros oak forest	35	1	18	9	6	5
Upper savanna	15	1	17	10	13	15
Plateau steppe	2	0	10	1	45	23

Expressed as mean percentages of total pollen with selective exclusions.

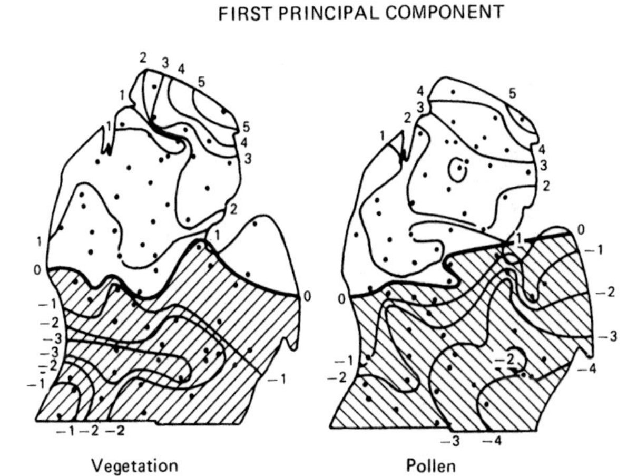

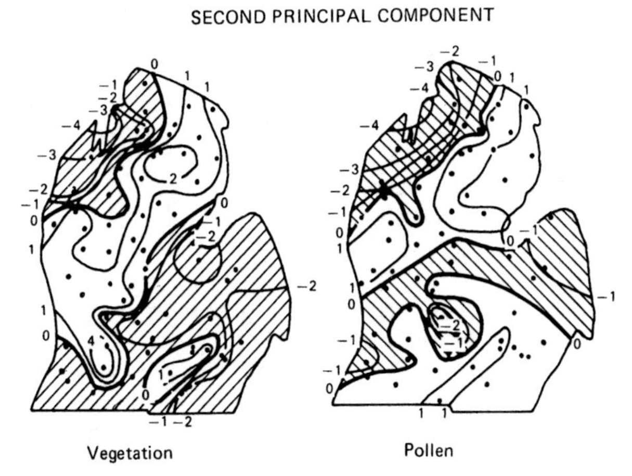

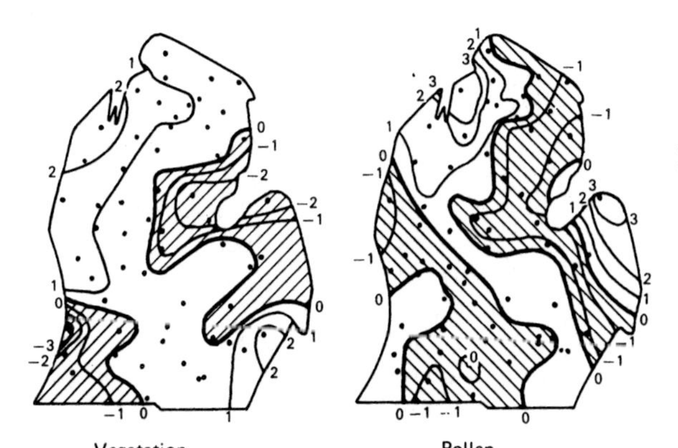

Fig. 11.12 Comparison of the results of principal components analysis of modern pollen spectra and corresponding vegetational data from 64 lakes in Lower Michigan. Negative values for the principal components are hatched. The positions of the lakes are indicated by dots. (After Webb, 1974a.)

in general Webb found a good correspondence between a taxon's abundance in the vegetation and in the modern pollen spectra. Independent principal components analysis summarized the two data sets (Fig. 11.12). The first component followed closely the formation boundary. The second component subdivided each formation into its two major forest types, on both vegetation and on pollen data. Webb concluded that this close correspondence between vegetation and the regional pollen rain provided a factual basis for reconstructing past vegetation at the regional level. Webb (1974b) showed a similar correspondence in northern Wisconsin. He applied these data to the interpretation of a pollen diagram from Lake Mary, Wisconsin by interpolating the fossil spectra from zones 3 and 4 into a principal components analysis of the modern pollen spectra from Wisconsin.

3. *The use of numerical methods in modern pollen surveys*

Many of the studies we have described have utilized numerical methods of data analysis to simplify the data and aid in its interpretation. Birks *et al.* (1975) tested the effectiveness of a whole range of numerical methods on one set of data, that of Lichti-Federovich and Ritchie (1968) from Canada. These data are summarized in Fig. 11.13, and can be briefly described as follows:

Low arctic shrub-tundra – high values of Cyperaceae and shrub birch pollen.

Forest-tundra – high conifer pollen values, similar to boreal forest, but high values of *Alnus* pollen. Not very distinct.

Boreal forest – high values of *Picea* and *Pinus* pollen, together with *Betula* and *Alnus*. The varying importance of *Populus* in the different boreal forest types is not registered, because of its poor pollen preservation; it is a 'blind spot' (*sensu* Davis, 1967).

Mixed forest – lower conifer pollen values are associated with higher non-arboreal pollen and deciduous tree pollen values.

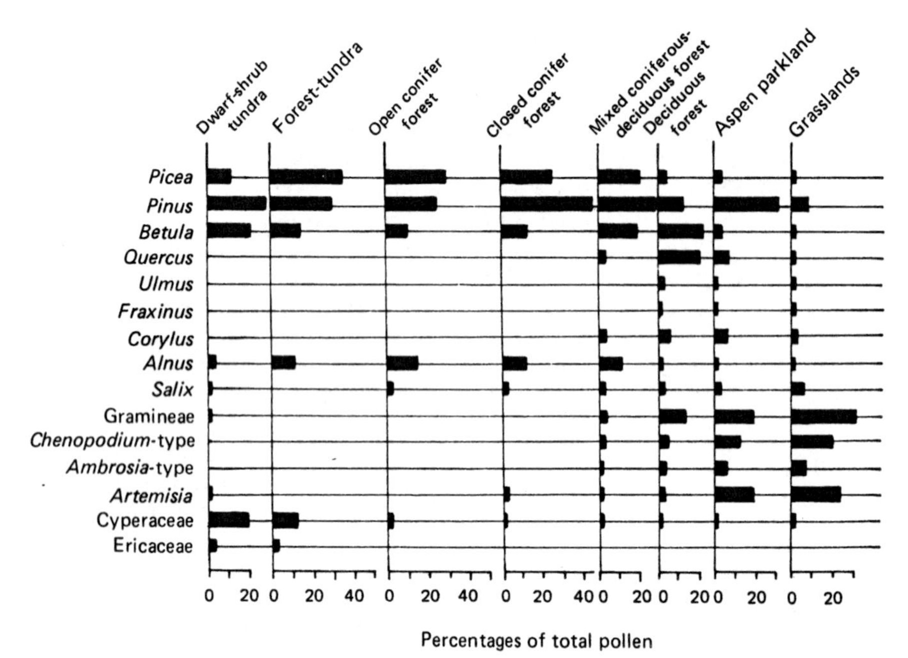

Fig. 11.13 Summary pollen diagram of mean pollen percentages in surface samples from lake muds for each main vegetation type in the western interior of Canada. (Data of Lichti-Federovich and Ritchie (1968) redrawn after Birks, 1973b.)

Deciduous forest	– high values of *Quercus*, *Betula*, *Ulmus*, and *Fraxinus* pollen, and relatively low amounts of conifer pollen. *Populus* again forms a 'blind spot'.
Aspen parkland	high values of Gramineae, *Artemisia*, and Chenopodi, aceae pollen. *Populus* is again a 'blind spot'.
Grassland	– similar to aspen parkland

Birks *et al.* (1975) analysed the data using a variety of numerical methods to see whether any further distinctions could be made. A mapping procedure showed a good correlation between distributions of vegetation types and the pollen percentages of individual taxa. A principal components analysis of all the taxa together resulted in the separation of grassland, parkland, and deciduous forest from the coniferous forests and tundra along the first axis. The second axis differentiated vegetation types within the latter areas into tundra and forest-tundra, and the closed coniferous and mixed coniferous–deciduous forests. The third axis separated the grasslands and parklands from the deciduous forests. These results suggest that the pollen assemblages correspond closely in area to the vegetation types. A similar result was obtained using canonical variates analysis. Again similar results were obtained using minimum-variance or sum-of-squares cluster analysis which distinguished fourteen groups of samples which corresponded broadly with the main vegetation types, but showed some overlap. These results supported Lichti-Federovich and Ritchie's conclusion that there is a broad correlation between different vegetation-landform units within the formation, and, as proposed by Janssen (1973), the regional pollen assemblages in each formation have a characteristic pollen composition which differs from that of adjacent formations. The cluster technique gave the most detailed correspondence with the vegetation-landform units, whereas the ordination procedures only distinguished the major vegetation formations.

4. *The use of modern pollen spectra in vegetational reconstruction*

Many of the studies we have described have used the modern pollen spectra to interpret fossil spectra, in order to reconstruct the vegetational history of an

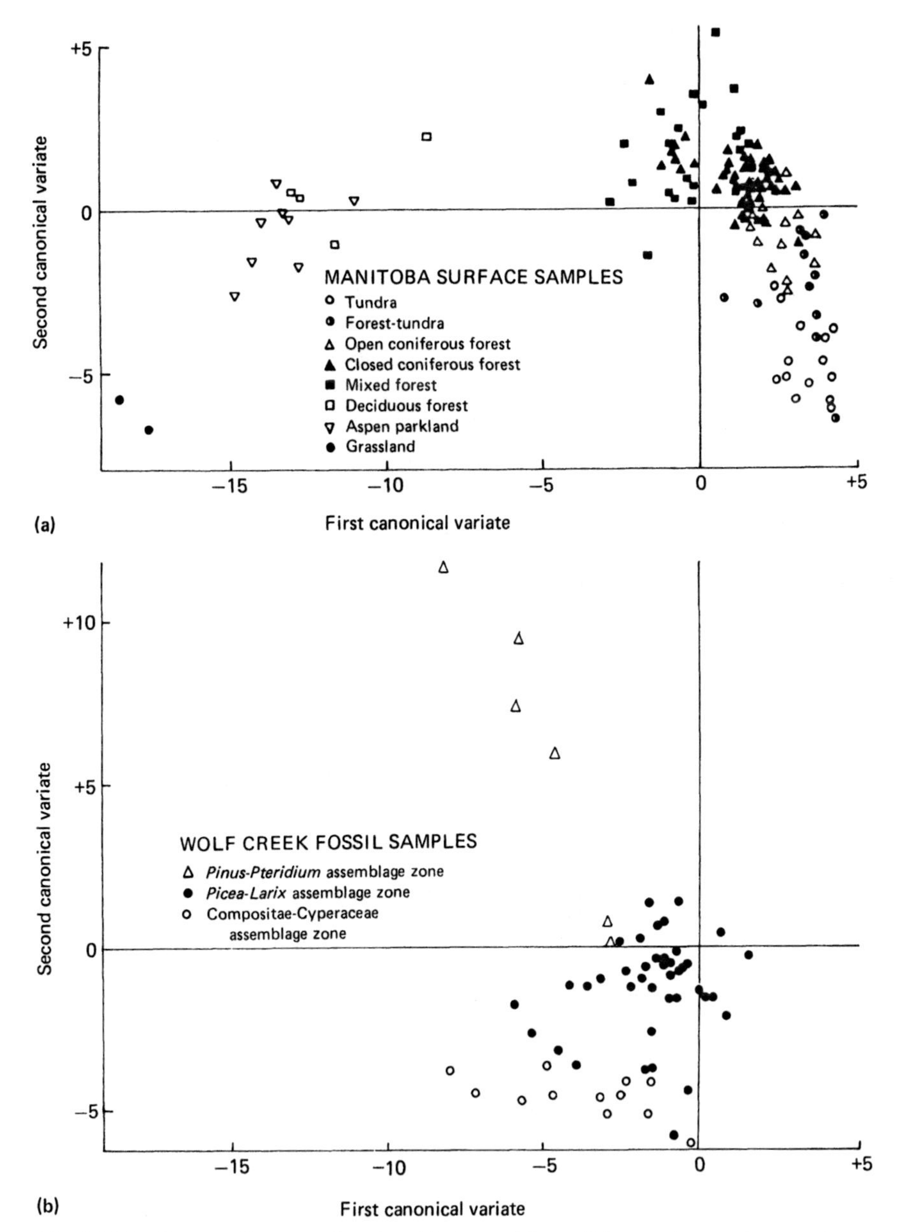

Fig. 11.14 Scatter diagrams showing (**a**) the positions of the modern pollen spectra from Canada of Lichti-Federovich and Ritchie (1968) on the first and second canonical variate axes, and (**b**) the position of 58 fossil samples from Wisconsinan late-glacial sediments at Wolf Creek, Minnesota, on the same axes. (After Birks (1976). © 1976 by the Ecological Society of America.)

area. In most cases, the modern pollen spectra were collected with a specific purpose in mind. Other studies have been only concerned with the relationships of the modern pollen and the modern vegetation. It is encouraging that sets of modern pollen spectra are beginning to be used by different authors to interpret fossil pollen diagrams for which they were not specifically collected. The following two studies both utilize the data of Lichti-Federovich and Ritchie (1968) from arctic Canada.

Birks (1976) used the Lichti-Federovich and Ritchie (1968) data to interpret the past vegetation at Wolf Creek, Minnesota, using canonical variates analysis. Figure 11.14 shows the comparison of the positions of the modern spectra with the positions of the fossil spectra on the first two canonical variate axes. Samples of the Compositae–Cyperaceae zone (WO-1) and of the *Pinus–Pteridium* zone (WO-3) did not match the modern spectra. Some, but by no means all of the samples from the *Picea–Larix* zone (WO-2) match modern samples from the coniferous forests of Manitoba. Birks concluded that there are, so far, no known modern vegetation analogues for these zones, and that vegetation interpretation would have to rely on knowledge of the present ecological and sociological relationships of the taxa, particularly of indicator species with narrow ecological tolerances today.

Ritchie and Yarranton (1978a,b) also compared the modern spectra of Lichti-Federovich and Ritchie (1968) with fossil sequences from central Canada, using principal components analysis. The mean position of each fossil assemblage zone was plotted on a PCA ordination of the total fossil data, and individual modern surface samples were positioned on the principal components axes, as in Fig.

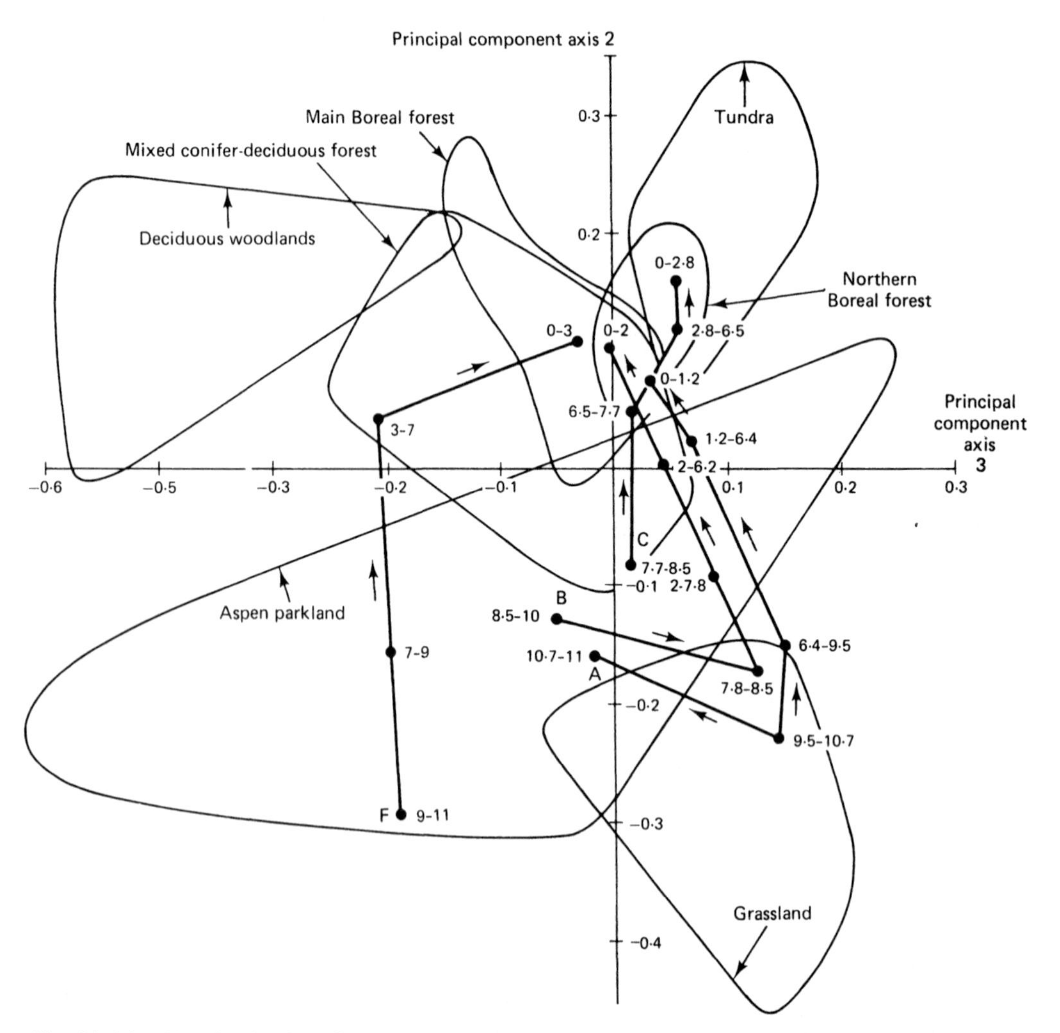

Fig. 11.15 The distribution of the major vegetation formations in Canada on the second and third axes of a principal components analysis of the 110 modern pollen spectra of Lichti-Federovich and Ritchie (1968). Mean weightings for each fossil pollen assemblage zone from four sites, A, B, C, and F, are shown as solid dots, and the trends by continuous lines. ^{14}C-year age ranges (in thousands of years) are shown for each zone. (After Ritchie and Yarranton, 1978a.)

11.15. Sites A and B start with a *Picea*-dominated assemblage which is then replaced by grassland during the prairie expansion. The modern boreal forest developed at about 6000 B.P. A grassland phase was absent from the more north-eastern sites C and F (modern samples from the aspen parkland tend to contain high percentages of *Picea* and *Pinus* pollen as a result of recent planting). The warm dry climatic phase is reflected by an increase in deciduous trees, after which vegetation resembling the modern boreal forest developed here also. This study indicates that the present northern boreal forest of central Canada has developed from several types of past vegetation, and not simply from one type in the past.

Disadvantages and limitations of the comparative approach

1. Each basin of deposition may have different properties as a pollen trap (see Chapter 9). If the modern spectra for comparison come from the same basin as the fossil spectra, the comparison would be relatively straightforward. However, past and present spectra are rarely the same, and therefore modern analogues have to be sought elsewhere. The comparison has to be separated in space as well as in

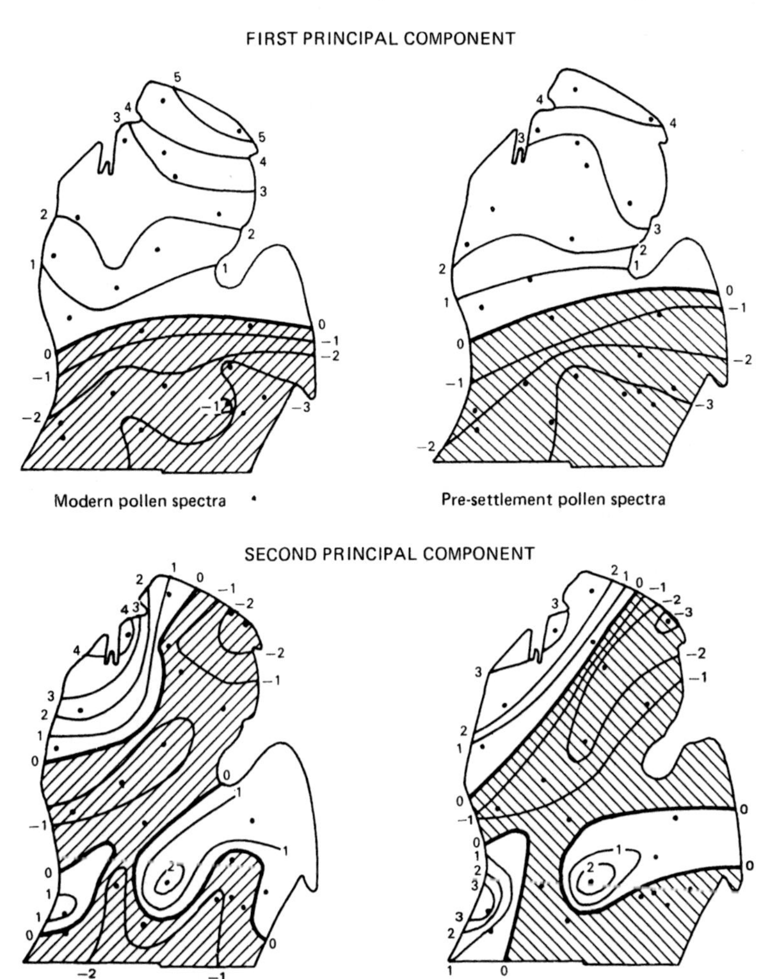

Fig. 11.16 Comparison of the results of principal components analysis of modern and pre-settlement pollen spectra from 23 lakes in Michigan, U.S.A. The scores on the first and second components are mapped. Areas with negative scores are hatched. (After Webb, 1973.)

time, and thus different pollen collecting properties of different lakes may well become important. A certain amount of inexactness must therefore be allowed for when making a match.

2. Not all modern vegetation types have been examined in terms of their modern pollen rain. It may not always be justified to conclude that no modern analogue exists, although the coverage of the major temperate vegetation formations is now becoming fairly good in the northern hemisphere.

3. Most modern vegetation has been disturbed by man, and therefore it can never exactly match fossil vegetation. Webb (1973) tested the degree to which disturbance had altered the spatial patterns in pollen rain in southern Michigan, U.S.A. Even when obvious indicators of human activity such as *Ambrosia* were removed from the pollen sum, the tree pollen assemblage still differed from presettlement assemblages in the greater quantities of *Betula*, *Ulmus*, and *Salix* pollen, and the smaller quantities of pollen of timber trees, such as *Fagus*, *Acer*, *Tsuga*, and *Pinus strobus*. However, PCA of modern and presettlement pollen percentages from all his 23 lakes showed a very similar disposition of the component scores (Fig. 11.16) showing that the over-all vegetational pattern had not been destroyed, and with specific and defined reservations, modern pollen spectra from Michigan could be usefully compared with fossil spectra (see Brubaker, 1975).

4. Many modern pollen spectra are derived from moss cushions or surface litter or soils, whereas most fossil pollen spectra are derived from lakes or peat bogs. Consequently, the modern spectra have a high local representation and may be used to

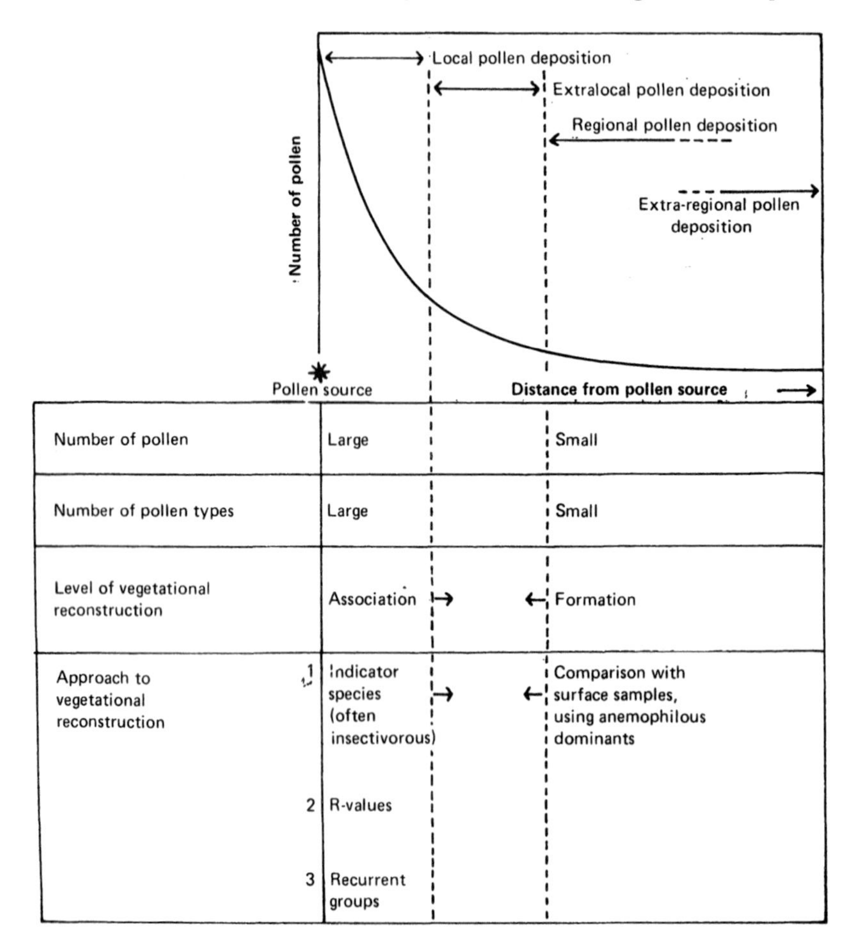

Fig. 11.17 Summary diagram of the distribution of pollen from a source, and the different approaches available for studying the pollen spectra at different distances from the source. (Modified from Janssen, 1980.)

distinguish individual communities (see Janssen, 1966, 1973). However, lake spectra are an amalgamation of pollen from all the surrounding communities. The representation of pollen from a point source at various distances away is discussed by Janssen (1980) and summarized in Fig. 11.17.

A moss cushion, soil sample, or a sample from a small (20–40 m) diameter bog contains a large proportion of locally produced pollen, with a large number of pollen types, some of which may be indicator species for the local community. The indicator species approach can be used to reconstruct the local community, probably at the association level (e.g. Janssen, 1972; Birks, 1973b). The R-value approach to reconstructing populations can also be profitably used (Chapter 10) at this scale and the delimitation of recurrent groups will probably result in assemblages representing local communities. In contrast, a pollen assemblage from the centre of a lake will contain pollen from the source at only a regional level, and it will also contain pollen from all surrounding sources at this level and lower. The number of pollen types will be small, because only abundant, wind-dispersed pollen will tend to reach the site, and consequently the reconstruction of the vegetation will be at a regional or formation level, by comparison with the anemophilous dominants in surface samples.

A practical demonstration of Janssen's generalization was shown by Ritchie (1974), when he compared moss and lake samples from the same region (p. 243), and Birks (1977) when he compared moss, sedge swamp, and lake samples from the south-west Yukon (p. 243).

5. In theory, the same pollen assemblage may be derived from two or more different vegetation types, because some ecologically important species produce low amounts of pollen, or poorly preserved pollen, which represent 'blind spots' in the vegetation. For example, aspen parkland samples are indistinguishable from grassland samples in Manitoba because of the poor preservation of *Populus* pollen; the poor representation of *Larix* pollen caused M. B. Davis problems in the estimation of R-values (see Chapter 10); and montane communities often contain major proportions of bryophytes and lichens, viviparous grasses, and Juncaceae with poorly preserved pollen (Birks, 1973b).

6. Difficulties arise in evaluating the significance of variations in pollen percentages between samples. For example, a small variation in percentages of a taxon with low pollen representation will represent greater differences in terms of vegetation composition than large variations in percentages of a taxon with high pollen representation (see Faegri and Iversen, 1975, p. 158). The analyst has to assess subjectively the variability in the percentages of each taxon which can be tolerated before the samples are declared to be different. For example, quite large variations in *Pinus* percentages mean relatively little in terms of vegetational differences, whereas variations of a similar magnitude for a taxon such as *Salix* with a low pollen representation would be of major importance in vegetational terms (see discussion of correction factors in Chapter 10).

7. Bias may occur when fossil and modern spectra are compared, usually in order to find a 'fit'. Comparisons using numerical methods avoid this bias, and emphasize situations where there is no modern analogue. Ogden (1969) used Spearman's rank correlation to compare modern and fossil spectra from North America. However, this is a relatively insensitive statistic, and more powerful methods are available, using multivariate techniques. We have already discussed the use of canonical variates analysis by Birks (1976) at Wolf Creek, Minnesota (p. 250), and of principal components analysis by Ritchie (1977) and Ritchie and Yarranton (1978a,b) in central Canada, and Webb (1974b) in Wisconsin. Adam (1970) was one of the pioneers of the use of numerical methods in Quaternary palynology. He used canonical variates analysis to find the optimal separation between groups of surface samples from oak grassland, desert grassland, and desert in south Arizona, U.S.A. He then positioned the fossil samples on the canonical variates axes, and was able to use their correspondence or non-correspondence to interpret the vegetational history of the area.

R. B. Davis *et al.* (1975) used 406 modern pollen spectra from eastern North America collated by Davis and Webb (1975) and calculated Pearson product-moment correlation coefficients (Yarranton and Ritchie, 1972) between them and pollen subzone average spectra from Moulton Pond, Maine, U.S.A. They then produced maps of the correlations. Figure 11.18 shows examples which can be compared with the distribution of the present vegetation. Between 13 400 and 12 400 B.P., Moulton Pond was surrounded by vegetation resembling the tundra and forest tundra of Canada, whereas between 4700 and 3900 B.P. it was surrounded by vegetation most resembling the northeastern transitional conifer–hardwood forests north-east of Lake

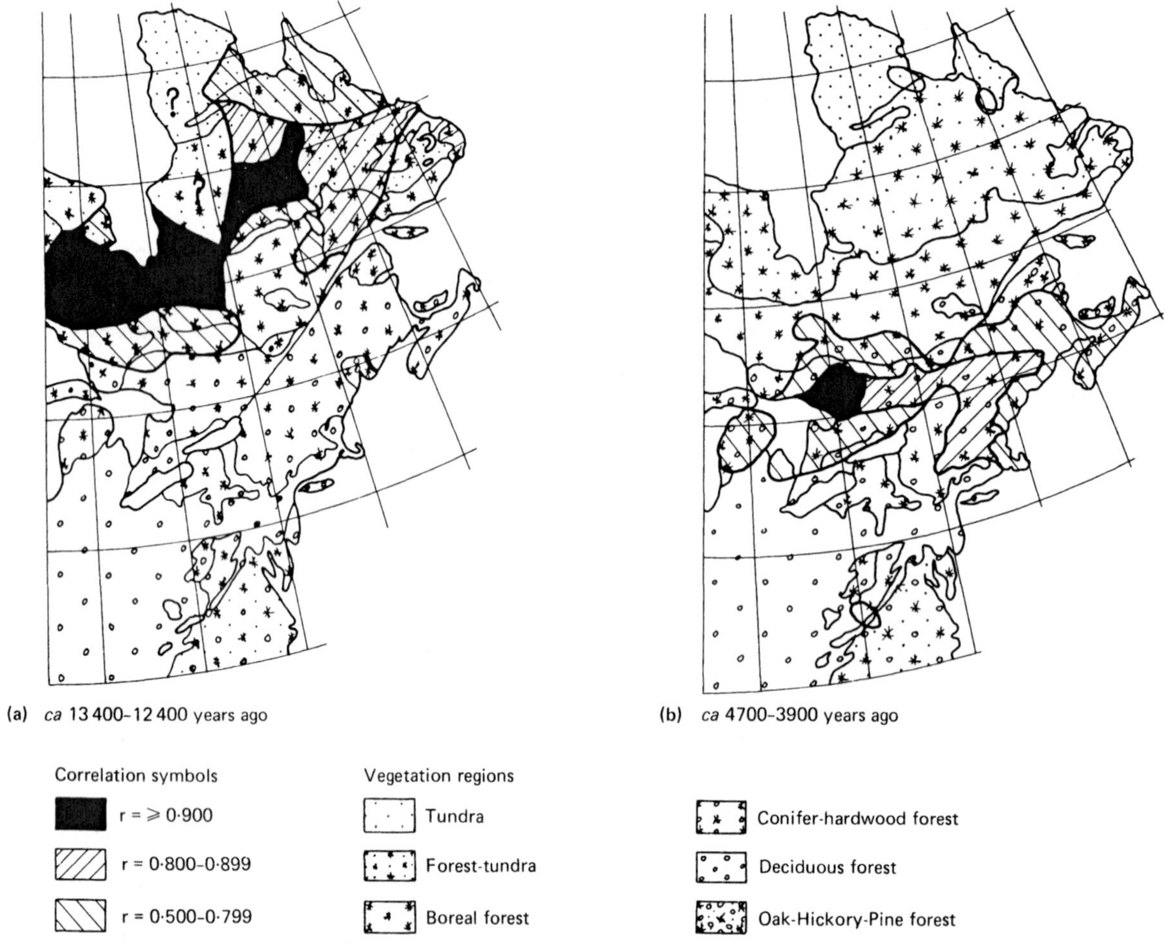

Fig. 11.18 Correlation maps of fossil pollen spectra from Moulton Pond, Maine, with 406 surface sample spectra in different vegetation types in eastern North America. (**a**) Comparison of fossil spectra 13 400–12 400 years old. (**b**) Comparison of specta 4700–3900 years old. The present vegetational regions are indicated (After Davis *et al.*, 1975.)

Huron. This method encourages interpretations in terms of modern vegetation to be made, although pollen zones with relatively over-all low correlation coefficients may indicate that no true modern analogue exists. However, the mapping approach is useful in that it allows the reader to see geographical relationships at a glance, and appreciate vegetational changes in the past by the distance the vegetation type has moved since it surrounded the site.

8. It has long been appreciated (Firbas, 1949; Szafer, 1935; Davis, 1976) that plants have migrated to their present ranges during the Holocene or post-glacial at different rates. Therefore modern analogues of early post-glacial forests probably do not now exist at all. The past ecological behaviour of taxa may also have been different in the absence of competition from their present associates. Therefore it may be difficult to reconstruct past communities by extending their present ecological tolerances back into the past (Janssen, 1970). However, groups of taxa have tended to occur together in the past which have ecological association at the present, for example in the last glaciation (Birks, 1973a; Andersen, 1961), in the Eemian (last) interglacial (Andersen, 1964, 1966), and even in the Pliocene (Zagwijn, 1967). The fact that such reconstructions are possible suggests that the ecology of many species has remained relatively constant since the Tertiary. Indeed, many vicariant species in N. America and Europe have comparable ecologies (e.g. *Fraxinus excelsior*, *Fraxinus nigra*, both on flood plains; *Pinus sylvestris*, *Pinus banksiana* on sandy soils; *Ledum palustre*, *Ledum groenlandicum* on bogs), suggesting that their ecological amplitude was established before the continuous distribution of species was broken up in the early Tertiary and subsequent taxonomic evolution occurred (Janssen, 1980). Therefore it may not be unreasonable to attempt to reconstruct vegetation types from a knowledge of the present autoecology and synecology of individual taxa.

The value of the comparative approach

The comparative approach is most useful when fossil spectra can be matched with modern spectra which are

i) distinctive from all other modern spectra
ii) refer to characteristic and homogeneous vegetation types
iii) derived from comparable sites of deposition in topographically similar areas.

Often surface samples are obtained from within an area of homogeneous vegetation, whereas the fossil spectra amalgamate pollen from a mosaic of communities surrounding the site. Modern spectra are most useful for reconstructions at the formation or vegetation-landform unit level. More detailed reconstructions can be made with the use of indicator species, which can reduce the level of vegetational reconstruction to that of the association if the fossil samples are from small basins or soils.

If no modern analogue can be found after an extensive search, it must be concluded that no modern analogue exists. Gleason's (1939) individualistic concept of the nature of vegetation implies that vegetation varies continuously in space and constantly in time, and it has been shown to have some foundation in the past behaviour of trees in particular. West (1964, 1970) has demonstrated how different combinations of the same tree taxa came together during the series of interglacials in Europe. This led him to remark '... we may conclude that our present plant communities have no long history in the Quaternary, but are merely temporary aggregations under given conditions of climate, other environmental factors, and historical factors' (West, 1964, p. 55). Hence there is no reason why past communities should be the same as modern ones. In this case, the investigator has to rely on estimates of past abundance of the taxon, as we discussed in Chapter 10.

Nevertheless, the comparative approach is probably the soundest method currently available for the reconstruction of past communities from pollen diagrams. In most cases it can be combined with the other approaches. A palaeoecologist frequently considers the pollen representation of the taxa concerned in order to make a rough estimate of their abundance in the past vegetation. From a knowledge of their present autecology and synecology, he can suggest how they may have been grouped into communities. In some cases indicator species can be employed to yield more precise information about particular plant communities, depending upon the site investigated and the taxonomic precision of the identifications. Groups of taxa are often delimited from a knowledge of their present ecological behaviour, such as aquatics, reed-swamp species, forest species, species of open ground, etc. (Iversen, 1954; Janssen, 1967). From their behaviour through time, the sequence of ecological events in the succession can be reconstructed.

The distribution of past vegetation

Because pollen percentages correspond in a broad way with the vegetation which produced them, it is possible to reconstruct not only the vegetation around a site, but also the vegetation over an area at a particular time. If the geographical patterns of the pollen percentages change through time, the rate and direction of the movement of individual species and whole vegetation types can be followed.

Birks and Saarnisto (1975) discuss the construction of fossil isopollen maps. The percentages of each taxon can be mapped at a particular point in time. The individual maps can then be combined by an ordination technique to show the main discontinuities in the combined spectra, which should correspond to the major vegetation divisions. Figure 11.19 shows the distribution of sample scores on the first principal component of pollen percentage data from Finland. At 8000 B.P. the main division lies between the *Betula*-dominated spectra in the north and the *Pinus*-dominated spectra in the south. At 6000 B.P. the first component distinguishes pollen spectra with high percentages of deciduous tree pollen in the south from those with high *Pinus* values further north, and *Betula*-dominated spectra in Lappland. At 4000 B.P. the major distinction is

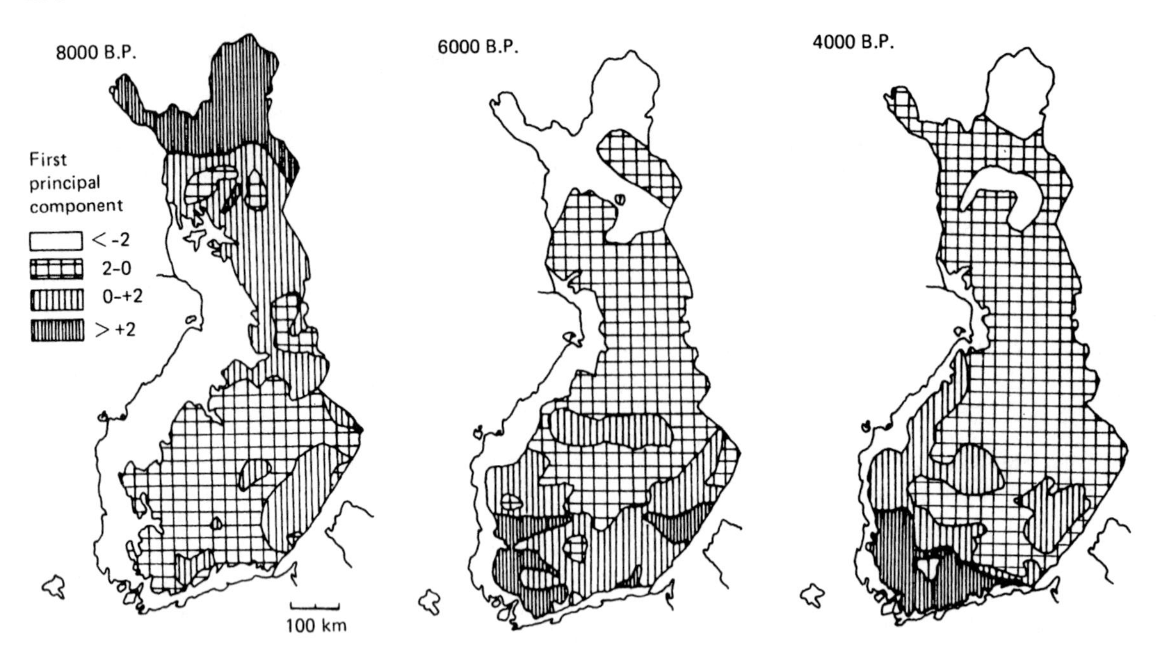

Fig. 11.19 Maps of the scores on the first principal component of a principal components analysis of Finnish fossil pollen data at 8000, 6000, and 4000 B.P. (After Birks and Saarnisto, 1975.)

between spectra with high values of deciduous tree pollen in the south-west, and spectra dominated by *Pinus* and *Picea* over most of the rest of the country. If the major boundaries in the maps of the PCA scores correspond with major vegetation boundaries we can see how these boundaries have changed through time. There are two major boundaries. The one in the south delimiting deciduous from coniferous forest has remained fairly static from 8000 to 4000 B.P. In the north, the boundary between conifer-dominated forests and birch forests and tundra has moved progressively northwards during this time. The relationship between tree percentages and pollen percentages in Finland has been studied by Prentice (1978) (see p. 239). With these data it may be possible to reconstruct past tree percentages, and produce more detailed maps of past vegetation.

Birks *et al.* (1975) constructed isopollen maps for 5000 B.P. in the British Isles, just before the onset of forest clearance by Neolithic man. Maps of the individual tree taxa showed striking patterns, which were summarized using PCA, and mapping the component scores as before (Fig. 11.20). The first component distinguished northern Scotland and south-west Ireland with high values of *Betula* and *Pinus* pollen from the rest of Britain with lower *Betula* and *Pinus* percentages and high percentages of deciduous trees. The second component strongly distinguished central Ireland, with high *Ulmus* and *Corylus* percentages from the rest of Britain. The third component distinguished central and southern England from the north and west. This division follows the well known present phytogeographical division of England, and indicates that this division separating the warmer south with more fertile Mezozoic rocks from the cooler wetter north and west with more acid Palaeozoic rocks, acted on the vegetation in the past. As Godwin (1940) emphasized, the Flandrian pollen record convincingly demonstrates the ancient and permanent character of vegetation differentiation within the British Isles, which the presence of man has not changed.

Bernabo and Webb (1977) mapped pollen percentages of major tree taxa at specified times in the Holocene of eastern North America. They emphasized the migration and changing abundance of taxa by constructing percentage difference maps over particular time intervals and isochrone maps of the boundaries of major vegetation types at different time intervals. For example, Fig. 11.21 illustrates the movement of the prairie–forest border through the Holocene. Their maps illustrate the progressive shrinkage of the *Picea*-dominated boreal forest after 11 000 B.P. as it became

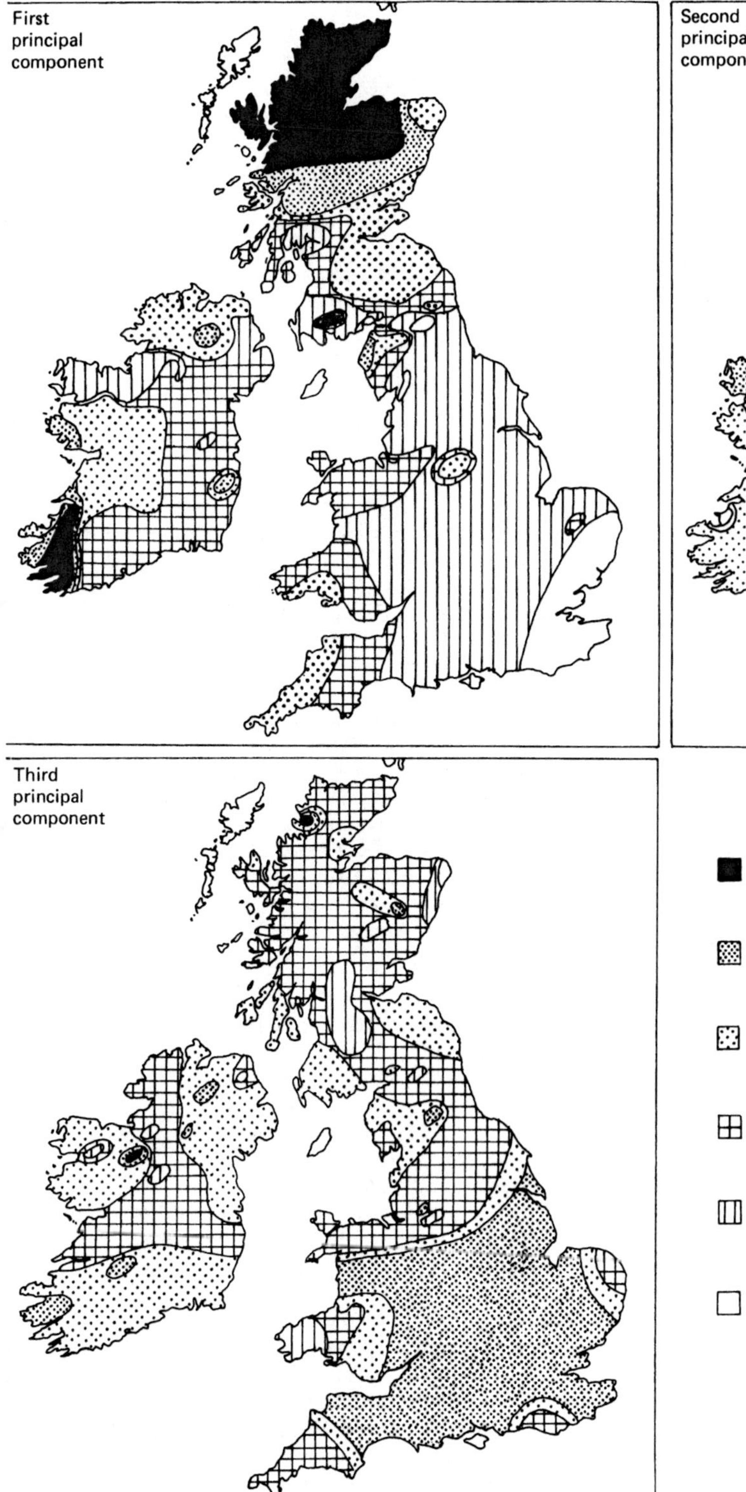

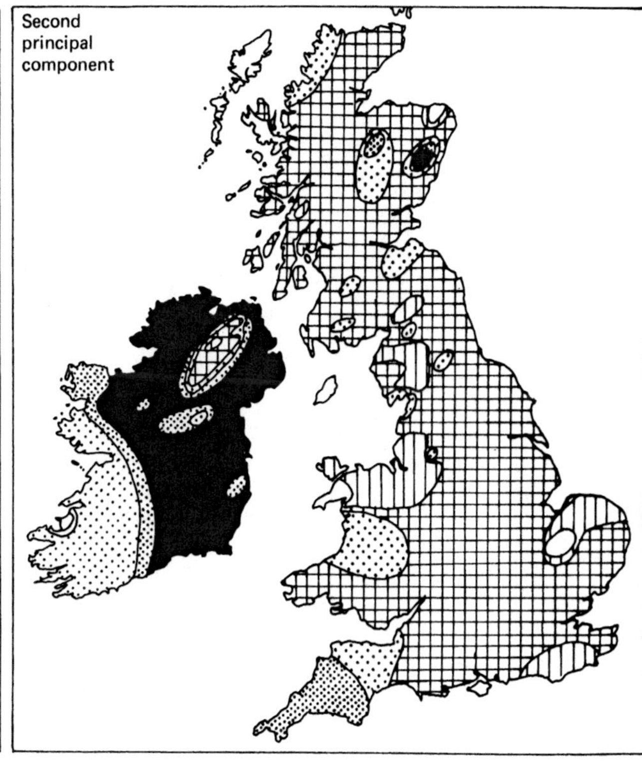

Fig. 11.20 Maps of the scores on the first three principal components of a principal components analysis of fossil pollen spectra from the British Isles 5000 years old. (After Birks *et al.*, 1975.)

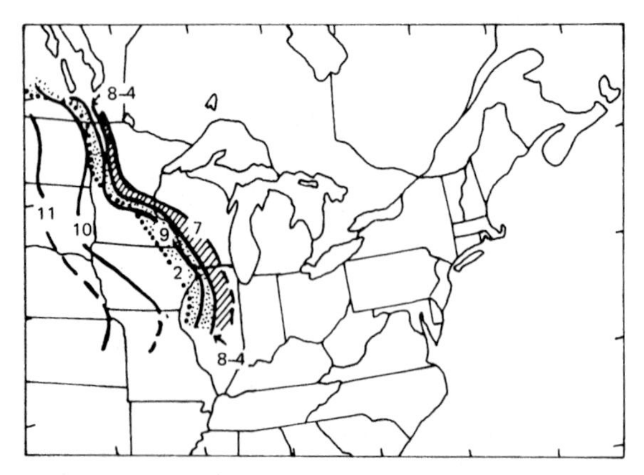

Fig. 11.21 Isochrones, in thousands of years B.P. illustrating the eastward expansion and subsequent westward retreat of the prairie/forest ecotone, based on isopoll maps of herb pollen from north-east American pollen diagrams. The shaded area shows the region over which the prairie retreated after its maximum extent 7000 years ago. (After Bernabo and Webb, 1977.)

compressed between the ice margin and the invasion of *Pinus* from the south and east. Simultaneously, deciduous forest moved northward and prairie expanded from the west. The prairie reached its easternmost extent at about 7000 B.P. and then started a westward retreat, while the boreal and deciduous zones moved slowly southwards again. The effect of European man was well marked by an increase in herb pollen percentages.

In a detailed local study of the Dinkel valley, Netherlands, Van der Hammen and Bakker (1971) used pollen data combined with a knowledge of the present distribution of soil types and vegetation to construct tentative maps of the past vegetation of the valley in Allerød time and in the late post-glacial (sub-Boreal) at the association level. Kral (1971) similarly constructed maps of the past vegetation of the Dachstein mountains of Austria. All these studies emphasize the value of mapping fossil pollen data in the way pioneered by Szafer (1935) as a means of displaying geographical variations in past and present pollen assemblages, and by inference, in past vegetation.

References

Reconstruction of past communities – general

BIRKS, H. J. B. (1973a). *Past and Present Vegetation of the Isle of Skye – a Palaeoecological Study*. Cambridge University Press (Chapters 11 and 13).

DAVIS, M. B. (1969). Palynology and environmental history during the Quaternary Period. *Amer. Sci.*, **57**, 317–32.

JANSSEN, C. R. (1970). Problems in the recognition of plant communities in pollen diagrams. *Vegetatio*, **20**, 187–98.

JANSSEN, C. R. (1980). On the reconstruction of past vegetation by pollen analysis: a review. Proc. IVth International Palynological Conference, Lucknow (in press).

OLDFIELD, F. (1970). Some aspects of scale and complexity in pollen-analytically based palaeoecology. *Pollen Spores*, **12**, 163–71.

WALKER, D. (1972). Quantification in historical plant ecology. *Proc. Australian Ecol. Soc.*, **6**, 91–104.

Reconstruction of communities preserved *in situ*

BIRKS, H. H. (1975). Studies in the vegetational history of Scotland. IV. Pine stumps in Scottish blanket peats. *Phil. Trans. R. Soc.* B, **270**, 181–226.

BIRKS, H. H. AND MATHEWES, R. M. (1978). Studies in the vegetational history of Scotland. V. Late Devensian and early Flandrian pollen and macrofossil stratigraphy at Abernethy Forest, Inverness-shire. *New Phytol.*, **80**, 455–84.

BIRKS, H. J. B. (1970). Inwashed pollen spectra at Loch Fada, Isle of Skye. *New Phytol.*, **69**, 807–20.

CASPARIE, W. A. (1972). *Bog Development in Southeastern Drenthe (The Netherlands)*. Dr. Junk.

GODWIN, H. (1968). Terneuzen and buried forests of the East Anglian Fenland. *New Phytol.*, **67**, 733–8.

RYBNÍČEK, K. AND RYBNÍČKOVÁ, E. (1968). The history of flora and vegetation on the Blátomire in southeastern Bohemia, Czechoslovakia. *Folia Geobot. Phytotax.*, **3**, 117–42.

Statistical approaches to the reconstruction of past communities

BUZAS, M. A. (1969). On the quantification of biofacies. *Proc. N. Amer. Paleont. Convention B*, 101–16.

CLAPHAM, W. B. (1970a). Nature and paleogeography of Middle Permian floras of Oklahoma as inferred from their pollen record. *J. Geol.*, **78**, 153–71.

CLAPHAM, W. B. (1970b). Evolution of Upper Permian terrestrial floras of Oklahoma as determined from pollen and spores. *Proc. N. Amer. Paleont. Convention E*, 411–27.

CLAPHAM, W. B. (1972). Numerical analysis and group formation in palynology. *Geoscience and Man*, **4**, 73–85.

HARRIS, W. F. AND NORRIS, G. (1972). Ecologic significance of recurrent groups of pollen and spores in Quaternary sequences from New Zealand. *Palaeogeography, Palaeoclimatology, Palaeoecology*, **11**, 107–24.

MARTIN, A. R. H. (1968). Pollen analysis of Groenvlei lake sediments, Knysna, South Africa. *Rev. Palaeobot. Palynol.*, **7**, 107–44.

MARTIN, P. S. AND MOSIMANN, J. E. (1965). Geochronology of Pluvial Lake Cochise, Southern Arizona. III. Pollen statistics and Pleistocene metastability. *Am. J. Sci.*, **263**, 313–58.

MOSIMANN, J. E. (1964). Null models and their use in interpreting fossil pollen frequencies. Unpubl. manuscript.

OLTZ, D. F. (1971). Cluster analysis of Late Cretaceous–Early Tertiary pollen and spore data. *Micropaleontology*, **17**, 221–32.

ORLOCI, L. (1967). An agglomerative method for classification of plant communities. *J. Ecol.*, **55**, 193–206.

REYMENT, R. A. (1963). Multivariate analytical treatment of quantitative species associations: an example from palaeoecology. *J. Anim. Ecol.*, **32**, 535–47.

SHAFFER, B. L. AND WILKE, S. C. (1965). The ordination of fossil communities: an approach to the study of species interrelationships and communal structure. *Papers Michigan Acad. Sci. Arts & Letters*, **50**, 199–214.

The indicator-species approach to reconstruction of past communities

ANDERSEN, S. T. (1961). Vegetation and its environment in Denmark in the Early Weichselian Glacial. *Danm. geol. Unders.* Ser. II, **75**, 175 pp.

DAVIS, M. B., BRUBAKER, L. B. AND WEBB, T. (1973). Calibration of absolute pollen influx. In *Quaternary Plant Ecology* (eds H. J. B. Birks and R. G. West). Blackwell.

IVERSEN, J. (1954). The late-glacial flora of Denmark and its relation to climate and soil. *Danm. geol. Unders.* Ser. II, **80**, 87–119.

IVERSEN, J. (1964). Plant indicators of climate, soil, and other factors during the Quaternary. *Rep. VI Intern. Congr. Quater. Warsaw 1961*, Sect. **2**, 421–8.

JANSSEN, C. R. (1967). A post-glacial pollen diagram from a small *Typha* swamp in Northwestern Minnesota, interpreted from pollen indicators and surface samples. *Ecol. Monogr.*, **37**, 145–72.

JANSSEN, C. R. (1972). The palaeoecology of plant communities in the Dommel Valley, North Brabant, Netherlands. *J. Ecol.*, **60**, 411–37.

RYBNÍČKOVÁ, E. AND RYBNÍČEK, K. (1971). The determination and elimination of local elements in pollen spectra from different sediments. *Rev. Palaeobot. Palynol.*, **11**, 165–76.

The comparative approach to reconstruction of past communities

AARIO, L. (1940). Waldgrenzen und subrezenten pollenspektren in Petsamo, Lappland. *Ann. Acad. Sci. Fenn.*, A**54**(8), 120 pp.

ADAM, D. P. (1970). *Some Palynological Applications of Multivariate Statistics.* University of Arizona Ph.D. thesis.

ANDERSEN, S. T. (1964). Interglacial plant successions in the light of environmental changes. *Rep. VI Intern. Congr. Quater. Warsaw 1961*, Sect. **2**, 359–68.

ANDERSEN, S. T. (1966). Interglacial vegetational succession and lake development in Denmark. *Palaeobotanist*, **15**, 117–27.

AUER, V. (1927). Untersuchungen über die Waldgrenzen und Torfböden in Lappland. *Comm. Inst. Quaest. Forest. Finlandiae*, **12**, 1–52.

BIRKS, H. J. B. (1973b). Modern pollen rain studies in some arctic and alpine environments. In *Quaternary Plant Ecology* (eds H. J. B. Birks and R. G. West). Blackwell.

BIRKS, H. J. B. (1976). Late-Wisconsinan vegetational history at Wolf Creek, Central Minnesota. *Ecol. Monogr.*, **46**, 395–429.

BIRKS, H. J. B. (1977). Modern pollen rain and vegetation of the St. Elias Mountains, Yukon Territory. *Can. J. Bot.*, **55**, 2367–82.

BIRKS, H. J. B., WEBB, T. AND BERTI, A. A. (1975). Numerical analysis of surface samples from central Canada: a comparison of methods. *Rev. Palaeobot. Palynol.*, **20**, 133–69.

BRUBAKER, L. B. (1975). Post-glacial forest patterns associated with till and outwash in Northcentral Upper Michigan. *Quat. Res.*, **5**, 499–527.

COLINVAUX, P. A. (1964). The environment of the Bering Land Bridge. *Ecol. Monogr.*, **34**, 297–329.

DAVIS, M. B. (1967). Late-glacial climate in Northern United States. A comparison of New England and the Great Lakes Region. In *Quaternary Paleoecology* (eds E. J. Cushing and H. E. Wright). Yale University Press.

DAVIS, R. B., BRADSTREET, T. E., STUCKENRATH, R. AND BORNS, H. W. (1975). Vegetation and associated environments during the past 14 000 years near Moulton Pond, Maine. *Quat. Res.* **5**, 435–65.

DAVIS, R. B. AND WEBB, T. (1975). The contemporary distribution of pollen in Eastern North America: a comparison with vegetation. *Quat. Res.*, **5**, 395–434.

DONNER, J. J. (1972). Pollen frequencies in the Flandrian sediments of Lake Vakojärvi, south Finland. *Comm. Biol.*, **53**, 19 pp.

FAEGRI, K. AND IVERSEN, J. (1975). *Textbook of Pollen Analysis* (3rd edition). Blackwell.

FIRBAS, F. (1934). Über die Bestimmung der Walddichte und der Vegetation Waldloser Gebiete mit Hilfe der Pollenanalyse. *Planta*, **22**, 109–45.

FREDSKILD, B. (1969). A postglacial standard pollen diagram from Peary Land, North Greenland. *Pollen Spores*, **11**, 573–83.

FREDSKILD, B. (1973). Studies in the vegetational history of Greenland. *Meddr. om Grønland*, **198**(4), 245 pp.

GLEASON, H. A. (1939). The individualistic concept of the plant association. *Amer. Midl. Nat.*, **21**, 92–108.

HICKS, S. (1977). Modern pollen rain in Finnish Lappland investigated by analysis of surface moss samples. *New Phytol.*, **78**, 715–34.

JANSSEN, C. R. (1966). Recent pollen spectra from the deciduous and coniferous deciduous forests of north-eastern Minnesota: a study in pollen dispersal. *Ecology*, **47**, 804–25.

JANSSEN, C. R. (1967). A post-glacial pollen diagram from a small *Typha* swamp in northwestern Minnesota, interpreted from pollen indicators and surface samples. *Ecol. Monogr.*, **37**, 145–72.

JANSSEN, C. R. (1973). Local and regional pollen deposition. In *Quaternary Plant Ecology* (eds H. J. B. Birks and R. G. West). Blackwell.

JANSSEN, C. R. (1980). On the reconstruction of past vegetation by pollen analysis: a review. *Proc. IVth Int. Palynol. Conference, Lucknow, India* (in press).

LICHTI-FEDEROVICH, S. AND RITCHIE, J. C. (1968). Recent pollen assemblages from the Western Interior of Canada. *Rev. Palaeobot. Palynol.*, **7**, 297–344.

LIVINGSTONE, D. A. AND ESTES, A. H. (1967). A carbon-dated pollen diagram from the Cape Breton Plateau, Nova Scotia. *Can. J. Bot.*, **45**, 339–59.

MATTHEWS, J. V. (1970). Quaternary environmental history of interior Alaska: pollen samples from organic colluvium and peats. *Arctic and Alpine Research*, **2**, 241–51.

MATTHEWS, J. V. (1974). Quaternary environments at Cape Deceit (Seward Peninsula, Alaska): evolution of a tundra ecosystem. *Bull. Geol. Soc. Amer.*, **85**, 1353–84.

MCANDREWS, J. H. (1966). Postglacial history of Prairie, Savanna, and Forest in Northwestern Minnesota. *Mem. Torrey Bot. Club*, **22**(2), 72 pp.

MCANDREWS, J. H. (1967). Pollen analysis and vegetational history of the Itasca region, Minnesota. In *Quaternary Paleoecology* (eds E. J. Cushing and H. E. Wright). Yale University Press.

OGDEN, J. G. (1969). Correlation of contemporary and Late Pleistocene pollen records in the reconstruction of Postglacial environments in Northeastern North America. *Mitt. Int. Verein. Limnol.*, **17**, 64–77.

PRENTICE, I. C. (1978). Modern pollen spectra from lake sediments in Finland and Finmark, north Norway. *Boreas*, **7**, 131–53.

RAMPTON, V. (1971). Late Quaternary Vegetational and Climatic History of the Snag-Klutlan area, Southwest Yukon Territory, Canada. *Bull. Geol. Soc. Amer.*, **82**, 959–78.

RITCHIE, J. C. (1974). Modern pollen assemblages near the arctic tree line, Mackenzie Delta region, Northwest Territories. *Can. J. Bot.*, **52**, 381–96.

RITCHIE, J. C. (1977). The modern and late Quaternary vegetation of the Campbell-Dolomite uplands, near Inuvik, N.W.T. Canada. *Ecol. Monogr.*, **47**, 401–23.

RITCHIE, J. C. AND LICHTI-FEDEROVICH, S. (1967). Pollen dispersal phenomena in Arctic-Subarctic Canada. *Rev. Palaeobot. Palynol.*, **3**, 255–66.

RITCHIE, J. C. AND YARRANTON, G. A. (1978a). The late-Quaternary history of the Boreal Forest of central Canada, based on standard pollen stratigraphy and principal components analysis. *J. Ecol.*, **66**, 199–212.

RITCHIE, J. C. AND YARRANTON, G. A. (1978b). Patterns of change in the late-Quaternary vegetation of the Western Interior of Canada. *Can. J. Bot.*, **56**, 2177–83.

WEBB, T. (1973). A comparison of modern and pre-settlement pollen from southern Michigan, U.S.A. *Rev. Palaeobot. Palynol.*, **16**, 137–56.

WEBB, T. (1974a). Corresponding patterns of pollen and vegetation in Lower Michigan: a comparison of quantitative data. *Ecology*, **55**, 17–28.

WEBB, T. (1974b). A vegetational history from Northern Wisconsin: evidence from modern and fossil pollen. *Amer. Midl. Natur.*, **92**, 12–34.

WEBB, T. AND MCANDREWS, J. H. (1976). Corresponding patterns of contemporary pollen and vegetation in Central North America. *Geol. Soc. Amer. Memoir*, **145**, 267–99.

WEST, R. G. (1964). Inter-relations of ecology and Quaternary palaeobotany. *J. Ecol.*, **52** (Suppl.), 47–57.

WEST, R. G. (1970). Pleistocene history of the British flora. In *Studies in the Vegetational History of the British Isles* (eds D. Walker and R. G. West). Cambridge University Press.

WRIGHT, H. E. (1967). The use of surface samples in Quaternary pollen analysis. *Rev. Palaeobot. Palynol.*, **2**, 321–30.

WRIGHT, H. E., MCANDREWS, J. H. AND VAN ZEIST, W. (1967). Modern pollen rain in Western Iran, and its relation to plant geography and Quaternary vegetational history. *J. Ecol.*, **55**, 415–43.

YARRANTON, G. A. AND RITCHIE, J. C. (1972). Sequential correlations as an aid in placing pollen zone boundaries. *Pollen Spores*, **14**, 213–23.

ZAGWIJN, W. H. (1967). Ecologic interpretation of a pollen digram from Neogene beds in the Netherlands. *Rev. Palaeobot. Palynol.*, **2**, 173–81.

Palaeovegetation and fossil isopollen maps

BERNABO, J. C. AND WEBB, T. (1977). Changing patterns in the Holocene pollen record of northeastern North America: a mapped summary. *Quat. Res.*, **8**, 64–96.

BIRKS, H. J. B., DEACON, J. AND PEGLAR, S. M. (1975). Pollen maps for the British Isles 5000 years ago. *Proc. R. Soc.* B, **189**, 87–105.

BIRKS, H. J. B. AND SAARNISTO, M. (1975). Isopollen maps and principal components analysis of Finnish pollen data for 4000, 6000, and 8000 years ago. *Boreas*, **4**, 77–96.

DAVIS, M. B. (1976). Pleistocene biogeography of temperate deciduous forests. *Geoscience and Man*, **13**, 13–26.

FIRBAS, F. (1949). *Waldgeschichte Mitteleuropas, 1: Allgemeine Waldgeschichte.* Gustav Fischer, Jena.

GODWIN, H. (1940). Pollen analysis and forest history of England and Wales. *New Phytol.*, **39**, 370–400.

KRAL, F. (1971). Pollenanalytische Untersuchungen zur Waldgeschichte des Dachsteinmassivs. *Veröff. Inst. Waldbau Wien*, 145 pp.

SZAFER, W. (1935). The significance of isopollen lines for the investigation of the geographical distribution of trees in the post-glacial period. *Bull. l'Acad. Sci. Polon.* B, 1935, 235–9.

VAN DER HAMMEN, T. AND BAKKER, J. A. (1971). Former vegetation, landscape and man in the Dinkel valley. *Med. Rijks Geol. Dienst.*, N.S. **22**, 147–58.

12
The reconstruction of past environments

The reconstruction of plant communities in space

Having discussed ways of reconstructing floras, plant populations, and plant communities in the past, we can now ask: what space did each community occupy? This is the fourth question in the series of questions we asked in the interpretation of pollen diagrams on p. 195. It is a very difficult question to answer, particularly from pollen analytical evidence, because of the variable dispersal of pollen from the parent plants. Tauber's (1965) model of pollen dispersal (see Chapter 9) indicates that, of the pollen sedimented into a pond 100–200 m diameter, about 80% would have come from the trunk space (Janssen's (1966) local and extra-local components) and the remaining 20% would have come from the canopy and rainout components (Janssen's regional and extra-regional components). In contrast, of the pollen in the sediments of a large lake several kilometres in diameter, about 10% would have come from the trunk space, about 70% from the canopy component, and the remaining 20% from the rainout component. Hence the size of the basin of deposition is important in the reconstruction of local and regional vegetation (see Chapter 11 and Fig. 11.17) and it is important to realize the considerable distances from which pollen in lake sediments can be derived. For example, of the pollen in a small lake, about 50% comes from within 250 m of the site, and about 75% from within 5 km. Only about 5% originates beyond about 10 km. However, the opposite is the case for a large lake. For a sampling site 1 km from the forest edge, about 50% of the pollen is derived from a radius of 10 km, and for a site 5 km from the forest edge, about 50% of the pollen is derived from a radius of 30 km.

The distribution of pollen from a source

The rapid fall off of pollen from a source, predicted by Tauber's (1965) model, has been tested by Raynor and Ogden (1965) and Raynor *et al.* (1968). They showed that *Ambrosia* pollen concentrations reached levels indistinguishable from the background regional deposition within 100–200 m of the source. Similarly, Turner (1964) demonstrated a fall of *Pinus* pollen percentages to a regional level within about 300 m (see Chapter 9 and Fig. 9.3). Janssen (1966, 1973) also effectively demonstrated the fall off of pollen from the source in transects of surface samples across bogs in Minnesota. The local pollen component from the trunk space usually declined very steeply with distance, because of effective filtration by neighbouring shrubby vegetation. Herb pollen from the forest understorey was rarely dispersed beyond the community in which it was produced.

There are two aspects to our question of what space did each community occupy? Firstly, we can consider vegetational differentiation over a small geographical area, and secondly, we can attempt to assess the area which each community occupied.

Vegetational differentiation in a small geographical area

There are several studies which have attempted to describe vegetational differentiation in the past. We have already discussed McAndrews' (1966) investigation along the Itasca transect (Chapter 11). Birks' (1973) study of the Isle of Skye revealed vegetational differences in Devensian late-glacial times equivalent to differences observable at the present day, which could be explained in terms of geology and soil type, and also local climate as affected by altitude and wind exposure. In the Pennine hills of northern England, Turner and Hodgson (1979) also demonstrated local differentiation. The slight differences in their pollen diagrams were effectively identified by principal components analysis. The differences could be explained in terms of soil type, drainage, and altitude.

We have already discussed in Chapter 11 how

Lichti-Federovich and Ritchie (1968) and others have demonstrated that regional vegetation at the formation or vegetation-landform level can be reflected by surface pollen samples from lake sediments. The reconstruction of past vegetational differences are also usually at this level; a change in the pollen assemblage indicates a change in the vegetation-landform or formation type. Because of the rapid fall-off of pollen from a source, the transition between two vegetation types in pollen terms will be quite narrow. For example, Smith (1964) demonstrated differences between two sites in Northern Ireland (Cannons Lough and Fallahogy) which were only 5 km apart, Pennington (1970, 1973) showed differences between sites in the English Lake District depending upon their altitude, and Brubaker (1975) showed how soil differences on adjacent outwash and till in Michigan influenced the vegetation in the past (see Chapter 10 and Fig. 12.1).

The area covered by a plant community

Very few people have attempted to assess the area covered by a community using evidence from pollen

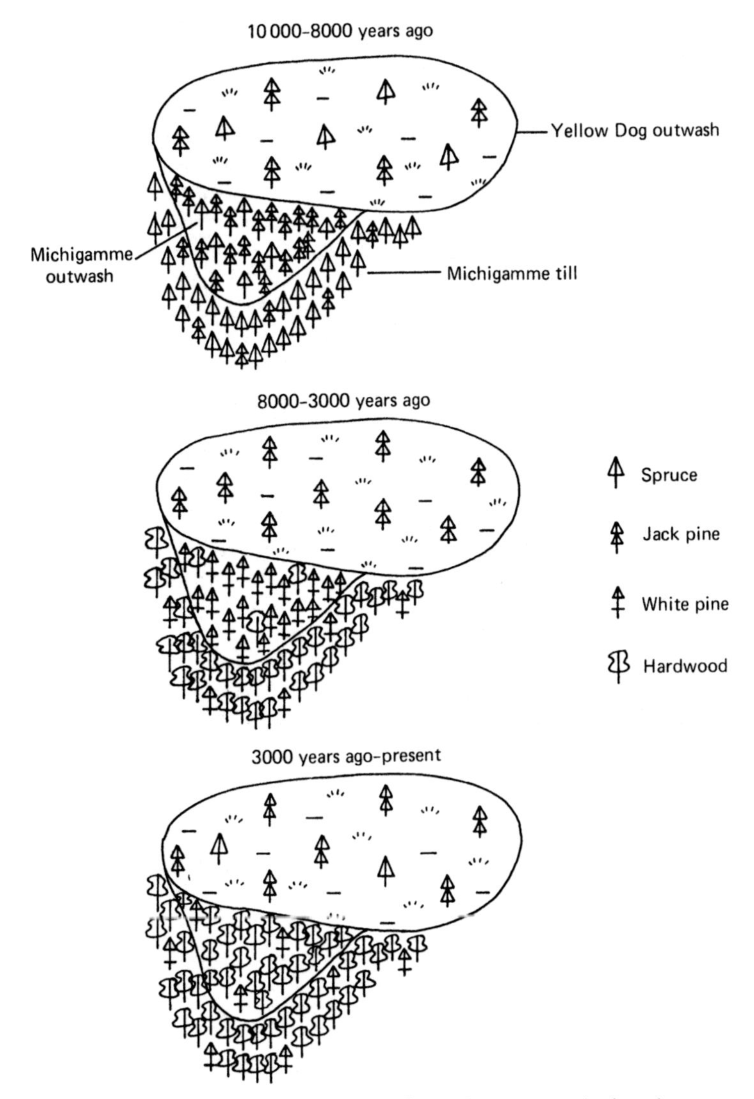

Fig. 12.1 The inferred distribution of past forest types during the Holocene on till and outwash soils in Michigan. (After Brubaker, 1975.)

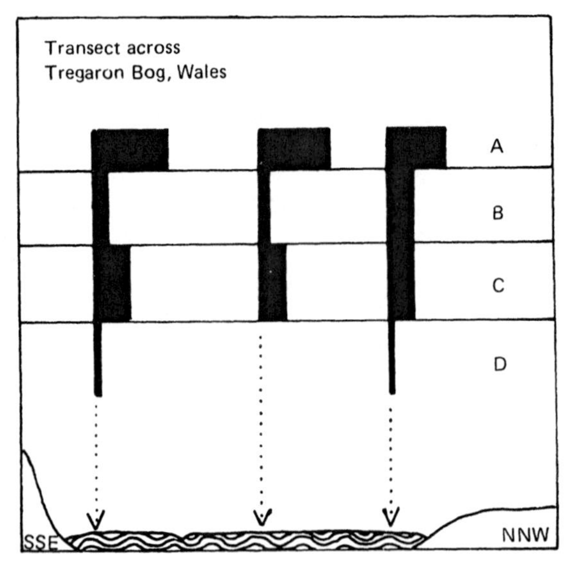

Fig. 12.2 Schematic grass pollen curves from three pollen diagrams across Tregaron Bog, Wales. A, B, C, and D, are clearance phases, discussed in the text. (After Turner, 1965.)

diagrams, because it is so difficult. The nearest approach has been made by Turner (1970) who considered the question: how large were the areas of forest cleared by prehistoric man? Two types of forest clearance are detectable in pollen diagrams from southern Scotland and Wales. Small temporary clearances ('landnam' of Iversen, 1941, 1949) lasted about 50 years (Turner, 1965; Smith and Willis, 1961) and were followed by forest regeneration as the people moved on elsewhere. Extensive clearances involved larger areas which were maintained for several hundred years. Small temporary clearances can only be detected in pollen diagrams near to the clearances. Otherwise they are masked by the regional pollen rain. For example, Turner (1965, 1970) detected such temporary clearances in Tregaron Moss and Bloak Moss from sites near the edge, but not in Flanders Moss, where the diagram was about 1.5 km from the edge. Therefore these clearances must have been sufficiently small not to have affected the regional pollen rain.

Three diagrams in a transect across Tregaron Bog, Wales (Turner, 1965) showed a difference of land use around the bog. Figure 12.2 shows schematic grass pollen curves from the south-south-east edge, the centre, and the north-north-west edge. The central diagram only extends to the base of phase C. Phase D is the phase of small temporary clearances, and it is followed by phase C of extensive clearance where the land was used mainly for pasture, and grass pollen values were high. Subsequently, during phase B, agriculture was practiced generally, resulting in lower grass frequencies, except in the north-west, where pasture continued to predominate. This suggests that arable activities were only on the south and south-east side of the bog. The local pollen production is clearly important in affecting pollen diagrams near to the forest or grassland edge. Phase A reflects recent complete clearance to form the modern landscape.

Turner (1970) followed up this work at Bloak Moss in south-west Scotland. Here the first extensive clearance lasted a maximum time of 350 years. She constructed 'three-dimensional pollen diagrams' using three sites across Bloak Moss (Fig. 12.3). She used grass pollen frequencies to indicate forest clearance. Consecutive samples from each diagram are shown, from equivalent subzones, with depth increasing to the right, and percentages increasing upwards. In Fig. 12.3a grass pollen is higher at the edges of the bog (sites A and K) than in the centre (site B), indicating a more intense forest clearance around the edges of the bog than in the region as a whole. This is confirmed by the equal percentages of tree pollen across the bog. During the following forest regeneration phase (b), local and regional grass pollen values were the same, and, as expected on Tauber's model, tree pollen values were higher near the margins because of the greater representation of the trunk space component (Fig. 12.3b). Local pollen produced by the bog vegetation was correspondingly higher in the centre. During the second extensive clearance (Fig. 12.3c) grass pollen values decrease from site A to K across the bog. Tree pollen values at site A fall and then rise again, complementing the rise and fall in the grass pollen there. This effect is visible but less marked in the centre, and not statistically significant at the opposite edge. This indicates that the clearance was at one side of the bog only. Even though she had successfully located areas of forest clearance, Turner was unable to estimate the size of the areas, although she could say that the small and extensive clearances were likely to have been on the edge of the bog, and that the fluctuations in the pollen curves were not reflecting large clearances some distance away.

Janssen (1973) also demonstrated very local changes in the area of pine forest in north-west Minnesota. Table 12.1 shows the percentages of

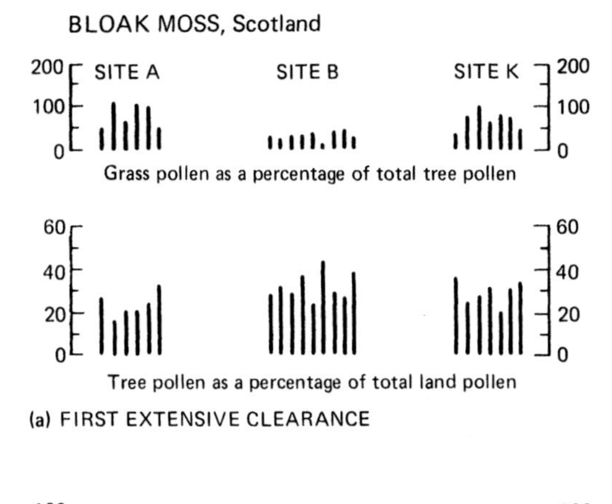

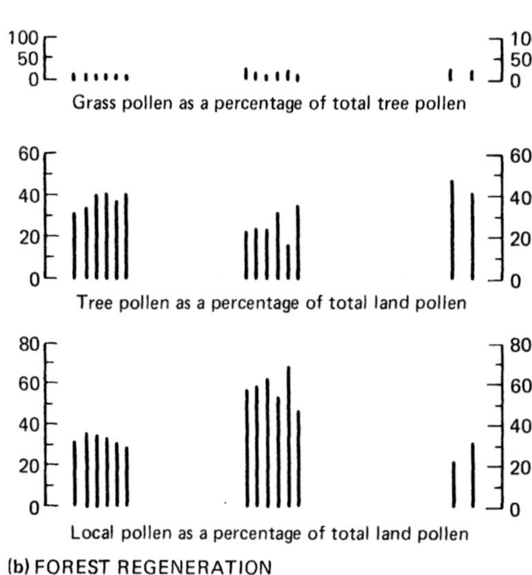

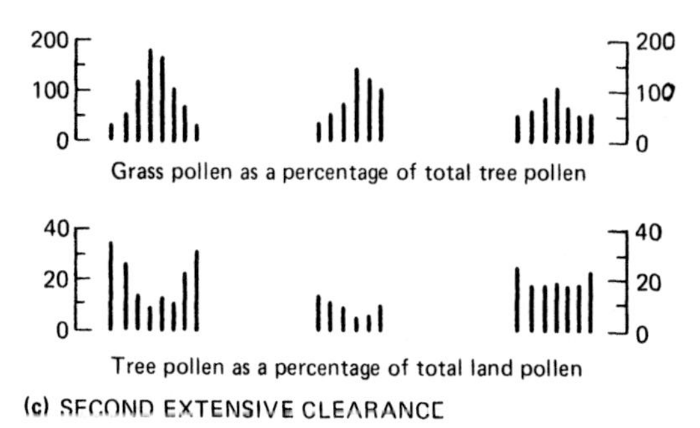

Fig. 12.3 Bloak Moss, south-west Scotland. 'Three-dimensional' pollen diagram from three sites, A, B, and K, in a transect across the bog. Consecutive samples for the equivalent subzones are shown, with depth increasing to the right, and percentages increasing upwards. See text for explanation. (After Turner, 1970.)

pine pollen in the centre and at the margin of a large lake (Bog D) and in two very small basins, Nicolet Creek and Stevens Pond, for pre- and post-settlement times.

The values at Bog D illustrate the difference between the local component at the margin and the regional component at the centre. Nicolet Creek is a narrow valley, and it shows local values similar to those at the edge of Bog D. Stevens Pond is very small, and also shows local values in pre-settlement times. However, these decrease to regional values after settlement, similar to those in the centre of Bog D. As the basin has stayed the same size, the distance of pine forest from the pond must have increased. Therefore the pine forest had been logged immediately around the site. There was no logging at the other two sites, and the pine pollen values are unchanged.

The reconstruction of plant communities in time

In order to answer our fifth question, what time did past communities occupy?, a detailed and independent chronology is necessary. The study of environmental changes through time is an essential feature of palaeoecology, and one of its main contributions to ecological theory is the addition of the time dimension beyond measurement at the present day.

The main method of constructing an independent chronology is by radiocarbon dating, which we have discussed on p. 208. Radiocarbon dating cannot extend with confidence beyond about 50 000 years at present, and it is too inaccurate to determine young ages. Young sediments can be dated using other, relatively short-lived isotopes, such as caesium-137, lead-210, and thorium-228:232 (see Chapter 6). Pollen horizons can also be used to date recent sediments. The age of the *Ambrosia* pollen rise is historically documented in North America, and it has been used extensively. The recent death of *Castanea* trees in eastern North America is another useful pollen horizon. In north-west Europe, the decline of *Ulmus* pollen seems to be a synchronous horizon at about 5000 B.P.

The radiocarbon method has drawbacks before 2000 B.P. due to the fluctuations in the ratio of ^{12}C : ^{14}C in the atmosphere. These fluctuations are gradually being calibrated by comparison with dates from tree rings of known ages, and from annually laminated sediments. Annually laminated lake sedi-

Table 12.1 Average pine pollen percentages based upon an upland tree pollen sum in a restricted area in northern Minnesota in four different locations. (From Janssen, 1973.)

	Bog D (centre of lake)	Bog D (margin of lake)	Nicolet Creek (narrow valley)	Stevens Pond (small pond)
Post-settlement time	83.6	91.5	over 90	84.7
Pre-settlement time	82.7	92.0	92–95	92.0
Present vegetation around site	Primeval pine forest sites in protected area			Fields, logging at settlement time

ments provide the most reliable chronology, and beyond the range of radiocarbon, they may be the only means of creating any chronology, even though it is not anchored in time, but only relative. Annual laminations provide close time control for dating events (see Chapter 6) and for estimating the length of time taken for changes in populations and communities to occur (Watts, 1973).

Other useful time markers in sediments may be abrupt stratigraphic changes. The minerogenic sedimentation during the Younger Dryas cold stadial in the Devensian late-glacial of north-west Europe is very characteristic. Layers of volcanic ash are widespread in western North America. When these have been dated by radiocarbon, they can be used to make definite time correlations between sites, providing that care is taken that the ash layer is correctly identified.

A further means of creating an independent chronology is palaeomagnetism which we discussed in Chapter 6. Palaeomagnetism can be used to check radiocarbon dates suspected of being erroneous due to the incorporation of old ^{14}C-deficient carbonates in calcareous sediments or of old ^{14}C from inwashed soils, etc. (e.g. Dickson *et al.*, 1978).

Post-glacial (Holocene, Flandrian) sediments can usually be dated satisfactorily by one or several of these methods. The dating of inorganic glacial sediments becomes more difficult, and sediments older than the limit of radiocarbon cannot so far be dated satisfactorily, although new methods are continually being tested, such as thermoluminescence in clays. The youngest limit of the potassium-argon method is not young enough to include the Quaternary. In older Quaternary sediments, only a relative chronology can be established. In marine sediments the $^{18}O:^{16}O$ ratio reflects the ice volume, and it can be used to correlate cores, and these cores can be also dated by measuring palaeomagnetic reversals. West (1977) discusses chronology and dating of Quaternary sediments in detail.

The reconstruction of the environment

We have now considered the reconstruction of the biotic component of the past ecosystem, the reconstruction of past communities, and an assessment of the space and time they occupied. We can now turn to our final question: what were the other factors operating in the ecosystem in that time and space? The reconstruction of the whole palaeoecosystem is the aim of many palaeoecological investigations. It can only be attempted after all the observations and analyses of the biota and the lithologies have been assembled. Evidence of the physical environment, particularly the climate can be drawn from both the organisms and the sediments. Abiotic evidence of past environments can be derived from sediment lithology and chemistry, site geomorphology, isotope content, etc. Biotic evidence of past environments can be derived from a knowledge of the present ecological tolerances and requirements of the organisms, assuming that these have not changed during the intervening time interval. Individual taxa can be used, in an indicator-species approach, or whole assemblages can be used.

In practice, a palaeoecologist uses evidence from as wide a variety of sources as possible, both biotic and abiotic, to provide the fullest and most coherent reconstruction of past environments.

Abiotic sources of environmental evidence

1. The lithology of sediments may enable inferences to be made about the environment in which the sediments were laid down. For example, coarse minerogenic sediments in a lake suggest active inwashing of material, possibly due to the action of low temperatures in the catchment, or to the activities of man.

2. The chemistry of lake sediments yields information on changes in the soils in the catchment (e.g. Mackereth, 1966). Phases of soil stability, leaching, and erosion can be detected (see Chapter 6). Pennington *et al.* (1972) and Pennington and Sackin (1975) compared chemical and pollen analytical data from lake sediments using principal components analysis. Figure 12.4 is a stratigraphic plot of the component scores of the chemical data, which shows that major changes in the sediment chemistry correspond with major changes in the pollen stratigraphy. A common cause can be inferred affecting both sediments and vegetation, namely the soils of the catchment. In the early Flandrian these were fertile and stable, as shown by high values of carbon, magnesium, and calcium in the sediments. Subsequently they became leached and podsolized releasing significant amounts of iron and manganese as peats developed.

3. Stable isotopes can provide abiotic evidence for past terrestrial environments, particularly climatic changes. In general, the proportions of different isotopes of the same element depend upon the temperature at which the sample was formed.

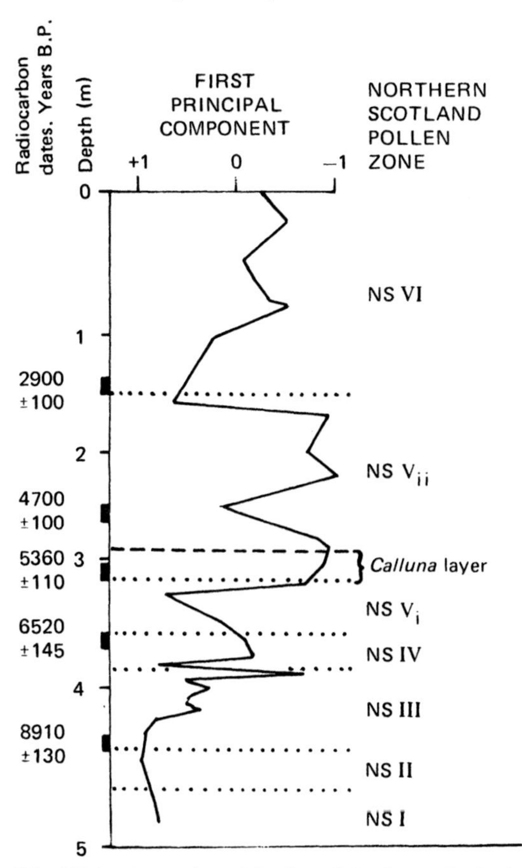

Fig. 12.4 A stratigraphic plot of the first principal component of a PCA of the sediment chemistry of Loch Clair, north-west Scotland, compared with the positions of the pollen zones, which correspond to changes in the pollen proportions. Radiocarbon dates are also shown. (After Pennington *et al.*, 1972.)

a) $^{18}O:^{16}O$ ratios in freshwater carbonate sediments

The equilibrium between the ^{16}O and ^{18}O isotopes of oxygen depends on temperature. For a drop of 1 °C, the proportion of ^{18}O in rainwater decreases by 0.70‰. Stuiver (1970) demonstrated this temperature effect in a north–south transect of lakes in eastern North America. He also demonstrated that seasonal changes in the ratio in rainwater were reflected in the surface and the deep water of a lake. Oxygen reaches the lake sediments as carbonate, precipitated by water plants, and also in carbonate in mollusc shells. Once formed, the ratio is preserved in the carbonate. Therefore an analysis of $^{18}O:^{16}O$ in carbonate in a lake sediment can give some indication of temperature changes through time. The ratio cannot be calibrated directly in terms of absolute temperature, but it can be used to trace temperature changes (Stuiver, 1970). Stuiver (1968, 1970) produced curves for changes in ^{18}O through the Holocene and into the Wisconsinan late-glacial of North America. The main rise of temperature appears to have occurred during the *Pinus* pollen maximum, at about 9000 B.P. After a maximum between 9000 and 5000 B.P., reflecting the hypsithermal interval inferred previously from other palaeoecological evidence, temperatures fell slowly to those of the present day.

Similar work was done on two Swiss lakes by Eicher and Seigenthaler (1976). Figure 12.5 shows the deviation of ^{18}O ($\delta^{18}O$) from the PDB (a standard measured on the Pee Dee belemnite), and the associated pollen curves. During the Oldest Dryas (zone Ia) temperatures were low. There was a rapid rise at the opening of the Bølling–Allerød period (zones Ib, Ic, and II), followed by a slow decline throughout this period. There was little pollen analytical evidence for an Older Dryas cold phase (Ic) and no indication at all in the ^{18}O curve. There is a sharp fall in temperature during the Younger Dryas stadial, although this period is not well delimited in the pollen diagram (zone III). This is followed by a rapid rise of temperature to zone Ib levels at the opening of the Holocene period (Zones IV and V). This temperature pattern for the Weichselian late-glacial is reminiscent of that inferred from evidence from Fossil Coleoptera (see Chapter 7, and Lang, 1970).

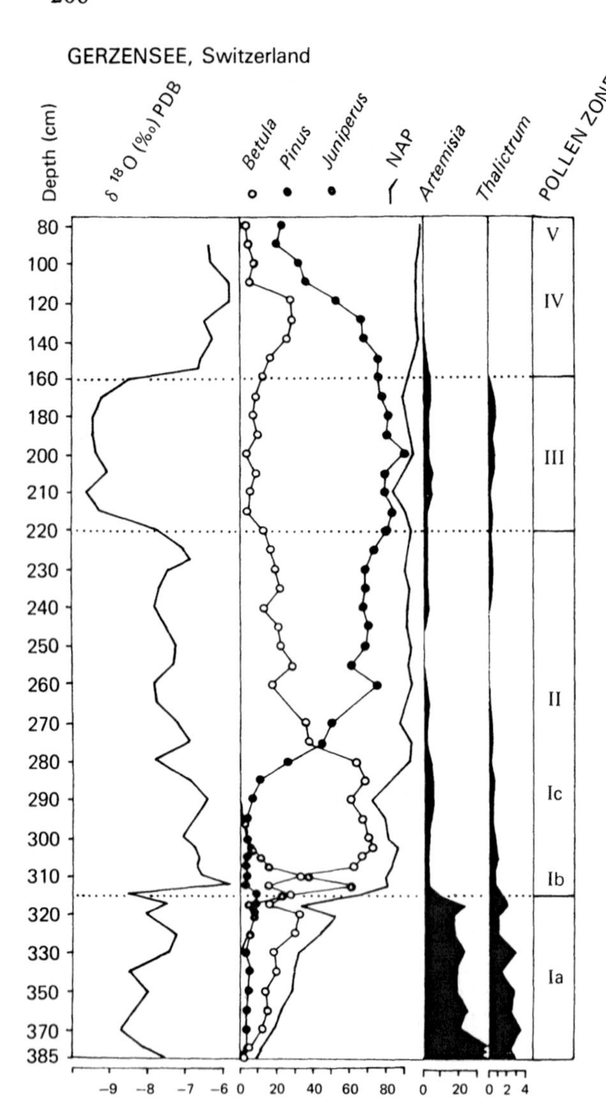

Fig. 12.5 The deviation of $\delta(^{18}O)_{PDB}$ in Weichselian late-glacial and early Holocene marl sediments from Gerzensee, Switzerland, together with the major pollen curves. The standard pollen zones are indicated. (After Eicher and Seigenthaler, 1976.)

b) $^{18}O:^{16}O$ ratios in the Greenland ice cap

Dansgaard *et al.* (1975) measured $\delta^{18}O$ from the Standard Mean Ocean Water (SMOW) down a core through the Greenland ice cap which extended back to 554 A.D. The fluctuations are shown in Fig. 12.6. Temperature fluctuations reconstructed from Iceland and from England have been similar since 1700 A.D. However, the warm twelfth and thirteenth centuries in England were relatively cold in Greenland and Iceland, and the 'Little Ice Age' in England, between 1450 and 1700 A.D. was not particularly cold in Greenland or Iceland. It is, however, detectable in North America. It would appear that climatic oscillations in Greenland are correlated with the pattern of changes in England 250 years later. This may be explained by features of atmospheric circulation.

Dansgaard *et al.* (1975) related Norse history to the climatic fluctuations revealed in the ice core. The first Norseman colonized Iceland in 865 A.D. during a short cold period. He lost all his cattle, and the fjord became blocked by sea ice. He called the country Iceland, and returned home. Other settlers soon recolonized Iceland, as climatic conditions improved rapidly. In 982, Erik the Red found land west of Iceland at the end of a period of warmer climate than any known since. He called it Greenland, and founded the first Norse community there. These short term climatic changes explain how Iceland and Greenland came by their apparently contradictory names. By 1350, the climate in Greenland became very cold, and this coincided with the extinction of the Norse colonies, showing how human populations can be dramatically affected directly by climate.

c) Stable isotopes in tree rings

Lerman (1973) has indicated that the reliability of isotope thermometers in wood is considerably less than in other organic material such as peat, and in carbonates, phosphates, etc. However, Libby *et al.* (1976) and Libby and Pandolfi (1977) have analysed tree rings for ^{18}O and deuterium, and plotted the changes against age. Using the mathematical method of Fourier transform analysis, they have shown periodicities in the results which they claim conform with periodicities in the Greenland ice cores, suggesting that they are reflections of a world-wide climatic periodicity. Libby *et al.* (1976) demonstrated that changes in $\delta^{18}O$ and δD in recent trees correlate well with measured temperatures, and they suggest that these isotopes can be used as relative thermometers. A similar claim, that changes in the ^{13}C content of tree rings is related to temperature, has been made by Pearman *et al.* (1976), but the theory behind this claim is controversial (Lerman, 1973).

d) Other stable isotope studies

i) Stuiver (1975) measured ^{13}C in freshwater carbonates. Amongst the complex factors control-

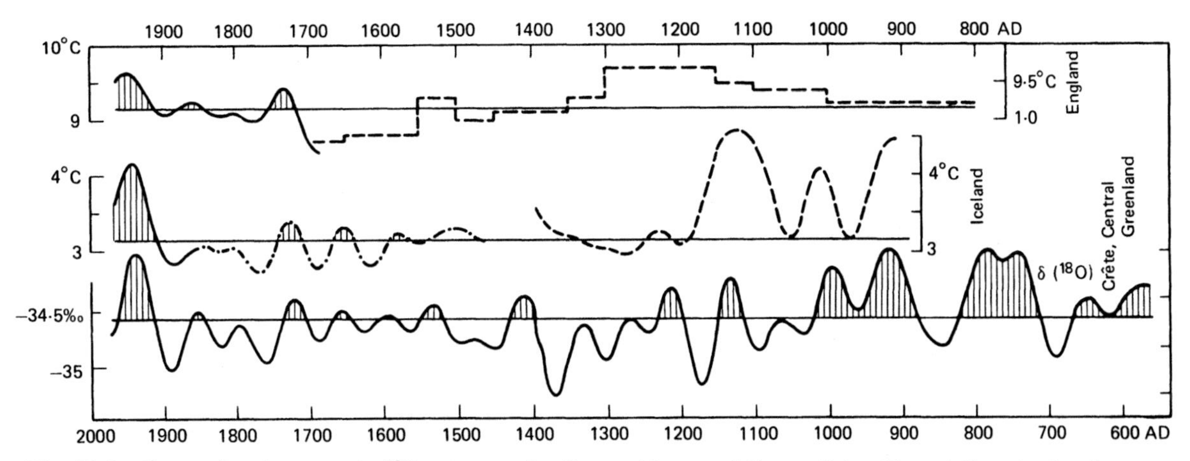

Fig. 12.6 Comparison between the ^{18}O concentration (bottom) in snow fallen at Crête, Central Greenland and preserved as ice (δ scale at left), and temperatures for Iceland and England. The curves are smoothed by a 60-yr low pass digital filter, except for England 800–1700 A.D. Dashed curves depend on indirect evidence. Dashed/dotted curve for Iceland from systematic sea-ice observations. (After Dansgaard *et al.*, 1975.)

ling ^{13}C incorporation, the most important are changes in the hardness of the water and changes in organic productivity. However, changes in ^{13}C values tend to show some correlation with climatic changes, but they are not well marked.

ii) Schiegl (1972) used the fact that the deuterium content of rainwater decreases with decreasing temperature to reconstruct past temperature changes from the deuterium content of the organic matter in peat. In Fig. 12.7 δD_{SMOW} from fossil peat rises considerably at the beginning of the Flandrian 10 000 years ago to a maximum, and then shows a gradual, irregular fall. He compared the curve with temperature variations deduced from tree-line studies. Values from wood on the same graph show widely ranging values, usually much higher than the peat values of δD.

iii) The ^{12}C:^{13}C ratio is different in plants using the C_4 or the C_3 pathway of photosynthesis. The recently discovered C_4 pathway predominates in plants with crassulacean acid metabolism (CAM) found in succulents and cacti, and also found in grasses, sedges, and Chenopodiaceae. The proportion of C_3 and C_4 photosynthesis varies with environmental conditions (Lerman, 1973). Thus measurements of ^{13}C from remains of these plants can be converted to the proportion of C_4 photosynthesis, which is related primarily to temperature and water availability. Measurements by Troughton *et al.* (1974) showed that the proportion of C_4 photosynthesis in *Opuntia polycantha* and *Atriplex confertifolia* increased substantially at the opening of the Holocene in the deserts of south-west U.S.A. indicating increases in temperature and aridity at that time.

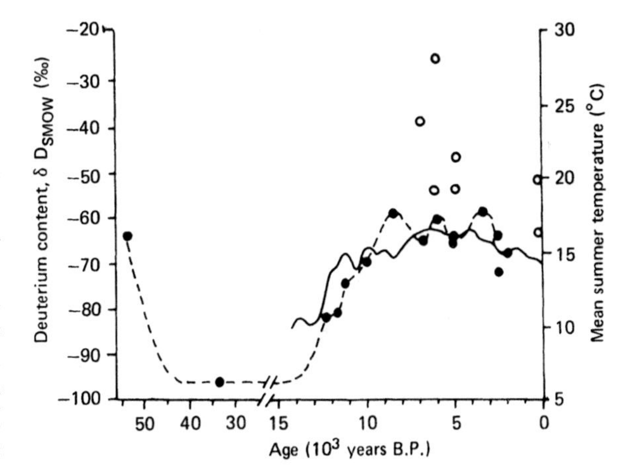

Fig. 12.7 Deuterium content in peats plotted against radiocarbon years B.P. (Schiegl, 1972), compared with deuterium content of wood samples, and the temperatures inferred from timber-line fluctuations. (After Lerman, 1973.)

Biotic sources of environmental evidence

If we assume that the environmental tolerances of taxa have not changed over the time of the palaeoecological investigation, we can use them to reconstruct past environments. As with community reconstruction, we can use a single species indicator-species approach, and a multivariate assemblage approach (see Birks, 1979).

1. *Indicator-species approach to environmental reconstruction*

This approach requires autecological knowledge of factors which control the distribution of the organism at the present day. Coope (see Chapter 7) has followed this approach in reconstructing past temperature conditions from fossil coleopteran remains. The same approach can be taken using plants, by taking the distribution of the plant concerned, and trying to fit it to the distribution of selected climatic parameters.

a) Conolly and Dahl (1970) estimated past temperature conditions by fitting maximum summer temperature isotherms to plant distributions in Britain and transferring these temperatures to areas containing fossil records. From the present and fossil distribution of *Salix herbacea*, they deduced that the over-all temperature depression during the Devensian late-glacial of northern Britain and Ireland compared with the present day was about 3°C, and during earlier Devensian times, it was about 5°C. Using *Betula nana*, they deduced that the Devensian late-glacial temperature depression in East Anglia was 8°C. Conolly and Dahl assumed a causal relationship between maximum summer temperature and species distribution, although experimental evidence demonstrating the validity of this relationship was lacking. They also assumed that the limiting effect of summer temperature was the same in the Devensian, when conditions of soil and competition were presumably very different.

b) It is obviously important to look at species distribution in relation to a range of physiologically important variables, rather than one simply measurable climatic factor. The interplay between factors in different parts of a species' range can be quite striking. Iversen (1944) showed this in his classic study on the factors limiting the distributions of *Viscum album*, *Hedera helix*, and *Ilex aquifolium*. Using ecological observations on how each species was affected by climatic factors at the edges of its range, Iversen chose to relate the performance of these three species to the mean temperature of the coldest month and of the warmest month. Figure 12.8 shows the results for the tree-climbing *Hedera* *helix*. It shows that *Hedera* can tolerate colder winter temperatures up to a certain limit if the maximum summer temperatures are higher. It can grow beyond this limit, but it is confined as sterile plants on the ground in woods. The thermal limits of all three species are compared in Fig. 12.9. *Ilex* is the most oceanic, tolerating the lowest summer

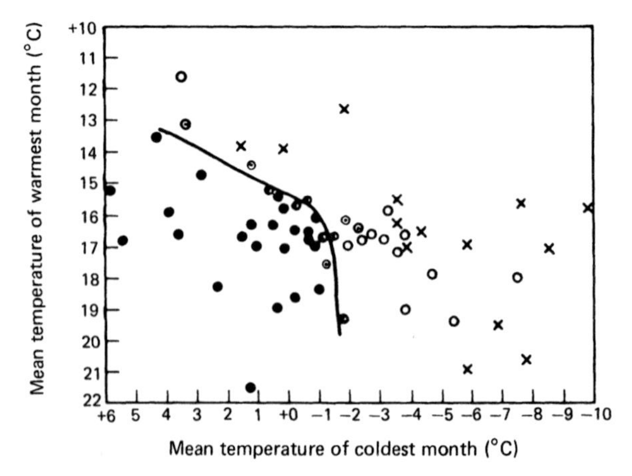

Fig. 12.8 The performance of *Hedera helix* related to mean temperatures of the warmest and coldest months. (After Iversen, 1944.)

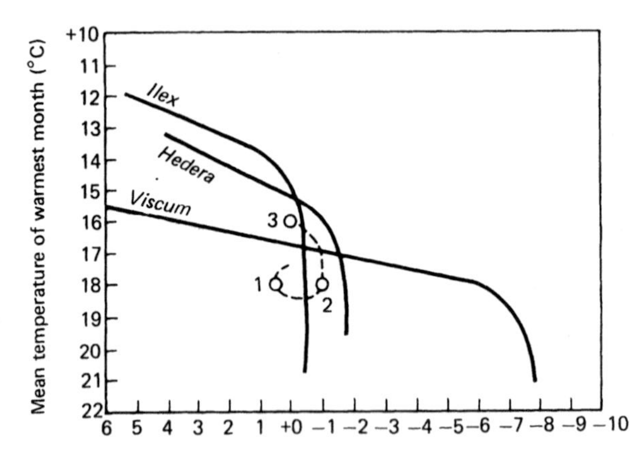

Fig. 12.9 The thermal limits for *Ilex aquifolium*, *Hedera helix*, and *Viscum album*. The approximate temperatures from Atlantic time (**1**) through Sub-boreal time (**2**) to the present day (**3**) for Denmark have been deduced from the relative proportions of fossil pollen of the three species. (After Iversen, 1944.)

temperatures but being the most intolerant of low winter temperatures. *Viscum* is the most continental, seeming to require high summer temperatures, but with a relative indifference to winter cold. Because of these precise thermal limits, Iversen felt justified in using the plants as indicator species, to reconstruct the temperature regime of Denmark during the Flandrian. The inferred temperature changes are also shown on Fig. 12.9, and they clearly indicate an increase in temperature range from the Atlantic (point 1), through sub-Boreal times (2), with summer temperatures 2–3°C warmer than present, and a decrease to more oceanic conditions through sub-Atlantic times (3) to the present day.

c) Churchill (1968) adopted a similar approach to that of Iversen, using three *Eucalyptus* species in Australia. Each is an important forest tree, and can be readily identified by its pollen. Their distribution is correlated with the mean rainfall of the driest and wettest month (Fig. 12.10), and there is no apparent relationship with temperature. *E. calophylla* and *E. marginata* are closely similar in their limits, and Churchill therefore used the ratio of *E. diversicolor* to *E. calophylla*, because *E. marginata* pollen was not so abundant. An increase in the proportion of *E. diversicolor* indicated wetter conditions, and from dated pollen diagrams Churchill inferred that the climate was relatively wet until 5000 B.P., after which it became dry until 2500 B.P., with a maximum dryness about 3100 B.P. Conditions then became wetter until 1400 B.P., followed by a short dry phase until about 700 B.P., since when the climate has become wetter once more.

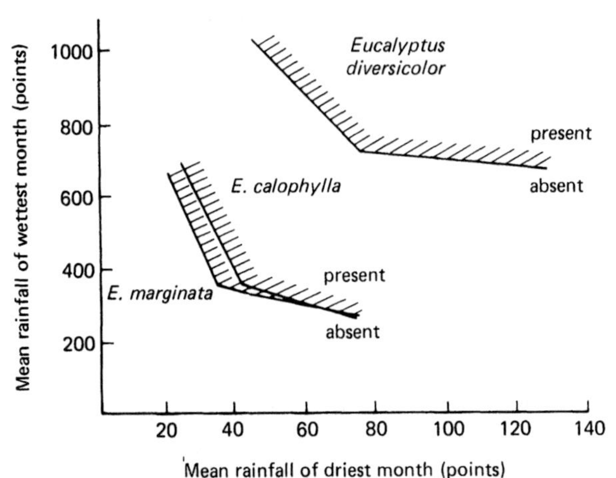

Fig. 12.10 Rainfall limits of *Eucalyptus calophylla*, *E. marginata*, and *E. diversicolor* in Australia. The first two species can tolerate drier conditions than the third. (After Churchill, 1968.)

d) Ratcliffe (1968) points out that gross measurements of temperature and rainfall are not necessarily realistic expressions of the factors limiting plant distributions. He shows how closely the distribution of the liverwort *Adelanthus decipiens* in the British Isles is delimited by humidity and temperature expressed as wet days and days of snowfall. Figure 12.11 shows how *Adelanthus* is confined to areas with high atmospheric humidity (>190 wet days a year) and freedom from low winter temperatures (<23 days of snow lie in the lowlands). A whole group of liverworts which grow together forming the mixed North Atlantic hepatic mat community are even more dependent on atmospheric humidity (Fig. 12.11), but within this limit they are, as a group, indifferent to winter temperatures. The outliers in central Scotland are in places with prolonged snow-lie.

e) Grichuk (1969) attempted to reconstruct aspects of the climate of the northern hemisphere at 5500 B.P. (±200 years) using the indicator-species approach. He considered January mean temperature, July mean temperature, total annual precipitation, and number of frost-free days. For each variable, he plotted the sphere of tolerance of the taxa found in pollen diagrams. He gives an example (Fig. 12.12) with a pollen diagram containing *Ulmus*, *Alnus*, *Picea*, and *Myrica*. When the present distributions of these plants are plotted against January and July temperature, the area where all their limits coincide may represent the climate at the time of the pollen sample, namely the climatic conditions which permit the growth of all four taxa.

Grichuk made estimates of climatic variables from twenty fossil pollen sites in the northern hemisphere at 5500 B.P. The temperature results are shown in Fig. 12.13. He took the 90°W meridian to represent North America, and the 35°E meridian to represent eastern Europe. Mean January temperatures were elevated most at 5500 B.P. during the so-called 'climatic optimum', or hypsithermal interval, at about 58–59°N in northern Europe and Canada. Further south, there was less deviation, and south of 40°N mean January temperatures were lower than at present, especially in the Middle East and south-east U.S.A. The July temperature curve is slightly different. Temperatures were not raised so much in northern Europe and northern North America, and there was no depression towards the North Pole.

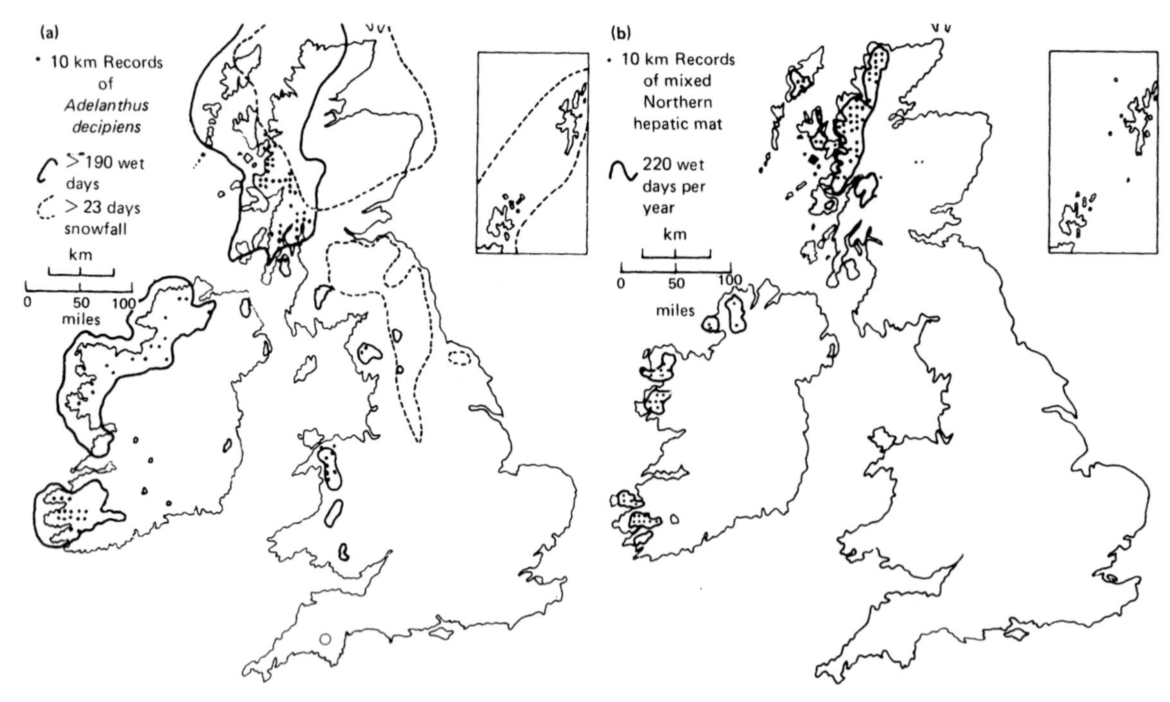

Fig. 12.11 The correlation of liverwort distributions with climatic measurements. (**a**) Map of *Adelanthus decipiens* in relation to indices of limiting humidity and temperature, which suggests that *Adelanthus* is confined to areas of both high atmospheric humidity and freedom from low winter temperatures. Absences or discontinuities within the area of suitable climate are largely due to destruction of the original woodland habitat. (**b**) Map of the mixed northern hepatic mat community, in relation to an index of limiting humidity. Absences within the area of suitable climate are largely due to destruction of the boulder-strewn dwarf-shrub-dominated habitat by grazing or burning.
(After Ratcliffe, 1968.)

However, July temperatures were 1–2°C cooler in more southern latitudes. Over-all, precipitation tended to be higher throughout the northern hemisphere at 5500 B.P. than at present, although data from present arid regions are generally lacking.

The indicator-species approach to climatic reconstruction is limited by the fact that it assumes that particular taxa are controlled by one or a few environmental factors only. In general, only a few taxa are considered, usually on a presence/absence basis (Birks, 1979). The alternative approach is to use the whole fossil assemblage and a large number of environmental variables, to try to find some correlations between the two. Because of the volume of data, multivariate methods are required for the analyses. The two approaches should complement one another.

2. *Assemblage approach to environmental reconstruction*

This approach assumes that an assemblage of organisms is related to the environment by some complex function, called a ***transfer function.*** If this function can be determined for modern situations, it should be applicable backwards in time, unless the reaction of the whole assemblage to its environment has changed. If X is an assemblage of organisms, E is the environment, and R is the transfer function:

$$E = XR$$

In palaeoecology, the task is usually to estimate E from X. If we know R from modern studies, and assume it has not changed with time, we can use it to estimate E from X.

This approach involves the use of multivariate numerical analyses for establishing the relationships between biological data and environmental variables. Such methods were devised for pollen data by Webb and Bryson (1972), for foraminiferal data from the oceans by Imbrie and Kipp (1971), and for tree-ring data by Fritts *et al.* (1971). Recently, Webb and Clark (1977) and Sachs *et al.*

(1977) have reviewed the various methods, and they have shown that they are all variants of the basic multiple linear regression model.

a) *Climatic estimates from pollen analytical data.* Webb and Bryson (1972) outline the process of relating pollen data to climate by a series of equations. If P represents the pollen data, C the climatic estimates, and B is the transfer function:

$$P \times B = C$$

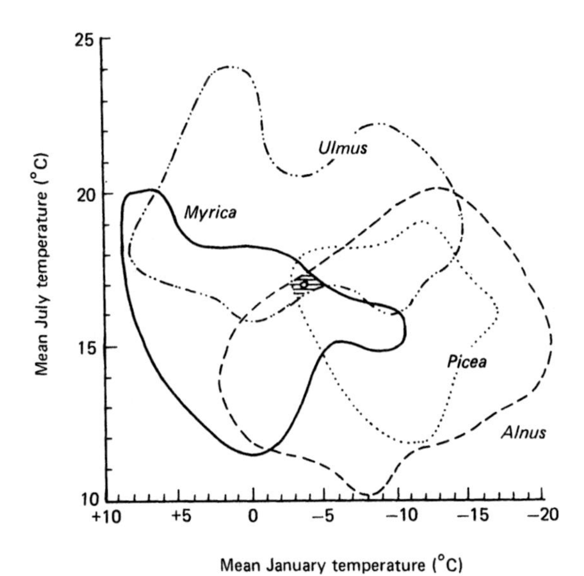

Fig. 12.12 The thermal limits of *Ulmus*, *Picea*, *Alnus*, and *Myrica* in northern Europe. The range at which the distributions overlap can be taken as an indication of the climate during which a fossil pollen sample was deposited which contained pollen of all the species.
(After Grichuk, 1969.)

Pollen analysts generally approach this formulation in two steps. First;

$$P \times R = V$$

where V is the vegetation which has produced the pollen and R is the function relating the two (this is the R-value we have discussed in Chapter 10). Second;

$$V \times D = C$$

where D is the function relating vegetation to climate.

These equations can be combined into:

$$P(R \times D) = C$$

Pollen analysts infer environmental conditions qualitatively using these steps of argument. They establish transfer functions non-analytically, as mental qualitative judgements to reconstruct past vegetation and climate. This process can be compared with the above equation:

Fossil pollen	.	Qualitative reasoning	→	
P	$\times$	R	$=$	V

Past vegetation	.	Qualitative reasoning	→	Past climate
V	$\times$	D	$=$	C

The qualitative reasoning to relate pollen and vegetation has, in some instances, been quantified as R-values (Chapter 10) thus:

If $P_m \times R_m = V_m$ m = modern
then $P_f \times R_m = \hat{V}_f$ f = fossil

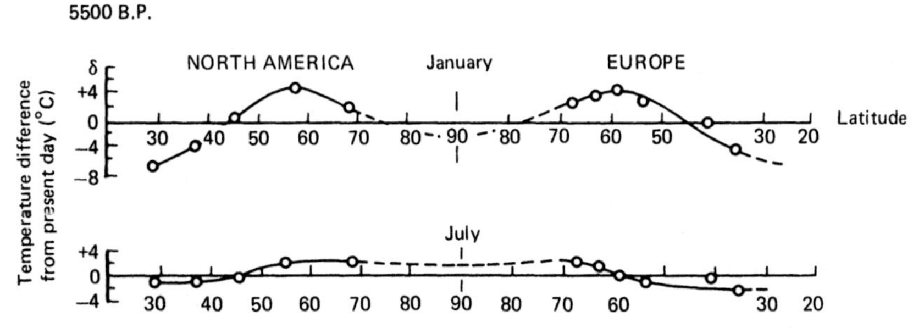

Fig. 12.13 Northern hemisphere temperature deviations from the present day at 5500 B.P. in two north–south transects, down the 90°W meridian through North America, and the 35°E meridian through eastern Europe.
(After Grichuk, 1969.)

However, the function D relating vegetation and climate has been quantified in very few cases (e.g. Iversen, 1944; Churchill, 1968). The function D is the factors limiting the plants' growth.

Because of the difficulty of defining D, Webb and Bryson (1972) took the alternative approach of relating pollen data to climate directly by the transfer function B.

$$\text{If} \quad P_m \times B_m = C_m$$
$$\text{then} \quad P_f \times B_m = \hat{C}_f$$

In order to define B_m, Webb and Bryson assembled modern pollen data from as wide an area as possible, and related it to climatic data using canonical correlation analysis (see Chapter 2). For B_m to be applicable to fossil data they assumed (1) that climate is the ultimate cause of changes in pollen stratigraphy, (2) that the vegetation, and hence the pollen assemblages have responded in a constant way to climatic changes, and (3) that linear relationships can adequately approximate to the relationships between climatic variables and pollen. These assumptions are, according to Webb and Bryson, probably valid for the relatively small areas of space and time that they were concerned with.

For the numerical analyses, Webb and Bryson excluded several pollen taxa from the calculations, for various reasons.

i) Taxa of local origin, e.g. Cyperaceae.
ii) Taxon favoured by recent human disturbance, e.g. *Ambrosia*.
iii) Taxa not present in all samples, and in very low amounts, e.g. *Populus*, Tubuliflorae, Liguliflorae, Leguminosae.
iv) Taxa with values 2–5 times higher in the fossil samples than in the modern samples; *Ulmus*, *Larix*, *Fraxinus*, *Corylus*, and *Ostrya*/*Carpinus*.
v) Taxa excluded for no apparent reason except to increase the ratio of observations to pollen types, and hence increase the statistical significance; Gramineae, *Alnus*, *Abies*, *Tilia*, and *Salix*.

The remaining eight pollen taxa were *Pinus*, *Picea*, *Quercus*, *Carya*, *Juglans*, *Acer*, *Betula*, and Chenopodiaceae. They were expressed as percentages of their sum plus the sum of *Abies*, *Salix*, and *Artemisia*. These manoeuvres were intended to reduce the inherent variability of the pollen data, and ensure that fossil samples from all the various zones in the Late Wisconsinan and the Holocene could be used.

The climatic variables chosen included months' duration of major air masses, and factors thought to be important to plant growth, such as July mean temperature, rainfall during the growing season, length of growing season, moisture stress, hours of sunshine, and snowfall.

Canonical correlations were calculated between both data sets. The first two canonical correlations were used, and the canonical regression loadings for each variable were used as the transfer functions. The accuracy of the transfer functions was tested by applying them to an independent set of modern pollen data, and comparing the predicted climate with the recorded modern climate. There was a close correspondence, confirming the transfer functions were valid to within 0.6°C and 2 cm precipitation during the growing season.

The transfer functions were then applied to three fossil sites, Kirchner Marsh on the prairie/deciduous forest boundary in Minnesota, Disterhaft Farm Bog on the deciduous/coniferous forest boundary, and Lake Mary in the coniferous forest to the east in Wisconsin. The changes in some climatic variables at three sites are shown in Fig. 12.14. There is a sharp rise in temperature over 200–300 years at the opening of the Holocene at about 10 000 years ago at all the sites. During the Wisconsinan late-glacial, precipitation decreased gradually. However, about 9000 years ago, precipitation decreased rapidly at Kirchner Marsh, with a resulting greater moisture stress. This corresponds with the eastward expansion of the prairie at this time. Disterhaft Farm Bog shows a much smaller decrease in precipitation, and the effect is only slight at Lake Mary at this time. After about 5000 B.P. precipitation increases gradually at Kirchner Marsh, which corresponds to a small decrease in temperature and less moisture stress. These diagrams emphasize that the prairie expansion was largely due to increased aridity rather than higher temperatures. Indeed, temperatures show little variation through the Holocene with a slight over-all rise up till 6000 B.P., followed by a very slow fall. This pattern corresponds to the climatic changes discovered by Grichuk (1969) for these latitudes (p. 271).

The results obtained by Webb and Bryson are generally consistent with climatic reconstructions inferred qualitatively by previous workers. However, their numerical approach has the advantages that (1) all taxa and all samples that they selected can be considered at once in a repeatable way, and all the quantitative information is used, (2)

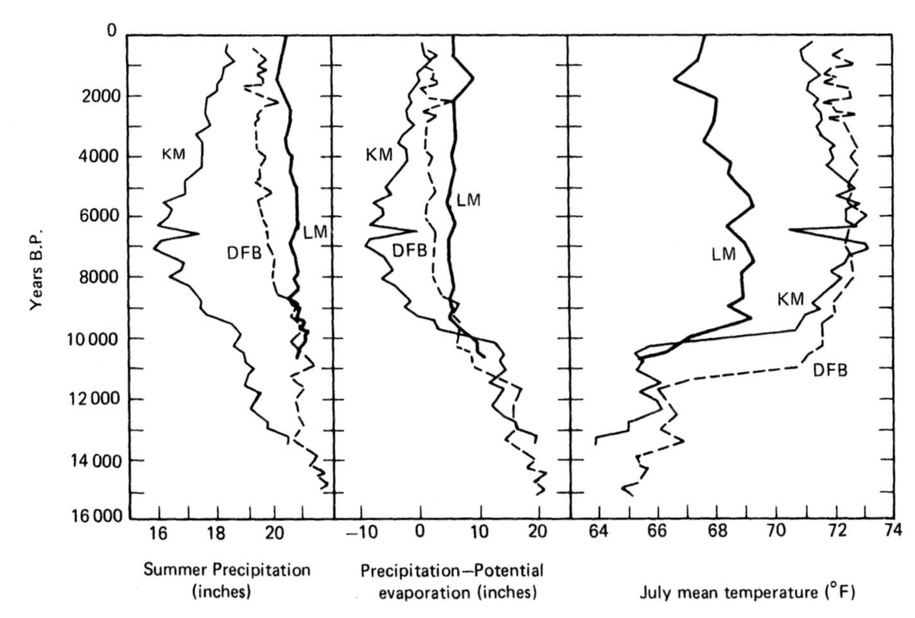

Fig. 12.14 Plots of the reconstructed values of the precipitation during the growing season, the precipitation minus potential evaporation, and July mean temperature for three sites in midwestern U.S.A., using transfer functions on fossil pollen data. KM, Kirchner Marsh; DFB, Disterhaft Farm Bog; LM, Lake Mary. (After Webb and Bryson, 1972.)

the authors are forced to be explicit about their methods, and (3) new working hypotheses can be generated and unsuspected features may be revealed. This method can be refined by using more contemporary pollen data, more detailed climatic information, and more versatile and refined numerical techniques (e.g. Bryson and Kutzbach, 1974; Howe and Webb, 1977; Webb and Clark, 1977). Maps of palaeoclimate derived from pollen data may eventually be produced. The main problems of the use of pollen analogues arises in the late-glacial, particularly in North America, where no satisfactory modern pollen analogues have yet been found.

A different approach to the reconstruction of past climate from fossil pollen data from that of Webb and Bryson was used by Harris *et al.* (1976) in New Zealand. They identified some samples as representative of warm conditions, and some others as representative of cold conditions using independent geological criteria. They calculated a linear discriminant function between the two groups (see Chapter 2), and then positioned the samples of unknown climatic significance along the discriminant function. The discriminant function can be calibrated in absolute climatic terms (temperature), by placing on it modern samples originating from areas of known climate. Then quantitative estimates of the past climate could be produced.

b) *Environmental estimates from marine-core data.* We shall briefly mention this aspect of marine palaeoecology because the assemblage approach using transfer functions originated here, and estimates of climate from marine data, particularly during glacial and interglacial periods are applicable to the continents.

The approach taken by Imbrie and Kipp (1971) is essentially the same as that of Webb and Bryson (1972) (see Webb and Clark, 1977). Imbrie and Kipp analysed foraminiferal assemblages from 61 core tops, mostly from the Atlantic Ocean, as the present biological expression of the present sea conditions. The core-top data were then analysed by varimax factor analysis into five assemblages. These assemblages represented polar, sub-polar, sub-tropical, tropical, and gyre margin conditions. The first four assemblages corresponded closely with winter sea-surface isotherms. The environmental parameters of the ocean (winter and summer surface temperature,

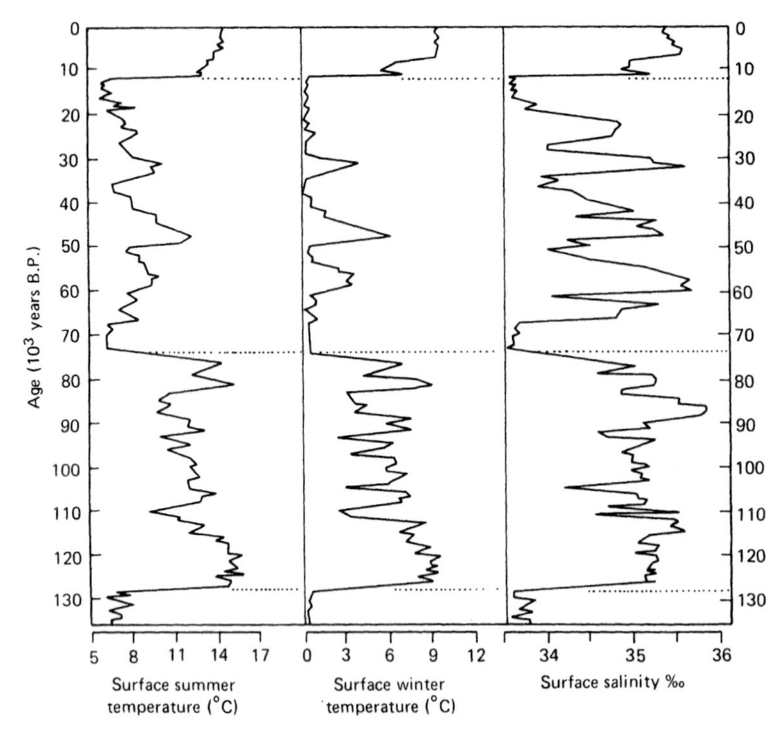

Fig. 12.15 Sea surface summer temperature, surface winter temperature, and surface salinity reconstructed from fauna down a long core from the North Atlantic near Ireland, using transfer functions. (After Sancetta *et al.*, 1973.)

surface salinity) were related by a set of palaeoecological equations to the foraminiferal assemblages by least-squares regression techniques. Fossil data from a core from the Caribbean Sea were then described in terms of the modern assemblages, and the palaeoecological equations, or transfer functions, were used to estimate past environmental conditions. The Caribbean core was correlated with others from nearby using the $\delta^{18}O$ curve, and temperature changes deduced from this method corresponded well with those obtained from foraminifers.

The mathematical method was refined by Imbrie *et al.* (1973). It was applied to a core from near the Irish coast by Sancetta *et al.* (1973). As expected, this area was more sensitive to glacial–interglacial fluctuations than the Caribbean, where tropical or sub-tropical conditions predominated throughout. Figure 12.15 illustrates the trends in winter and summer surface temperatures and surface salinity since about 130 000 years ago. The rapid warming at about 127 000 years ago marks the beginning of the penultimate interglacial (Eemian in Europe), and temperatures reached a maximum about 124 000 years ago. The beginning of the last glaciation (Weichselian or Devensian in Europe), is marked by a temporary temperature drop at about 110 000 years ago representing a temporary cooling in the North Atlantic. Full-glacial cooling started at about 73 000 years ago. There are short warm intervals detectable during the last glaciation at 59 000, 48 000, and 31 000 years ago, and the coldest conditions occurred between 30 000 and 11 000 years ago. An abrupt warming at the beginning of the Holocene (Flandrian) is interrupted by a short cold phase, particularly in winter temperatures, about 10 000 years ago, which corresponds to the Younger Dryas stadial of north-west Europe. Sancetta *et al.* compared their results favourably with those from the Greenland ice cap, and indicated some differences with the reconstruction of the last glaciation in Europe using pollen and soil evidence.

This general approach of transfer functions for

marine fossil assemblages has now been appplied to radiolarian, coccolith, and diatom assemblages to derive estimates of sea-surface temperatures and salinities at many places in the oceans (see the papers in Cline and Hays, 1976).

c) *Environmental estimates from tree rings*. The width of an annual ring in the trunk of a tree is an expression of the growth of the tree, which is related to climate. Therefore ring width can be related to climatic variables by transfer functions. Fritts *et al.* (1971) pioneered this approach. They calibrated the width of annual rings with climatic observations over the recent past for which climatic data are available. Then they used the calibrations to reconstruct climate for the period of time covered by the tree-ring data but for which no climatic records were available.

Fritts *et al.* (1971) simplified 28 climatic variables from 28 recording stations over 31 years by principal components analysis into 20 components. They then related these to tree-ring data for the 31 years from 120 forest stands in the same area by multiple regression, and multiplied the regression coefficients by the principal component eigenvectors to obtain transfer functions. The transfer functions varied between localities, but they were quite consistent among different species in the same site. Fritts *et al.* (1971) also calculated transfer functions between tree-ring widths and seasonal atmospheric pressure patterns, and used them to reconstruct large-scale pressure anomaly maps from 1521 to the present. These reconstructions compared favourably with reconstructions using other lines of evidence.

La Marche (1974) demonstrated the great sensitivity of tree-ring width at the limits of the tree's ecological tolerance. In one area of the White Mountains, California, U.S.A., he showed that bristle-cone pines (*Pinus aristata*, *P. longaeva*) were most sensitive to precipitation at their lower altitudinal limit, where they bordered on desert, but they were most sensitive to temperature at high altitudes, where precipitation was not so critically limiting. These factors could be combined to give a detailed reconstruction of past climate in the area. Figure 12.16 shows the effects on temperature and precipitation on tree-ring width at high and low altitudes. Figure 12.17 shows how this information could be used to reconstruct the climate of the White Mountains over the past 1000 years.

Climatic reconstructions using tree rings in both modern and fossil timber have been made by Pilcher and his associates in Northern Ireland (Hughes *et al.*, 1978). They have also made an important contribution to the calibration of radiocarbon dating from their tree-ring work (Pearson *et al.*, 1977; Pilcher *et al.*, 1977; reviewed by Moore, 1978). The whole field of dendrochronology has been reviewed in a recent book by Fritts (1976), and is one of very considerable importance in future palaeoclimatology. In theory it will be ultimately possible to derive yearly climatic estimates from fossil tree-ring data for the last 8000 years.

Global palaeoclimatic reconstructions

The use of transfer functions on pollen data, marine-core data, and tree-ring data has all been directed to the reconstruction of past climate. This is a logical end point to a palaeoecological investigation because climate and geology ultimately determine all other processes in the ecosystem. The earth's atmosphere is one system, and a change in any one place will produce corresponding changes in the rest of the system. Hence, from a knowledge of climatic changes in western U.S.A. from tree-ring data, Fritts *et al.* (1971) could reconstruct pressure

Precipitation	Temperature: HIGH	Temperature: LOW	
HIGH	Wide	Narrow	High altitude
HIGH	Wide	Wide	Low altitude
LOW	Wide	Narrow	High altitude
LOW	Narrow	Narrow	Low altitude

Fig. 12.16 The effects of climate on tree-ring widths at high and low altitudes in the White Mountains of California, U.S.A.

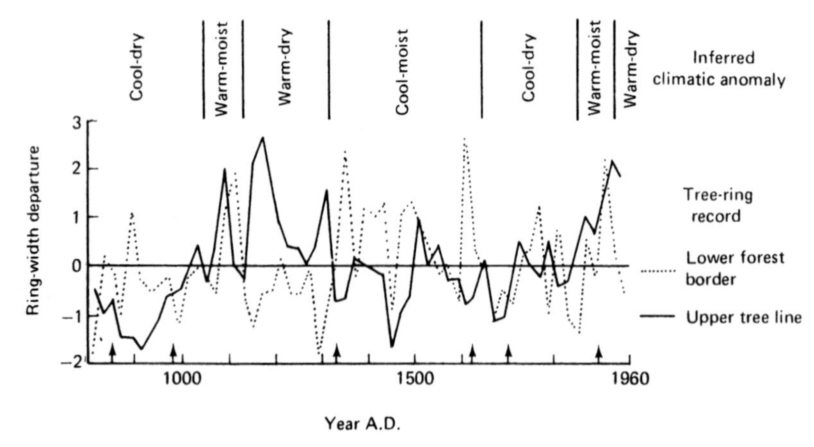

Fig. 12.17 Departure from mean tree-ring width in trees from ecologically contrasting sites in the White Mountains, California, and inferred climatic anomalies. Arrows show dates of glacial moraines in the nearby Sierra Nevada; all except the youngest were formed during periods judged to be relatively cool from the tree-ring evidence. Glacial advances of the early 1300s and early 1600s also coincide with unusually wet periods. (After La Marche (1974). © 1974 by the American Association for the Advancement of Science.)

anomaly maps for the northern hemisphere since 1521.

Recently, a synthetic programme of palaeoclimatic research has been set up to reconstruct past global climates. CLIMAP (Climatic Long Range Mapping and Prediction) seeks to reconstruct earth's climate during the Pleistocene, using all available evidence from the oceans and continents (Cline and Hays, 1976). To determine a model of the present atmosphere, several boundary conditions have to be described. To produce a model for the past, CLIMAP first reconstructed the boundary conditions which prevailed at 18 000 B.P., which is the time of the maximum extent of ice in the last glaciation. The climate could be modelled from four boundary conditions.

1. The geography of the continents. This can be estimated by assuming a minimum sea level lowering of 85 m.
2. The land albedo. This can be estimated for non-glaciated areas from the type of vegetation, which can be reconstructed from pollen analysis.
3. The extent and thickness of the ice sheets. This can be reconstructed from geomorphological evidence on land, and by the presence of clay rather than diatoms in marine sediments.
4. Sea-surface temperatures. These can be estimated using the transfer function techniques of Imbrie and Kipp (1971) on foraminiferan, coccolith, and radiolarian data from modern samples and fossils 18 000 years old. The fluctuations in the $^{16}O:^{18}O$ ratio can be used to determine the age of the sediments, because it depends largely on the volume of the ice caps.

The resulting map of sea surface temperatures, ice sheets, and land distribution published by CLIMAP (1976) is shown in Fig. 12.18. It shows several interesting features.

1. The ice and snow cover is double that of the present day.
2. In the North Atlantic, the Gulf Stream was displaced southwards, only reaching as far north as Portugal and Spain. To the north, Europe, including Britain, was surrounded by polar water masses.
3. There was a marked steepening of the thermal gradients, especially along the North Polar Front. As a result, the main westerly wind track and its associated depressions were displaced southwards.
4. There was a global cooling by an average of −2.3°C.
5. There was a cooling in equatorial regions of the ocean as a result of increased upwelling of cool polar water. However, there was little change in sea-surface temperatures in mid latitudes, as shown in a map of temperature differences between present and 18 000 B.P. in Fig. 12.19.

The first stage of the CLIMAP project described above then allowed the modelling of the climate at

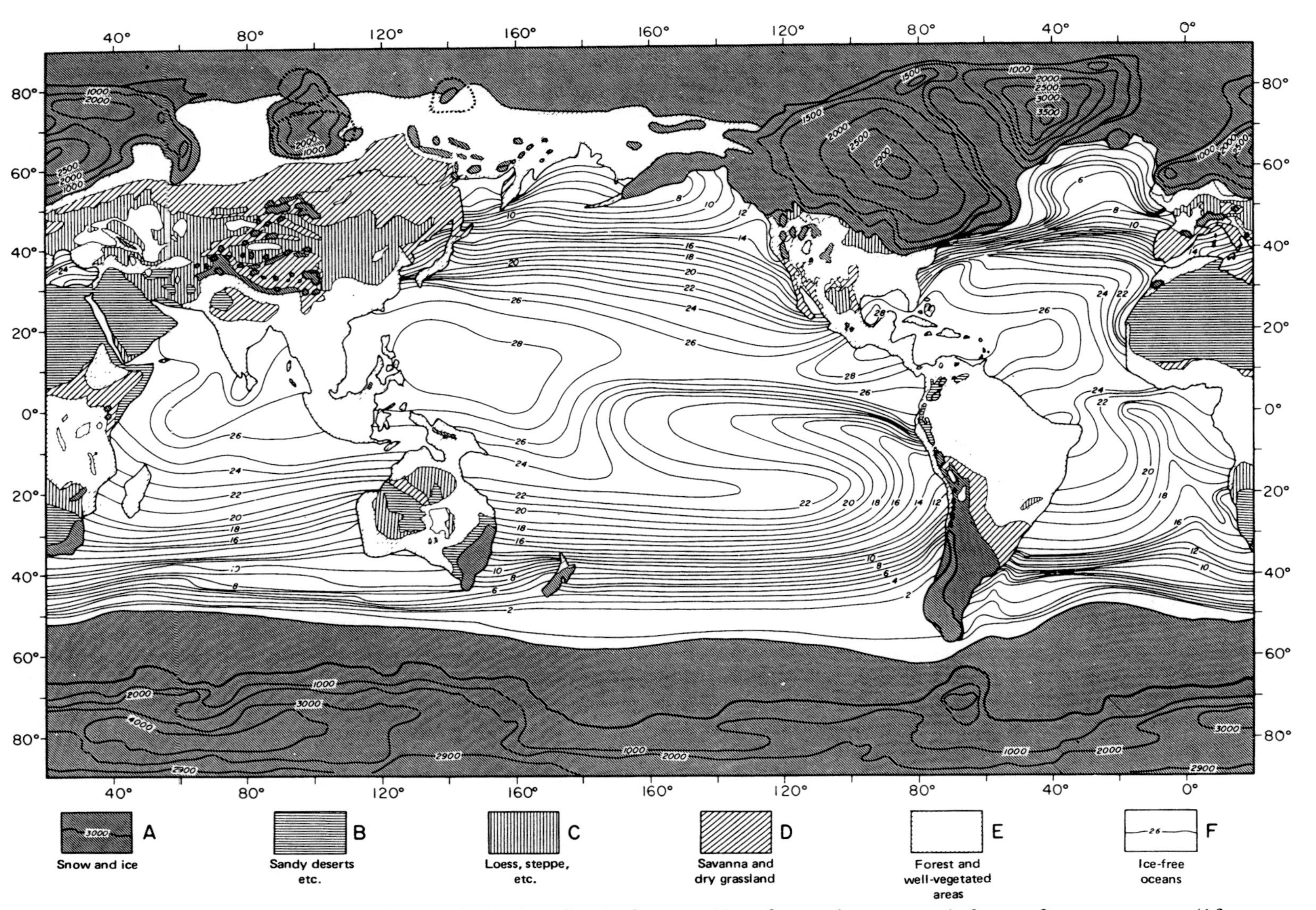

Fig. 12.18 World map at 18 000 B.P. showing the distribution of land, of snow and ice, of vegetation types, and of sea-surface temperatures. (After CLIMAP (1976). © 1976 by the American Association for the Advancement of Science.)

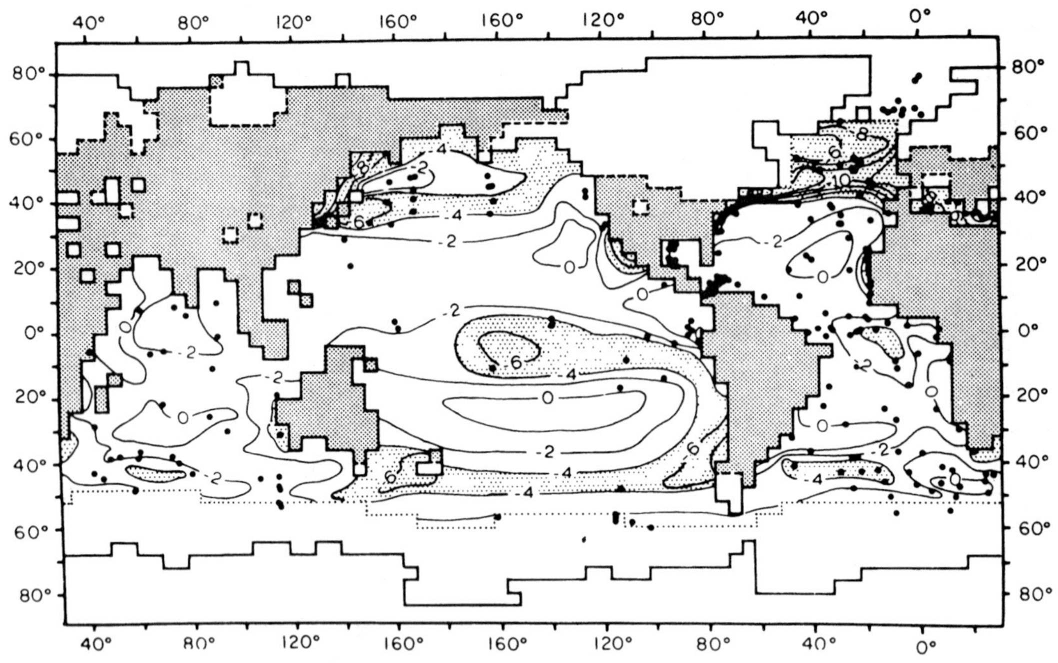

Fig. 12.19 World map showing differences between August sea-surface temperatures 18 000 years ago, and modern values. Continental and ice outlines conform to a grid spacing of 4° latitude by 5° longitude. Areas where temperature differences were greater than 4°C are shown in light stippling. Note the large differences in the northern North Atlantic. Dots are sites of cores used to reconstruct sea-surface temperaturees 18 000 years ago. (After CLIMAP (1976). © 1976 by the American Association for the Advancement of Science.)

18 000 B.P. (Gates, 1976). The conditions of the earth's surface reconstructed at 18 000 B.P. were used as boundary conditions in a global circulation model, which included the continents as well as the oceans. Gates' model simulated July climate at 18 000 B.P., and it predicted that the ice-age climate was substantially drier and cooler over the unglaciated continents. His estimates of July temperatures on land coincide well with estimates from pollen and other sources of data. Deficiencies in the model require the refinement of modelling techniques and of palaeoclimatic data bases. Manabe and Hahn (1977), using a more refined general circulation model, used the CLIMAP data base to simulate the climate of tropical regions during the last ice age. Their results predicted widespread tropical continental aridity, as did Gates (1976). The reason for this appears to have been the stronger surface air flow from the tropical continental regions to the oceans that resulted from the greater reduction in atmospheric temperature over the continents than over the oceans.

The third stage of the CLIMAP project will be to test the model of climate at 18 000 B.P. using independently derived data. If the model is satisfactory, it can be used with different boundary conditions pertaining to other times in the Pleistocene. As yet, this stage has not been reached.

Numerical analysis in Quaternary palaeoecology

Throughout this book we have followed the palaeoecological trail in various directions. But there has always been an over-riding aim – to reconstruct the organisms and the environment in which they lived. We have not hesitated to discuss the use of numerical methods of data analysis at all stages along the trail. The computer age is here to stay, and numerical methods have become an essential tool in many palaeoecological investigations. Indeed some, such as the CLIMAP project, could not proceed without the use of computers. Numerical methods have several advantages in Quaternary palaeoecology.

1. Palaeoecology is an intellectual challenge to reconstruct the past. The more quantitative is the reconstruction desired, the greater is the challenge.
2. Numerical methods provide a common language between different biological and environmental variables, and they are therefore of great importance in the reconstruction of the past environment. They also provide a communications link between palaeoecologists and other scientists in related fields, such as ecologists, geologists, and climatologists. Information on past processes and events in quantitative terms is needed as input for simulation models of population structure, climate, etc.
3. Numerical methods provide a useful means of processing large data sets, and of detecting 'structure' or 'pattern' in the data.
4. The use of numerical methods forces an investigator to be explicit about his aims, his assumptions, and his methods. Therefore it is possible for another investigator to follow his arguments exactly in order to take them further or prove them wrong. It is preferable to be explicit but wrong rather than inexplicably right.
5. Numerical methods often expose areas where our knowledge is weakest, such as the processes of pollen transport, the relationships of organisms to climate, and their response to climatic change, etc.
6. Numerical methods require data sets in standard terms, such as a standard taxonomy, a standard preparation method, a standard sampling method etc. Standardization in data collection and data presentation is generally a good thing, as the quality of the practical research is usually raised.
7. The use of numerical multivariate methods allows one to consider all samples and all taxa in a rapid and precise way, and to produce results which are repeatable by the same or different investigators.
8. Numerical methods may reveal unsuspected features or patterns in the data. New ways of looking at the data may lead to the generation of new hypotheses, which may point to further lines of research.

Conclusions

No numerical method should become an end in itself. The use of each method should be carefully evaluated, and its value to the project in hand should be assessed, so that the maximum efficiency of data handling is obtained. It is the interpretation of the data which is the essence of palaeoecology, and no amount of computerization, however useful, can compensate for a first-hand knowledge of the ecology of the organisms being studied. So often, a palaeoecologist is stopped short by the lack of such knowledge. Although this is unavoidable in most pre-Quaternary research, most of the organisms found in Quaternary sediments are still alive today. A lack of modern ecological knowledge has frequently stimulated research into the palaeoecological characteristics of the present day, such as the modern representation of organisms that have been found as fossils, and of processes occurring at the present day which are believed to have been active and relevant in the past. Most of this type of experimental palaeoecology has been directed towards the interpretation of pollen data, but recently studies of other organisms found as fossils have been initiated.

Quaternary palaeoecology continues to provide challenges. Much basic descriptive work still needs to be done on all groups of organisms from all parts of the world. As more problems are investigated and wholly or partially solved, new ones appear over the horizon, which demand new techniques and inspiration from enthusiastic palaeoecologists. Although its substance is fossils and sediments laid down long ago, Quaternary palaeoecology, as we have tried to show in this book, is very much an active, living science.

References

Distribution of communities in space

BIRKS, H. J. B. (1973). *Past and Present Vegetation of the Isle of Skye – a Palaeoecological Study*. Cambridge University Press (Chapter 14).

BRUBAKER, L. B. (1975). Post-glacial forest patterns associated with till and outwash in Northcentral Upper Michigan. *Quat. Res.*, **5**, 499–527.

IVERSEN, J. (1941). Landnam i Danmarks Stenalder. *Danm. geol. Unders.* Ser. II, **66**, 68 pp.

IVERSEN, J. (1949). The influence of prehistoric man on vegetation. *Danm. geol. Unders.* Ser. IV, **3**, 1–25.

JANSSEN, C. R. (1966). Recent pollen spectra from the deciduous and coniferous deciduous forests of northeastern Minnesota: a study in pollen dispersal. *Ecology*, **47**, 804–25.

JANSSEN, C. R. (1973). Local and regional pollen deposition. In *Quaternary Plant Ecology* (eds H. J. B. Birks and R. G. West). Blackwell.

LICHTI-FEDEROVICH, S. AND RITCHIE, J. C. (1968). Recent pollen assemblages from the Western Interior of Canada. *Rev. Palaeobot. Palynol.*, **7**, 297–344.

MCANDREWS, J. H. (1966). Postglacial history of Prairie, Savanna, and Forest in Northwestern Minnesota. *Mem. Torrey Bot. Club*, **22**(2), 72 pp.

PENNINGTON, W. (1970). Vegetational history in the north-west of England: a regional synthesis. In *Studies in the Vegetational History of the British Isles* (eds D. Walker and R. G. West), Cambridge University Press.

PENNINGTON, W. (1973). Absolute pollen frequencies in sediments of lakes of different morphometry. In *Quaternary Plant Ecology* (eds H. J. B. Birks and R. G. West). Blackwell.

RAYNOR, G. S. AND OGDEN, E. C. (1965). Twenty-four-hour dispersion of ragweed pollen from known sources. *Brookhaven National Laboratory, Publ.*, **957**, 201–11.

RAYNOR, G. S., OGDEN, E. C. AND HAYES, J. V. (1968). Effect of a local source on ragweed pollen concentrations from background sources. *J. Allergy*, **41**, 217–25.

SMITH, A. G. (1964). Problems in the study of the earliest agriculture in Northern Ireland. *Rep. VI Intern. Congr. Quater. Warsaw, 1961*, Sect. **2**, 461–71.

SMITH, A. G. AND WILLIS, E. H. (1961). Radiocarbon dating of the Fallahogy landnam phase. *Ulster J. Archaeol.*, **24/25**, 16–24.

TAUBER, H. (1965). Differential pollen dispersion and the interpretation of pollen diagrams. *Danm. geol. Unders.* Ser. II, **89**, 69 pp.

TURNER, J. (1964). Surface sample analyses from Ayrshire, Scotland. *Pollen Spores*, **6**, 583–92.

TURNER, J. (1965). A contribution to the history of forest clearance. *Proc. R. Soc.* B, **161**, 343–54.

TURNER, J. (1970). Post-Neolithic disturbance of British vegetation. In *Studies in the Vegetational History of the British Isles* (eds D. Walker and R. G. West). Cambridge University Press.

TURNER, J. AND HODGSON, J. (1979). Studies in the vegetational history of the Northern Pennines. I. Variations in the composition of the early Flandrian forests. *J. Ecol.*, **67**, 629–46.

Distribution of communities in time

DICKSON, J. H., STEWART, D. A., THOMPSON, R., TURNER, G., BAXTER, M. S., DRNDARSKY, N. D. AND ROSE, J. (1978). Palynology, palaeomagnetism and radiometric dating of Flandrian marine and freshwater sediments of Loch Lomond. *Nature*, **274**, 548–53.

WATTS, W. A. (1973). Rates of change and stability in vegetation in the perspective of long periods of time. In *Quaternary Plant Ecology* (eds H. J. B. Birks and R. G. West). Blackwell.

WEST, R. G. (1977). *Pleistocene Geology and Biology* (2nd edition). Longman. (Chapter 9).

Environmental reconstructions from terrestrial sediment chemistry and stable isotopes

DANSGAARD, W., JOHNSEN, S. J., REEH, N., GUNDESTRUP, N., CLAUSEN, H. B. AND HAMMER, C. U. (1975). Climatic changes, Norsemen and modern man. *Nature*, **255**, 24–8.

EICHER, U. AND SEIGENTHALER, U. (1976). Palynological and oxygen isotope investigations on Late-Glacial sediment cores from Swiss lakes. *Boreas*, **5**, 109–17.

LANG, G. (1970). Florengeschichte und mediterran-mitteleuropäische Florenbeziehungen. *Feddes Repert.*, **81**, 315–35.

LERMAN, J. C. (1973). Isotope 'paleothermometers' on continental matter: assessment. *Colloques Internationoux du C.R.N.S.*, No. **219**, 163–81.

LIBBY, N. M. AND PANDOLFI, L. J. (1977). Climatic periods in tree, ice and tides. *Nature*, **266**, 415–17.

LIBBY, N. M., PANDOLFI, L. J., PAYTON, P. H., MARSHALL, J., BECKER, B. AND GIERTZ-SIEBENLIST, V. (1976). Isotopic tree thermometers. *Nature*, **261**, 284–6.

MACKERETH, F. J. H. (1966). Some chemical observations on post-glacial lake sediments. *Phil. Trans. R. Soc.* B, **250**, 165–213.

PEARMAN, G. I., FRANCEY, R. J. AND FRASER, P. J. B. (1976). Climatic implications of stable carbon isotopes in tree rings. *Nature*, **260**, 771–3.

PENNINGTON, W., HAWORTH, E. Y., BONNY, A. P. AND LISHMAN, J. P. (1972). Lake sediments in northern Scotland. *Phil. Trans. R. Soc.* B, **264**, 191–294.

PENNINGTON, W. AND SACKIN, M. J. (1975). An application of principal components analysis to the zonation of two Late-Devensian profiles. *New Phytol.*, **75**, 419–53.

SCHIEGL, W. E. (1972). Deuterium content of peat as a paleoclimatic recorder. *Science*, **175**, 512–13.

STUIVER, M. (1968). Oxygen-18 content of atmospheric precipitation during last 11,000 years in the Great Lakes Region. *Science*, **162**, 994–7.

STUIVER, M. (1970). Oxygen and carbon isotope ratios of fresh-water carbonates as climatic indicators. *J. Geophys. Res.*, **75**, 5247–57.

STUIVER, M. (1975). Climate versus changes in ^{13}C content of the organic component of lake sediments during the late Quaternary. *Quat. Res.*, **5**, 251–62.

TROUGHTON, J. E., WELLS, P. V. AND MOONEY, H. A. (1974). Photosynthetic mechanisms and paleoecology from carbon isotope ratios in ancient specimens of C_4 and CAM plants. *Science*, **185**, 610–12.

Univariate or bivariate approaches to environmental reconstruction from biotic evidence

BIRKS, H. J. B. (1979). The use of pollen analysis in the reconstruction of past climates: a review. In *Proceedings of the International Conference on Climate and History, Norwich*, July 1969 (eds. H. H. Lamb *et al.*). University of East Anglia.

CHURCHILL, D. M. (1968). The distribution and prehistory of *Eucalyptus diversicolor* F. Muell., *E. marginata* Donn ex Sm., and *E. calophylla* R. Br. in relation to rainfall. *Aust. J. Bot.*, **16**, 125–51.

CONOLLY, A. P. AND DAHL, E. (1970). Maximum summer temperature in relation to the modern and Quaternary distributions of certain arctic-montane species in the British Isles. In *Studies in the Vegetational History of the British Isles* (eds D. Walker and R. G. West). Cambridge University Press.

GRICHUK, V. P. (1969). An attempt to reconstruct certain elements of the climate of the Northern Hemisphere in the Atlantic Period of the Holocene. In *Golotsen* (ed. M. I. Neishtadt). Moscow, Izd-vo Nauka.

IVERSEN, J. (1944). *Viscum*, *Hedera* and *Ilex* as climatic indicators. *Geol. Fören. Förhandl. Stock.*, **66**, 463–83.

RATCLIFFE, D. A. (1968). An ecological account of Atlantic bryophytes in the British Isles. *New Phytol.*, **67**, 365–439.

Multivariate approaches to environmental reconstruction from biotic evidence

BRYSON, R. A. AND KUTZBACH, J. E. (1974). On the analysis of pollen-climate canonical transfer functions. *Quat. Res.*, **4**, 162–74.

FRITTS, H. C. (1976). *Tree Rings and Climate*. Academic Press.

FRITTS, H. C., BLASING, T. J., HAYDEN, B. P. AND KUTZBACH, J. E. (1971). Multivariate techniques for specifying tree-growth and climate relationships and for reconstructing anomalies in paleoclimate. *J. Appl. Met.*, **10**, 845–64.

HARRIS, W. F., DARWIN, J. H. AND NEWMAN, M. J. (1976). Species clustering and New Zealand Quaternary climate. *Palaeogeography, Palaeoclimatology, Palaeoecology*, **19**, 231–48.

HOWE, S. AND WEBB, T. (1977). Testing the statistical assumptions of paleoclimatic calibration functions. *Preprint volume Fifth Conference on Probability and Statistics in Atmospheric Sciences. Am. Met. Soc.*, 152–7.

HUGHES, M. K., GRAY, B., PILCHER, J., BAILLIE, M. AND LEGGETT, P. (1978). Climatic signals in British tree-ring chronologies. *Nature*, **272**, 605–6.

HUTSON, W. H. (1977). Transfer functions under no-analog conditions: experiments with Indian Ocean planktonic foraminifera. *Quat. Res.*, **8**, 355–67.

IMBRIE, J. AND KIPP, N. G. (1971). A new micropaleontological method for quantitative paleoclimatology: application to a late Pleistocene Caribbean core. In *The Late Cenzoic Glacial Ages* (ed. K. K. Turekian). Yale University Press.

IMBRIE, J., VAN DONK, J. AND KIPP, N. G. (1973). Paleoclimatic investigation of a late Pleistocene Caribbean deep-sea core: comparison of isotopic and faunal methods. *Quat. Res.*, **3**, 10–38.

LA MARCHE, V. C. (1974). Paleoclimatic influences from long tree-ring records. *Science*, **183**, 1043–8.

MOORE, P. D. (1978). Tree-ring chronology. *Nature*, **272**, 578–9.

PEARSON, G. W., PILCHER, J. R., BAILLIE, M. G. L. AND HILLAM, J. (1977). Absolute radiocarbon dating using a low altitude European tree-ring calibration. *Nature*, **270**, 25–8.

PILCHER, J. R., HILLAM, J., BAILLIE, M. G. L. AND PEARSON, G. W. (1977). A long sub-fossil oak tree-ring chronology from the North of Ireland. *New Phytol.*, **79**, 713–29.

SACHS, H. M., WEBB, T. AND CLARK, D. R. (1977). Paleoecological transfer functions. *Ann. Rev. Earth Planet. Sci.*, **5**, 159–78.

SANCETTA, C., IMBRIE, J. AND KIPP, N. G. (1973). Climatic record of the past 130 000 years in North Atlantic Deep-Sea Core V23-82: correlation with the terrestrial record. *Quat. Res.*, **3**, 110–16.

WEBB, T. AND BRYSON, R. A. (1972). Late- and postglacial climatic change in the Northern mid west, U.S.A.: Quantitative estimates derived from fossil pollen spectra by multivariate statistical analysis. *Quat. Res.*, **2**, 70–115.

WEBB, T. AND CLARK, D. R. (1977). Calibrating micropaleontological data in climatic terms: a critical review. *Annals New York Acad. Sci.*, **288**, 93–118.

Global palaeoclimatic reconstructions

CLIMAP (1976). The surface of the Ice-Age Earth. *Science*, **191**, 1131–7.

CLINE, R. M. AND HAYS, J. D. (1976). Investigation of Late Quaternary Paleoceanography and Paleoclimatology. *Geol. Soc. Amer. Memoir*, **145**, 464 pp. (especially papers by N. G. Kipp and A. McIntyre, and N. G. Kipp).

GATES, W. L. (1976). Modelling the Ice-Age climate. *Science*, **191**, 1138–44.

MANABE, S. AND HAHN, D. G. (1977). Simulation of the tropical climate of an ice age. *J. Geophys. Res.*, **82**, 3889–911.

Index

Printed in the United Kingdom
by Lightning Source UK Ltd.
129251UK00002B/17/A